Springer Series in Computational Physics

Editors: R. Glowinski M. Holt P. Hut
H.B. Keller J. Killeen S.A. Orszag V.V. Rusanov

Springer Series in Computational Physics

Editors: R. Glowinski M. Holt P. Hut H. B. Keller J. Killeen
S. A. Orsag V. V. Rusanov

E.V. Vorozhtsov
N.N. Yanenko

Methods for the Localization of Singularities in Numerical Solutions of Gas Dynamics Problems

With 112 Illustrations

Springer-Verlag
New York Berlin Heidelberg
London Paris Tokyo Hong Kong

E.V. Vorozhtsov
Institute of Theoretical and Applied Mechanics
U.S.S.R. Academy of Sciences
Siberian Division
Novosibirsk 630090 U.S.S.R.

Editors

R. Glowinski
Institut de Recherche d'Informatique
et d'Automatique (IRIA)
Domaine de Voluceau, Rocquencourt, B.P. 105
F-78150 Le Chesnay, France

P. Hut
The Institute for Advanced Study
School of Natural Sciences
Princeton, NJ 08540, U.S.A.

J. Killeen
Lawrence Livermore Laboratory
P.O. Box 808
Livermore, CA 94551, U.S.A.

N.N. Yanenko
(deceased)

M. Holt
College of Engineering and
Mechanical Engineering
University of California
Berkeley, CA 94720, U.S.A.

H.B. Keller
Applied Mathematics 101-50
Firestone Laboratory
California Institute of Technology
Pasadena, CA 91125, U.S.A.

S.A. Orszag
Department of Mechanical and
Aerospace Engineering
Princeton University
Princeton, NJ 08544, U.S.A.

V.V. Rusanov
Keldysh Institute of Applied Mathematics
4 Miusskaya pl.
SU-125047 Moscow, U.S.S.R.

Library of Congress Cataloging-in-Publication Data
Vorozhtsov, E. V. (Evgeniɪ Vasil'evich)
  [Metody lokalizatsii osobennostei pri chislennom reshenii zadach
gazodinamiki. English]
  Methods for the localization of singularities in numerical
solutions of gas dynamics problems / E.V. Vorozhtsov, N.N. Yanenko.
  p. cm. — (Springer series in computational physics)
  Translation of: Metody lokalizatsii osobennostei pri chislennom
reshenii zadach gazodinamiki.
  Includes bibliographical references.
  ISBN-13: 978-3-642-64770-3
  1. Gas dynamics. 2. Numerical analysis. I. IAnenko, N. N.
(Nikolai Nikolaevich) II. Title.III. Series.
  QA930.V6613 1990
  533' .2—dc20

Title of the original Russian edition: Metody lokalizatsii osobennostey pri chislennom reshenii zadach gazodinamiki © "Nauka" Publishing House (Siberian Division), Novosibirsk, 1985.

ISBN-13: 978-3-642-64770-3   e-ISBN-13: 978-3-642-61271-8
DOI: 10.1007/978-3-642-61271-8

© Springer-Verlag Berlin Heidelberg 1990
Softcover reprint of the hardcover 1st edition 1990

Typesetting: Asco Trade Typesetting Ltd., Hong Kong; printing and binding: R.R. Donnelley & Sons, Harrisonburg, Virginia.
2156/30-543210—Printed on acid-free paper

Preface

As a result of the numerical simulation of multidimensional gas dynamics problems on a computer, the output information is obtained in the form of immense arrays of numerical data. In this connection, there arises the problem of extracting the actually needed information from these arrays; in other words, it is necessary to solve the problem of information compression. In particular, the numerical solution of gas dynamics problems often aims at the information on the solution singularities—the shock waves, contact interfaces, slip lines, etc. Our book is devoted to the development and investigation of accuracy of the algorithms for the localization of such singularities. In addition, the questions of development of the algorithms for the classification of singularities into several types (on the basis of shock-capturing numerical solutions of two-dimensional gas dynamics problems) are considered for the first time in the monographic literature. For this purpose, some ideas and methods of the modern theory of digital-image processing and of the pattern recognition theory are used. The information obtained at the output of the systems of the singularities classification presented in this book is rich in content, because it contains both physical and geometrical characteristics of recognized objects. Therefore, such "intellectual" systems of information extraction may be used in the expert systems of automated design of aero-dynamic bodies which meet some optimality requirements. This is, in our opinion, very attractive from the point of view of applications.

The methods of differential approximation, variational calculus, and numerical optimization have been used in the studies of accuracy of the well-known algorithms for the localization of singularities, as well as the new algorithms proposed by the present authors.

We have aimed at a balanced presentation of the material, therefore, the applications of developed algorithms of the singularities localization to the analysis of various two-dimensional fluid mechanics problems have been included in the book along with the theoretical results. In particular, we have considered problems of high-velocity impact, transonic flow around an airfoil, hypersonic flow around a nonconvex body, etc.

We express our gratitude to the research workers of the Department for Numerical Methods of Continuum Mechanics of the Computing Center of the U.S.S.R. Academy of Sciences, Siberian Branch, and to professional colleagues from the Institute of Theoretical and Applied Mechanics of the U.S.S.R. Academy of Sciences, Siberian Branch, in whose collectives the work had been discussed. Our opinions and points of view were also affected by the interaction with collectives headed by A.N. Tikhonov, L.V. Ovsyannikov. The discussions with Yu.A. Berezin, Yu.M. Davydov, V.M. Fomin, A.N. Konovalov, B.G. Kuznetsov, V.A. Novikov, V.V. Pikalov, N.G. Preobrazhenskiĭ, B.L. Roždestvenskiĭ, V.V. Rusanov, Yu.I. Shokin, A.F. Voyevodin, Yu.S. Zavyalov were especially useful to us. We are very grateful to the editor of the English language edition of this book, V.V. Rusanov, for his careful editing of the manuscript.

We also express our thanks to Professor K.G. Roesner from Darmstadt, F.R.G., who immediately recommended our book to Springer-Verlag in Heidelberg, having read the Russian edition of the book. We are grateful to Professor W. Beiglböck, the editor of the Springer Series in Computational Physics, for allowing us to publish in this series, and to T.A. Alexandrova for typing the manuscript.

Contents

Introduction and Necessary Notions from the Theory of Difference Schemes for Gas Dynamics Problems

The use of electronic computers in numerical simulations of gas flows with singularities started practically immediately after the first such computers had appeared in the 1940s [1.1], [1.2]. Presently, a wide application of computers for the solution of various fluid dynamics problems has become possible, owing to the development of powerful computers and efficient numerical methods—and the corresponding field of science has been termed "computational fluid dynamics" [1.3], [1.4].

In the course of mathematical modeling of many fluid dynamics problems one often has to deal with the solutions containing singularities of various types. For example, in the problems of inviscid compressible gas dynamics there are singularities of shock wave and contact discontinuity type [1.5], in the filtration problems there are saturation fronts [1.6]. In combustion problems one has to deal with flame fronts [1.7], in turbulence theory with coherent structures [1.8], in meteorology problems with atmospheric fronts [1.9], in magnetohydrodynamics with magnetohydrodynamic shock waves [1.10], [1.11], etc. At present, finite-difference shock-capturing schemes are widely used for the numerical investigation of such problems. In the numerical solutions obtained with the aid of such schemes the discontinuities are approximated by some transition regions, the size of which (in the direction of a normal to the discontinuity surface) is usually equal to several intervals of a spatial computing mesh. As a result of this it proves difficult to effectively use and interpret the numerical data obtained and, in addition, there arises the problem of increasing the accuracy of difference solutions in the neighborhood of discontinuities. In particular, a research worker dealing with shocked gas flows is in many cases interested primarily in the information on shock surfaces: their disposition, shape, propagation speed, etc.

In connection with the foregoing there exists a need, in the development and foundation of specialized algorithms, to process the numerical results of solving fluid dynamics problems which are intended for the localization of discontinuity surfaces in a flow and for their classification into several types (shock waves, contact interfaces, etc.).

A problem of development of the singularities localization techniques on the basis of finite-difference solutions is closely related to increases in the accuracy of numerical solutions in the vicinity of discontinuities. Localization of strong discontinuities is substantially facilitated if one uses, in computations, a finite-difference method which enables one to reduce the width of a zone of discontinuity "smearing" to the size of one mesh interval. In this case it is possible to use, for shock localization, the already existing simplest procedures, for example, by maximum coalescence of isolines (for example, isochors) or by maximum gradients of any of the functions sought which undergo a discontinuity.

The structure of shock localization algorithms will also be affected by a further increase in the performance of computers. For example, a considerable increase in the computer core memory, as well as the use of parallel processors, enables one to use substantially finer meshes. Then the accuracy of determining the location of a discontinuity within the zone of its "smearing" is not so important as in the case of crude meshes, and it will then be possible to successfully apply the simplest procedures for the numerical shock localization. On the other hand, a manual processing of the results (and even a simple survey of them) becomes difficult with the increase in the number of grid points. Computer methods of processing can substantially aid the researcher in the interpretation of numerical results in this case. They also facilitate substantially the computer generation of pictures of temporal evolution of various singularities in cases when such pictures are of primary importance. If the solution accuracy in the neighborhood of discontinuities, which is achieved in the case of using a specific difference scheme, is insufficient, then the informaton on the singularities locaton may be incorporated directly into a computational algorithm to achieve an increase in the accuracy of computation (see Chapters 2, 3, and 6).

In addition, the localization of discontinuities, in particular, the ones arising in the process of computation, gives an opportunity of an active control of this process which may include, for example, the alteration of some boundary conditions, a switch to another construction of a difference grid or to another difference scheme, etc.

Mathematical models of the above-listed various problems of fluid dynamics are characterized by different levels of complexity. In our monograph we analyze a wide spectrum of algorithms for the localization and classification of strong discontinuities in the numerical solutions of inviscid compressible non-heat-conducting gas dynamics. This has been done for two reasons. First, the flows with singularities of different types (being different from the ones known in gas dynamics) are at present treated in much the same way as the gas-dynamical shocked flows (see, for example, [1.7], [1.12], [1.13]). Second, the results of the investigation of various algorithms for the strong discontinuities localization (presented in Chapters 2–7) were obtained by the present authors only for gas dynamics problems described by the Euler equations. Taking the above into account, the domain of applicability of the methods for the localization and classification of singularities on the basis of shock-capturing numerical solutions presented below goes beyond the scope of inviscid gas dynamics problems. The essence of a general approach (presented in our book) to the development and investigation of shock localization methods based on shock-capturing computations is in the maximal use of information on the structure of a finite-difference solution in the zone of "smearing" of a strong discontinuity, while constructing the algorithms for locating a true discontinuity within a zone of its numerical smearing. Aiming at brevity of presentation, we emphasize our own results; therefore, other algorithms of the singularities localization which are known in the literature are mentioned only briefly at the beginning of Chapters 2 and 4. We make no claim to completeness in the list of references where the localization algorithms developed by other authors are presented, although we hope that we have presented in this list the basic ideas and trends in the construction of the above algorithms.

The methods of the singularities localization presented in Chapters 2–6 can be united into one big group of methods whose realization is related substantially to the use of *a priori* information on the orientation of shock surfaces with respect to the axes of spatial coordinates. However, in some cases, such informaton is absent. There are, for example, such problems, the investigation of which is difficult to carry out by other techniques (for example, by experimental techniques), and then mathematical modeling becomes the only method of studying such complicated phenomena or processes [1.14]. The methods for the localization and classification of singularities which are presented in Chapter 7 may prove to be very useful in the analysis of computational results of such problems. These methods do not require for their implementation any *a priori* information on the presence or absence of singularities in the problem under consideration, as well as on their approximate orientation with respect to the spatial coordinate axes. The methods of Chapter 7 use substantially the ideas and algorithms of the digital-image processing theory and the theory of pattern recognition, and are very versatile and universal, and which is shown in a number of examples. Since the data obtained (which is presented in Chapter 7) at the output of a system of extraction of information from the results of two-dimensional gas-dynamical computations is rich in content—it contains both physical and geometrical characteristics of recognized objects—it can be used for controlling the process of the numerical solution of the basic problem as well as for decision-making in the expert systems of aerodynamic automatic design [1.15], [1.16].

Results of investigation of the accuracy of methods for the localization of singularities are illustrated by numerical computations of model problems having exact solutions. In addition, there are demonstrated examples of those complicated fluid mechanics problems, in the analysis of which the developed localization algorithms have been used. These are the high-velocity impact problems, a transonic flow around an airfoil, supersonic flows in annular nozzles and jets, hypersonic flow around a nonconvex body, interaction of jets with obstacles, etc.

This monograph represents the first systematic presentation of the results of accuracy analysis of the methods for the localization of singularities on the basis of the shock-capturing computation of gas dynamics problems. A number of new results obtained by the present authors is presented for the first time.

1.1. Original Equations. Jump Conditions in the Case of One-Dimensional Gas Flow

1.1.1. Divergence and Nondivergence Form of Equations

The system of differential equations governing the plane one-dimensional flow of an inviscid compressible non-heat-conducting gas, which depends on time t and on one Cartesian coordinate x, has the following divergence form [1.17], [1.18]:

$$\partial\rho/\partial t + \partial\rho u/\partial x = 0; \tag{1.1}$$

$$\partial\rho u/\partial t + \partial(p + \rho u^2)/\partial x = 0; \tag{1.2}$$

$$\partial\rho(\varepsilon + u^2/2)/\partial t + \partial[\rho u(\varepsilon + u^2/2) + pu]/\partial x = 0. \tag{1.3}$$

Here ρ is the density, p is the pressure, ε is the internal energy per unit mass of the gas, and u is the velocity in the direction of the x-axis. The four functions ρ, u, p, ε sought enter into the system (1.1)–(1.3). Therefore, one more equation is necessary to complete this system. As is known, among the thermodynamical quantities describing the gas state only the two quantities are independent, the remaining quantities can be expressed in terms of two chosen independent functions with the aid of an equation of state [1.18]. In particular, let the equations of state be given in the form

$$p = G(V, S), \qquad T = T(V, S), \tag{1.4}$$

where V is the specific volume, $V = 1/\rho$, S is the entropy, and T is the gas temperature. Then the specific internal energy ε may be calculated as a function of the variables V, S, with the aid of a thermodynamical identity $d\varepsilon + p\,dV = T\,dS$. Knowing the dependencies of the quantities p and ε on V and S, we can compute the pressure p as a function of ρ, ε:

$$p = F(\rho, \varepsilon). \tag{1.5}$$

Thus, in the case when the equation of state can be given in the form (1.5), the system of equations (1.1)–(1.3) is closed without using the entropy S. In the following we shall assume the presence of a dependency (1.5) or of a dependency

$$\varepsilon = f(p, \rho). \tag{1.6}$$

Of course, the functions f and F are such that the identity

$$p \equiv F(\rho, f(p, \rho)) \tag{1.7}$$

takes place. The ideal gas equation of state

$$p = (\gamma - 1)\rho\varepsilon \tag{1.8}$$

is one of the simplest equations of state, where the quantity γ is the ratio of specific heat, usually $\gamma = \text{const} > 1$.

In the following we shall often use a vector notation of the system (1.1)–(1.3). Introduce the column vectors

$$\mathbf{u} = \begin{pmatrix} \rho \\ \rho u \\ \rho E \end{pmatrix}, \qquad \boldsymbol{\varphi}(\mathbf{u}) = \begin{pmatrix} \rho u \\ p + \rho u^2 \\ pu + \rho uE \end{pmatrix}, \tag{1.9}$$

where

$$E = \varepsilon + u^2/2, \tag{1.10}$$

that is, E is the total energy per unit mass of the gas. Then the system (1.1)–(1.3) may be written in the form

$$\partial\mathbf{u}/\partial t + \partial\boldsymbol{\varphi}(\mathbf{u})/\partial x = 0. \tag{1.11}$$

The system (1.11) is the divergence, or conservative, form of the Euler equations. We shall also need a nondivergence form of the system (1.1)–(1.3). Set

$$\mathbf{u} = \begin{pmatrix} u_1 \\ u_2 \\ u_3 \end{pmatrix}, \qquad \boldsymbol{\varphi} = \begin{pmatrix} \varphi_1 \\ \varphi_2 \\ \varphi_3 \end{pmatrix},$$

where

$$u_1 \equiv \rho, \qquad u_2 \equiv \rho u, \qquad u_3 \equiv \rho E;$$

$$\varphi_1 = \rho u = u_2,$$

$$\varphi_2 = p + \rho u^2 = F(u_1, (u_3/u_1) - 0.5(u_2/u_1)^2) + u_2^2/u_1, \tag{1.12}$$

$$\varphi_3 = F(u_1, u_3/u_1 - 0.5(u_2/u_1)^2)(u_2/u_1) + u_2 u_3/u_1,$$

$F(\rho, \varepsilon)$ is the function entering the equation of state (1.5). Then the elements of the Jacobi matrix

$$A = \partial\boldsymbol{\varphi}/\partial\mathbf{u} \tag{1.13}$$

are determined by the formulas $a_{lm} = \partial\varphi_l/\partial u_m$. In the case when the equation of state (1.5) is employed to complete the system (1.1)–(1.3), the elements a_{lm} have the following expressions (see, for example, [1.19])

$$a_{11} = 0; \qquad a_{12} = 1; \qquad a_{13} = 0;$$
$$a_{21} = \theta + rz - u^2; \qquad a_{22} = u(2 - z); \qquad a_{23} = z; \tag{1.14}$$
$$a_{31} = -u(E + m - \theta - rz); \qquad a_{32} = E + m - u^2 z; \qquad a_{33} = uz_1.$$

In formulas (1.14) $r = u^2 - E$, $z = (1/\rho)\, \partial p/\partial \varepsilon$, $z_1 = 1 + z$, $\theta = \partial p/\partial \rho$, and $m = p/\rho$. With the use of the matrix A (1.13) the nondivergence form of the system (1.11) may obviously be written as

$$\partial \mathbf{u}/\partial t + A(\mathbf{u})\, \partial \mathbf{u}/\partial x = 0. \tag{1.15}$$

As is known, the eigenvalues λ_1, λ_2, λ_3 of the matrix A have the form

$$\lambda_1 = u - c, \qquad \lambda_2 = u, \qquad \lambda_3 = u + c, \tag{1.16}$$

where c is the adiabatic speed of sound. In the case of the equation of state (1.5) the square of the speed of sound is calculated by the formula

$$c^2 = (p/\rho^2)\, \partial F/\partial \varepsilon + \partial F/\partial \rho.$$

Thus, if the function $F(\rho, \varepsilon)$ in the equation of state (1.5) satisfies the inequality

$$(p/\rho^2)\, \partial F/\partial \varepsilon + \partial F/\partial \rho > 0, \tag{1.17}$$

then the equation system (1.15) is of hyperbolic type. It is assumed in the following that the equations of state employed satisfy the inequality (1.17). The matrix A whose elements are determined by formulas (1.14) also possesses the following property [1.18]:

$$\begin{aligned} (A - uI)^{2k+1} &= c^{2k}(A - uI), \\ (A - uI)^{2k+2} &= c^{2k}(A - uI)^2, \qquad k = 0, 1, 2, \ldots, \end{aligned} \tag{1.18}$$

where I is the unit matrix and c is the speed of sound.

1.1.2. Jump Conditions

Let Ω be an arbitrary subdomain with the boundary Γ in the (x, t)-plane which is in the domain of definition of the system (1.11) solution. Then the integral conservation laws for the system (1.11) have the form

$$\oint_\Gamma \mathbf{u}\, dx - \boldsymbol{\varphi}(\mathbf{u})\, dt = 0. \tag{1.19}$$

Unlike the system (1.1)–(1.3), the relationships (1.19) are also valid for discontinuous solutions. Let us derive the conditions which are to be satisfied along the discontinuity lines of the solutions of gas dynamics equations as consequences of the integral conservation laws. Let $x = x_s(t)$ be the equation of one of the lines of the jump in hydrodynamic quantities, and let the function $f(x, t)$ undergo a jump across the line $x = x_s(t)$. Denote by

$$\begin{aligned} f_1(t) &= f(x_s(t) - 0, t); \qquad f_2(t) = f(x_s(t) + 0, t); \\ [f] &= f_2(t) - f_1(t). \end{aligned} \tag{1.20}$$

Let the discontinuity propagate at a speed $dx_s/dt = D$. Consider in the (x, t)-

plane a closed contour, two lines of which adhere with an infinite proximity to some segment of the discontinuity line $x_s(t)$. It follows from the conservation laws written for this contour that along the discontinuity line

$$\int ([\mathbf{u}]D - [\varphi(\mathbf{u})])\, dt = 0,$$

where the integrals are taken along any segment of the discontinuity line. By virtue of an arbitrary choice of the integration domain, the relationships

$$[\mathbf{u}]D = [\varphi(\mathbf{u})] \tag{1.21}$$

are valid at each point of a discontinuity which relate the jumps of hydrodynamic quantities across the discontinuity line $x = x_s(t)$ and the speed $D = x_s'(t)$ of the discontinuity line. In the case of the Euler equation system (1.1)–(1.3) equations (1.21) may be written with regard to (1.9), (1.10) as the following three algebraic relations

$$D[\rho] = [\rho u]; \tag{1.22}$$

$$D[\rho u] = [p + \rho u^2]; \tag{1.23}$$

$$D[\rho(\varepsilon + u^2/2)] = [\rho u(\varepsilon + p/\rho + u^2/2)]. \tag{1.24}$$

The relations (1.22)–(1.24) are called the Rankine–Hugoniot conditions. Taking into account the notation (1.20), we can rewrite the Rankine–Hugoniot conditions (1.22)–(1.24) in the form of equations

$$\rho_2(u_2 - D) = \rho_1(u_1 - D) = m; \tag{1.25}$$

$$p_2 + \rho_2(u_2 - D)^2 = p_1 + \rho_1(u_1 - D)^2; \tag{1.26}$$

$$\rho_2(u_2 - D)(\varepsilon_2 + p_2/\rho_2 + (u_2 - D)^2/2)$$
$$= \rho_1(u_1 - D)(\varepsilon_1 + p_1/\rho_1 + (u_1 - D)^2/2). \tag{1.27}$$

If $m(t) = 0$ in equation (1.25), then this kind of discontinuity will be called contact; if $m(t) \neq 0$, then the discontinuity will be called a shock wave. In the case of a contact discontinuity it follows from (1.25) that

$$D = u_1 = u_2 = x_s'(t),$$

that is, the discontinuity line coincides with the particle trajectory. Assuming $u_1 = D$, $u_2 = D$, we obtain from (1.26) that $p_1 = p_2$. The condition (1.27) is satisfied identically at $u_1 = u_2 = D$. Thus, the pressure and the speed of the flow are continuous across a contact discontinuity in the one-dimensional gas flow. In particular, a contact discontinuity may be an interface between two different gases satisfying different equations of state.

In the case of a shock wave, that is, when $m \neq 0$, the Hugoniot adiabatic equation is obtained from (1.25)–(1.27) as an algebraic consequence [1.18]

$$\varepsilon_2 - \varepsilon_1 = (1/2)(p_2 + p_1)(V_1 - V_2), \tag{1.28}$$

where V is the specific volume, $V = 1/\rho$. Zemplén's theorem is valid for stable shock waves. This theorem asserts that the shock wave speed is subsonic with respect to a gas behind the shock front, and is supersonic with respect to a gas before the shock front.

Let the subscript 1 in (1.20) refer to a gas state behind the shock wave front, and let the subscript 2 refer to a state before the front. Then the above assertion (Zemplén's theorem) may be written in the form of the following inequalities:

$$|u_2 - D| > c_2; \qquad |u_1 - D| < c_1. \tag{1.29}$$

1.1.3. Riemann Problem

Concluding this section let us briefly consider the Riemann problem. An arbitrary discontinuity is an initial state of two infinite masses of gas characterized by constant parameters $u_1, p_1, V_1, \varepsilon_1, T_1$ and $u_2, p_2, V_2, \varepsilon_2, T_2$ adjoining along the plane $x = 0$ at the initial time $t = 0$. Here the magnitudes of the discontinuity to the left and right are arbitrary and subject only to the equations of state of the gases which may be different for the neighboring gases.

The determination of the flow arising for $t > 0$ with these initial conditions is called the Riemann problem, or the breakdown-of-discontinuity problem.

If an arbitrary discontinuity is not a contact discontinuity or a shock wave, it decomposes by forming some configuration of stable discontinuities and continuous gas-dynamical flows. All possible configurations of a flow arising in the process of a breakdown of a discontinuity in the gas have been considered in [1.18], [1.20]. Here the configuration A contains a rarefaction wave propagating into the gas "1", and a contact discontinuity and a shock wave propagating into the gas "2" (Figure 1.1(a)). The configuration B contains the shock waves propagating to the left and to the right of the point $x = 0$, and a contact discontinuity (Figure 1.1(b)). In configuraton C there are two rarefaction waves and a contact discontinuity (Figure 1.1(c)). Critical values of the parameters separating one configuraton from another have been derived in [1.18]. Certain flow configurations, which may be called intermediate configurations between the main configurations A, B, C, correspond to these critical values. For example, an intermediate configuration between configurations A and B is the configuration consisting of one shock wave and a contact discontinuity (Figure 1.1(d)). An intermediate configuration between configurations B and C is the configuration consisting of a stagnant contact boundary (Figure 1.1(e)) and a rarefaction wave. In the particular case of configuration C there can occur a separation of the gases "1" and "2" from one another, and then the rarefaction waves are separated by a region of vacuum in which $\rho = p = c = 0$ (Figure 1.1(f)).

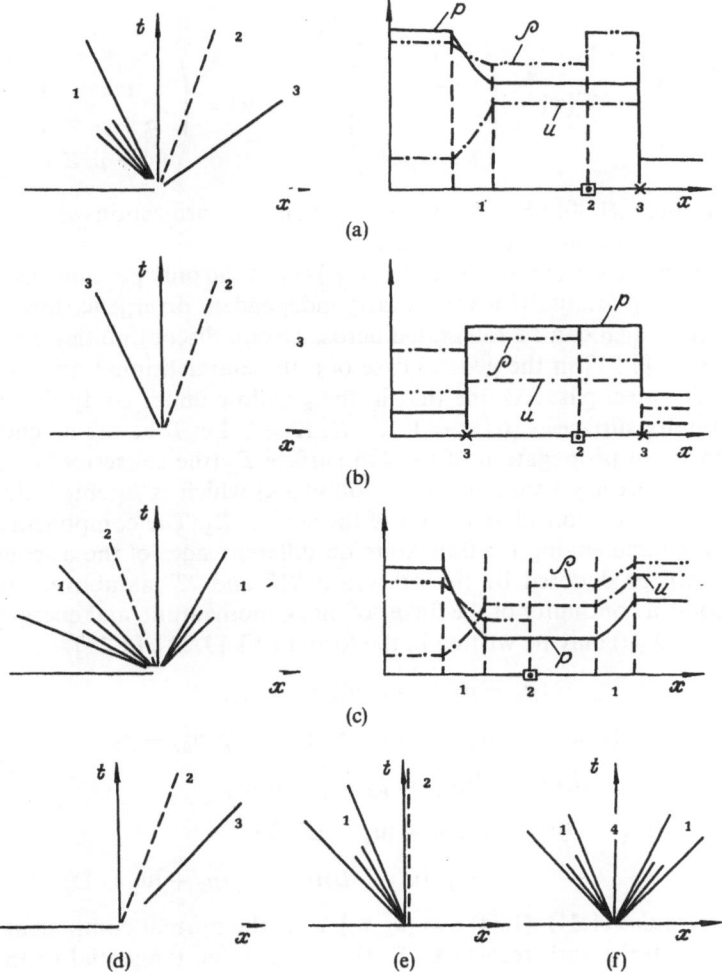

Figure 1.1. Flow configurations in the breakdown of the discontinuity problem (1—rarefaction wave; 2—contact discontinuity; 3—shock wave; 4—the domain of vacuum).

1.2. Jump Conditions in the Case of Two-Dimensional Gas Flow

The system of Euler equations governing the flow of an inviscid compressible non-heat-conducting gas in the case of two spatial variables x, y and time t may be written in the form [1.17]

$$\partial \mathbf{w}/\partial t + \partial \mathbf{F}(\mathbf{w})/\partial x + \partial \mathbf{G}(\mathbf{w})/\partial y = 0, \qquad (1.30a)$$

where

$$\mathbf{w} = \begin{pmatrix} \rho \\ \rho u \\ \rho v \\ \rho E \end{pmatrix}, \qquad \mathbf{F(w)} = \begin{pmatrix} \rho u \\ p + \rho u^2 \\ \rho u v \\ pu + \rho u E \end{pmatrix}, \qquad \mathbf{G(w)} = \begin{pmatrix} \rho v \\ \rho u v \\ p + \rho v^2 \\ pv + \rho v E \end{pmatrix}. \quad (1.30b)$$

In the formulas (1.30b) $E = \varepsilon + 0.5(u^2 + v^2)$, u, v are projections of the velocity vector on the x- and y-axis, respectively.

Note that the system of equations (1.30) is not the only possible divergence form of the equations. Thirteen linearly independent divergence forms of the equations, which can be integrated across strong discontinuities, have been obtained in [1.21] in the general case of a three-dimensional nonstationary flow of a perfect gas. Assume that in the gas flow under study there are K discontinuity surfaces $\Sigma_k(t)$, $k = 1, \ldots, K$, $K \geq 1$. Let D be the magnitude of the velocity of propagation of the kth surface Σ_k (the subscript "k" at D is omitted for brevity here and in the following) which is oriented along the normal to Σ_k at a considered point of the surface Σ_k. The components of the vector $\mathbf{w}$ characterizing the fluid state on different sides of the discontinuity surface will be denoted by the subscripts "1" and "2" as above. Then the conditions of continuity of the fluxes of mass, momentum, and energy across the surface $\Sigma_k(t)$ may be written in the form [1.5], [1.22], [1.23]

$$(\rho_1 - \rho_2)D = \rho_1 u_{n1} - \rho_2 u_{n2}; \quad (1.31)$$

$$(\rho_1 u_{n1} - \rho_2 u_{n2})D = \rho_1 u_{n1}^2 + p_1 - \rho_2 u_{n2}^2 - p_2; \quad (1.32)$$

$$\rho_1(u_{n1} - D)u_{\tau 1} = \rho_2(u_{n2} - D)u_{\tau 2}; \quad (1.33)$$

$$\rho_1(u_{n1} - D)(\varepsilon_1 + p_1/\rho_1 + (\mathbf{u}_1 - \mathbf{D})^2/2)$$
$$= \rho_2(u_{n2} - D)(\varepsilon_2 + p_2/\rho_2 + (\mathbf{u}_2 - \mathbf{D})^2/2). \quad (1.34)$$

In the formulas (1.31)–(1.34) $\mathbf{u} = (u_n, u_\tau)$, u_n is the normal component of the velocity vector $\mathbf{u}$ with respect to $\Sigma_k(t)$, and u_τ is the tangential to the $\Sigma_k(t)$ velocity component. As was pointed out in [1.5], the only case considered systematically was when the sets of the points of discontinuity form piecewise smooth discontinuity surfaces of the first kind of the flow parameters. In the following we shall assume that $\Sigma_k(t)$ is a smooth surface of a discontinuity in the gas flow. Then all the types of discontinuities in the two-dimensional inviscid gas flows are described by the solutions of the algebraic system (1.31)–(1.34). Consider now the solutions of this system for specific types of discontinuities.

Shock Waves. In this case $\hat{j} = \rho_1(u_{n1} - D) = \rho_2(u_{n2} - D)$, and it follows from (1.33) that

$$u_{\tau 1} = u_{\tau 2}, \quad (1.35)$$

that is, there is no jump in the tangential velocity component. The conditions (1.31)–(1.34) on the shock wave $\Sigma_k(t)$ are reduced to three equations: (1.31), (1.32), and

$$\varepsilon_1 + p_1/\rho_1 + (u_{n1} - D)^2/2 = \varepsilon_2 + p_2/\rho_2 + (u_{n2} - D)^2/2, \qquad (1.36)$$

which are called Rankine–Hugoniot conditions. If $u_{\tau 1} = u_{\tau 2} = 0$, then the shock wave is called normal shock wave, otherwise oblique shock wave. The elimination of the quantities $(u_{n1} - D)$, $(u_{n2} - D)$ from (1.31), (1.32), and (1.36) yields the relationship (1.28).

Let the surface $\Sigma_k(t)$ propagate in the direction of $\mathbf{D}$ from left to right. Then in the case $\hat{j} < 0$ the shock wave moves to the right relative to the medium, and in the process of wave motion the material intersects the shock wave $\Sigma_k(t)$ moving from right to left with respect to $\Sigma_k(t)$. Conversely, if $\hat{j} > 0$, then the shock wave moves to the left with respect to the medium. The medium to the right of the shock wave at $\hat{j} < 0$ and to the left of the shock wave at $\hat{j} > 0$ is called the medium before the shock wave, and another medium is called the medium behind the shock wave. The parameters of the medium before and behind the shock wave will be denoted (by analogy with Section 1.1.2) by u_{n2}, $u_{\tau 2}, \rho_2, p_2, \varepsilon_2, V_2$ and $u_{n1}, u_{\tau 1}, \rho_1, p_1, \varepsilon_1, V_1$, respectively. The entropy increases in a stable shock wave, so that

$$S_1 > S_2. \qquad (1.37)$$

The inequalities similar to (1.29) may be rewritten in the multidimensional case as [1.5]

$$|u_{n2} - D| > c_2, \qquad |u_{n1} - D| < c_1. \qquad (1.38)$$

In the case of the equation of state (1.8) the formula

$$u_{n2} - u_{n1} = 2(\text{sign } \hat{j})c_2(M_2 - 1/M_2)/(\gamma + 1) \qquad (1.39)$$

is valid ([1.5], [1.18]), where $M_2 = |u_{n2} - D|/c_2 > 1$.

Contact Discontinuities. In this case the fluid (the gas) does not flow across the discontinuity surface, that is, $\hat{j} = 0$, and we then obtain from (1.31), (1.32) that $p_1 = p_2$ on $\Sigma_k(t)$, and the quantities ρ, ε, u_τ can have an arbitrary jump across $\Sigma_k(t)$. If $\hat{j} = 0$ and

$$u_{\tau 1} - u_{\tau 2} \neq 0, \qquad (1.40)$$

then the discontinuity is called tangential; if

$$\hat{j} = 0, \qquad u_{\tau 1} = u_{\tau 2}, \qquad p_1 = p_2, \qquad (\rho_1 - \rho_2)^2 + (\varepsilon_1 - \varepsilon_2)^2 \neq 0, \qquad (1.41)$$

then such a contact discontinuity is called (in [1.5]) a purely contact discontinuity.

The relationships (1.31)–(1.34) are also obviously satisfied at the (x, y, t) points where the solution $(\rho, u, v, p, \varepsilon)$ is continuous. In particular, these relationships are satisfied in compression waves and rarefaction waves.

1.3. Homogeneous Difference Schemes and Their Differential Approximations

1.3.1. Two Approaches to Construction of Schemes for the Computation of Discontinuous Solutions

Consider the Cauchy problem

$$\partial \mathbf{u}/\partial t + \partial \boldsymbol{\varphi}(\mathbf{u})/\partial x = 0, \tag{1.11}$$

$$\mathbf{u}(x, 0) = \mathbf{u}_0(x), \qquad 0 \le t \le \bar{t}. \tag{1.42}$$

The corresponding difference Cauchy problem may be formulated as follows:

$$(1/\tau)(T_0 - I)\mathbf{u}^n = \Lambda(T_1, \mathbf{u}^n, \mathbf{u}^{n+1}); \tag{1.43}$$
$$\mathbf{u}^0(x) = \mathbf{u}_0(x),$$

where Λ is a vector-matrix difference operator and T_0, T_1 are the shift operators,

$$T_1 \equiv T_{+1}; \qquad T_0 \equiv T_{+0}; \qquad T_{+1}f(x, t) = f(x + h, t);$$
$$T_{+0}f(x, t) = f(x, t + \tau);$$
$$T_1^m f(x, t) = f(x + mh, t); \tag{1.44}$$
$$T_0^m f(x, t) = f(x, t + m\tau), \qquad m \in R,$$

where h is the step of a uniform grid on the x-axis; I is the identity operator,

$$I\mathbf{u}^n \equiv \mathbf{u}^n, \qquad \mathbf{u}^n(x) = \mathbf{u}(x, \tau n), \qquad n = 0, 1, 2, \ldots,$$

τ is the time step. The equation system (1.43) relating the functions $\mathbf{u}^{n+1}$, $\mathbf{u}^n$ at the two time levels $t^{n+1} = (n + 1)\tau$ and $t^n = n\tau$, is usually called a two-level difference scheme [1.18].

Introduce the difference operator Ω by the formula

$$\Omega\mathbf{u}^n = [(1/\tau)(T_0 - I)\mathbf{u}^n - \Lambda(T_1, \mathbf{u}^n, T_0\mathbf{u}^n)] = 0.$$

We shall say that the difference scheme (1.43) approximates the system (1.11) if the condition

$$\|\Omega\mathbf{u} - \partial\mathbf{u}/\partial t - \partial\boldsymbol{\varphi}(\mathbf{u})/\partial x\| \to 0 \qquad (\tau, h \to 0)$$

is satisfied in some class of the functions $\mathbf{u}(x, t)$ (these can be spline functions of a sufficiently high smoothness) where the symbol $\|\cdot\|$ denotes some appropriately chosen norm. If, for arbitrary n and at $\tau \to 0$, $h \to 0$, the condition $\|\mathbf{u}^n - \mathbf{u}(x, t^n)\| \to 0$ is satisfied, then one says that the solution of a difference problem converges to the solution of a differential problem. In the case when a difference scheme of the form, for example, (1.43), approximates a system of nonlinear equations (1.11), (1.9), it is very difficult to prove the convergence of the scheme, although it is not impossible. A review of the results, on the convergence of some class of the Cauchy problem $du/dt = Au$, $u|_{t=0} = u_0$

approximations where A is some linear or nonlinear operator, has been presented in [1.24]. Let K_τ be an operator of solution advancement by one time step τ which corresponds to a difference approximation of the equation $du/dt = Au$. If $\tau = t/n$ and we make n steps, then it is supposed that the operator $K_{t/n}^n = K_{t/n} \circ \cdots \circ K_{t/n}$ approximates the evolution operator of the equation $du/dt = Au$. Denote now by F_t the evolution operator of the equation, that is, $F_t(u_0)$ is the solution with initial value u_0. Then the convergence of the algorithm is written (in [1.24]) in the form $F_t = \lim_{n\to\infty} K_{t/n}^n$. The convergence theory presented in [1.24] generalizes substantially the well-known equivalence theorem of Lax [1.2] which was proved for linear systems. The theory of Lax yielded the conditions ensuring the convergence of the computational algorithm. One of these conditions, the stability, was studied in detail in [1.2] for the case when a two-level difference scheme (1.43) has the form $B_1 u^{n+1} = B_0 u^n$ where B_1 and B_0 are linear operators. The stability consideration in [1.2] was based on the use of the Fourier transformation of the operators B_1 and B_0 entering into the system of difference equations $B_1 u^{n+1} - B_0 u^n = 0$. The stability investigation of nonlinear difference schemes approximating the Euler equation system (1.11), (1.9) entails great difficulties. A stability analysis of the linearized difference equations (the presentation of a corresponding technique may be found, for example, in [1.2]) may lead to false conclusions on the scheme stability. There are known examples showing that a scheme which is unstable in accordance with the linear analysis is nonlinearly stable. For example, the schemes of the HARLOW particle-in-cell method [1.25], [1.26], the FLIC method [1.27], and the "coarse particles" method [1.28] are unstable in accordance with linear analysis in the computation of flows where the fluid velocity u is small; however, computation by these schemes does not result in an infinite increase in the amplitude of the difference solution oscillations with increasing t, but some auto-oscillatory regime of the difference solution in the subdomains with small fluid velocity does take place; this phenomenon was studied for the first time in [1.29]. Examples of computations by difference schemes approximating a nonlinear equation with sign-changing viscosity coefficient have been presented in [1.30]. These schemes also prove to be unstable from the point of view of linear analysis; at the same time, these schemes are nonlinearly stable.

Difference schemes whose formulas are of the same type at mesh points, regardless of the presence and character of singularities of the solution in a neighborhood of a point, have been called homogeneous. The simplicity of implementation of such schemes has stimulated their wide acceptance in computations of shocked gas flows.

The scheme homogeneity naturally makes added demands on the computational algorithm, since it must now, at least in principle, "equally well" describe both smooth and discontinuous flows.

There are two general approaches to the construction of homogeneous difference schemes for the computation of discontinuous solutions. The first approach is based on the fact that the integral conservation laws (1.19) should be satisfied for any distinguished portion of the gas or space. The conservation laws (1.19) must be augmented by requiring a nondecrease of entropy of any fixed mass of gas to exclude the appearance of unstable discontinuities. After that, the difference scheme is constructed by means of a direct approximation of integral conservation laws. A good difference approximation of the conservation laws should contain "positive" viscosity, otherwise the errors introduced by a difference approximation may destabilize the numerical solution. The Lax, Lax–Wendroff, and Godunov schemes are examples of stable schemes approximating the integral conservation laws without the explicit introduction of pseudoviscosity.

A characteristic feature of the direct approximation of integral conservation laws is the so-called conservative or divergence property of the difference schemes obtained by it. This property consists of the fact that the equations of the difference scheme can be interpreted as the integral conservation laws (1.19) for a cell of the mesh formed by the intersection of the lines $t = n\tau$ and $t = (n + 1)\tau$ with the lines $x = ih$, $x = (i + 1)h$, with some approximation (or interpolation) of the quantities contained in the conservation laws on the boundaries of the cell. The form of this approximation remains constant on the given boundary of the cell, that is, it is the same for the conservation laws in neighboring cells. This ensures the property of additivity of a conservative difference scheme which consists in the fact that, in summing the difference equations over neighboring cells, we obtain a new equation which can be considered as the conservation laws for the outer boundary of the region formed from the union of these cells. Thus, the difference equations possess the same properties as the contour integrals (1.19).

If the difference scheme converges, that is, the family of difference solutions has a limit as τ, $h(\tau) \to 0$ (in some weak norm), then this limit satisfies precisely the required conservation laws (1.19), and no others [1.18]. This property of conservative difference schemes is similar to the property of a system of differential equations of parabolic type

$$\partial \mathbf{u}_\mu/\partial t + \partial \varphi(\mathbf{u}_\mu)/\partial x = \mu\, \partial/\partial x[B(\mathbf{u}_\mu, x, t)\, \partial \mathbf{u}_\mu/\partial x], \qquad (1.45)$$

in which the "viscosity"—the right-hand side of (1.45)—is contained in divergence form as a derivative of the expression $B\, \partial \mathbf{u}_\mu/\partial x$, where B is a rather arbitrary matrix.

For the system (1.45) there is a similar property. If as $\mu \to 0$ the solution $\mathbf{u}_\mu(x, t)$ has a limit $\mathbf{u}(x, t)$ in some appropriate sense, then this limit satisfies the conservation laws (1.19), that is, it is a generalized solution of the system (1.11) [1.18]. Thus stable approximations of the integral conservation laws bear a resemblance to the method of "vanishing viscosity" for constructing discontinuous solutions of systems of quasi-linear equations.

We now proceed to the second and most widespread approach to the construction of homogeneous difference schemes—an approach based on the introduction of the artificial viscosity or "pseudoviscosity". By introducing a pseudoviscosity into the equations of gas dynamics we approximately describe shock waves as a smooth shock transition [1.1]. The first method of introducing the pseudoviscosity, which was proposed by von NEUMANN and RICHTMYER [1.1], consisted in the fact that the artificial viscosity q was introduced additively into the pressure. In more recent investigations other authors have proposed more complicated pseudoviscosities having a vector-matrix form and which was introduced into all the equations of the system (1.1)–(1.3). A review of a number of such investigations may be found in [1.3].

1.3.2. Two Forms of Differential Approximations

The difference solution behavior in the neighborhood of strong discontinuities depends substantially on the form of introduced pseudoviscosity, but is not determined by it completely. The "smearing" of discontinuities is also affected by the scheme viscosity which is implicitly present because of the errors in approximation of integral conservation laws. Thus, it is necessary to simultaneously take into account both the approximation and the artificial viscosity in a difference scheme for a detailed study of difference solution behavior. The differential approximation method [1.18], [1.19] is now widely used for this purpose. Let us introduce the notion of the first differential approximation (f.d.a.) of a difference scheme. We make use of operator representations [1.18], [1.19]

$$T_0 = \exp(\tau D_0) = \exp(\tau \partial/\partial t) = \sum_{l=0}^{\infty} \frac{\tau^l}{l!} D_0^l;$$

$$T_1 = \exp(h D_1) = \exp(h \partial/\partial x) = \sum_{l=0}^{\infty} \frac{h^l}{l!} D_1^l. \tag{1.46}$$

Substituting formulas (1.46) into scheme (1.43) instead of T_0 and T_1, we obtain a system of partial differential equations of an infinite order

$$\sum_{l=1}^{\infty} \frac{\tau^{l-1}}{l!} \frac{\partial^l \mathbf{u}}{\partial t^l} = \Lambda[\exp(h D_1), \mathbf{u}, \exp(\tau D_0)\mathbf{u}]. \tag{1.47}$$

Suppose that the difference scheme (1.43) approximates the original system $\partial \mathbf{u}/\partial t = L\mathbf{u}$ with an order $O(\tau^{\gamma_1}) + O(h^{\gamma_2})$. Then, neglecting in (1.47) the terms of the order $O(\tau^\alpha)$, $O(h^\beta)$, where $\alpha = \gamma_1 + 1, \gamma_1 + 2, \ldots, \beta = \gamma_2 + 1, \gamma_2 + 2, \ldots$, we obtain

$$\partial \mathbf{u}/\partial t + (\tau/2)\, \partial^2 \mathbf{u}/\partial t^2 + \cdots + [\tau^{\gamma_1}/(\gamma_1 + 1)!]\, \partial^{\gamma_1+1} \mathbf{u}/\partial t^{\gamma_1+1} = L\mathbf{u} + L_1(D)(\mathbf{u}), \tag{1.48}$$

where $L_1(D)$ is some differential operator whose coefficients have the orders $O(\tau, \tau^2, \ldots, \tau^{\gamma_1}, h, h^2, \ldots, h^{\gamma_2})$. The equation (1.48) will be called the H-form of the f.d.a. of a difference scheme (1.43). Let us replace the derivatives $\partial^l \mathbf{u}/\partial t^l$

($l \geq 2$) in (1.48) by the x-derivatives using the f.d.a. H-form. For example, in order to obtain such an expression for $\partial^2 \mathbf{u}/\partial t^2$ we differentiate both sides of the f.d.a. (1.48) with respect to t. As a result, we obtain the f.d.a. P-form of a difference scheme (1.43):

$$\partial \mathbf{u}/\partial t = L\mathbf{u} + \tilde{L}_1(D)(\mathbf{u}),$$

where $\tilde{L}_1(D)$ is a differential operator involving only the operators of x-differentiation $\partial^k/\partial x^k$, $k = 0, 1, \ldots, 1 + \max(\gamma_1, \gamma_2)$.

EXAMPLE. Consider an explicit difference scheme with one-sided differences

$$[u^{n+1}(x) - u^n(x)]/\tau + a[u^n(x) - u^n(x - h)]/h = 0 \qquad (1.49)$$

approximating the equation with a constant coefficient $a > 0$

$$\partial u/\partial t + a\,\partial u/\partial x = 0. \qquad (1.50)$$

The f.d.a. H-form of the difference scheme (1.49) is easily computed with the aid of expansions (1.46)

$$\partial u/\partial t + a\,\partial u/\partial x = (ah/2)\,\partial^2 u/\partial x^2 - (\tau/2)\,\partial^2 u/\partial t^2. \qquad (1.51)$$

In order to find the f.d.a. P-form from (1.51) let us express the derivative u_{tt} in terms of the x-derivatives. Differentiating both sides of (1.51) with respect to x we have

$$\partial^2 u/\partial t^2 = -a\,\partial^2 u/\partial x\,\partial t + (ah/2)\,\partial^3 u/\partial x^2 \partial t - (\tau/2)\,\partial^3 u/\partial t^3. \qquad (1.52)$$

Let us express the mixed derivative u_{xt} in (1.52) by the x-derivatives. For this purpose, we differentiate both sides of the equation (1.51) with respect to x:

$$\partial^2 u/\partial t\,\partial x = -a\,\partial^2 u/\partial x^2 + O(\tau) + O(h). \qquad (1.53)$$

Employing formula (1.53), we rewrite equation (1.52) in the form

$$\partial^2 u/\partial t^2 = a^2\,\partial^2 u/\partial x^2 + O(\tau) + O(h). \qquad (1.54)$$

Substituting the right-hand side of (1.54) instead of u_{tt} into (1.51), we obtain the f.d.a. P-form of scheme (1.49):

$$\partial u/\partial t + a\,\partial u/\partial x = (a/2)(h - a\tau)\,\partial^2 u/\partial x^2. \qquad (1.55)$$

1.4. On the Applicability of Progressive Wave-Type Solutions of the First Differential Approximation Equations

As was shown in Section 1.1, in the one-dimensional inviscid compressible gas flow there may take place strong discontinuities of two kinds: shock waves and contact discontinuities. The propagation speed D of a discontinuity is generally a function of time. In the case when a discontinuity moves at a constant speed the quantity D does obviously not depend on t, that is, $D = $ const. As is known, there are among the solutions of the system (1.1)–(1.3), (1.5) solutions containing a stationary shock wave or a stationary contact discontinuity [1.18], and in the problem on the breakdown of discontinuity there may take place simultaneously both types of discontinuities moving at constant speeds.

Of course, the position of a strong discontinuity within the zone of its "smearing", in the case of using homogeneous difference schemes, depends on the structure of the difference solution in the above zone. This structure substantially depends on the approximation and artificial viscosity present in a difference scheme [1.3], [1.18], [1.19]. Therefore, the construction of the algorithms for shock localization within the limits of zones of shock smearing depends on the properties of approximation and artificial viscosity of a scheme. A mathematical study of the shock transition structure in a fluid possessing physical viscosity and heat conduction is usually accomplished in the example of a stationary shock wave [1.18]. Thereby it is assumed that all the functions in the equations of gas flow depend on the only variable

$$\xi = x - Dt, \tag{1.56}$$

where D is the speed of a stationary shock wave. The solutions depending on the variable ξ (1.56) are usually called the progressive wave-type solutions. The use of such solutions enables one to reduce the analysis of shock transition structure to a study of a system of ordinary differential equations. In this book we utilize the same approach, while studying the properties of shock waves smearing which takes place as the result of action of the approximation or artificial viscosity of the difference scheme. In this study we use the system of equations of the f.d.a. of a difference scheme as a mathematical model describing the difference solution structure in a smeared shock wave. Of course, there arises the question of the degree of proximity of the progressive wave-type solutions of the f.d.a. equations to the grid solutions obtained by a finite-difference scheme. Three ways of investigating this question are possible. The first way consists in obtaining an explicit analytical solution of the f.d.a. equations. For many known scalar artificial viscosities introduced additively into the pressure this way proved to be realistic, as was demonstrated for the first time in [1.1]. In more complicated cases, when the scheme viscosity has a matrix-vector form, it is impossible to solve the f.d.a. equations explicitly because of their substantial nonlinearity. Two methods of investigation are possible in these cases. The first is to obtain in some norm, say, in the L_2 space norm, the estimates for the difference $\mathbf{w} - \mathbf{w}_h$, where $\mathbf{w}$ is the solution of the f.d.a. equations and $\mathbf{w}_h$ is the solution of the difference equations. It appears that the first estimates of this kind were presented in the first edition of the book [1.18] which appeared in 1968. These estimates, as well as the estimates published in [1.34], [1.35], addressed the case of a scalar linear equation (1.50).

However, in the case of a finite-difference approximation of a scalar nonlinear equation

$$\partial u/\partial t + \partial(u^2/2)/\partial x = 0 \qquad (1.57)$$

such estimates are absent in the literature. In this case (for investigation of the proximity of the f.d.a. solutions to the solutions of difference equations) one applies the technique of direct comparison of difference solutions, obtained by integrating the ordinary differential f.d.a. equations, and the grid solutions of finite-difference equations approximating the original system of partial differential equations. For the case of a finite-difference approximation of the nonlinear equation (1.57) such a direct comparison was apparently accomplished for the first time in [1.36]. Note that in [1.36] the LAX scheme [1.31] was taken as a difference scheme, and then the f.d.a. *P*-form of this scheme was integrated numerically. In [1.37] a finite-difference solution obtained by a "corner" scheme approximating a nonlinear inhomogeneous equation

$$\partial u/\partial t + \partial\varphi(u)/\partial x = \Psi(u)$$

was compared to the numerical solution of the f.d.a. equations written as a system of ordinary differential equations for two functions depending only on the variable (1.56). These two equations were integrated by an *A*-stable method from the GEAR package. As a result it was found that the relative deviation of the f.d.a. equation solution from the difference solution did not exceed 4%.

Indirect data on the applicability of progressive wave-type solutions of the f.d.a. equations (for the description of difference solution properties in the zone of smearing of a strong discontinuity) are also given by a qualitative analysis of the ordinary differential f.d.a. equations. This analysis gives an opportunity of elucidating the monotone or nonmonotone character of the numerical solution in the neighborhood of the discontinuities. As a rule, the results of such a f.d.a. analysis agree with the results of computations by finite-difference schemes of various orders of accuracy. In a series of works [1.38]–[1.42] this analysis has been carried out for the f.d.a. of equations (1.50) and (1.57), and in [1.43], [1.44] for (1.57).

1.4.1. Singular Points of the First Differential Approximation Equations

Consider the problem of the propagation of a stationary shock wave in a gas. Let W_1, W_2 be constant vectors describing the gas state behind and before the shock front, respectively. To study the properties of shock-front smearing in numerical solutions of the above problem, let us consider in more detail the case when the Euler equation system (1.11), (1.9) is approximated by first-order finite-difference schemes. To complete this system we make use, as in [1.45], of the ideal gas equation of state (1.8). Let the *H*-form of the f.d.a. of a divergence difference scheme approximating the system (1.11) have the form

$$\partial w/\partial t + \partial \varphi(w)/\partial x = -(\tau/2)\, \partial^2 w/\partial t^2 + \partial/\partial x[C(w, h, \tau)\, \partial w/\partial x]. \quad (1.58)$$

In (1.58) we have introduced the notation $w(x, t)$ for the solution of the f.d.a. equations to emphasize its difference from the solution $u(x, t)$ of the original system (1.11), (1.9). Further, in (1.58) C is some 3×3 matrix. Then the f.d.a. P-form may be written as follows:

$$\partial w/\partial t + \partial \varphi(w)/\partial x = \partial/\partial x\{B(w, h, \tau)\, \partial w/\partial x\}, \quad (1.59)$$

where

$$B(w, h, \tau) = B_P(w, h, \tau)$$

$$= C(w, h, \tau) - (\tau/2)A^2(w), \qquad A = \partial\varphi/\partial w. \quad (1.60)$$

Consider the solutions of the systems (1.58), (1.59) of the form $w = w(\xi)$ where ξ is a variable determined by (1.56). The system (1.58) then goes over to a system of ordinary differential equations

$$\varphi(w) - Dw + G = B(w, h, \tau)\, dw/d\xi, \quad (1.61)$$

where

$$B = B_H(w, h, \tau) = C(w, h, \tau) - (\tau/2)D^2 I, \quad (1.62)$$

I is the unit matrix, and G is a constant vector. For the determination of the vector G we require that the solution of the system (1.61) satisfies the conditions

$$w(\xi) = \begin{cases} W_1, & \xi \to -\infty; \\ W_2, & \xi \to +\infty. \end{cases} \quad (1.63)$$

The conditions of the form (1.63) are conventional in the studies of shock transition structure in the gases possessing physical viscosity and heat conduction (see, for example, [1.18]). For the existence of the solution $w(\xi)$ satisfying the conditions (1.63) it is necessary [1.18] for the points W_1, W_2 to be stationary points of the system (1.61) which leads to the relations

$$G = DW_1 - \varphi(W_1) = DW_2 - \varphi(W_2). \quad (1.64)$$

It is easy to see that the relations (1.64) coincide with the Hugoniot conditions (1.21).

Following [1.45], consider two first-order difference schemes for the system (1.11): the LAX scheme [1.31] and the "breakdown of discontinuity" scheme [1.33], [1,46]. The f.d.a. H-form of the Lax scheme has the form (1.58) where $C = (h^2/2\tau)I$, therefore,

$$B_P(w, h, \tau) = [h^2/(2\tau)]I - (\tau/2)A^2;$$

$$B_H(w, h, \tau) = [h^2/(2\tau)]I - (\tau/2)D^2 I. \quad (1.65)$$

In the case when the flow under study is supersonic and $u > c > 0$ the scheme

([1.46]) is a difference scheme with one-sided differences in x and in t. Hence the f.d.a. H-form of this scheme in the case when $u > c > 0$ is obtained easily, and has the form (1.58) where $C = (h/2)A$. Then

$$B_P(\mathbf{w}, h, \tau) = A[(h/2)I - (\tau/2)A];$$
$$B_H(\mathbf{w}, h, \tau) = (h/2)A - (\tau/2)D^2 I. \tag{1.66}$$

Denote singular points at which $\mathbf{w} = \mathbf{W}_1$, $\mathbf{w} = \mathbf{W}_2$, by O_1, O_2, respectively. Let us determine the type of the singular points O_1, O_2 when using the matrices B of the form (1.65), (1.66). For this purpose we linearize the system (1.61) in the vicinity of the singular point $O_j, j = 1, 2$. Let us set

$$\mathbf{w}(\xi) = \mathbf{W}_j + \delta\mathbf{w}'(\xi), \tag{1.67}$$

where δ is a small quantity. For $\mathbf{w}'(\xi)$ we obtain a linear system

$$d\mathbf{w}'/d\xi = \mathscr{F}_j\mathbf{w}', \tag{1.68}$$

where

$$\mathscr{F}_j = B^{-1}(\mathbf{W}_j)[A(\mathbf{W}_j) - DI], \qquad j = 1, 2. \tag{1.69}$$

It is assumed here and in the following that the time step τ satisfies the inequality $\tau < h/\max(|u| + c)$ known as the Courant–Friedrichs–Lewy stability condition [1.2], [1.3]. Employing formulas (1.16) for the eigenvalues of the matrix A, entering formulas (1.65), (1.66) as well as the inequalities (1.29), it is easy to show that in the case of the matrices B_P, B_H determined by (1.65), (1.66) det $B(\mathbf{W}_j) \neq 0$ at $B = B_H$ and $B = B_P$, thus, an inverse matrix $B^{-1}(\mathbf{W}_j)$ entering the formula (1.69) exists. For example, in the case of the matrix B_P from (1.66) we have at a singular point $\mathbf{w} = \mathbf{W}_j$:

$$\det B_P = \prod_{k=1}^{3} \lambda_{jk}[(h/2) - (\tau/2)\lambda_{jk}],$$

where λ_{jk} is the kth eigenvalue of the matrix $A(\mathbf{W}_j)$. Let $\mu_{jk}^P, \mu_{jk}^H, k = 1, 2, 3$, be the eigenvalues of the matrix $\mathscr{F}_j$ obtained from (1.69) at $B = B_P$ and $B = B_H$, respectively. In the case of the Lax scheme we obtain the following expression from (1.69) with regard for (1.65):

$$\mu_{jk}^P = 2(\lambda_{jk} - D)/(h^2/\tau - \tau\lambda_{jk}^2); \tag{1.70}$$

$$\mu_{jk}^H = 2(\lambda_{jk} - D)/(h^2/\tau - \tau D^2). \tag{1.71}$$

For definiteness consider a shock wave moving from left to right. Then $D > 0$, and the inequalities

$$u_1 + c_1 > D > u_2 + c_2 \tag{1.72}$$

take place by virtue of the Zemplén theorem (1.29). Taking into account (1.16), (1.72) we obtain from the relationships (1.70), (1.71) that the inequalities

$$\mu_{11}^{P,H} < 0; \qquad \mu_{12}^{P,H} < 0; \qquad \mu_{13}^{P,H} > 0;$$

take place at the singular point O_1 in the case of the matrix B_P as well as the matrix B_H from (1.65). Thus, O_1 is a saddle of the first kind [1.47]. It is stated analogously that at the right singular point O_2 the inequalities $\mu_{2k}^{P,H} < 0, k = 1$, 2, 3, take place when using the matrices (1.65), thus the point O_2 is a stable node [1.47]. Consequently, in the case of the Lax scheme a stable direction of numerical integration of the system (1.61) will be the integration direction from the singular point O_1 to the singular point O_2.

When using the matrices B_P, B_H (1.66) we obtain

$$\mu_{jk}^P = 2(\lambda_{jk} - D)/[\lambda_{jk}(h - \tau\lambda_{jk})];$$
$$\mu_{jk}^H = 2(\lambda_{jk} - D)/(h\lambda_{jk} - \tau D^2). \tag{1.73}$$

It is easily stated, analogously to the foregoing, that in this case also, the singular point O_1 is a saddle of the first kind and the singular point O_2 is a stable node when using the matrix B_P from (1.66). As will be shown below in Section 1.4.2, in the case when the direction of the stable integration of the system (1.61) is known, one can use one-step numerical integration methods, namely, the Runge–Kutta methods. For the realization of the Runge–Kutta methods it is necessary to transform the system (1.61) to a form resolved with respect to the derivatives $d\mathbf{w}/d\xi$:

$$d\mathbf{w}/d\xi = (\det B)^{-1}B^*(\mathbf{w})[\boldsymbol{\varphi}(\mathbf{w}) - D\mathbf{w} + \mathbf{G}], \tag{1.74}$$

where $B^*(\mathbf{w})$ is the adjoint of the matrix B [1.48]. We show that the system (1.61) can have several singular points within the smeared shock wave zone when using the matrix B_H from (1.66). These points will be called inner singular points unlike the singular points O_1, O_2. Suppose that in the smeared shock wave zone the inequalities

$$0 < |d\mathbf{w}/d\xi| < \infty \tag{1.75}$$

take place. Let the smeared shock wave zone possess such a point ξ_0 at which $\mathbf{w}(\xi_0) = \mathbf{w}_0$ and $\det B(\mathbf{w}_0, h, \tau) = 0$. Consider $\lim_{\xi \to \xi_0} d\mathbf{w}/d\xi$. By virtue of (1.75) we should suppose that

$$\lim_{\xi \to \xi_0} B^*(\mathbf{w})[\boldsymbol{\varphi}(\mathbf{w}) - D\mathbf{w} + \mathbf{G}] = 0,$$

thus there is a singularity of the system (1.75) at the point ξ_0. With regard to (1.60), (1.12) we obtain that

$$\det B_H(\mathbf{w}, h, \tau) = (h^3/8)(u - \kappa D^2)(u - c - \kappa D^2)(u + c - \kappa D^2),$$

where $\kappa = \tau/h$. Consider in the (u, c)-plane a rectangular domain such that the singular points O_1, O_2 lie in this domain (Figure 1.2). Let $u = u(\xi)$, $c = c(\xi)$, be the integral curve of the system (1.61) passing through the singular points O_1 and O_2. Considering various cases of this integral curve position with respect to the lines $u = \kappa D^2$, $u - c = \kappa D^2$, $u + c = \kappa D^2$ it is easy to find that the number of inner singular points O_j, $j \geq 3$, can vary from 0 to N where $N \geq 4$. For example, in Figure 1.2(a) one of four cases for the absence of inner singular points is shown (appropriate numerical results will be presented

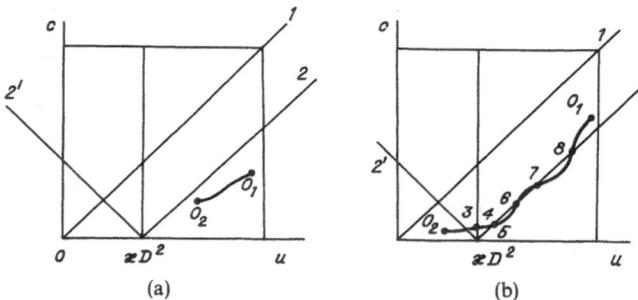

Figure 1.2. Singular points in the (u, c)-plane: (a) in the absence of inner singular points; (b) in the presence of such points.

below). In Figure 1.2(b) a case for the presence of eight singular points of the system (1.61) is shown where inner singular points are enumerated by the figures 3–8. The presence of inner singular points creates additional serious difficulties in the numerical integration of the system (1.61) [1.49]. However, even in the case when there are no inner singular points for the system (1.61), one-step numerical methods can prove to be inapplicable because of the absence of the stable integration direction. The last thing may take place when using the matrix B_H from (1.66), as follows from (1.73) (see also [1.10], [1.45]). In such situations it is necessary to apply iterative methods (as described in [1.49]–[1.52]) for the numerical solution of the problem (1.61), (1.63).

1.4.2. Numerical Study of the Smeared Shock Wave Structure

Let us describe the computational algorithm for solving the problem (1.61), (1.63) which we have created for the case of the absence of inner singular points and the presence of the stable integration direction. Let us nondimensionalize (1.61) using the formulas

$$\bar{p} = p/p_1; \qquad \bar{\rho} = \rho/\rho_1; \qquad \bar{u} = u/u_a; \qquad \bar{\varepsilon} = \varepsilon\rho_1/p_1;$$
$$\bar{\xi} = \xi/(0.5h); \qquad \bar{\tau} = \tau/t_a; \qquad u_a = \sqrt{p_1/\rho_1}; \qquad t_a = 0.5h/u_a.$$

It is easy to check that such a nondimensionalization does not change the form of system (1.61) when using matrices B from (1.65), (1.66). In what follows the bars over the dimensionless quantities will be omitted for brevity. To reduce the system (1.61) to the Cauchy form, let us calculate the matrix B^* in (1.74) in the case when employing the matrix B_P from (1.65). Using the results of [1.48] we have

$$B^*(\mathbf{w}) = (\tau^2/4)[A^4 - 2r_1 A^3 + (\kappa^{-2} - 2r_2 + r_1^2)A^2$$
$$+ 2r_1 r_2 A + (\kappa^{-2} - \kappa^{-1}r_1 - r_2)(\kappa^{-2} + \kappa^{-1}r_1 - r_2)\varGamma], \quad (1.76)$$

where $\kappa = \tau/h$ and r_1, r_2, r_3 are the coefficients of the matrix A characteristic equation

$$\lambda^3 - r_1\lambda^2 - r_2\lambda - r_3 = 0;$$

$$r_1 = 3u, \qquad r_2 = c^2 - 3u^2, \qquad r_3 = u(u^2 - c^2).$$

Let us simplify (1.76) using formulas (1.18)

$$A^3 = r_1 A^2 + r_2 A + r_3 I;$$

$$A^4 = (r_1^2 + r_2)A^2 + (r_1 r_2 + r_3)A + r_1 r_3 I.$$

As a result formula (1.76) takes the form

$$B^*(\mathbf{w}) = (\tau^2/4)(d_1 A^2 + d_2 A + d_3 I), \tag{1.77}$$

where

$$d_1 = \kappa^{-2} - r_2, \qquad d_2 = r_3 + r_1 r_2,$$
$$d_3 = d_1^2 - (r_1/\kappa)^2 - r_1 r_3. \tag{1.78}$$

The matrix B^* corresponding to the matrix B_P from (1.66) is calculated analogously. In this case one must set in (1.77)

$$d_1 = \kappa^{-2} - \kappa^{-1}r_1 - r_2; \qquad d_2 = r_3 - r_1 d_1; \qquad d_3 = r_3(\kappa^{-1} - r_1) - r_2 d_1.$$

Finally, the matrix B^* corresponding to the matrix B_H from (1.65) is computed by (1.77) where

$$d_1 = \kappa^{-2}; \qquad d_2 = \kappa^{-1}D^2 - \kappa^{-2}r_1; \qquad d_3 = D^4 - \kappa^{-1}r_1 D^2 - \kappa^{-2}r_2.$$

Consider the case when the singular point O_1 is a saddle of the first kind and O_2 is a stable node. Then according to the foregoing one should begin the integration from the singular point O_1. As is known ([1.47]) one can use the solution of the linearized system (1.68) for the description of the system (1.61) solution $\mathbf{w}(\xi)$ in the vicinity of the singular point O_1 in the case of a saddle-like singularity. Describe the algorithm for departure from the singular point O_1 based on (1.68). Let $\omega_k = \{\omega_{1k}, \omega_{2k}, \omega_{3k}\}$ be the eigenvector of the system (1.68) in the vicinity of the singular point O_1 where the index $k = 1, 2, 3$ corresponds to the eigenvalue μ_{1k}. According to [1.52], [1.10] the integral curves leave the singular point O_1 in the directions of the eigenvectors $\omega_1, \omega_2, \omega_3$. Let us arrange the eigenvalues $\mu_{11}, \mu_{12}, \mu_{13}$ of the matrix $\mathscr{F}_1$ in increasing sequence, that is, $\mu_{11} < \mu_{12} < 0, \mu_{13} > 0$. In the planes containing the singular point O_1 and the pairs of vectors ω_1, ω_3 and ω_2, ω_3, the point O_1 is a saddle, and in the plane of vectors ω_1 and ω_2 it is a node. Therefore, a unique integral curve leaves the point O_1 in the direction ω_3 [1.10]. Let us write an algebraic system for the determination of the eigenvector ω_3 components corresponding to the eigenvalue μ_{13}:

$$(\mathscr{F}_1 - \mu_{13}I)\omega_3 = 0. \tag{1.79}$$

Since the determinant of this system is equal to zero, one can only express two components of the eigenvector ω by the third one using (1.79). Let $\mathscr{F}_j = \| f_{ml} \|$, $m, l = 1, 2, 3$. Then from (1.79) we find

$$\omega_{13} = \omega_{13}; \qquad \omega_{23} = -(\Delta_1/\Delta_2)\omega_{13}; \qquad \omega_{33} = (\Delta_3/\Delta_2)\omega_{13}; \quad (1.80)$$

where

$$\Delta_1 = \begin{vmatrix} f_{11} - \mu_{13} & f_{13} \\ f_{21} & f_{23} \end{vmatrix}, \qquad \Delta_2 = \begin{vmatrix} f_{12} & f_{13} \\ f_{22} - \mu_{13} & f_{23} \end{vmatrix},$$

$$\Delta_3 = \begin{vmatrix} f_{11} - \mu_{13} & f_{12} \\ f_{21} & f_{22} - \mu_{13} \end{vmatrix}.$$

Taking into account the qualitative behavior of the integral curves in the vicinity of the first-type saddle, we have in (1.67) that $\delta w'(\xi) \cong \omega_3$ [1.47]. In computations we have set $\omega_{13} = -\varepsilon_0$ in (1.80) where ε_0 is a small positive number. In order that the functions $w_2(\xi)$ and $w_3(\xi)$ also decrease when moving along the integral curve from the singular point O_1 to the singular point O_2, we should then require that

$$-\Delta_1/\Delta_2 > 0; \qquad \Delta_3/\Delta_2 > 0. \qquad (1.81)$$

The numerical integration of the system (1.74) was carried out by the fourth-order Runge–Kutta method with a constant step $\Delta\xi$. When departing from the singular point O_1 with the help of formulas (1.80) the inequalities (1.81) were checked. In all the computations using f.d.a.s of the Lax scheme and the scheme from [1.46] these inequalities were satisfied. To check the correctness of the matrix $B^{-1}(\mathbf{w}, h, \tau)$ computation entering (1.74) the relation

$$\det(B^{-1}) = \prod_{j=1}^{3} \mu_j(B^{-1}) \qquad (1.82)$$

was employed where the determinant in the left-hand side of the equation was computed by expansion in the column, $\mu_j(B^{-1})$ is the jth eigenvalue of the matrix B^{-1} which is easily expressed by the jth eigenvale of the matrix A in the case of formulas (1.65), (1.66). Full coincidence of both sides of equation (1.82) was obtained in the computations. The influence of the choice of the constant $\varepsilon_0 = -\omega_{13}$ entering (1.80) was also studied in [1.45] where the quantity ε_0 was varied in the range $10^{-4} \leq \varepsilon_0 \leq 10^{-2}$. A very insignificant discrepancy (less than 0.1%) between the results obtained for different ε_0 is in agreement with the qualitative behavior of the integral curves in the neighborhood of the first-type saddle [1.47].

Let us introduce some notations before proceeding to a direct confrontation of the numerical solutions of ordinary differential f.d.a. equations and grid solutions of finite difference equations. Let $\rho(x, t)$ be the gas density obtained by the solution of difference equations of a scheme for some t, and let $\rho_P(\xi)$, $\rho_H(\xi)$ be the progressive wave-type solutions of the P-form and H-form of the f.d.a. of a difference scheme under study. For the numerical solution of a problem on the propagation of a stationary shock wave for the Euler equation system (1.1)–(1.3), (1.8), the Cauchy initial data of a "step" form was set for each of the functions sought, and the constant values of the solution components on both sides of a discontinuity satisfied the Hugoniot conditions (1.22)–(1.24), and the discontinuity itself was smeared over one cell of a computing mesh on the x-axis.

Since the progressive wave-type solutions are determined with accuracy up to translation along the ξ-axis for the comparison of $\rho_P(\xi)$ and $\rho_H(\xi)$ with $\rho(x, t)$, such a node was sought in these profiles at which the value of $|u_i + c_i - D|$ was minimum where i is the index of a node of the grid with a step h on the x-axis. This point was taken as an origin $i = 0$ for ρ_P, ρ_H, $\rho(x, t)$.

In [1.2] was indicated the existence of an initial unsteady interval when calculating the motion of a steady shock wave by a shock-capturing scheme. In this connection there arises a question on the choice of the moment of time t for the comparison of the difference solution $\rho(x, t)$ with the solutions $\rho_P(\xi)$, $\rho_H(\xi)$.

This question is closely related to a question of the existence of solutions $w_h(\eta)$ of the systems of difference equations approximating the Euler equation system (1.1)–(1.3) and depending only on the variable $\eta = (x - Dt)/h$. Indeed, in the co-moving coordinate system the profile of such a difference solution will be stationary, and then it will be reasonable to compare this solution with solutions $\rho_P(\xi)$ and $\rho_H(\xi)$.

Theoretical studies of the questions on the existence of difference solutions of the form $w_h(\eta)$ have been carried out up to now in [1.31], [1.53]–[1.60] only for the case of difference schemes approximating a scalar quasi-linear equation

$$\partial u/\partial t + \partial \varphi(u)/\partial x = 0, \tag{1.83}$$

where the function $\varphi(u)$ is assumed to satisfy the requirements

$$\varphi''(u) > 0, \qquad (u_1 - u_2)(\varphi'(u_1) - \varphi'(u_2)) > 0,$$

u_1, u_2 are constants characterizing the solution behind and before the shock wave front, respectively. The question on the existence of the solutions of the form $w_h(\eta)$ of the difference equations was formulated for the first time by P. Lax [1.31]. In the scalar case the solution $w_h(\eta)$ sought for satisfies a nonlinear finite-difference equation

$$w_h(\eta - \delta) = F(w_h(\eta + r), \ldots, w_h(\eta - l)), \tag{1.84}$$

where $\delta = \tau D/h, r$, and l are nonnegative integers, $r + l > 0$, and the nonlinear function F is determined by the form of a difference scheme approximating equation (1.83). One seeks for the solution of equation (1.84) having the limits

$$w_h(-\infty) = u_1, \qquad w_h(+\infty) = u_2. \tag{1.85}$$

It is possible to prove the existence of the solution to the problem (1.84), (1.85) by considering a question on the convergence of the iteration process

$$w_h^{n+1}(\eta - \delta) = F(w_h^n(\eta + r), \ldots, w_h^n(\eta - l)), \tag{1.86}$$

where n is the iteration number, $n = 0, 1, \ldots$, and the function w_h^0 is given and has the limits $w_h^0(-\infty) = u_1$, $w_h^0(+\infty) = u_2$. V.V. RUSANOV showed in [1.53] that it is necessary for the convergence of the process (1.86) that the initial function w_h^0 satisfies certain special conditions. G. JENNINGS [1.54] proved, in the case $r = l$, the existence of a monotone function satisfying equation (1.84) and the limit relationships (1.85). A similar result was obtained in [1.55] for three-point schemes by a different method based on using auxiliary functions. A generalization of these results for the case of arbitrary nonnegative r and l in (1.84) has been presented in [1.56]–[1.58].

A question on the existence of a limiting stationary profile of the shock wave in the solution of the difference Cauchy problem for a quasi-linear equation in the multidimensional case

$$\partial u / \partial t + \sum_{j=1}^{s} \partial \varphi_j(u) / \partial x_j = 0$$

has also been considered in [1.58], where $x_1, \ldots, x_s$ are spatial variables, $s > 1$. A class of explicit second-order TVD-schemes, which do not increase the total variation of the grid functions and which approximate equation (1.83), has been considered in [1.59], [1.60]. A theorem on the existence of a limit profile $w_h(\eta)$ has been proved in [1.59], [1.60] for these schemes (see, for further discussion of the TVD-schemes, the beginning of Chapter 6).

Let us assume that the limit stationary solution $w_h(\eta)$ at $n \to \infty$ (n is the number of the time level) is also realized in the computations of a problem on the propagation of a stationary shock wave in an inviscid gas by the LAX scheme [1.31] and by the GODUNOV scheme [1.46] being considered in this section. Then it is reasonable to assume that at sufficiently large n the difference solution $\mathbf{w}^n$ differs little from the stationary solution $\mathbf{w}_h(x/h - n\delta)$. Let us now take some integer m, $0 \le m < n$. It is necessary for the satisfaction of the equality $\mathbf{w}_n^m = \mathbf{w}(\eta_i^n)$ that

$$\eta_j^m = \eta_i^n, \tag{1.87}$$

where $\eta_i^n = (x_i - Dn\tau)/h$ and x_i is the abscissa of a node of the uniform mesh on the x-axis at which the components of the vector $\mathbf{w}$ are computed. We find from (1.87):

$$j = i + (m - n)\delta. \tag{1.88}$$

It follows from (1.88) that it is necessary for the quantity $(m - n)\delta$ to be integer if the quantity j is to be integer. This can easily be achieved on the corresponding choice of an integer m (n is assumed to be fixed in (1.88)), if δ is set to be a rational number, that is, $\delta = p/q$ where p and q are some natural numbers. Setting $\delta = p/q$, let us introduce the quantity

$$\Delta \rho^m = \log_{10} \left\{ \max_j |\rho_j^m - \rho(\eta_{j+(n-m)\delta})| \right\} \tag{1.89}$$

for such an integer m, that the quantity $(n - m)\delta$ is an integer. If there is a convergence of the difference solution to the limit stationary solution of the form $\mathbf{w}_h(\eta)$, then the quantity $\Delta\rho^m$ should decrease as m increases. The dimensionless time t/τ is measured in Figures 1.3 and 1.4 along the abscissa axis. Along the ordinate axis in Figure 1.3 the value $\Delta\rho^m$ is measured and which was obtained with $n = 500$ in (1.89). The results presented in Figures 1.3 and 1.4 have been obtained at $p_1 = 5$, $u_2 = 0$, $p_2 = \rho_2 = 1$, in (1.25)–(1.27), $\gamma = 2$, in the equation of state (1.8). By analogy with [1.2], 10 grid nodes with the outer pressure p_2 ahead of the shock were added periodically and, respectively, 10 grid nodes beginning from the left boundary $x = 0$ were removed. Thus, the propagation of the shock can be followed for an arbitrarily long time, although only 100 grid nodes on the x-axis are used in the computations at each moment of time. Such a logic of the computer code enables one to reduce substantially the amplitude of disturbances introduced along the characteristics through the boundary $x = 0$ inside the computational domain and caused

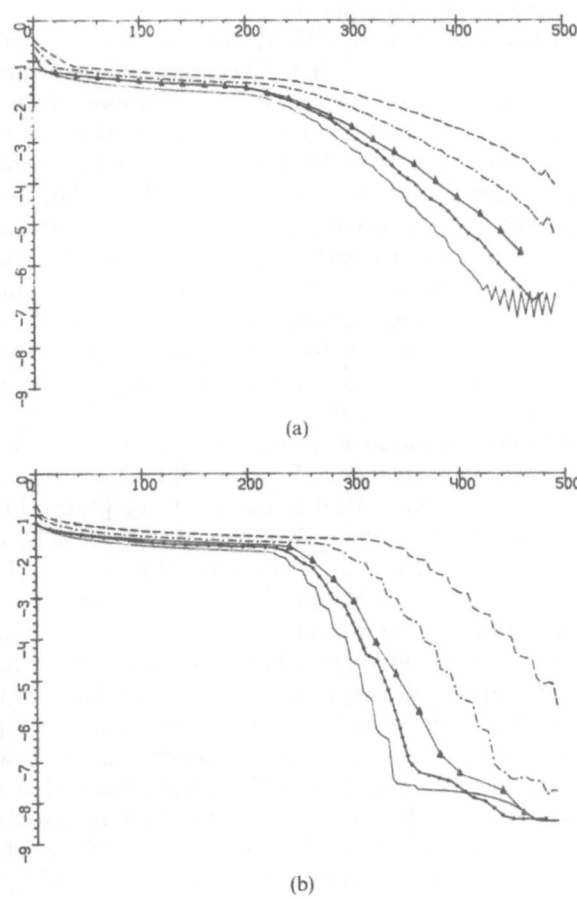

Figure 1.3. The quantity $\Delta\rho^m$ as a function of $m = t/\tau$; (a) the Lax scheme; (b) the "breakdown of discontinuity" scheme. $(----)$—$\delta = 1/2$; $(-\cdot-\cdot-)$—$\delta = 3/5$; $(\blacktriangle\!\!-\!\!\blacktriangle\!\!-\!\!\blacktriangle)$—$\delta = 13/20$; $(\bullet\!\!-\!\!\bullet\!\!-\!\!\bullet\!\!-\!\!\bullet)$—$\delta = 7/10$; (———)—$\delta = 3/4$.

by simple difference boundary formulas used at the left end $x = 0$: at this boundary the components of the flux vector $\boldsymbol{\varphi}(\mathbf{u})$ in the system (1.9)–(1.11) were computed with the use of the exact values u_1, p_1, ρ_1 characterizing the gas state behind the shock front.

It may be seen in Figure 1.3 that the quantity $\Delta\rho^m$ actually decreases as m increases. In the case of the "breakdown-of-discontinuity" scheme the value of ρ_j^m (for some δ and at sufficiently large m) differs from the limit stationary solution $\rho(\eta_{j+(n-m)\delta})$ by a very small magnitude of the order $10^{-8} \div 10^{-9}$. In the case of the Lax scheme the convergence of a difference solution to the stationary discrete profile is slower than in the case of the "breakdown-of-discontinuity" scheme [1.46] (compare Figure 1.3(a) and (b)). Comparing the graphs of the quantity (1.89) at different δ, one can observe that the speed of convergence of the difference solution to the stationary discrete profile $\mathbf{w}_h(\eta)$ increases with δ.

The evolution of the difference solution to a stationary regime may also be followed by observing the behavior of the width after Prandtl $X(w_j)$ [1.61], [1.18] as a function of time, where $w_1 \equiv \rho, w_2 \equiv \rho u, w_3 \equiv \rho(\varepsilon + u^2/2)$. It can be seen in Figure 1.4 that X becomes a periodic function of t beginning from some $t = t_* > 0$. The period of oscillations of the quantity X is equal to $v\tau$ where v is the least natural number, for which $v\delta$ is an integer. For example, $v = 2$ at $\delta = 1/2$; $v = 20$ at $\delta = 13/20$ (see also Figure 1.4). This periodicity of the width X may be explained as follows. The convergence of the iteration process (1.86) is usually proved for arbitrary real η and δ in (1.86). Suppose that $\delta = p/q < 1$ where p and q are reciprocals. Then it is easy to see that the fractional part of the index j (1.88) is a periodic function of m with a period being equal to q. A specific numerical value of the solution $\mathbf{w}_h(\eta)$ corresponds to each noninteger value of the quantity η. It follows from Figure 1.4 that at a fixed δ the "breakdown-of-discontinuity" scheme goes over to the stationary regime faster than the Lax scheme. In the "starting" interval $[0, t_*]$ the difference solution $\mathbf{w}(x, t)$ is substantially nonstationary in the co-moving coordinate system (see also [1.62], [1.63]).

Thus, the theoretical results of [1.31], [1.53]–[1.60], as well as the results of numerical experiments presented in Figures 1.3 and 1.4 for the case of two first-order schemes approximating the system (1.1)–(1.3), give an additional argument in favor of considering the progressive wave-type solutions of both difference equations and the f.d.a. differential equations while studying the properties of difference solutions in the neighborhood of shock wave fronts.

Taking the foregoing into account, the time t for the comparison of the grid solution for the density $\rho(x, t)$ with the solutions $\rho_P(\xi), \rho_H(\xi)$ was chosen from the requirement $t > t_*$. For example, in the case of the Lax scheme the stationary value $X(\rho) \approx 4.7927h$ was obtained at $t = 500\tau$ and at $\delta = 7/10$. This value of δ corresponds to the Courant number $K \approx 0.9034$. The value $X(\rho)$ found in [1.45] at $K = 0.9000$ on the basis of the numerical integration of the equation system of the f.d.a. P-form (1.74), (1.77) (1.78) was equal to

Figure 1.4. The width after
Prandtl $X(\rho)/h$ as a function of $m = t/\tau$; (a) the Lax
scheme; (b) the "breakdown
of discontinuity" scheme.
1—$\delta = 1/2$; 2—$\delta = 3/5$;
3—$\delta = 13/20$; 4—$\delta = 7/10$;
5—$\delta = 3/4$. (----)—the
line $X(\rho(x, 500\,\tau))/h$.

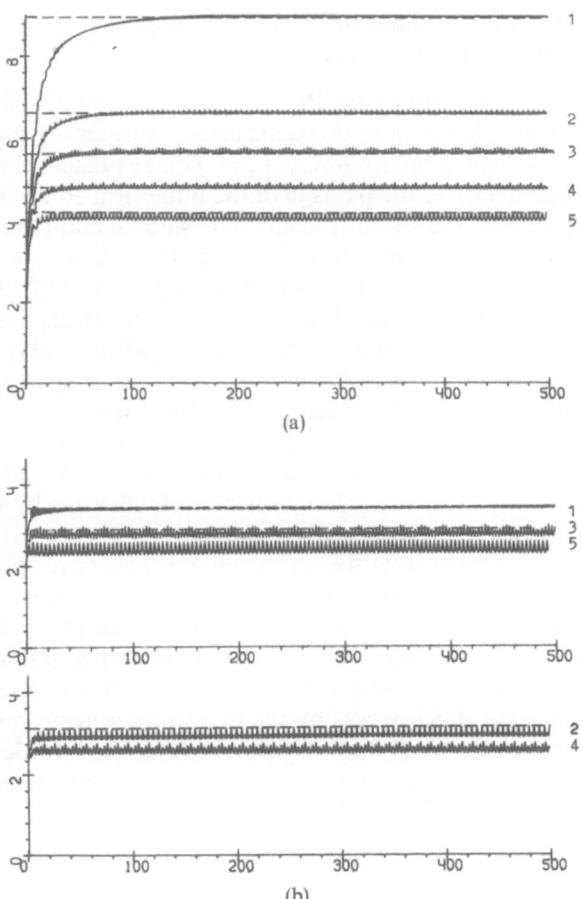

(a)

(b)

$\approx 4.66h$. A direct comparison of the finite-difference solution $\rho(x, t)$ with the
solutions $\rho_P(\xi)$, $\rho_H(\xi)$ was also carried out in [1.45] for the cases of the LAX
scheme [1.31] and the GODUNOV scheme [1.46]. This comparison showed that
the corresponding relative errors do not exceed 4% in the case of shock waves
of finite intensity.

When solving some gas dynamic problems, weak shock waves arise in the
flow. Such waves should generally also be calculated effectively by finite-
difference schemes. To elucidate the applicability of the progressive wave-type
solutions of the f.d.a. equations to the description of a smeared shock wave
of weak intensity, a series of computations by the Lax scheme and the
"breakdown-of-discontinuity" scheme [1.46] was carried out in [1.45]. As a
parameter characterizing the shock wave intensity, the quantity $\tilde{\eta} = p_1/p_2$ was
taken where it was assumed that $p_2 \neq 0$; p_1, p_2 are the gas pressures behind
and before the shock wave front, respectively. The computations were carried

out for a number of values of $\bar{\eta}$, such that $\bar{\eta} \to 1$. As a result the following conclusions were drawn.

1. The width after Prandtl X of the smeared shock wave depends on the wave intensity and this dependence is generally nonmonotonic.

2. The property $\lim_{\bar{\eta} \to 1} t_* = \infty$ takes place where the moment of time t_* characterizes the passage of the numerical solution to the stationary regime. At $t < t_*$ the width X is an increasing function of t. This dependence of X on t cannot be described by a simple formula $X \sim t^\alpha$ where $\alpha = $ const, as takes place in the case of a linear discontinuity [1.65], [1.66].

3. When the shock wave intensity increases, less time steps are needed (at fixed Courant number $K = \max[(\tau/h)(|u| + c)]$) to achieve the stationary regime of numerical solution. The value of time t_* at fixed $\bar{\eta}$ is different for different difference schemes and shock waves.

4. Thus, at $t < t_*$, the progressive wave-type solutions of the f.d.a. equations of a difference scheme cannot be used for the description of numerical solution properties in the smeared shock wave by virtue of a purely unsteady character of the numerical solution.

5. A more intensive smearing of shock waves of weak intensity takes place in comparison with shock waves of finite intensity. Note that this property takes place not only for the Lax scheme and the "breakdown-of-discontinuity" scheme of [1.46]. In [1.67] a strong smearing of shock waves of weak intensity was indicated when using the scheme [1.68]. This effect of intensive smearing of weak shock waves by the first-order schemes must be taken into account in the numerical studies of problems in which the laws of propagation of weak shock waves are investigated, in particular, geometric acoustics problems [1.11], [1.69], [1.70].

CHAPTER 2

Differential Analyzers of Shock Waves in One-Dimensional Gas Flows

At present the finite-difference shock-capturing methods are one of the most efficient means of mathematical modeling of shocked gas flows. One of the characteristic features of these methods, arising in the course of their realization on an electronic computer, consists in the fact that the amount of numerical information obtained as a result of the solution of a problem, exceeds by several orders of magnitude that which is of real interest to a research worker [2.1], [2.2]. Another feature of the shock-capturing techniques is their low accuracy in the vicinity of strong discontinuities, which are smeared thereat within several intervals of the computing mesh. These circumstances give rise to two practical problems: one on interpretation of the numerical results obtained, and the other on the increase in accuracy of the numerical solution in the vicinity of strong discontinuities. The problem of interpretation includes the localization (that is, determination of the location) of various singularities in gas flows and their classification (shock waves, contact discontinuities, etc.).

On the other hand, the information on shock location within the zone of its "smearing" may be directly used to increase the numerical solution accuracy in the vicinity of discontinuities. It has been proposed in [2.3] to use the information on shock location to improve the quality of smeared profiles, without violating the accuracy in smooth regions on the basis of a passage from homogeneous schemes to nonhomogeneous schemes of variable orders of accuracy. It will be shown in Chapter 6 how information on shock location can be used for purposes of difference solution refinement.

Thus, the questions of the development and of the foundation of the differential analyzers of strong discontinuities (that is, algorithms for the localization of singularities within the computational domain by shock-capturing computational results), are of present interest. An automation of shock localization on the basis of the application of these algorithms enables one to facilitate substantially the process of extracting really needed information from the immense numerical arrays produced by a computer in the process of solving steady and, especially, unsteady multidimensional gas dynamics problems. Note one more aspect of using information on the location of shocks, in particular, of shocks which emerge in the process of calculation: it can be used for the active control of computational process including, for example, the alteration of some boundary condition, a passage to another structure of difference grid or to another difference scheme, etc. Such an active control of computation may be accomplished by means of a corresponding computer code or through a dialogue regime based on interactive computer graphics [2.4]–[2.6].

Let us enumerate the known algorithms of shock wave localization by shock-capturing computational results. They can be subdivided into two large groups. The first group of methods for locating shock waves is characterized by the fact that a computation of discontinuity surfaces is organized alongside the computation of the overall flow field which includes shock waves. In the second group of shock localization techniques only the instantaneous (for a given moment of time) numerical values of the flow parameters are used. Let us, at first, enumerate the methods of explicit extraction and tracking of shock front surfaces. It was proposed in [2.7] to approximate the discontinuity curve by a certain number of points. The temporal propagation of the discontinuity was then found as a motion of the above points along a normal to the discontinuity front. Another simple technique for representing the discontinuity line presented in [2.8] consists in the introduction of a "height function" $h = f(x, t)$ of the points of this line over the chosen line, in the present case, the line $y = 0$. As was pointed out in [2.8], this technique works poorly when the slope dh/dx exceeds the relation h_y/h_x, and does not work at all in the case of non-single-valued functions $h = f(x, t)$ when more than one value of h corresponds to a single value of x; h_x, h_y are the steps of a rectangular grid in the plane of spatial coordinates x, y.

In the method of front tracking proposed by J. GLIMM et al. [2.9], and developing the ideas presented earlier in [2.7] (see above), the motion of shock wave fronts is computed on the basis of the Rankine–Hugoniot conditions. In the interior domains between the shock surfaces the flow variables of the method in [2.9] are computed in the nodes of a rectangular grid. If the grid point P is close to the shock front, then the flow parameters at this point are computed by linear interpolation involving the values in the interior nodes and the values on that side of the shock front which refers to the same flow subdomain in which the point P is located. To calculate the temporal evolution of the flow in the interior subdomains between the shock surfaces (that is, in the subdomains of continuous flow), in [2.9] the two-step Lax–Wendroff scheme is used. The modular structure of the code realizing the method in [2.9] will make it possible to take into account various effects, for example, the disappearance of decaying shock waves, with comparatively moderate additional effort. Impressive examples presented in [2.9] demonstrate the efficiency of the method.

A technique for shock line representation as an ensemble of rectilinear segments was employed by the authors of [2.10]–[2.14]. In [2.10]–[2.12], within the framework of a computer code realizing the relaxation computation by the Godunov scheme, the shock wave was extracted by using a special procedure which was based on the Huygens principle and which helped to avoid the sawtooth-like structure of the shock surface. In this procedure the shock front surface was approximated by a broken line consisting of rectilinear segments. In [2.13] the shock fronts were also approximated by rectilinear segments. Therein the speeds of motion of the broken line segments approximating the shock front were found from the solution of a Riemann problem. In [2.14] the rectilinear shock front segments were at first determined in the rectangular grid cells. Then the ends of the segments were moved in two steps: in the first step along the x-axis, and in the second step along the y-axis.

Due to errors in the numerical solution in the vicinity of a discontinuity the motion of broken line segments may become irregular as time increases, and then a sawtooth-like broken line will be obtained. In this connection, a procedure for computing the motion of two neighboring segments of the broken line may be useful, in which the speed of these segments is computed as some mean value of the speeds of these two segments [2.15].

The random choice method of A. CHORIN [2.16]–[2.21] enables one to reduce to a minimum the width of shock wave smearing. However, it was pointed out in [2.22] that this method produces errors in the shock location of magnitude $\sim \pm 2.5h$. In this connection, it was proposed in [2.22] to use, in the context of the random choice method, two techniques of shock front tracking: the method of grid coalescence in the vicinity of a discontinuity and the method of shock fitting. In [2.23] it has been proposed to use explicitly the Rankine–Hugoniot condition on a shock and the condition of entropy increase across the shock wave front, while calculating the breakdowns of discontinuities in the Chorin random choice method. Such modification of the random choice method, as was shown in [2.23] in computational examples, increases substantially the accuracy of shock localization while solving the Buckley–Leverett equation from the filtration theory. In [2.24] the discontinuities were tracked explicitly, and in the continuous flow domains between discontinuities the computation was carried out by the MACCORMACK scheme [2.25]. In [2.26] a method of computing shocked flows has been proposed in which the position $s(t)$ of a shock is considered as a dependent variable. The quasi-one-dimensional flow equations are transformed to equations in the new coordinates (ξ, τ) in which the shock wave position proves to be fixed. For this purpose a substitution, $\xi = \xi(x, t) = x/s(t)$, is made. It was shown in [2.27], [2.28] in the example of the Burgers equation that a problem on the study of the temporal evolution of an initially given discontinuity can be reduced to some variational problem, and the flow in a spatial domain can be computed by a shock-capturing scheme. In Chapter 5 we shall investigate in detail the properties of the Miranker–Pironneau method and present a number of generalizations and modifications of this method.

Let us now enumerate the known algorithms of the shock wave localization by shock-capturing computation results that belong to the second of the above two groups, when the position of shocks is determined only on the basis of

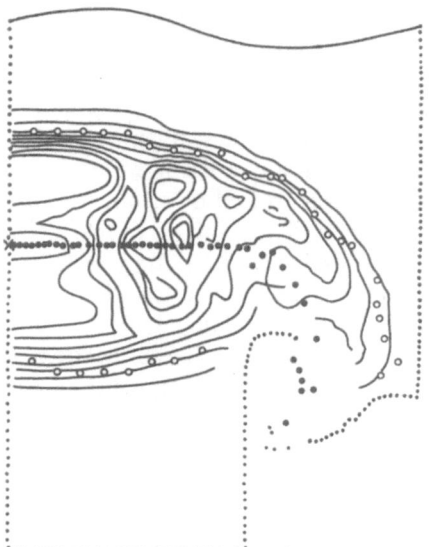

Figure 2.1. High-velocity impact of a cylinder onto an obstacle. Solid circles represent marker particles located on the contact discontinuity. Open circles show the points of shock wave fronts determined by the differential analyzer of Section 3.4.

"smeared" profiles at a given moment in time. Here the most widespread technique is a visual determination of the location of shock fronts and contact interfaces by the coalescence of different isolines. As isolines we can employ, for example, the lines of constant pressure (isobars), the lines of constant density (isochors), and the lines of constant Mach number [2.29], [2.30]. In Figure 2.1, taken from [2.31], we illustrate this technique by a picture of isobars obtained in a computation by the HARLOW method [2.32] of a problem on the high-velocity impact of a cylindrical projectile onto a semi-infinite target. The location of curved shock waves propagating in the target and in the projectile can be approximately determined in Figure 2.1 by the coalescence of isobars. In some works (see, for example, [2.33]–[2.35]), the shock front was determined by maximum gradients of the flow parameters. In Figure 2.2 the circles refer to the shock wave points determined in [2.35]

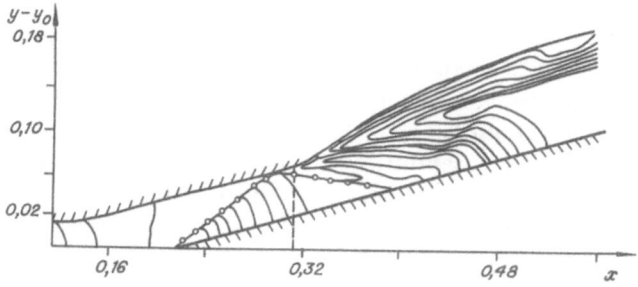

Figure 2.2 Computation of a two-dimensional flow of a chemically nonequilibrium gas mixture in an annular nozzle and in the jet by the method of [2.36]. Solid lines are isobars.

by $\max |\partial p/\partial y|$ on the lines $x = $ const. In [2.36], [2.37] the shock wave position was determined by an inflection point in the pressure profile along the beam $x = $ const; in [2.38]–[2.40] by a sudden change in the direction of stream lines; and in [2.41]–[2.44] by the maximum of the artificial viscosity. In [2.45] the values of the saturation function $s(x, y, t)$ in the process of computation of a problem on two-phase filtration with a jump in s varied from $s = 0$ (pure oil) to $s = 1$ (pure water). The position of a saturation jump front for fixed t was determined as a level curve with $s(x, y, t) = 0.01$. A method for locating a self-similar shock wave front in a smeared profile, based on the fact that the quantity

$$2(\gamma + 1)^{-1} \cdot [\gamma a_-^{-2} t^{-2} (\mathbf{rn} - \mathbf{u}_- \mathbf{n}t)^2 - (\gamma - 1)/2] - p_+/p_-$$

changes its sign at the front of such a wave, has been proposed in [2.46]. Here $\mathbf{r}$ is the radius vector of a point of the shock surface; $\mathbf{n}$ is the normal to the shock surface; γ is the ratio of specific heats c_p/c_v; and $\mathbf{u}_-$, a_-, p_-, are, respectively, the gas speed, the sound speed, and the pressure behind the shock wave front. The results of application of this localization algorithm to the computation of a two-dimensional shocked flow by the FLIC method have been compared with experimental data in [2.46]. This comparison shows the efficiency of the developed localization technique.

The major part of the above techniques of shock wave localization by shock-capturing computational results that belong to the second group had, until recently, no theoretical background. In [2.47] the present authors have apparently proposed the first technique for studying the accuracy of some techniques of shock wave localization by the results of the shock-capturing computation of one-dimensional gas dynamics problems. The basic notion used in this technique is the notion of a smeared shock wave center. Let us elucidate the meaning of the basic notion in one simple example before beginning with the systematic presentation of the material of differential analyzer studies.

2.1. An Introductory Example

Consider the Burgers equation

$$\partial u_\mu/\partial t + u_\mu \, \partial u_\mu/\partial x = \mu \, \partial^2 u_\mu/\partial x^2, \qquad \mu = \text{const} > 0, \qquad (2.1.1)$$

with initial conditions

$$u_\mu(x, 0) = \begin{cases} u_1, & x < x_0, \\ u_2, & x > x_0, \end{cases} \qquad (2.1.2)$$

where u_1, u_2, x_0 are constants and $u_1 > u_2$, x is the spatial coordinate, and t is time. The exact solution of the problem (2.1.1), (2.1.2) has the form [2.48]

$$u_\mu(x, t) = u_2 + (u_1 - u_2) \cdot \{1 + g(x, t) \exp[(u_1 - u_2)(x - x_0 - Dt)/(2\mu)]\}^{-1}, \tag{2.1.3}$$

where $D = (u_1 + u_2)/2$,

$$g(x, t) = \left[\int_{-(x-x_0-u_2t)/\sqrt{4\mu t}}^{\infty} \exp(-\zeta^2)\, d\zeta \right] \left[\int_{(x-x_0-u_1t)/\sqrt{4\mu t}}^{\infty} \exp(-\zeta^2)\, d\zeta \right]^{-1}.$$

(2.1.4)

Indicate the following properties of the solution (2.1.3) [2.49]:

1. For arbitrary fixed $t > 0$ there exists a point (x_c, t) at which the graphs of the solutions (2.1.3), (2.1.4), obtained for the different values of the constant $\mu > 0$, intersect (see Figure 2.3).

2.
$$\max_x |\partial u_\mu/\partial x| = \partial u_\mu/\partial x|_{x=x_c}.$$

3. It was shown in [2.48] that in the limit $t \to \infty$ the solution of the problem (2.1.1), (2.1.2) converges in the domain $u_1 > (x - x_0)/t > u_2$ to the progressive wave-type solution $U_\mu(\xi)$ of equation (2.1.1) where $\xi = x - x_0 - 0.5(u_1 + u_2)t$,

$$U_\mu(\xi) = 0.5(u_1 + u_2) - 0.5(u_1 - u_2)th[(u_1 - u_2)\xi/(4\mu)].$$

A number of the results of studies of the solution $u_\mu(x, t)$ convergence to the progressive wave-type solution may also be found in [2.50]–[2.52].

4. The generalized (discontinuous) solution of the Cauchy problem

$$\partial u/\partial t + u\, \partial u/\partial x = 0;$$

$$u(x, 0) = \begin{cases} u_1, & x < x_0, \\ u_2, & x > x_0, \end{cases} \quad u_1 > u_2,$$

(2.1.5)

can be constructed as a limit at $\mu \to 0$ of the solution to the Cauchy problem (2.1.1), (2.1.2) [2.41], [2.53].

5. The locus of points (x_c, t) forms in the (x, t)-plane a line which is the discontinuity trajectory in the solution $u(x, t)$ of the problem (2.1.5). Thus, despite the fact that the discontinuity in the solution of the problem (2.1.1), (2.1.2) is smoothed at $t > 0$, there exists a point at which the quantity $u_\mu(x_c, t_0)$, $t_0 > 0$, does not depend on μ in the profiles of $u_\mu(x, t_0)$ obtained for different μ, and the abscissa of this point coincides with the exact position of the discontinuity in the solution of the problem (2.1.5). This point was called a smeared shock wave center in [2.47], [2.49], [2.54]–[2.58].

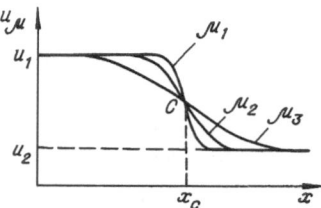

Figure 2.3. Solution graphs of the Burgers equation for different μ.

Let us show that, in the case of an arbitrary profile with shock wave, the properties 1–3, and 5 of the Cauchy problem solution for the Burgers equation (2.1.1) generally do not take place. For this purpose consider the N-wave solution of the Burgers equation (2.1.1). Following [2.48] let us make use of the Cole–Hopf substitution

$$u_\mu(x, t) = -2\mu\varphi_x/\varphi \qquad (2.1.6)$$

to obtain this solution where the function $\varphi(x, t)$ is chosen in the case under study as a source function solution

$$\varphi = 1 + \sqrt{a/t} \, \exp(-x^2/(4\mu t)) \qquad (2.1.7)$$

of the heat equation with $a = \text{const} > 0$. By virtue of formula (2.1.6) the corresponding u_μ solution is

$$u_\mu = -2\mu\varphi_x/\varphi = (x/t)\cdot[\sqrt{a/t} \, \exp(-x^2/(4\mu t))]$$
$$\times [1 + \sqrt{a/t} \, \exp(-x^2/(4\mu t))]^{-1}. \qquad (2.1.8)$$

The function φ (2.1.7) at $t \to 0$ behaves like the delta function. In this connection it is difficult to interpret the expression (2.1.8) as a solution of the Cauchy problem for u_μ with initial data given at $t = 0$ [2.48]. However, for any $t > 0$, the solution (2.1.8) is represented by a graph resembling the letter N, with positive and negative phases, and as an initial profile one can take a profile for any $t = t_0 > 0$. The area of the positive phase of the profile is equal to

$$\int_0^\infty u_\mu \, dx = -2\mu[\ln \varphi]_0^\infty = 2\mu \ln(1 + \sqrt{a/t}). \qquad (2.1.9)$$

If we denote the value of the integral (2.1.9) at the initial moment of time t_0 by A, then we can introduce the Reynolds number

$$R_0 = A/(2\mu) = \ln(1 + \sqrt{a/t_0}). \qquad (2.1.10)$$

From formula (2.1.10) we find the following expression for a: $a = t_0 \times [\exp(R_0) - 1]^2$, therefore, formula (2.1.8) may be rewritten as

$$u_\mu = (x/t)\cdot\{1 + (t/t_0)^{0.5}[\exp(x^2/(4\mu t))]/[\exp(R_0) - 1]\}^{-1}. \qquad (2.1.11)$$

If $\mu \to 0$, then $R_0 \gg 1$ in accordance with (2.1.10), because $A = \text{const}$. Therefore, one can expect that the solution of the corresponding Cauchy problem for the equation $u_t + uu_x = 0$ will be a good approximation during some time t. However, at $t \to \infty$, the diffusive term will dominate [2.48].

Now consider the solution (2.1.11) for finite x, t at $R_0 \gg 1$ ($t_0 \ll a$, respectively). According to [2.48], at fixed t and $R_0 \to \infty$,

$$u_\mu \cong \begin{cases} x/t, & -\sqrt{2At} < x < \sqrt{2At}, \\ 0, & |x| > \sqrt{2At}. \end{cases} \qquad (2.1.12)$$

The result (2.1.12) completely coincides with the solution of an "inviscid" equation $u_t + uu_x = 0$. Let $x = \xi(t)$ be the equation of the trajectory of a shock moving to the right in the solution of the "inviscid" equation. According to (2.1.12),

$$x = \xi(t) = \sqrt{2At}. \qquad (2.1.13)$$

As in the foregoing example (2.1.1), (2.1.2), with an initial function $u_\mu(x, 0)$ of a "step" form, let us try to elucidate a question as to whether the value of the solution (2.1.11) along the trajectory (2.1.13) will not depend on μ. For this purpose let us substitute the quantity $\xi(t)$ defined by formula (2.1.13) into the solution (2.1.11) instead of x. Write the result of this substitution as

$$u_\mu(\xi(t), t, v) = \sqrt{2A/t}/\{1 + \sqrt{t/t_0}\, \exp(v)/[\exp(v) - 1]\}, \qquad (2.1.14)$$

where we have introduced the notation $v = R_0 = A/(2\mu)$. It is evident that for the independence of the value $u_\mu(\xi(t), t, v)$ of v it is necessary that the derivative $\partial u_\mu(\xi(t), t, v)/\partial v$ is identically equal to zero. We find from (2.1.14):

$$\partial u_\mu(\xi(t), t, v)/\partial v = \sqrt{2A/t_0}\, \exp(v)[\exp(v) - 1 + \sqrt{t/t_0}\, \exp(v)]^{-2}$$
$$\neq 0, \qquad (2.1.15)$$

thus, along the "inviscid" trajectory (2.1.13) the value of the solution (2.1.11) depends on μ. However, at sufficiently small μ, that is, when the value v is large, $\partial u_\mu(\xi(t), t, v)/\partial v \sim \exp(-v)$, which means a weak dependence of the quantity $u_\mu(\xi(t), t, v)$ on μ at small μ, in other words, at small $\mu > 0$ the value $u_\mu(\xi(t), t, v)$ is approximately constant. This property can also be established by direct use of formula (2.1.14). Introduce the notation $\varepsilon = \exp(-v)$. The value $\varepsilon \to 0$ as $R_0 \to \infty$ or as $\mu \to 0$. Let us take two small positive values ε_1 and ε_2 and estimate the difference

$$u_\mu(\xi(t), t, \varepsilon_1) - u_\mu(\xi(t), t, \varepsilon_2).$$

By the Lagrange theorem,

$$u_\mu(\xi(t), t, \varepsilon_1) - u_\mu(\xi(t), t, \varepsilon_2)$$
$$= (\varepsilon_1 - \varepsilon_2)(\partial u_\mu/\partial \varepsilon)|_{\varepsilon = \varepsilon^*}$$
$$= (\varepsilon_1 - \varepsilon_2)(-\sqrt{2A/t_0})(1 - \varepsilon_* + \sqrt{t/t_0})^{-2}, \qquad (2.1.16)$$

where $\varepsilon_* = \varepsilon_1 + \theta(\varepsilon_2 - \varepsilon_1), 0 < \theta < 1$. It follows from formula (2.1.16) that at small $\varepsilon_1, \varepsilon_2$ the values $u_\mu(\xi(t), t, \varepsilon_1)$ and $u_\mu(\xi(t), t, \varepsilon_2)$ differ little from each other along the "inviscid" trajectory (2.1.13), so that one can say that the properties of the u_μ solution that are analogous to the above-indicated properties 1 and 5 at the initial function $u_\mu(x, 0)$ of the "step" form (2.1.2) are satisfied approximately for the initial function $u_\mu(x, t_0)$ considered above, and being different from a "step".

Note that an analogue of property 3, the property of stabilization of the $u_\mu(x, t)$ solution to a stationary solution, in the example on the N-wave is absent simply because the shape of the N-wave profile changes with t in such a way that the area of the positive phase, that is, $\int_0^\infty u_\mu(x, t) \, dx$, tends to zero as $t \to \infty$, according to (2.1.9) (the negative phase has the same area).

2.2. Existence and Uniqueness of the Smeared Wave Center in the Solution of the System with Artificial Viscosity

2.2.1. General Considerations

The stationary gas dynamic equation system derived from the equation system (1.1)–(1.3) by substituting the pressure p by the quantity $p + q$ where q is a pseudoviscosity, or an artificial viscosity which was proposed by NEUMANN and RICHTMYER [2.59], seems to be the most simple mathematical model describing the effects of "smearing" of a gas dynamic shock wave resulting from the introduction of an artificial viscosity. In this connection it seemed natural at first to carry out an investigation of a question on the existence of the smeared shock wave center in the solution of the above-mentioned system. Consider the problem of a steady shock wave in which u, p, ρ, ε are constant quantities in the regions behind and before the shock front. We shall denote these values for the states behind and before the front by subscripts "1" and "2", respectively. Following [2.59] we shall study the solutions of the system (1.1)–(1.3) which depend only on the variable $x' = x - Dt$, that is, the progressive wave-type solutions where D is the speed of a stationary shock wave. Let us set $u' = u - D$. Integrating the system once, we obtain

$$\rho u = C_1 = m, \qquad (2.2.1)$$

$$p + q + mu = C_2, \qquad (2.2.2)$$

$$m[p/(\rho(\gamma - 1)) + u^2/2] + (p + q)u = C_3, \qquad (2.2.3)$$

where C_1, C_2, C_3 are integration constants, and the primes at the quantities x, u are omitted for brevity of notation. For definiteness, the shock wave is assumed to move from left to right; then $D > 0$, $m < 0$. Let us assume analogously to [2.59], [2.41], [2.44] that the artificial viscosity q entering equations (2.2.1)–(2.2.3) has the form

$$q = \begin{cases} F(h \, du/dx, h \, dp/dx, h \, d\rho/dx, p, \rho), & du/dx < 0, \\ 0 & du/dx \geq 0, \end{cases} \qquad (2.2.4)$$

where h is the step of a uniform computing grid on the x-axis.

Considering the equation system (2.2.1)–(2.2.3) as a system of three equations with respect to four functions ρ, u, p, q, we find u, p, q as functions of the specific volume $\vartheta = 1/\rho$,

$$u = m\vartheta, \qquad p = (\gamma - 1)[C_3/(m\vartheta) + m^2\vartheta/2 - C_2],$$

$$q = (1/\vartheta)[C_2\gamma\vartheta - (\gamma + 1)m^2\vartheta^2/2 - (\gamma - 1)C_3/m]. \qquad (2.2.5)$$

By virtue of (2.2.4) $q(\vartheta_1) = q(\vartheta_2) = 0$. From these conditions we find the constants C_2, C_3, assuming the value m given,

$$C_2 = (\gamma + 1)m^2(\vartheta_1 + \vartheta_2)/(2\gamma), \qquad C_3 = (\gamma + 1)m^3\vartheta_1\vartheta_2/[2(\gamma - 1)]. \quad (2.2.6)$$

Taking into account (2.2.6) we have from (2.2.5)

$$q = (\gamma + 1)m^2(\vartheta_2 - \vartheta)(\vartheta - \vartheta_1)/(2\vartheta), \quad (2.2.7)$$

$$p(\vartheta) = (\gamma - 1)(m^2/2)[(\gamma + 1)\vartheta_1\vartheta_2/((\gamma - 1)\vartheta)$$

$$+ \vartheta - (\gamma + 1)(\vartheta_1 + \vartheta_2)/\gamma]. \quad (2.2.8)$$

Substituting an expression (2.2.4) for q in the zone of the smeared shock wave into formula (2.2.7) we obtain with regard to (2.2.5), (2.2.7), (2.2.8) an ordinary differential equation with respect to $\vartheta(x)$

$$\tilde{F}(\vartheta(x), h\, d\vartheta/dx) = (\gamma + 1)m^2(\vartheta_2 - \vartheta)(\vartheta - \vartheta_1)/(2\vartheta), \quad (2.2.9)$$

where

$$\tilde{F}(\vartheta(x), h\, d\vartheta/dx) = F(hm\, d\vartheta/dx, h(\gamma - 1)(m^2/2)$$

$$\times [1 - (\gamma + 1)\vartheta_1\vartheta_2(\gamma - 1)^{-1}\vartheta^{-2}]\, d\vartheta/dx, -\vartheta^{-2}h\, d\vartheta/dx, p(\vartheta), 1/\vartheta). \quad (2.2.10)$$

In the following equation (2.2.9) will be considered under the boundary conditions

$$\lim_{x \to -\infty} \vartheta(x) = \vartheta_1, \qquad \lim_{x \to +\infty} \vartheta(x) = \vartheta_2. \quad (2.2.11)$$

similarly to (1.63).

Note that (2.2.9) is an autonomous differential equation, that is, the variable x does not enter this equation explicitly. According to the general properties of the solutions of autonomous equations the variable x can enter the solution $\vartheta(x)$ only in the form of a combination $x - x_0$ where x_0 is an arbitrary constant, that is, $\vartheta = \vartheta(x - x_0)$ (see, for example, [2.60], [2.61]).

Denote by G_s a finite interval on the x-axis which is located in the zone of "smearing" of a shock wave and in which the most significant variation of the $\vartheta(x - x_0)$ solution takes place with increasing x.

Definition 2.1. The point $x_* \in G_s$, at which the value of the solution $\vartheta(x_* - x_0)$ does not depend on the value of the mesh step h at a fixed value of the constant x_0, will be called the center of a smeared shock wave in the solution $\vartheta(x - x_0)$ of the problem (2.2.9), (2.2.11).

We shall say further that the smeared shock wave center is unique if there exists the only point x_* at which $\vartheta(x_* - x_0)$ does not depend on h at a fixed value of the constant x_0, and, in addition, $\vartheta(x_* - x_0)$ does not depend on the choice of the shift constant x_0.

Since the material is subject to compression in a shock wave (see, for example, [2.41]), $\rho_1 > \rho_2$, consequently, $\vartheta_1 < \vartheta_2$. In this connection we shall be interested in a root $h\,d\vartheta/dx = f(\vartheta)$ of equation (2.2.9) which is positive in the interval $(\vartheta_1, \vartheta_2)$. However, equation (2.2.9) can, in fact, have a few roots $h\,d\vartheta/dx = f_i(\vartheta)$, $i = 1, \ldots, N$. This problem of nonuniqueness of the solution of equation (2.2.9), which is an algebraic or trancendental equation with respect to $h\,d\vartheta/dx$, arises already in the case of the quadratic Neumann–Richtmyer viscosity (see an example below). A q viscosity of the form (2.2.4) has been proposed in [2.62], and the corresponding function F has the form

$$F = ah^{3/2}\rho(p/\rho)^{1/4}|du/dx|^{3/4}|[-(dp/dx)(d\vartheta/dx)]^{1/2}|^{3/4}.$$

It is easy to see that the equation for determining the roots $h\,d\vartheta/dx = f_i(\vartheta)$ of equation (2.2.9) is in this case an algebraic equation of degree 6. In the general case let equation (2.2.9) have N roots, among which there are K real roots $f_i(\vartheta)$, where $K \geq 1$. Among these roots $f_i(\vartheta)$ there can be present those roots which change their sign in the interval $\vartheta_1 < \vartheta < \vartheta_2$. We are interested in the roots $f_i(\vartheta)$ which are positive in the above interval; we shall show in the following that the positiveness of $f_i(\vartheta)$ is one of the conditions which ensures the uniqueness of a smeared shock wave center. Let there be among K real roots $f_i(\vartheta)$, K_1 roots $f_i(\vartheta)$ ($K_1 > 1$), such that $f_i(\vartheta) > 0$ for $\vartheta_1 < \vartheta < \vartheta_2$, $f_i(\vartheta_1) = f_i(\vartheta_2) = 0$. Solving each one of the K_1 equations $h\,d\vartheta/dx = f_i(\vartheta)$ we obtain K_1 integral curves $\vartheta = \vartheta(x)$ which satisfy the conditions (2.2.11); that is, in this case there is a nonuniqueness of the equation (2.2.9) solution under conditions (2.2.11).

Note that it is possible to investigate the integral curves of the problem (2.2.9)–(2.2.11) by a different technique. In this technique equation (2.2.9) is directly investigated, that is, this equation is considered in a form which is not resolved with respect to the derivative $d\vartheta/dx$. Here one can try to use the qualitative techniques and methods presented in [2.63]. Such a study was not carried out until now.

In the case when one is able to solve equation (2.2.9) with respect to $h\,d\vartheta/dx$, and there is only one positive root at $\vartheta_1 < \vartheta < \vartheta_2$ among the roots $f_i(\vartheta)$, it is not difficult to consider the question of the existence and uniqueness of the smeared shock wave center in the solution of the problem (2.2.9), (2.2.11). Let $C^l[\vartheta_1, \vartheta_2]$ ($l = 0, 1, \ldots$) be a set of the functions which are continuous in $[\vartheta_1, \vartheta_2]$ and which have continuous derivatives of order up to l in $[\vartheta_1, \vartheta_2]$.

Theorem 2.1. *If the following conditions are satisfied:*

(a) *equation (2.2.9), where the function $\tilde{F}$ is determined by formulas (2.2.8), (2.2.10), has the unique positive root $h\,d\vartheta/dx = f(\vartheta)$;*
(b) *$f(\vartheta) > 0$ at $\vartheta_1 < \vartheta < \vartheta_2$, $f(\vartheta_1) = f(\vartheta_2) = 0$; and*
(c) *$f(\vartheta) \in C^1[\vartheta_1, \vartheta_2]$,*

then there exists the unique smeared shock wave center in the solution of the problem (2.2.9), (2.2.11).

PROOF. The solution of the equation $h \, d\vartheta/dx = f(\vartheta)$ has the form

$$x - x_0 = h\Phi(\vartheta), \tag{2.2.12}$$

where

$$\Phi(\vartheta) = \int d\vartheta/f(\vartheta), \tag{2.2.13}$$

x_0 is an integration constant. Note that the function $\Phi(\vartheta)$ is continuously differentiable at $\vartheta_1 < \vartheta < \vartheta_2$ with regard to condition (c), and $\Phi'(\vartheta) > 0$ at $\vartheta_1 < \vartheta < \vartheta_2$ by virtue of condition (b). Taking into account formulas (2.2.11), (2.2.12) we obtain that $\Phi(\vartheta)$ changes its sign in the interval $(\vartheta_1, \vartheta_2)$. Therefore, the equation $\Phi(\vartheta) = 0$ has the only root $\vartheta_0 \in (\vartheta_1, \vartheta_2)$. Since the function $\Phi(\vartheta)$ does not depend on h and x_0, the quantity ϑ_0 does not depend on h and x_0. In accordance with the known theorems on the existence and uniqueness of solutions of autonomous differential equations there exists the unique solution of the problem (2.2.9), (2.2.11) which satisfies the condition $\vartheta((x - x_0)/h)|_{x=x_0} = \vartheta_0$. Thus, the smeared shock wave center is unique, as was to be proved.   □

Corollary 1. *If the artificial viscosity q of the form* (2.2.4) *is chosen in such a way that the conditions of Theorem 2.1 are satisfied, then the quantity* max q *does not depend on the specific form of q.*

In fact, the solution $\vartheta(x - x_0) \in C^2(-\infty, \infty)$ if conditions (a)–(c) are satisfied [2.60]; in addition, a transition from the singular point $\vartheta = \vartheta_1$ to the singular point $\vartheta = \vartheta_2$ with increasing x occurs monotonously by virtue of condition (b). Consequently, in this case, the solution ϑ of the problem (2.2.9), (2.2.11) possesses the property $\vartheta_1 \le \vartheta \le \vartheta_2$. Let us find the maximum of the function $q(\vartheta)$ defined by (2.2.7) in the interval $[\vartheta_1, \vartheta_2]$:

$$\max q = q(\sqrt{\vartheta_1 \vartheta_2}) = (\gamma + 1)m^2(\sqrt{\vartheta_2} - \sqrt{\vartheta_1})^2/2. \tag{2.2.14}$$

This fact was indicated earlier in [2.64].

Corollary 2. *If $\Phi(\sqrt{\vartheta_1 \vartheta_2}) < 0$ $(\Phi(\sqrt{\vartheta_1 \vartheta_2}) > 0)$, then the point of maximum of the artificial viscosity is to the left (to the right) of the smeared shock wave center. If, besides, $B = |\Phi(\sqrt{\vartheta_1 \vartheta_2})|$, then the abscissa of the point of maximum of the artificial viscosity q is at the distance $B \cdot h$ from the smeared shock wave center abscissa.*

Corollary 3. *The inequality $u_1 > u_2$ takes place at the front of a stable shock wave* [2.41]. *Therefore, in the case when the function $u(x)$ is monotone in the zone of smearing of a shock wave, the inequality $du/dx < 0$ takes place in this zone. It may easily be seen from equations* (2.2.1)–(2.2.3) *that the functions ρ, $p + q$, ϑ will also be monotone in the above zone, and the relationships*

$$du/dx < 0, \qquad \mathrm{sign}[d(p + q)/dx] = \mathrm{sign} \, m,$$

$$\mathrm{sign}(d\vartheta/dx) = -\mathrm{sign} \, m,$$

take place.

Corollary 4. *If the function $f(\vartheta)$ in (2.2.13) is representable in the form*

$$f(\vartheta) = f_1(\vartheta)(\vartheta - \vartheta_1)^{\beta_1}(\vartheta_2 - \vartheta)^{\beta_2},$$

where $\beta_1 < 1$, $\beta_2 < 1$, and the function f_1 is bounded and positive in the interval $\vartheta_1 \leq \vartheta \leq \vartheta_2$, then the width X of the zone of the smeared shock wave is finite.

Indeed, the width X with (2.2.12), (2.2.13) in view is determined by the formula

$$X = h[\Phi(\vartheta_2) - \Phi(\vartheta_1)] = h \int_{\vartheta_1}^{\vartheta_2} d\vartheta/f(\vartheta) \tag{2.2.15}$$

and the integrand has integrable singularities at the points ϑ_1 and ϑ_2.

Remark. The function F in (2.2.4) is chosen independently of u in order to provide the invariance of viscosity (2.2.4) under the Galilean transformation.

Corollary 5. *If the viscosity (2.2.4) is such that the conditions of Theorem 2.1 are satisfied, and, besides, the equality $\Phi''(\vartheta_0) = 0$ holds, where ϑ_0 is the value of the function ϑ at the smeared shock wave center $x = x_0$, then the center of the smeared shock wave coincides with the point of extremum of the functions*

$$|du/dx|, \quad d\vartheta/dx, \quad |d(p + q)/dx|. \tag{2.2.16}$$

If, besides, $\Phi'''(\vartheta_0) > 0$ ($\Phi'''(\vartheta_0) < 0$), then the smeared shock wave center is the point of maximum (local minimum) of functions (2.2.16).

Differentiate twice, with respect to x, both sides of equation (2.2.12), as a result the formula

$$\Phi''(\vartheta)(d\vartheta/dx)^2 + \Phi'(\vartheta)\, d^2\vartheta/dx^2 = 0 \tag{2.2.17}$$

is obtained. $\Phi'(\vartheta) > 0$, taking into account condition (b), therefore, from (2.2.17) it follows that $(d^2\vartheta/dx^2)|_{x=x_0} = 0$. Differentiating twice, with respect to x, both sides of equation (2.2.17) we find that at the point x_0 there occurs the formula

$$(d\vartheta/dx)^3 \Phi'''(\vartheta(x_0)) + \Phi'(\vartheta(x_0))(d^3\vartheta/dx^3)|_{x=x_0} = 0,$$

from which the second assertion follows relative to $\vartheta(x)$; the same conclusion is also valid for the rest of the functions in (2.2.16), taking into account (2.2.1), (2.2.2). In addition, if x_0 is the point of minimum of the functions (2.2.16), then this minimum is local because $d\vartheta/dx > 0$ within the smeared shock wave zone according to Theorem 2.1, and the absolute minimum of the functions (2.2.16), equal to zero, is reached at the ends of this zone.

EXAMPLE. Consider the viscosity q from [2.59], [2.65]

$$q = \begin{cases} a\rho(h\, du/dx)^2, & du/dx < 0, \\ 0 & du/dx \geq 0. \end{cases} \tag{2.2.18}$$

Taking into account (2.2.5), (2.2.7), (2.2.18), equation (2.2.9) may be written in the case under study as follows:

$$a(h\,d\vartheta/dx)^2 = (1/2)(\gamma + 1)(\vartheta_2 - \vartheta)(\vartheta - \vartheta_1). \qquad (2.2.19)$$

Solving equation (2.2.19) as an algebraic quadratic equation with respect to $h\,d\vartheta/dx$, we find that in the interval $\vartheta_1 < \vartheta < \vartheta_2$ it has the only positive root of the form

$$h\,d\vartheta/dx = \{[(\gamma + 1)/(2a)](\vartheta_2 - \vartheta)(\vartheta - \vartheta_1)\}^{0.5}. \qquad (2.2.20)$$

It is easy to find with regard to (2.2.20) that the function $\Phi(\vartheta)$ in (2.2.12) has the form

$$\Phi(\vartheta) = [2a/(\gamma + 1)]\arcsin[(2\vartheta - \vartheta_1 - \vartheta_2)/(\vartheta_2 - \vartheta_1)]. \qquad (2.2.21)$$

The equation $\Phi(\vartheta) = 0$ has in the interval $\vartheta_1 \le \vartheta \le \vartheta_2$ the only root $\vartheta_0 = 0.5(\vartheta_1 + \vartheta_2)$. Thus, there exists a center of the smeared shock wave. According to (2.2.12), (2.2.21), $\vartheta = \vartheta_0$ at $x'_0 = x_0 \pm k\pi$, k is an integer, $k = 0, 1, 2, \ldots$. We get from (2.2.12), with (2.2.21) in view, that

$$\vartheta(x - x'_0) = \frac{\vartheta_1 + \vartheta_2}{2} + \frac{\vartheta_2 - \vartheta_1}{2}\sin\left(\sqrt{\frac{\gamma + 1}{2a}}\frac{x - x'_0}{h}\right). \qquad (2.2.22)$$

We are interested only in that part of the curve (2.2.22) in which $d\vartheta/dx > 0$. This gives the interval

$$-\pi/2 < [(\gamma + 1)/(2a)]^{0.5}(x - x'_0)/h < \pi/2.$$

Thus, in the case under consideration the solution $\vartheta(x)$ of the problem (2.2.9), (2.2.11) has the form (see also [2.7])

$$\vartheta = \begin{cases} \dfrac{\vartheta_1 + \vartheta_2}{2} + \dfrac{\vartheta_2 - \vartheta_1}{2}\sin\left(\sqrt{\dfrac{\gamma + 1}{2a}}\dfrac{x - x'_0}{h}\right), \\ \qquad\qquad |x - x'_0| \le (\pi/2)\sqrt{2a/(\gamma + 1)}\,h, \qquad (2.2.23) \\ \vartheta_1, \qquad x - x'_0 \le -(\pi/2)\sqrt{2a/(\gamma + 1)}\,h, \\ \vartheta_2, \qquad x - x'_0 \ge (\pi/2)\sqrt{2a/(\gamma + 1)}\,h. \end{cases}$$

It follows from (2.2.21) that at $x = x'_0$ $\Phi''(\vartheta(x'_0)) = 0$, that is, x'_0 is the point of maximum of the functions (2.2.16) in accordance with Corollary 5 of Theorem 2.1; in addition, at $x = x'_0$

$$(d^{2n}/dx^{2n})(\vartheta, p + q, u) = 0, \qquad n = 1, 2, \ldots.$$

Let x_* be the abscissa of the point of maximum of the artificial viscosity q (2.2.18). It is easy to be sure of the fact that $\Phi(\sqrt{\vartheta_1\vartheta_2}) < 0$, therefore, $x_* < x_0$ in accordance with Corollary 2 of Theorem 2.1. It follows from (2.2.21) that the value $\Phi(\sqrt{\vartheta_1\vartheta_2})$ increases as the shock wave intensity increases. As is known, in the case of an infinitely strong shock wave $\vartheta_2/\vartheta_1 = (\gamma + 1)/(\gamma - 1)$, therefore,

$$|x_* - x_0| \leq h\sqrt{2a/(\gamma + 1)}\ \arcsin[(\gamma + \sqrt{\gamma^2 - 1})^{-1}]. \qquad (2.2.24)$$

Formula (2.2.24) can provide additional information for the choice of a non-dimensional coefficient a if one requires that the inequality $|x_* - x_0| \leq h$ be satisfied. Then we obtain from (2.2.24) the following limitation on a:

$$a \leq 0.5(\gamma + 1)\{\arcsin[(\gamma + (\gamma^2 - 1)^{1/2})^{-1}]\}^{-2}. \qquad (2.2.25)$$

For example, at $\gamma = 1.4$ we find from (2.2.25): $a \leq 6.34$, thus, the inequality $|x_* - x_0| \leq h$ is satisfied at usually used values of a (usually one takes $a \leq 4$, see [2.59]).

It follows from formula (2.2.20) that the width X of a zone of "smearing" of the shock in the solution of the problem (2.2.9), (2.2.11) is finite while using the quadratic viscosity (2.2.18),

$$X = \sqrt{2a/(\gamma + 1)}\ \pi h. \qquad (2.2.26)$$

It follows from (2.2.26) that $X \to 0$ as $h \to 0$, so that in the limit $h = 0$ we have a discontinuous solution. If the width X determined by formula (2.2.15) proves to be infinite, then one can consider the width after Prandtl X_1,

$$X_1 = |\vartheta_2 - \vartheta_1|\bigg/\bigg(\max_x |d\vartheta/dx|\bigg). \qquad (2.2.27)$$

Formula (2.2.27) determines the size along the x-axis of a region in which the most substantial change of the solution ϑ in the zone of smearing of a shock wave takes place. If the function $d\vartheta/dx$ is continuous and has a constant sign within the zone of smearing, then one can employ for the computation of the width X_1 the formula

$$X_1 = h|\vartheta_1 - \vartheta_2| \min_{\vartheta_1 \leq \vartheta \leq \vartheta_2} |\Phi'(\vartheta)| \qquad (2.2.28)$$

taking into account (2.2.12). It follows from (2.2.28) that $X_1 \to 0$ as $h \to 0$, thus, we again obtain a discontinuous solution in the limit (see also [2.66]). There arises the question as to whether this limiting discontinuous solution coincides with the generalized solution of the original Euler equations (1.1)–(1.3). A discussion of this question is deferred to Section 2.3 where we consider more complicated expressions for the dissipative terms than in this section. Here we only note that in the case when the above convergence takes place at $h \to 0$, then it is reasonable to use the above introduced notion of a smeared shock wave center, while estimating the accuracy of various techniques for shock front localization in the smeared profiles, since by virtue of Definition 2.1 the position of the smeared shock wave center does not depend on the stepsize h in the coordinate system which is co-moving with the shock wave. Consequently, the center of a smeared shock wave in the solution of the problem

(2.2.9), (2.2.11) propagates at an accurate speed of the shock wave. The determination of the smeared shock wave center coordinates can be used directly as the basis of the algorithms for locating shock fronts which was realized in [2.54], [2.55], [2.57], [2.58], [2.67]–[2.69]. The questions of the practical realization of such algorithms are considered in Section 2.4.3.

2.2.2. Analysis of Artificial Viscosities

Consider the question of existence of the smeared shock wave center in the solutions of the system (2.2.1)–(2.2.3) at a number of specific forms of the artificial viscosity.

Theorem 2.2 [2.55]. *If the dimensionless constants a and b entering the artificial viscosity*

$$q = -ahc\rho\left(1 + bc\rho\frac{du}{dx}\bigg/\frac{dp}{dx}\right)\min\left(\frac{du}{dx}, 0\right), \qquad (2.2.29)$$

where c is the local speed of sound, satisfy the inequalities

$$a > 0;$$

$$b > [\eta - (\gamma + 1)/(\gamma - 1)]\sqrt{\gamma - 1} \qquad (2.2.30)$$
$$\times \{2\gamma[(\gamma + 1)\eta/(\gamma - 1) + \eta^2 - (\gamma + 1)\eta(1 + \eta)/\gamma]\}^{-1/2},$$

with $\eta = \vartheta_2/\vartheta_1$, then the smeared shock wave center in the solution of the corresponding problem (2.2.9)–(2.2.11) exists and is unique.

PROOF. Employing formulas (2.2.7), (2.2.8), (2.2.9), (2.2.29) we obtain for $\vartheta(x)$ an ordinary differential equation

$$d\vartheta/dx = [(\gamma + 1)|m|(\vartheta_2 - \vartheta)(\vartheta - \vartheta_1)/2]/\{ah$$
$$\times [c - b(c^2/\vartheta)((\gamma - 1)(|m|/2)(1 - (\gamma + 1)\vartheta_1\vartheta_2/((\gamma - 1)\vartheta^2)))^{-1}]\}, \qquad (2.2.31)$$

where the right-hand side is positive in the interval $(\vartheta_1, \vartheta_2)$ by virtue of (2.2.30). The solution of equation (2.2.31) may be written in the form

$$\Phi(\vartheta) = J_1(\theta) - J_2(\theta) = (x - x_0)/h, \qquad (2.2.32)$$

where $\theta = \vartheta/\vartheta_1$,

$$J_1(\theta) = 2a[0.5\gamma(\gamma - 1)]^{1/2}[(\eta - 1)(\gamma + 1)]^{-1} \times \{(1 - \eta)\ln[2(R(\theta))^{0.5} + 2\theta + a_2]$$
$$- (R(1))^{1/2}\ln[2(R(1)(1 + (a_2 + 2)/(\theta - 1) + R(1)(\theta - 1)^{-2}))^{1/2}$$
$$+ a_2 + 2 + 2R(1)(\theta - 1)^{-1}]$$
$$+ [R(\eta)]^{1/2}\ln[2(R(\eta)(1 + (a_2 + 2\eta)|\theta - \eta|^{-1} \qquad (2.2.33)$$
$$+ R(\eta)(\theta - \eta)^{-2}))^{1/2} + 2R(\eta)/(\theta - \eta) + a_2 + 2\eta]\};$$

$$J_2(\theta) = (\gamma + 1)^{-1} 2ab \ln[(\theta - 1)^{\beta_2}(\eta - \theta)^{-\beta_1}(\theta - \sqrt{a_1})^{\beta_3}(\theta + \sqrt{a_1})^{\beta_4}];$$

$$R(\theta) = a_1 + a_2\theta + \theta^2;$$

$$a_1 = (\gamma + 1)\eta/(\gamma - 1); \qquad a_2 = -(\gamma + 1)(1 + \eta)/\gamma;$$

$$\eta = \vartheta_2/\vartheta_1; \qquad \beta_i = \Delta_i/\Delta, \qquad i = 1, \ldots, 4;$$

$$\Delta = 2(\eta - 1)a_1^{1/2}(1 - a_1)(a_1 - \eta^2);$$

$$\Delta_1 = 2a_1^{1/2}\eta(a_1 - 1)R(\eta); \qquad \Delta_2 = 2a_1^{1/2}R(1)(a_1 - \eta^2);$$

$$\Delta_3 = (\eta - 1)a_1(a_2 + 2\sqrt{a_1})(1 + \sqrt{a_1})(\eta + \sqrt{a_1});$$

$$\Delta_4 = (\eta - 1)a_1(a_2 + 2\sqrt{a_1})(\sqrt{a_1} - 1)(\eta - \sqrt{a_1}).$$

Let us find the conditions on dimensionless coefficients a, b in (2.2.29) under which the function $\Phi(\theta)$ determined by (2.2.32), (2.2.33) changes its sign in the interval $(1, \eta)$. For this purpose let us at first collect in (2.2.33) the terms containing the difference $\theta - 1$:

$$-a \ln\{[2(R(1)(1 + (a_2 + 2)/(\theta - 1) + R(1)(\theta - 1)^{-2}))^{1/2}$$
$$+ 2R(1)/(\theta - 1) + a_2 + 2]^{\lambda_1}(\theta - 1)^{\lambda_2}\} \quad (2.2.34)$$

where

$$\lambda_1 = [R(1)0.5\gamma(\gamma - 1)]^{1/2}(\eta - 1)^{-1}, \qquad \lambda_2 = b\beta_2\gamma.$$

Using the expression (2.2.34) it is easy to show that for the satisfaction of the relationship

$$\lim_{\theta \to 1} \Phi(\theta) = -\infty$$

it is necessary that $\lambda_2 - \lambda_1 < 0$ from where we get, with the inequality $\beta_2 < 0$ in view,

$$b > [R(1)0.5\gamma(\gamma - 1)]^{1/2}/[\beta_2\gamma(\eta - 1)] = b_2. \quad (2.2.35)$$

Considering analogously the case $\theta \to \eta$ we obtain from the requirement $\lim_{\theta \to \eta} \Phi(\theta) = +\infty$ the inequality

$$b > [R(\eta)0.5\gamma(\gamma - 1)]^{1/2}/[\beta_1\gamma(\eta - 1)] = b_1. \quad (2.2.36)$$

It is clear that both inequalities (2.2.35), (2.2.36) are satisfied if $b > \max(b_1, b_2)$. A simple analysis of the expressions for b_1 and b_2 shows that $\max(b_1, b_2) = b_1$ and that b_1 coincides with the right-hand side of the second of the inequalities (2.2.30). Thus, in the case of the viscosity (2.2.29), the condition of the smeared shock wave center existence proves to be coincident with one of the conditions for the positiveness of $d\vartheta/dx$ in the zone of a smeared shock wave. $\qquad \Box$

Remark 1. The artificial viscosity (2.2.29) is a particular case of the pseudo-viscosity of a more general form

$$q = -ahc\rho\left\{1 + \sum_{k=1}^{N} b_k\left(c\rho\left|\frac{du}{dx}\bigg/\frac{dp}{dx}\right|\right)^k\right\}\min\left(\frac{du}{dx}, 0\right),$$

where $N \geq 1$; b_k, $k = 1, \ldots, N$, are nonnegative constants.

Remark 2. A number of recommendations on the choice of the values of constants a and b in (2.2.29) have been given in [2.55]. It has been shown in computational examples (in [2.55]) that the viscosity (2.2.29) at $b \neq 0$ leads to a lesser smearing of the shock wave front than the viscosity (2.2.29) with $a \neq 0$, $b = 0$. At the same time, in the case where $b \neq 0$ the post-shock oscillations are effectively damped. A possible choice of the values of the coefficients a and b in (2.2.29) (which was indicated in [2.55] for the case $\gamma = 2$ in the equation of state (1.8)) is $a = 0.15$, $b = 0.77$.

Theorem 2.3 [2.55]. *If the dimensionless constants a, σ_1, and σ_2 entering the artificial viscosity*

$$q = -ah\rho(\sigma_1 c + \sigma_2 h |du/dx|) \min(du/dx, 0), \qquad (2.2.37)$$

where c is the local speed of sound, satisfy the inequalities

$$(1 + \eta)^2(\gamma + 1)[4\sigma_2 - a\sigma_1^2(\gamma - 1)]^2 - 4\eta$$
$$\times [a\sigma_1^2\gamma(\gamma - 1) - 4\sigma_2(\gamma + 1)](a\sigma_1^2\gamma - 4\sigma_2) \leq 0; \quad (2.2.38)$$
$$a > 0; \qquad 4a\sigma_2(\gamma + 1) < a^2\sigma_1^2\gamma(\gamma - 1); \qquad \sigma_2 \neq 0; \qquad (2.2.39)$$
$$a^2\sigma_1^2\gamma(\gamma - 1) < b, \qquad (2.2.40)$$

where

$$b = \min(b_1, b_2), \qquad \eta = \vartheta_2/\vartheta_1;$$
$$b_1 = \{[A_1(1)/(A_2(1))^\lambda]^{(\eta-1)}[B_2(\eta)/B_1(\eta)]^{\sqrt{s_2}}\}^{1/s_1};$$
$$b_2 = \{[A_1(\eta)/(A_2(\eta))^\lambda]^{-(\eta-1)}[B_2(1)/B_1(1)]^{\sqrt{s_1}}\}^{1/s_2};$$
$$A_i(\theta) = 2[a_iR_i(\theta)]^{0.5} + 2a_i\theta + b_i; \qquad s_1 = R_1(1);$$
$$s_2 = R_1(\eta); \qquad \lambda = [1 - 4a\sigma_2(\gamma + 1)/(a^2\sigma_1^2\gamma(\gamma - 1))]^{0.5};$$
$$B_i(\theta) = 2\{R_i(\theta)[a_i + (b_i + 2a_i\theta)/(\eta - 1) + R_i(\theta)$$
$$\times (\eta - 1)^{-2}]\}^{0.5} + 2R_i(\theta)/(\eta - 1) + b_i + 2a_i\theta;$$
$$R_i(\theta) = a_i\theta^2 + b_i\theta + c_i, \qquad i = 1, 2;$$
$$a_1 = 1; \qquad b_1 = -(\gamma + 1)(1 + \eta)/\gamma;$$
$$c_1 = (\gamma + 1)\eta/(\gamma - 1); \qquad a_2 = (a\sigma_1)^2\gamma(\gamma - 1) - 4a\sigma_2(\gamma + 1);$$
$$b_2 = (\gamma + 1)(1 + \eta)[4a\sigma_2 - a^2\sigma_1^2(\gamma - 1)];$$
$$c_2 = a(\gamma + 1)\eta(a\sigma_1^2\gamma - 4\sigma_2),$$

then the smeared shock wave center in the solution of the corresponding problem (2.2.9)–(2.2.11) exists and is unique.

The proof of this theorem is similar to the proof of Theorem 2.2, therefore it is not presented here.

Table 2.1

q	$ah^2\rho\left(\dfrac{du}{dx}\right)^2$	$-ahc_0\rho\dfrac{du}{dx}$	$-ahc\rho\dfrac{du}{dx}$	$-ah\rho\dfrac{du}{dx}(\sigma_1 c + \sigma_2 h|du/dx|)$	$-ahc\rho\left[1 + bc\rho\dfrac{du}{dx}\left|\dfrac{dp}{dx}\right|\dfrac{du}{dx}\right]$
Reference	[2.59]	[2.70]	[2.71]	[2.71]	[2.47]
Does the smeared shock wave center exist?	Yes	Yes	Yes	Yes, at (2.2.38)–(2.2.40)	Yes, at (2.2.30)
Practical criterion for finding the smeared shock wave center	by max$\|du/dx\|$ at $du/dx < 0$	by max$\|du/dx\|$ at $du/dx < 0$	by max q at $\|\Phi(\sqrt{\vartheta_1\vartheta_2})\| < 1$	by max q at $\|\Phi(\sqrt{\vartheta_1\vartheta_2})\| < 1$	by max q at $\|\Phi(\sqrt{\vartheta_1\vartheta_2})\| < 1$

Remark 1. The inequalities (2.2.38) and $a > 0$, $4a\sigma_2(\gamma + 1) < a^2\sigma_1^2\gamma(\gamma - 1)$ are the conditions for the existence of a continuous monotone transition in the zone of a smeared shock wave, whereas the inequality (2.2.40) is a condition for the existence of a smeared shock wave center.

Remark 2. The width X of a smeared shock wave at the values $a \neq 0$, $\sigma_1 \neq 0$, $\sigma_2 \neq 0$, satisfying the conditions (2.2.38)–(2.2.40), is computed by the formula

$$X = ha\sigma_1[\gamma(\gamma - 1)/2]^{0.5}[(\gamma + 1)(\eta - 1)]^{-1}$$
$$\times \{R_1(1)\ln b_1 + R_1(\eta)\ln b_2 - [(R_1(1))^{0.5}$$
$$+ (R_1(\eta))^{0.5}]\ln[a^2\sigma_1^2\gamma(\gamma - 1)]\}.$$

Remark 3. Consider a weak shock wave for which $\eta = 1 + \varepsilon$, ε is a small positive number. It is not difficult to show, by linearizing the expressions for b_1 and b_2 with respect to ε, that with accuracy up to the terms $O(\varepsilon)$ the relationship $b = 1 + \varepsilon \ln \varphi(\gamma, a, \sigma_1, \sigma_2)$ takes place where $\varphi > 1$. Then instead of (2.2.40) one can use a more restrictive, but at the same time very simple inequality $a^2\sigma_1^2\gamma(\gamma - 1) < 1$.

Remark 4. In the case where $\sigma_2 \equiv 0$ (linear in du/dx viscosity (2.2.37)) it is easy to show that the smeared shock wave center exists and is unique only under the condition $a\sigma_1 > 0$, and the width X of the smeared shock wave is infinite.

In Table 2.1 we summarize the results of the investigation of the smeared shock wave center existence when using a number of artificial viscosities of the form (2.2.4).

2.3. Scheme Viscosity and Smeared Shock Wave Center Existence

As was already demonstrated in Section 1.4, the first differential approximation (f.d.a.) is applicable to the description of the approximation viscosity structure of the difference schemes of gas dynamics. The structure of approximation viscosity, as determined by the f.d.a. equations, is much more complicated than in the case of a model system (2.2.1)–(2.2.4), which was considered in detail in the foregoing section. In addition, the only grid parameter in Section 2.2 was the step h of a mesh on the x-axis; in the general case, as follows from Sections 1.2 and 1.4, two grid parameters h and τ are present in a difference scheme approximating a nonstationary system (1.1)–(1.3) where τ is a time step.

Since at present the divergence difference schemes are basically used in gas dynamic computations [2.41], [2.44], [2.66], we shall restrict ourselves in the following to a consideration of some classes of such schemes.

Consider a class of difference schemes approximating the system (1.11), (1.9) with an rth order of accuracy, $r \geq 1$, and such that their f.d.a. H-form is representable in the form

$$\partial \mathbf{w}/\partial t + \partial \varphi(\mathbf{w})/\partial x = (\partial/\partial x)\left[\sum_{\substack{i,j \\ i+j=r}} h^i \tau^j \, \mathbf{F}_{ij}(\mathbf{w}, \partial \mathbf{w}/\partial x, \partial \mathbf{w}/\partial t, \ldots, \right.$$

$$\left. \partial^r \mathbf{w}/\partial x^r, \partial^r \mathbf{w}/\partial x^{r-1}\partial t, \ldots, \partial^r \mathbf{w}/\partial t^r) \right]. \qquad (2.3.1)$$

Let us express the derivatives $\partial^m \mathbf{w}/\partial x^k \, \partial t^{m-k}$, $m - k > 0$, entering into $\mathbf{F}_{ij}$, in terms of x-derivatives making use of an algorithm described in Section 1.2

$$\partial^m \mathbf{w}/\partial x^k \partial t^{m-k} = \mathbf{f}_{m,k}(\mathbf{w}, \partial \mathbf{w}/\partial x, \ldots, \partial^m \mathbf{w}/\partial x^m). \qquad (2.3.2)$$

As a result we obtain from the system (2.3.1) the f.d.a. P-form

$$\partial \mathbf{w}/\partial t + \partial \varphi(\mathbf{w})/\partial x = (\partial/\partial x)\left[\sum_{\substack{i,j \\ i+j=r}} h^i \tau^j \mathscr{F}_{ij}(\mathbf{w}, \partial \mathbf{w}/\partial x, \ldots, \partial^r \mathbf{w}/\partial x^r) \right]. \qquad (2.3.3)$$

We shall consider the progressive wave-type solutions of the equations of a f.d.a. (2.3.3) or (2.3.1), that is, the solutions depending only on the variable $\xi = x - Dt$ where D is the speed of a stationary shock wave. Substituting the function $w(\xi)$ instead of $w(x, t)$ into (2.3.1), (2.3.3) we obtain, with account being taken of the results of Section 1.4, the following two systems of ordinary differential equations

$$\sum_{\substack{i,j \\ i+j=r}} h^i \tau^j \mathscr{F}_{ij}(\mathbf{w}, d\mathbf{w}/d\xi, -D \, d\mathbf{w}/d\xi, \ldots, d^r\mathbf{w}/d\xi^r,$$

$$-D d^r\mathbf{w}/d\xi^r, \ldots, (-D)^r \, d^r\mathbf{w}/d\xi^r) = \varphi(\mathbf{w}) - \varphi(\mathbf{W}_1) - D(\mathbf{w} - \mathbf{W}_1); \qquad (2.3.4)$$

$$\sum_{\substack{i,j \\ i+j=r}} h^i \tau^j \mathscr{F}_{ij}(\mathbf{w}, d\mathbf{w}/d\xi, \ldots, d^r\mathbf{w}/d\xi^r) = \varphi(\mathbf{w}) - \varphi(\mathbf{W}_1) - D(\mathbf{w} - \mathbf{W}_1), \qquad (2.3.5)$$

where $\mathbf{W}_1$ is a constant vector entering the boundary conditions

$$\mathbf{w}(\xi) = \begin{cases} \mathbf{W}_1, & \xi \to -\infty, \\ \mathbf{W}_2, & \xi \to +\infty. \end{cases} \qquad (1.63)$$

At $r > 1$ we shall also impose alongside the boundary conditions (1.63) the following conditions

$$\lim_{\xi \to -\infty} d\mathbf{w}/d\xi = 0, \qquad r = 2;$$

$$\lim_{|\xi| \to \infty} d\mathbf{w}/d\xi = 0, \qquad r = 3; \qquad (2.3.6)$$

$$\lim_{\xi \to -\infty} d^2\mathbf{w}/d\xi^2 = \lim_{|\xi| \to \infty} d\mathbf{w}/d\xi = 0, \qquad r = 4.$$

In (2.3.6) $(r - 1)$ boundary conditions are written for the cases $r = 2$, $r = 3$, $r = 4$; $(r - 1)$ boundary conditions for $r > 4$ may be written from similar considerations.

As is known, the progressive wave-type solution is determined with an accuracy up to the translation along the ξ-axis. Let ξ_0 be a constant characterizing an arbitrary translation, or shift, along the ξ-axis. As in Section 2.2, we denote by G_s a finite interval on the ξ-axis within which the most significant variation of the w solution occurs.

Definition 2.2. The point $\xi_* \in G_s$, at which the values of the components of the progressive wave-type solution vector $w(\xi_*, \xi_0, h, \tau)$ of the system of equations of the first differential approximation (2.3.4) or (2.3.5) are obtained under the boundary conditions (1.63), (2.3.6) do not depend on h and τ at each fixed value of ξ_0, is called the smeared shock wave center.

Remark. It is possible to consider a wider class of schemes than the one for which the f.d.a. representation in the form (2.3.1) is possible while also studying the question on the existence of a smeared shock wave center on the basis of the f.d.a. H-form (2.3.1) which leads to the system (2.3.4). This is a class of schemes of the rth order of approximation whose f.d.a. H-form is representable in the form

$$\partial w/\partial t + \partial \varphi(w)/\partial x = (\partial/\partial x)\left[\sum_{\substack{i,j \\ i+j=r}} h^i \tau^j F_{ij}^{(1)}(w, \partial w/\partial x, \partial w/\partial t, \dots, \right.$$

$$\left. \partial^r w/\partial x^r, \partial^r w/\partial x^{r-1}\, \partial t, \partial^r w/\partial t^r \right]$$

$$+ (\partial/\partial t)\left[\sum_{\substack{i,j \\ i+j=r}} h^i \tau^j F_{ij}^{(2)}(w, \partial w/\partial x, \partial w/\partial t, \dots, \right.$$

$$\left. \partial^r w/\partial x^r, \partial^r w/\partial x^{r-1}\, \partial t, \dots, \partial^r w/\partial t^r) \right]. \quad (2.3.7)$$

Indeed, substituting $w = w(\xi)$ into (2.3.7) we obtain a system of ordinary differential equations which can be integrated once.

2.3.1. Relation Between the Divergence Property of Difference Schemes and the First Differential Approximation Divergence Property

We introduce some notations before formulating a class of divergence difference schemes for the system (1.11) which will be considered below from the point of view of the theory of differential analyzers. Let $T_{\pm 1}$, $T_{\pm 0}$ be the shift operators along the x- and t-axis, respectively, that is,

$$T_{\pm 1}\mathbf{w}(x, t) = \mathbf{w}(x \pm h, t); \qquad T_{\pm 0}\mathbf{w}(x, t) = \mathbf{w}(x, t \pm \tau);$$

$$T_0 \equiv T_{+0}, \qquad T_1 \equiv T_{+1}, \qquad T_{\pm 1/2}\mathbf{w}(x, t) = \mathbf{w}(x \pm h/2, t).$$

Theorem 2.4. *If the finite-difference scheme approximating the system* (1.11) *is representable in the form*

$$(1/\tau)C\{\beta_1[T_0 - (\alpha_3 T_1 + \alpha_4 I + \alpha_5 T_{-1})]$$

$$+ \beta_2[(\alpha_6 T_1 + \alpha_7 I + \alpha_8 T_{-1}) - T_{-0}]\}\mathbf{w}(x, t)$$

$$+ (1/h)[\alpha_1(T_1 - I) + \alpha_2(I - T_{-1})](\beta_3 T_0 + \beta_4 I + \beta_5 T_{-0})\varphi(\mathbf{w}(x, t))$$

$$= (1/h)(T_{1/2} - T_{-1/2})[(1/h)\Omega(\mathbf{w}(x, t))$$

$$\times (T_{1/2} - T_{-1/2})(\beta_6 T_0 + \beta_7 I + \beta_8 T_{-0})]\mathbf{w}(x, t), \tag{2.3.8}$$

and the following conditions are satisfied:

(a) *C is a three-point operator such that its second differential approximation $\tilde{C}$ is of the form*

$$\tilde{C} = I + \sum_{i=1}^{2} h^i(\partial^i/\partial x^i)a_i(\mathbf{w})I, \tag{2.3.9}$$

where $a_i(\mathbf{w})$, $\Omega(\mathbf{w})$ are matrices of 3×3 dimension; these matrices may depend not only on $\mathbf{w}$ but also on the derivatives of the functions $\mathbf{w}(x, t)$;
(b) *elements of the matrix Ω have magnitude of order $O(\tau) + O(h)$;*
(c) *$\alpha_i, \beta_i, i = 1, \ldots, 8$, are constants, such that*

$$\alpha_1 + \alpha_2 = 1, \qquad \alpha_3 + \alpha_4 + \alpha_5 = 1, \qquad \alpha_6 + \alpha_7 + \alpha_8 = 1;$$

$$\beta_1 + \beta_2 = 1, \qquad \beta_3 + \beta_4 + \beta_5 = 1, \qquad \beta_6 + \beta_7 + \beta_8 = 1; \tag{2.3.10}$$

$$0 \le \alpha_j \le 1, \qquad 0 \le \beta_j \le 1, \qquad j = 1, 2, \ldots, 8,$$

then the f.d.a. of the difference scheme (2.3.8) *can be presented in the divergence form*

$$\partial \mathbf{w}/\partial t + \partial \varphi(\mathbf{w})/\partial x = \partial \psi(\mathbf{w})/\partial x. \tag{2.3.11}$$

The proof of the theorem is reduced to expansion of the operators entering scheme (2.3.8) in Taylor series in powers of h and τ with respect to the point (x, t). As a result of trivial but rather cumbersome calculations, the following expression for the differential approximation is obtained:

$$\partial \mathbf{w}/\partial t + \partial \varphi(\mathbf{w})/\partial x = \sum_{i=1}^{3} \tau^{-1}(h^i/i!)[\beta_1(\alpha_3 + (-1)^i\alpha_5) - \beta_2(\alpha_6 + (-1)^i\alpha_8)]$$

$$\times \partial^i\mathbf{w}/\partial x^i - \sum_{j=2}^{3} (\tau^{j-1}/j!)[\beta_1 + (-1)^{j-1}\beta_2] \partial^j\mathbf{w}/\partial t^j$$

$$- \sum_{i=1}^{2} \sum_{j=1}^{2} (h^i \tau^{j-1}/j!)[\beta_1 + (-1)^{j-1}\beta_2]$$

$$\times (\partial^i/\partial x^i) a_i(\mathbf{w}) \, \partial^j \mathbf{w}/\partial t^j - \sum_{j=1}^{2} (\tau^j/j!)$$

$$\times [\beta_3 + (-1)^j \beta_5] \, \partial^{j+1} \varphi / \partial t^j \, \partial x$$

$$- \sum_{i=2}^{3} (h^{i-1}/i!)[\alpha_1 + (-1)^{i-1}\alpha_2](\partial^i/\partial x^i)$$

$$\times \left\{ \varphi + \sum_{j=1}^{2} (\tau^j/j!)[\beta_3 + (-1)^j \beta_5] \, \partial^j \varphi / \partial t^j \right\}$$

$$+ (\partial/\partial x)[\Omega(\mathbf{w})(\partial/\partial x)(\mathbf{w} + \tau(\beta_6 - \beta_8) \, \partial \mathbf{w}/\partial t)]. \quad (2.3.12)$$

To reduce (2.3.12) to the form (2.3.11) it is sufficient to change differentiation with respect to t in the right-hand side of (2.3.12) for differentiation with respect to x. Using the f.d.a., easily obtained from (2.3.12), the following expression can be found

$$\partial^j \mathbf{w}/\partial t^j = (-1)^j (\partial^{j-1}/\partial x^{j-1})(A^j \, d\mathbf{w}/\partial x) + \delta_j^1 \, \partial \varphi_1/\partial x + \delta_j^2 \, \partial \varphi_2/\partial x, \quad (2.3.13)$$

$j = 1, 2, (\partial^0/\partial x^0) \equiv I, A = \partial \varphi/\partial \mathbf{w}, \delta_j^1, \delta_j^2$ are Kronecker symbols,

$$\varphi_1(\mathbf{w}) = \tau(\beta_3 - \beta_5 - 0.5\beta_1 + 0.5\beta_2)A^2 \, \partial \mathbf{w}/\partial x + h a_1(\mathbf{w}) A(\mathbf{w}) \, \partial \mathbf{w}/\partial x$$

$$- (h/2)(\alpha_1 - \alpha_2)A \, \partial \mathbf{w}/\partial x + \Omega \, \partial \mathbf{w}/\partial x;$$

$$\varphi_2(\mathbf{w}) = -\tau(\beta_3 - \beta_5 - 0.5\beta_1 + 0.5\beta_2)(\partial/\partial x)(A^3 \, \partial \mathbf{w}/\partial x)$$

$$- (\partial/\partial x)(\Omega A \, \partial \mathbf{w}/\partial x) - hA[(\partial/\partial x)(a_1 A \, \partial \mathbf{w}/\partial x)$$

$$+ a_1(\partial A/\partial x)(\partial \mathbf{w}/\partial x)] + (h/2)(\alpha_1 - \alpha_2)(\partial/\partial x)(A^2 \, \partial \mathbf{w}/\partial x).$$

Making use of (2.3.12), (2.3.13), and (2.3.10), the function $\psi(\mathbf{w})$ entering (2.3.11) can be easily found

$$\psi(\mathbf{w}) = \sum_{i=1}^{3} h^i/(i! \, \tau)[\beta_1(\alpha_3 + (-1)^i \alpha_5) - \beta_2(\alpha_6 + (-1)^i \alpha_8)]$$

$$\times (\partial^{i-1} \mathbf{w}/\partial x^{i-1}) - \sum_{j=2}^{3} (\tau^{j-1}/j!)[\beta_1 + (-1)^{j-1}\beta_2]$$

$$\times [(-1)^j (\partial^{j-2}/\partial x^{j-2})(A^j \, \partial \mathbf{w}/\partial x) + \delta_j^1 \varphi_1 + \delta_j^2 \varphi_2]$$

$$- \sum_{i=1}^{2} \sum_{j=1}^{2} (h^i \tau^{j-1}/j!)[\beta_1 + (-1)^{j-1}\beta_2]$$

$$\times (\partial^{i-1}/\partial x^{i-1})[a_i(\mathbf{w})(-1)^j (\partial^{j-1}/\partial x^{j-1})$$

$$\times (A^j \, \partial \mathbf{w}/\partial x) + a_i(\mathbf{w})\delta_j^1 \, \partial \varphi_1/\partial x + a_i(\mathbf{w})\delta_j^2 \, \partial \varphi_2/\partial x]$$

$$- \sum_{j=1}^{2} (\tau^j/j!)[\beta_3 + (-1)^j \beta_5]$$

$$\times [(-1)^{j+1}(\partial^{j-1}/\partial x^{j-1})(A^{j+1} \, \partial w/\partial x) - \delta_j^1 \varphi_1 - \delta_j^2 \varphi_2]$$

$$- \sum_{i=2}^{3} (h^{i-1}/i!)(\alpha_1 + (-1)^{i-1}\alpha_2)(\partial^{i-1}/\partial x^{i-1})[\varphi - \tau(\beta_3 - \beta_5) \cdot$$

$$\times A^2 \, \partial w/\partial x] + \Omega(w)(\partial/\partial x)[w - \tau(\beta_6 - \beta_8)(A \, \partial w/\partial x - \partial \varphi_1/\partial x)],$$

that was required.

Remark. The scheme with an operator C of the form (2.3.9) for the system (1.11) was considered in [2.72].

The family of schemes (2.3.8) has a relatively simple structure, therefore, we were able to derive the differential approximation (2.3.12) by hand. However, the derivation of differential approximations needed for the analysis of properties of sufficiently complicated difference schemes, for example, the schemes with many fractional steps, proves to be very difficult because of an immense amount of analytic calculations. In these cases one can resort to automatic calculation of differential approximations of difference schemes on a computer by means of symbolic manipulation languages. Let us enumerate a number of works pertinent to this topic. In [2.73] a computer system has been mentioned which utilizes the FORMAC language for obtaining the f.d.a.s of difference schemes approximating the scalar linear partial differential equation. An analysis of some difference schemes for the equation $u_t + cu_x = 0$ ($c = $ const) by means of the f.d.a. obtained on a computer has been carried out in [2.73], [2.74]. A very brief presentation of an implementation of one more machine system for the f.d.a. calculation—the ALTRAN system—is contained in [2.75]. According to [2.75], this system enables one to automatically obtain the f.d.a. of difference schemes for nonlinear schemes for nonlinear systems in the case of two independent variables x and t. Unfortunately, [2.73]–[2.75] do not contain the information on computer algorithms, on code structure, or on the organization of symbolic manipulation on a computer. In addition, FORMAC and ALTRAN are barely accessible for users in the Soviet Union.

The realization of automatic derivation of the f.d.a. in the case of two-layer one-step schemes for linear and nonlinear scalar equations has been communicated in [2.76]–[2.78], and some examples have been presented. Basic principles underlying the automatic derivation of differential approximations in the case of the fractional step schemes, using the symbolic manipulation language REFAL, have been presented in [2.79]. It should be noted that the automation of the f.d.a. derivation in the case of fractional step schemes for a nonlinear equation is algorithmically the most complicated and has been realized for the first time in [2.79].

2.3.2. On the Existence of the Smeared Shock Wave Center

Consider the question of the existence of a smeared shock wave center for a multiparametric family of difference schemes (2.3.8) approximating the system of equations of gas dynamics in Eulerian variables (1.11), (1.9).

Suppose that the constant coefficients α_i, β_i in the scheme (2.3.8) and the matrices $a_1(\mathbf{w})$, $\Omega(\mathbf{w})$ satisfy the conditions

(a) $$\beta_1(\alpha_3 - \alpha_5) - \beta_2(\alpha_6 - \alpha_8) = 0;$$

(b) $hDa_1(\mathbf{w}) + \tau(\beta_3 - \beta_5)DA(\mathbf{w}) - (h/2)(\alpha_1 - \alpha_2)A(\mathbf{w}) + \Omega(\mathbf{w}) = \beta_9(D, h, \tau)I;$

(c) $\varepsilon = (h^2/(2\tau))[\beta_1(\alpha_3 + \alpha_5) - \beta_2(\alpha_6 + \alpha_8)] - (\tau/2)(\beta_1 - \beta_2)D^2 + \beta_9 > 0.$

Substituting $\mathbf{w} = \mathbf{w}(\xi)$ into (2.3.12) we obtain

$$\Psi(\mathbf{w}) = \sum_{i=1}^{3} [h^i/(i!\,\tau)][\beta_1(\alpha_3 + (-1)^i\alpha_5) - \beta_2(\alpha_6 + (-1)^i\alpha_8)]\,d^{i-1}\mathbf{w}/d\xi^{i-1}$$

$$- \sum_{j=2}^{3} (\tau^{j-1}/j!)[\beta_1 + (-1)^{j-1}\beta_2]\,d^{j-1}\mathbf{w}/d\xi^{j-1}(-D)^j$$

$$- \sum_{i=1}^{2}\sum_{j=1}^{2} h^i\tau^{j-1}[(-D)^j/j!][\beta_1 + (-1)^{j-1}\beta_2]$$

$$\times (d^{i-1}/d\xi^{i-1})a_i(\mathbf{w})\,d^j\mathbf{w}/d\xi^j - \sum_{j=1}^{2} (\tau^j/j!)[\beta_3 + (-1)^j\beta_5]$$

$$\times (d^j\varphi/d\xi^j)(-D)^j - \sum_{i=2}^{3} (h^{i-1}/i!)[\alpha_1 + (-1)^{i-1}\alpha_2]$$

$$\times (d^{i-1}/d\xi^{i-1})\left[\varphi + \sum_{j=1}^{2} (\tau^j/j!)(\beta_3 + (-1)^j\beta_5)(-D)^j(d^j/d\xi^j)\varphi\right]$$

$$+ \Omega(\mathbf{w})(d/d\xi)(\mathbf{w} + \tau(\beta_6 - \beta_8)(-D)\,d\mathbf{w}/d\xi). \qquad (2.3.14)$$

Employing formula (2.3.14) it is easy to obtain an expression for the f.d.a. of scheme (2.3.8) which satisfies the boundary conditions (1.63)

$$B(\mathbf{w}, h, \tau)\,d\mathbf{w}/d\xi = \varphi(\mathbf{w}) - D_1\mathbf{w} - \varphi(\mathbf{W}_1) + D_1\mathbf{W}_1, \qquad (2.3.15)$$

where

$$B(\mathbf{w}, h, \tau) = [h^2/(2\tau)][\beta_1(\alpha_3 + \alpha_5) - \beta_2(\alpha_6 + \alpha_8)]I - (\tau/2)(\beta_1 - \beta_2)D^2I$$

$$+ ha_1(\mathbf{w})D + \tau(\beta_3 - \beta_5)DA(\mathbf{w})$$

$$- (h/2)(\alpha_1 - \alpha_2)A(\mathbf{w}) + \Omega(\mathbf{w}); \qquad (2.3.16)$$

$$D_1 = D + (h/\tau)[\beta_1(\alpha_3 - \alpha_5) - \beta_2(\alpha_6 - \alpha_8)].$$

Note that $D_1 = D$ by virtue of condition (a). Taking into account condition (c) let us rewrite the system (2.3.15) in the form

$$\varepsilon\,d\mathbf{w}/d\xi = \varphi(\mathbf{w}) - \varphi(\mathbf{W}_1) - D(\mathbf{w} - \mathbf{W}_1). \qquad (2.3.17)$$

Since we are considering the system of Euler equations (1.11), (1.9), (1.8) we can write system (2.3.17) as follows:

$$dw_1/d\xi = (1/\varepsilon)g_1(w_1, w_2, w_3);$$
$$dw_2/d\xi = (1/\varepsilon)g_2(w_1, w_2, w_3); \qquad (2.3.18)$$
$$dw_3/d\xi = (1/\varepsilon)g_3(w_1, w_2, w_3);$$

where

$$\mathbf{w} = \{w_1, w_2, w_3\}, \qquad w_1 \equiv \rho, \qquad w_2 \equiv \rho u,$$
$$w_3 \equiv \rho\{p/[\rho(\gamma - 1)] + u^2/2\};$$
$$\mathbf{W}_1 = \{W_{11}, W_{12}, W_{13}\};$$
$$\mathbf{g} = \{g_1, g_2, g_3\}, \qquad \mathbf{g} = \boldsymbol{\varphi}(\mathbf{w}) - \boldsymbol{\varphi}(\mathbf{W}_1) - D(\mathbf{w} - \mathbf{W}_1); \qquad (2.3.19)$$
$$\boldsymbol{\varphi} = \{\varphi_1, \varphi_2, \varphi_3\}, \qquad \varphi_1 = w_2,$$
$$\varphi_2 = (\gamma - 1)(w_3 - w_2^2/(2w_1)) + w_2^2/w_1;$$
$$\varphi_3 = w_2 w_3/w_1 + (\gamma - 1)(w_3 - w_2^2/(2w_1))w_2/w_1;$$

here γ is a constant in the equation of state (1.8). The existence of a continuous solution of the problem (2.3.17), (1.63) was proved in [2.80] for the case of weak shock waves. If one assumes that a continuously differentiable solution of the problem (2.3.17), (1.63) exists in the case of shock waves of arbitrary finite intensity, then it is possible to prove the existence of a smeared shock wave center in the following way.

Since the system (2.3.18) is autonomous, it can be reduced to two equations with respect to the functions $w_2(w_1)$, $w_3(w_1)$

$$dw_2/dw_1 = r_1(w_1, w_2, w_3); \qquad dw_3/dw_1 = r_2(w_1, w_2, w_3), \qquad (2.3.20)$$

where

$$r_j(w_1, w_2, w_3) \equiv g_{j+1}(w_1, w_2, w_3)/g_1(w_1, w_2, w_3), \qquad j = 1, 2.$$

Solving system (2.3.20) we obtain the equation for determining $w_1(\xi)$

$$dw_1/d\xi = (1/\varepsilon)g_1(w_1, w_2(w_1), w_3(w_1)) = \tilde{g}_1(w_1)/\varepsilon. \qquad (2.3.21)$$

The solution of (2.3.21) has the form

$$\Phi(w_1) = (\xi - \xi_0)/\varepsilon, \qquad \Phi(w_1) = \int dw_1/\tilde{g}_1(w_1). \qquad (2.3.22)$$

The function $\Phi(w_1)$ does not depend on ε because ε does not enter the system (2.3.20) for determining $w_2(w_1)$, $w_3(w_1)$. In order to be convinced in monotonicity of the continuous function $w_1(\xi)$ in the zone of the "smeared" shock wave it is sufficient to show that system (2.3.17) has no stationary singular points within this zone. In fact, let $\mathbf{w}_*$ be one more singular point of system

(2.3.17) along with the singular points $\mathbf{W}_1$ and $\mathbf{W}_2$, that is,

$$\varphi(\mathbf{w}_*) - \varphi(\mathbf{W}_1) - D(\mathbf{w}_* - \mathbf{W}_1) = 0. \qquad (2.3.23)$$

This algebraic system, as is known [2.41], has only two solutions: $\mathbf{w}_* = \mathbf{W}_1$, $\mathbf{w}_* = \mathbf{W}_2$, that is, it is found that $\mathbf{w}_*$ coincides with one of the constant limit values of the function $\mathbf{w}(\xi)$. According to the general properties of autonomous systems of the form (2.3.17), [2.60], [2.81], the values $\mathbf{w}_* = \mathbf{W}_1$ and $\mathbf{w}_* = \mathbf{W}_2$ can be reached only at $\xi \to -\infty$ and at $\xi \to +\infty$, respectively. Taking into account the above established monotonicity of the function $w_1(\xi)$ in the zone of smearing of the shock wave and the fact that the material is compressed in a shock wave, so that $\rho_2 - \rho_1 < 0$, we have, in the zone under consideration, the inequality $1/\tilde{g}_1(w_1) < 0$. It is easy to show that the function $\Phi(w_1)$ changes its sign in the interval (W_{21}, W_{11}), $\mathbf{W}_2 = \{W_{21}, W_{22}, W_{23}\}$. Indeed, the argument ξ changes from $-\infty$ to $+\infty$ when moving along the integral curve $\mathbf{w}(\xi)$ from the singular point $\mathbf{w} = \mathbf{W}_1$ to the singular point $\mathbf{w} = \mathbf{W}_2$. Then on the basis of the first formula in (2.3.22) we arrive at a conclusion that the function $\Phi(w_1)$ changes its sign in the interval (W_{21}, W_{11}). In combination with the monotonicity and continuity of the function $1/\tilde{g}_1(w_1)$ this means that the function $\Phi(w_1)$ has only the root $w_{10} \in (W_{21}, W_{11})$. At the point ξ_0 the values of the functions $w_2 = w_2(w_{10})$, $w_3 = w_3(w_{10})$, where the functions $w_2(w_1)$, $w_3(w_1)$ represent the solution of system (2.3.20). With this we conclude the proof of the smeared shock wave center existence in the solution of the problem (2.3.17), (1.63).

In the case when the matrix of coefficients of the approximation viscosity $B(\mathbf{w}, h, \tau)$ differs in the system of ordinary differential equations from the diagonal matrix $B = \varepsilon(D, h, \tau)I$, then the investigation of the question on the existence and uniqueness of a smeared shock wave center in the solution of problems of the form (1.61), (1.63) can be carried out by the methods of the qualitative theory of ordinary differential equations (see, for example, [2.82], [2.83]). The investigation is then subdivided into two stages. At the first stage the existence and smoothness properties of the integral curve $\mathbf{w} = \mathbf{w}(\xi)$, connecting the singular points $\mathbf{w} = \mathbf{W}_1$ and $\mathbf{w} = \mathbf{W}_2$, are established by means of the afore-mentioned qualitative methods. At the second stage the existence and uniqueness of a smeared shock wave center is elucidated. While carrying out investigations on the first stage there can arise questions on the qualitative behavior of the integral curves of a system of form (1.61) in the phase space $\{w_1, w_2, w_3\}$ in the vicinity of the interior singular points (see Section 1.4 on these singular points). There are a number of qualitative techniques and methods for studying the solutions of three-dimensional autonomous systems in the neighborhood of saddle-like singular points (see, for example, [2.84]–[2.87]). In their time the present authors have undertaken an attempt to study the f.d.a. of a "breakdown-of-discontinuity" difference method [2.88] within the framework of the above scheme; however, this complicated study was unfortunately not carried to its conclusion.

Examples of the existence and nonexistence of a smeared shock wave center, for the case of using finite-difference shock-capturing schemes for the numerical solution of problems in which there are shock waves with relaxation, have been given in [2.89]. Such shock waves with a relaxation zone adhering to the wave front may be encountered, for example, in nonequilibrium gas dynamics and in the dynamics of multiphase media.

As in Section 2.2. since we want to take the position of a smeared shock wave center (in the solution of the problem (1.61), (1.63)) as the basis for estimating the accuracy of the localization of shock fronts within the zone of their "smearing", the stage of investigation of the convergence of the solution of a problem with approximation viscosity to a generalized solution of inviscid flow equations (1.1)–(1.3) should follow the elucidation of the question on the existence and uniqueness of a smeared shock wave center. In the case when the matrix $B(\mathbf{w}, h, \tau)$ of viscosity coefficients in system (1.61) has the form $B = \varepsilon I$, where $\varepsilon = \mathrm{const} > 0$ is a sufficiently small value, the convergence at $\varepsilon \to 0$ of the solution of the problem (1.61), (1.63) with viscosity to the solution of an "inviscid" hyperbolic system (1.11) was proved in [2.80]. A review of a number of similar investigations may be found in [2.90]. In [2.91] the system governing the flow of a viscous barotropic gas was taken together with the equation of state corresponding to the hypothetic Chaplygin gas for which the ratio of specific heats $\gamma = -1$, and then the convergence of the solution of this system (as the viscosity coefficient $\mu \to 0$) to a generalized solution of a corresponding system without viscosity was proved. A number of examples of such a choice of matrix B in system (1.61), that the nonuniqueness of the generalized solution takes place as a small parameter tends to zero, have been cited in [2.41]. The question of establishing general criteria, which the matrices B should satisfy in (1.61) in order that the convergence to the unique generalized solution takes place as $h \to 0$, $\tau \to 0$, still remains open.

A study of questions on the existence and uniqueness of a smeared shock wave center and on the convergence to a generalized solution as $h \to 0$, $\tau \to 0$, in the case of the difference schemes of the second and higher orders of accuracy, that is, when $r \geq 2$ in (2.3.4)–(2.3.6), is absent in the literature up to now. The proof of existence and uniqueness of a continuous solution for problems of type (2.3.4), (1.63), (2.3.6) at $r = 2$ is substantially hampered by an oscillatory behavior of the solution $\mathbf{w}(\xi)$ in the neighborhood of a shock wave. For the case of the approximation of a model equation $u_t + uu_x = 0$ the presence of such oscillations, as well as soliton-like solutions of the corresponding f.d.a. equations, has been demonstrated in a number of works (see, for example, [2.92]–[2.94]).

An available extensive practice of computations of discontinuous gas flows by well-established finite-difference shock-capturing schemes (as a rule, these schemes are conservative or in divergence form) indicates that the accuracy of numerical solutions increases when the spatial mesh is refined, that is, when step h is reduced. In these cases it is reasonable to expect that the direct application of a smeared shock wave center notion for the purposes of shock wave front localization will yield good results.

Similar to a technique of strong discontinuity localization within the zone of shock wave smearing, which is based on determining the abscissa of a point of maximum of the artificial viscosity q introduced additively into the pressure (see Section 2.2 above), one can consider a technique in which the position of a shock is determined by the maximum modulus of some of the components of a vector $B(\mathbf{w}, h, \tau) \, d\mathbf{w}/d\xi$ entering the system of the f.d.a. equations (1.61). Some preliminary results of investigation of such a localization technique have been presented in [2.58]. In particular, certain necessary conditions of extremum of the vector $B(\mathbf{w}, h, \tau) \, d\mathbf{w}/d\xi$ have been found.

Following [2.58] let us consider a hyperbolic system

$$\partial \mathbf{U}/\partial t + \partial \boldsymbol{\varphi}(\mathbf{U})/\partial x = 0; \qquad \mathbf{U} = (U_1, \ldots, U_n)^{\mathrm{T}};$$

$$\boldsymbol{\varphi} = (\varphi_1, \ldots, \varphi_n)^{\mathrm{T}}. \tag{2.3.24}$$

Taking into account the results of [2.95] we shall assume that along a discontinuity line the relationships

$$D[\mathbf{U}] = [\boldsymbol{\varphi}] \tag{2.3.25}$$

are satisfied where the symbol $[\boldsymbol{\varphi}]$ denotes a jump across the discontinuity line and is determined by (1.20), and $D = D(t)$ is the discontinuity propagation speed. The relationships (2.3.25) are a generalization of the Rankine–Hugoniot conditions (1.22)–(1.24). We shall assume further that the system (2.3.24) admits the presence of shock waves (in the sense of the definition given in [2.95]), and in particular, stationary shock waves. Approximate the system (2.3.24) by a divergence form finite-difference first-order scheme, such that its f.d.a. is representable in the form (1.59). As in Section 1.4, we search for a progressive wave-type solution of a system (1.59) that satisfies the conditions (1.63).

Suppose that the matrix $A = \partial \boldsymbol{\varphi}/\partial \mathbf{w}$ has real eigenvalues $\lambda_1(\mathbf{w}), \ldots, \lambda_n(\mathbf{w})$ and $\lambda_1(\mathbf{w}) < \lambda_2(\mathbf{w}) < \cdots < \lambda_n(\mathbf{w})$ in the zone of a smeared shock wave where the functions $\lambda_j(\mathbf{w})$, $j = 1, \ldots, n$, are assumed to be continuous and monotone in the zone of a smeared shock wave. Let us show that if $\mathbf{w}(\xi) \in C^2(-\infty, \infty)$, and the elements of the matrix $B(\mathbf{w}, h, \tau)$ are continuously differentiable with respect to the components of $\mathbf{w}$, and there exist a number k, $1 \le k \le n$, such that

$$\lambda_k(\mathbf{W}_1) > D > \lambda_k(\mathbf{W}_2), \tag{2.3.26}$$

and the components of the vector of the approximation viscosity $\mathbf{Q}$ in equation (1.61),

$$\mathbf{Q} = B \, d\mathbf{w}/d\xi; \qquad \mathbf{Q} = (Q_1, \ldots, Q_n)^{\mathrm{T}}; \tag{2.3.27}$$

reach their extremum at one point ξ_0, which is common for all the components of the $\mathbf{Q}$ vector, then the relation

$$\lambda_k(\mathbf{w}) = D \tag{2.3.28}$$

takes place at the point ξ_0.

It is evident that at the point of extremum of the vector components (2.3.27) the equality

$$(d/d\xi)[B(\mathbf{w}, h, \tau)\, d\mathbf{w}/d\xi] = 0$$

should take place. Differentiating both sides of equation (1.61) with respect to ξ we obtain, at the point under consideration,

$$[A(\mathbf{w}) - DI](d\mathbf{w}/d\xi) = 0, \tag{2.3.29}$$

where I is the unit matrix. Consider the system (2.3.29) as an algebraic system for determining the components of the vector $d\mathbf{w}/d\xi$. Let $|d\mathbf{w}/d\xi| \neq 0$ in the zone of the "smeared" shock wave. Then one should set $\det[A(\mathbf{w}) - DI] = 0$ for the existence of a nontrivial solution to the system (2.3.29).

Since

$$\det[A(\mathbf{w}) - DI] = \prod_{j=1}^{n} [\lambda_j(\mathbf{w}) - D],$$

we obtain, with regard to the inequalities (2.3.26), the relationship (2.3.28). Note that equation (2.3.28) is only a necessary condition of extremum of the quantity (2.3.27). For the satisfaction of a sufficient condition of extremum it is necessary that the derivative

$$(d^2/d\xi^2)[B(\mathbf{w}, h, \tau)\, d\mathbf{w}/d\xi] \tag{2.3.30}$$

is different from zero at the point $\xi = \xi_0$ of the sought for extremum. Before using the quantity (2.3.30) it is necessary to elucidate whether the function $B(\mathbf{w}, h, \tau)$ and $d\mathbf{w}/d\xi$ entering (2.3.30) possess the necessary degree of smoothness and differentiability. In the case when, for example, the elements of the B matrix are polynomial functions of the vector $\mathbf{w}$ component (see [2.42], [2.97]), the presence of the second continuous derivatives of these elements with respect to the components of $\mathbf{w}$ is obvious.

Now consider the case when the system (2.3.24) is the system of Euler equations governing the inviscid compressible fluid flow. Let the necessary and sufficient conditions of extremum of the quantities (2.3.27) be satisfied at some point $\xi = \xi_0$. Then this point moves at a shock wave speed D. Indeed, let us consider a shock wave in gas moving from left to right. Then $D > 0$ and, by virtue of Zemplén's theorem, the inequalities

$$u_1 + c_1 > D > u_2 + c_2 \tag{2.3.31}$$

take place where u and c are the gas speed and the sound speed, respectively. $\lambda_3 = u + c$ taking into account (1.16), consequently, $k = 3$ in (2.3.26), with (2.3.31) in view, and

$$u(\xi_0) + c(\xi_0) = D. \tag{2.3.32}$$

As was pointed out in [2.2], the characteristics of the numerical solution having the slope $(u + c)$ in the (x, t)-plane do not intersect at the shock wave, unlike the exact solution, but approach asymptotically the line corresponding to the shock front trajectory. From this it follows, with regard to (2.3.32), that the point of extremum of the quantity (2.3.27) moves along the shock wave trajectory.

2.4. An Analysis of Difference Schemes of Gas Dynamics

2.4.1. First-Order Schemes

Let us analyze a number of well-known difference schemes from the point of view of the theorems proved in Sections 2.2. and 2.3.

The Lax scheme [2.98] can be obtained as a particular case of the scheme (2.3.8) if one sets, in (2.3.8),

$$\beta_1 = 1; \quad \beta_2 = 0; \quad \alpha_3 = \alpha_5 = 0.5; \quad \alpha_4 = 0; \quad \alpha_1 = \alpha_2 = 0.5;$$
$$\beta_3 = \beta_5 = 0; \quad \beta_4 = 1; \quad \Omega(\mathbf{w}) = a_1(\mathbf{w}) = a_2(\mathbf{w}) = 0. \tag{2.4.1}$$

The system (2.3.15) takes the form (2.3.17) where $\varepsilon = h^2/(2\tau) - \tau D^2/2$, taking into account (2.4.1). The inequality $\varepsilon > 0$, by virtue of Zemplén's theorem (1.29), does not contradict the well-known stability condition of the Lax scheme. Thus, the smeared shock wave center for the Lax scheme exists at least for the shock wave of moderate intensity. In Figure 2.4 we present an example of the Lax scheme computation of a problem on the propagation of a stationary shock wave of finite intensity (pressures, $p_1 = 5$, $p_2 = 1$) [2.31]. The abscissa of a point of intersection of the finite-difference solution graphs obtained by the Lax scheme at different τ and h coincides with the exact position of a shock front within an error $\approx 0.1h$.

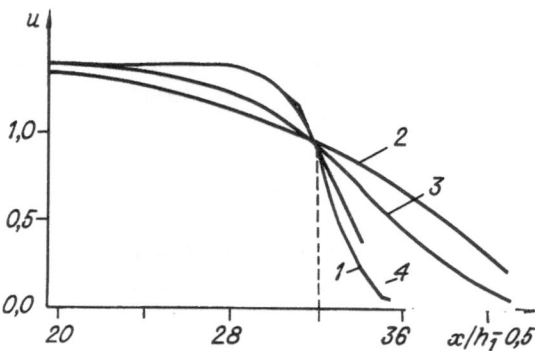

Figure 2.4. The Lax scheme. 1—h_1, $2\tau_1$; 2—h_1, $\tau_1/2$; 3—h_1, τ_1; 4—$h_1/3$, $\tau_1/3$.

The Rusanov Scheme [2.99]. In order to obtain this scheme from the family of schemes (2.3.8) it is sufficient to set

$$\beta_1 = 1, \qquad \beta_2 = 0, \qquad \alpha_3 = \alpha_5 = 0, \qquad \alpha_4 = 1,$$

$$C \equiv I, \qquad \alpha_6 = \alpha_7 = \alpha_8 = 0,$$

$$\alpha_1 = \alpha_2 = 0.5, \qquad \beta_3 = \beta_5 = 0, \qquad \beta_4 = 1, \qquad \Omega(\mathbf{w}) = (h/2)\omega(|u| + c)I,$$

$$\omega = \text{const}, \qquad \beta_6 = \beta_8 = 0, \qquad \beta_7 = 1,$$

where c is the local speed of sound. The f.d.a. H-form of the Rusanov scheme with (2.3.12) in view may be written as

$$\frac{\partial \mathbf{w}}{\partial t} + \frac{\partial \varphi(\mathbf{w})}{\partial x} = \frac{\partial}{\partial x}\left[\omega \frac{h}{2}(|u| + c)\frac{\partial \mathbf{w}}{\partial x}\right] - \frac{\tau}{2}\frac{\partial^2 \mathbf{w}}{\partial t^2}. \qquad (2.4.2)$$

Substituting $\mathbf{w} = \mathbf{w}(\xi)$ into the system (2.4.2) where ξ is determined by formula (1.56) we obtain a system

$$(h/2)[\omega(|u| + c) - \kappa D^2]\, d\mathbf{w}/d\xi = \varphi(\mathbf{w}) - D\mathbf{w} - \varphi(\mathbf{W}_1) + D\mathbf{W}_1, \quad (2.4.3)$$

where $\kappa = \tau/h$. It follows from (2.4.3) that the condition $\kappa = \text{const} > 0$ is a necessary condition for the smeared shock wave center existence, because in the case under consideration an analogue of the formulas (2.3.22) may be written in the form

$$\Phi(w_1) = (\xi - \xi_0)/h, \ \Phi(w_1) = \int [\omega(|u| + c) - \kappa D^2][2\tilde{g}_1(w_1)]^{-1}\, dw_1.$$

The Particle-In-Cell Method [2.32]. This method will be investigated in an asymptotic approximation when the number of particles in each cell tends to infinity. Then it is possible to obtain the f.d.a. H-form of the integral step scheme of the particle-in-cell method [2.5], [2.55]

$$\rho_t + (\rho u)_x = [(h/2)|u|\rho_x]_x + \tau(p_{xx} - 0.5\rho_{tt});$$

$$(\rho u)_t + (p + \rho u^2)_x = [(h/2)|u|(\rho u)_x]_x + \tau[2(p_x u)_x - 0.5(\rho u)_{tt}];$$

$$(\rho E)_t + (pu + \rho u E)_x = [(h/2)|u|(\rho E)_x]_x + \tau[(p_x E)_x + ((1/(2\rho))pp_x)_x$$

$$+ (u(pu)_x)_x - 0.5(\rho E)_{tt}], \qquad (2.4.4)$$

where $E = \varepsilon + 0.5u^2$. The FLIC method [2.100] has the same f.d.a. if one nullifies in it the artificial viscosities introduced into the Eulerian computing stage scheme. Consider a case when the step τ is related to h by a dependence

$$\tau = \kappa h, \qquad \kappa = \text{const} > 0, \qquad (2.4.5)$$

at different values of step h. Employing the vectors $\mathbf{w}$ and φ with components determined by the formulas (2.3.19) we can obtain from (2.4.4) the following system of ordinary differential equations

$$(h/2)B(\mathbf{w}, \kappa)\, d\mathbf{w}/d\xi = \varphi(\mathbf{w}) - D\mathbf{w} - \varphi(\mathbf{W}_1) + D\mathbf{W}_1 \qquad (2.4.6)$$

by substituting the progressive wave-type solution $\mathbf{w} = \mathbf{w}(\xi)$, $\xi = x - Dt$, into (2.4.4). The matrix B in (2.4.6) has the form

$$B = (|u| - \kappa D^2)I + \kappa(\gamma - 1)\tilde{B}(\mathbf{w}), \qquad (2.4.7)$$

$$\tilde{B}(\mathbf{w}) = \begin{pmatrix} u^2 & -2u & 2 \\ 2u^3 & -4u^2 & 4u \\ u^2(2u^2 - E + 0.5c^2/\gamma) & -3u^3 - c^2u/\gamma & 2(E + u^2 + c^2/(2\gamma)) \end{pmatrix},$$
$$(2.4.8)$$

where c is the local speed of sound. It follows from formula (2.4.7) that the matrix B differs from a diagonal one. Making, in (2.4.6), a change of independent variable ξ by the formula

$$\eta = (\xi - \xi_0)/h, \qquad (2.4.9)$$

we rewrite the system (2.4.6) as follows

$$(1/2)B(\mathbf{w}, \kappa)\, d\mathbf{w}/d\eta = \varphi(\mathbf{w}) - D\mathbf{w} - \varphi(\mathbf{W}_1) + D\mathbf{W}_1. \qquad (2.4.10)$$

It follows from (2.4.10) that if the solution of the system (2.4.10) at the boundary conditions (1.63) exists, then it is obviously a function of an independent variable η and the constants κ, D, $\mathbf{W}_1$:

$$\mathbf{w} = \mathbf{w}(\eta, \kappa, D, \mathbf{W}_1) \qquad (2.4.11)$$

It follows from the relationship (2.4.11) that different graphs of the solution $\mathbf{w}$ correspond to different values of the constant κ. Thus, if the condition $\kappa = $ const is not satisfied in computations on three different spatial grids, then the three solution graphs obtained will not intersect at the same point; that is, the smeared shock wave center does not exist. Thus, the condition $\tau/h = $ const is a necessary condition for the existence of a smeared shock wave center for the particle-in-cell method [2.33] and for the FLIC method [2.100]. On the other hand, it follows from (2.4.7) that

$$\lim_{\kappa \to 0} B(\mathbf{w}, \kappa) = |u| I,$$

that is, the matrix B at small κ differs little from the diagonal one.

In Figure 2.5 we present the results of the computation (by a modified Harlow method [2.55]) of the problem on the propagation of a stationary shock wave in a gas at rest, and which were obtained on three different grids on the x-axis. In the case of Figure 2.5(a) the condition $\tau/h = $ const was satisfied, but in the case of Figure 2.5(b) this condition was premeditatedly violated. These computations confirm the absence of a smeared shock wave center for different κ.

A consideration similar to the foregoing one, and carried out for the "coarse

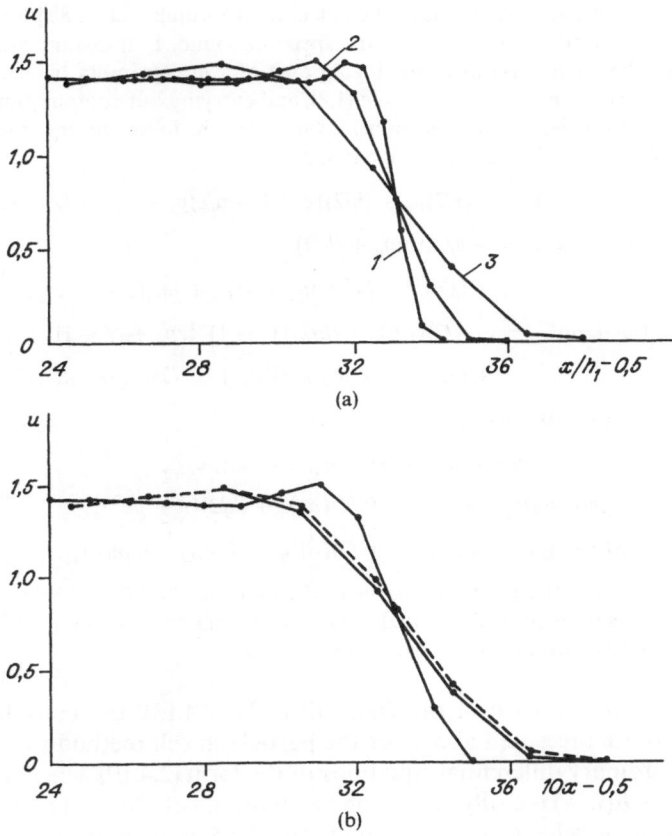

Figure 2.5. Results of computation by the Harlow method: (a) $\tau/h = $ const; (b) $\tau/h \neq$ const. $1—0.5h_1$; $2—h_1$; $3—2h_1$.

particles" method [2.5], leads to a matrix of coefficients of the approximation viscosity of the form (2.4.7) where the matrix $B(\mathbf{w})$ differs slightly from (2.4.8). Consequently, in the case of the "coarse particles" method, the condition (2.4.5) is also a necessary condition for the existence of a smeared shock wave center.

The "Breakdown-of-Discontinuity" Scheme [2.88]. If the waves under study are weak (sound waves), then one can use approximate formulas for the computation of a breakdown of a discontinuity, which were presented in [2.88]. In this case, the following lemma is valid [2.101].

Lemma 2.1. *Approximation viscosity of the "breakdown-of-discontinuity" scheme [2.88] changes continuously when passing from the subsonic flow regime to the supersonic flow regime.*

To prove this lemma we consider the calculation formulas of [2.88] for the two cases: $0 < u < c, u > c > 0$, where c is the speed of sound. Employing approximate formulas for the computations of the breakdown of a discontinuity (which are presented in [2.88]) for the equation of state (1.8), and carrying out computations by the use of an algorithm described in Section 1.3 we obtain the following equation system of the f.d.a. H-form in the case when $0 < u < c$

$$\rho_t + (\rho u)_x = -(\tau/2)\rho_{tt} + (h/2)[c^{-1}(1 - u/c)p_x + u\rho_x + (u/c)\rho u_x]_x;$$

$$(\rho u)_t + (p + \rho u^2)_x = -(\tau/2)(\rho u)_{tt} + (h/2)$$
$$\times [(2u/c - (u/c)^2)p_x + u^2\rho_x + \rho u_x(c + u^2/c)]_x; \quad (2.4.12)$$

$$(\rho E)_t + (pu + \rho uE)_x = -(\tau/2)(\rho E)_{tt} + (h/2)[(\gamma - 1)^{-1}cp_x + (\gamma - 1)^{-1}\gamma\rho cuu_x$$
$$+ (3u^2/(2c) - u^3/(2c^2))p_x + (u^3/2)(\rho_x + \rho u_x/c)]_x.$$

Similarly, for $u > c > 0$, we have

$$\rho_t + (\rho u)_x = -(\tau/2)\rho_{tt} + (h/2)(\rho u)_{xx};$$

$$(\rho u)_t + (p + \rho u^2)_x = -(\tau/2)(\rho u)_{tt} + (h/2)(p + \rho u^2)_{xx}; \quad (2.4.13)$$

$$(\rho E)_t + (pu + \rho uE)_x = -(\tau/2)(\rho E)_{tt} + (h/2)(pu + \rho uE)_{xx}.$$

It is easy to see that the right-hand sides of the systems (2.4.12), (2.4.13) coincide at $u = c$, which was to be proved. An analogous fact was stated earlier in [2.102] for the case of linearized equations of gas dynamics. $\qquad\square$

Repeating, in the case of the f.d.a.s (2.4.12), (2.4.13), the considerations similar to those presented above for the particle-in-cell method we obtain a system of ordinary differential equations of the form (2.4.10), where the form of the matrix $B(\mathbf{w}, \kappa)$ is easily determined with the aid of (2.4.12), (2.4.13). Thus, in the case of the scheme in [2.88], condition (2.4.5) is again necessary for the existence of a smeared shock wave center. In Figure 2.6 (taken from [2.58]) we present velocity profiles in the neighborhood of a shock wave that were obtained by the scheme of [2.88] on three different spatial meshes under the

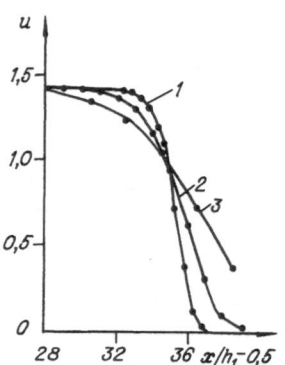

Figure 2.6. S.K. Godunov's scheme. 1—$h_1/2$; 2—h_1; 3—$2h_1$.

condition $\tau/h = \text{const}$. The abscissa of a point of intersection of difference solution graphs for different h is from the exact position of the shock front at a distance $< 0.3h$.

2.4.2. Second-Order Schemes

The family of schemes $\mathcal{S}_\beta^\alpha$ approximating the system (1.11) may be written in the form [2.103]–[2.105]

$$\tilde{w}_i = (1 - \beta)w_i^n + \beta w_{i+1}^n - \alpha\kappa[\varphi(w_{i+1}^n) - \varphi(w_i^n)];$$

$$w_i^{n+1} = w_i^n - [\kappa/(2\alpha)][(\alpha - \beta)\varphi(w_{i+1}^n) + (2\beta - 1)\varphi(w_i^n) \qquad (2.4.14)$$

$$+ (1 - \alpha - \beta)\varphi(w_{i-1}^n) + \varphi(\tilde{w}_i) - \varphi(\tilde{w}_{i-1})],$$

where α, β are constant weight coefficients and $\kappa = \tau/h$. The scheme (2.4.14) yields oscillations in the regions behind the shock wave fronts and contact discontinuities. Therefore, it is useful to introduce into equations (2.4.14) some dissipative mechanism to suppress these spurious oscillations. To preserve the second order of accuracy of scheme (2.4.14) let us introduce into equations (2.4.14) the quadratic viscosity q of the form (2.2.18) into the pressure additively as in the system (2.2.1)–(2.2.3). This slightly modified family will be denoted in the following by the symbol $\mathcal{S}_\beta^\alpha$. Taking into account the results of [2.103], the f.d.a. H-form of the family $\mathcal{S}_\beta^\alpha$ may be written in the form

$$\partial w/\partial t + \partial\varphi(w)/\partial x = (h^2/6)\{-\kappa^2\,\partial^3 w/\partial t^3 - \partial^3\varphi(w)/\partial x^3 + [3/(2\alpha)]$$

$$\times [\mathcal{B}(\varphi_x^0, \varphi_x^1)]_x\} + h^2(\partial/\partial x)[\Omega_1(w, w_x)\,\partial w/\partial x], \quad (2.4.15)$$

where Ω_1 is a matrix of the form

$$\Omega_1 = \begin{pmatrix} 0 & 0 & 0 \\ -gw_1^{-2}w_2 & g/w_1 & 0 \\ -gw_1^{-3}w_2^2 & gw_1^{-2}w_2 & 0 \end{pmatrix},$$

$$g = aw_1 \min[(\partial/\partial x)(w_2/w_1), 0],$$

$$\varphi^0 = (1 - \beta)w + \alpha\kappa\varphi(w),$$

$$\varphi^1 = \beta w - \alpha\kappa\varphi(w),$$

and $\mathcal{B}$ is a symmetric bilinear transformation, $\mathcal{B} = \varphi''(w)$. Substituting a progressive wave-type solution into (2.4.15) we obtain the system

$$h^2[(\kappa^2/6)D^3\,d^2 w/d\xi^2 - (1/6)\,d^2\varphi(w)/d\xi^2$$

$$+ [1/(4\alpha)]\mathcal{B}(\varphi_\xi^0, \varphi_\xi^1) + \Omega_1(w, w_\xi)\,dw/d\xi$$

$$= \varphi(w) - Dw - \varphi(W_1) + DW_1. \quad (2.4.16)$$

Making, in system (2.4.16), a substitution (2.4.9) we can rewrite (2.4.16) as

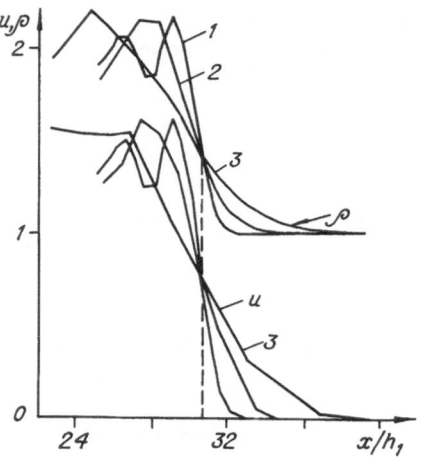

Figure 2.7. The Lax–Wendroff scheme. 1—
$h_1/2$; 2—h_1; 3—$2h_1$.

follows:

$$(\kappa^2/6)D^3\, d^2 w/d\eta^2 - (1/6)\, d^2\varphi(w)/d\eta^2$$

$$+ [1/(4\alpha)]\mathscr{B}(\varphi_\eta^0,\, \varphi_\eta^1) + \Omega_1(w,\, w_\eta)\, dw/d\eta$$

$$= \varphi(w) - Dw - \varphi(W_1) + DW_1. \quad (2.4.17)$$

It follows from system (2.4.17) that a necessary condition for the smeared shock wave center existence (while using the schemes from the family $\mathscr{S}_\beta^\alpha$) is the relationship $\tau/h = \text{const}$. The influence of various terms of the order $O(h^2)$ entering the f.d.a. P-form of the family $\mathscr{S}_\beta^\alpha$ on the dispersive and dissipative properties of the difference solutions has been studied in [2.103]. This study enabled the authors of [2.103] to find the terms responsible for the presence of spurious oscillations in the numerical solution. The introduction of the pseudoviscosity q into $\mathscr{S}_\beta^\alpha$ gives an additional dissipative effect, as was shown in Section 2.2, in the example of a model system (2.2.1)–(2.2.4). The family $\mathscr{S}_\beta^\alpha$

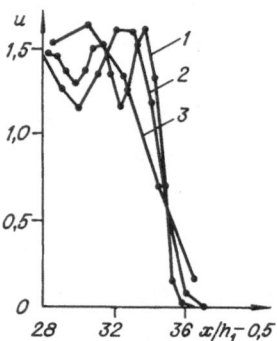

Figure 2.8. The MacCormack scheme (the notation is the same as in Figures 2.5–2.7).

includes as a particular case the two-step Lax–Wendroff scheme [2.7] at
$\alpha = \beta = 1/2$ and the MacCormack scheme [2.25] at $\alpha = 1$, $\beta = 0$ and at
$\alpha = 1$, $\beta = 1$. In Figures 2.7 and 2.8 we present graphs of the difference
solutions obtained by the two-step Lax–Wendroff scheme with an additional
viscosity q of the form (2.2.18) at $a = 3$, and by the MacCormack scheme $\mathscr{S}_0^1$,
respectively. The abscissa of the point of intersection of the solution graphs
for different h is from the exact position of a shock wave at a distance $< 0.3h$;
that is, the accuracy of the discontinuity localization on the basis of the wave
center notion is high, despite the presence of parasitic oscillations in the
numerical solution.

2.4.3. Practical Realization of the Shock Wave Differential Analyzer Algorithms (With Regard to Sections 2.2. and 2.3)

Unlike the smeared shock wave we understand the finite-difference shock
wave as a solution of the finite-difference equations approximating the shock
wave, and calculated at the discrete grid points in the (x, t)-plane. Analogous
to the notion of a smeared shock wave center, let us now introduce the notion
of the finite-difference shock wave center [2.56], [2.57].

Definition 2.3. By a finite-difference shock wave center (f.-d.s.w.c.) we mean a
point in the finite-difference shock wave zone belonging to the solutions
of finite-difference equations at the same t which are obtained at two dif-
ferent values of the constant steps h, τ: $h = h_1$, $\tau = \tau_1$, $h = h_2$, $\tau = \tau_2$, and
$|h_1 - h_2| + |\tau_1 - \tau_2| \neq 0$, and which are determined with the help of linear
interpolation at the (x, t) points that are not nodes.

Definition 2.4. By the differential analyzer of a shock wave we mean the
algorithm which enables us to find the finite-difference shock wave center
coordinates in a computational cell by shock-capturing calculation.

In Section 1.4.2 we have presented a survey of the results of theoretical
studies on a question of the existence of the solutions of the form $w_h(\eta)$ with
$\eta = (x - Dt)/h$ of the difference equations approximating the scalar quasi-
linear equation (1.83). In this case it is easy to prove that in the $w_h(\eta)$ profiles
obtained at different τ and h there exists a point—the discrete shock center—
at which the value of the solution $w_h(\eta)$ does not depend on h. This is obviously
the point $x = Dt$, since at $x = Dt$ we obtain that $\eta = 0$ independently of τ and
h. Since the equation $x = Dt$ is the equation of the exact trajectory of a
stationary shock wave, it is clear that the determination of the coordinates of
a finite-difference shock wave center may be put into the basis of the algorithms for the localization of a shock wave front, in the cases when there exist
solutions of the form $\mathbf{w}_h(\eta)$ of difference equations approximating the original

equation (1.83) or the system of Euler equations (1.1)–(1.3). However, for the case of the systems of quasi-linear hyperbolic equations, there are at present no theoretical results on the existence and uniqueness of solutions of the form $\mathbf{w}_h(\eta)$ of difference schemes approximating such systems. The results of the numerical experiments presented in Section 1.4.2 uniquely indicate the existence of the discrete shock-type solutions of difference equations approximating the Euler equation system (1.1)–(1.3).

Let us now discuss a question on the accuracy of the localization of the shock wave front abscissa on the basis of Definition 2.3. In the case of the stationary shock waves this accuracy should be higher for a larger number n of the time level because, in accordance with Section 1.4.2 (see Figure 1.3), the convergence of a difference solution to the stationary solution of the form $\mathbf{w}_h(\eta)$ takes place at $n \to \infty$. At a given finite n the localization accuracy will be higher in the case when the value $Dn\tau$ coincides with the value of the abscissa of some node of a spatial grid in which the difference solution $\mathbf{w}_h(\eta)$ components are computed. However, it is usually impossible to meet, in the actual problems of gas dynamics, the following three requirements: first, the speed D of shock waves usually depends on t; second, the computations are carried out only for finite n; and third, the value $D(t) \cdot n\tau$ usually does not coincide with the abscissas of the nodes of a spatial grid for natural n. This leads in practice to some errors in determining the location of the shock wave front on the basis of the notion of a finite-difference shock wave center. However, it should be noted that the size of these errors is usually insignificant $(< h)$ (see Figures 2.4, 2.5(a), and 2.6–2.8 and the discussion of these figures).

It follows from Definition 2.3 that it is necessary to carry out two calculations of the same problem with different steps in x or in t for the direct application of the finite-difference shock wave center notion. The results of the first of these calculations can be recorded on the external memory devices, and then a calculation on another mesh in x or in t can be carried out. After that a search for the domains of the smearing of shock waves is organized by means of a specialized computer subroutine. For this purpose the inequality $\partial u / \partial x < 0$ (see Section 2.2) is used in the one-dimensional case and, besides, a numerical check-up of the Rankine–Hugoniot conditions across shock waves is carried out. For this purpose one can make use of the formulas for the calculation of the error Δ for the satisfaction of the Rankine–Hugoniot conditions that were proposed in [2.69]; for example,

$$\Delta_1 = [\rho_1(u_1 - D) - \rho_2(u_2 - D)]/[\rho_2(u_2 - D)], \qquad (2.4.18)$$

where the shock wave speed D can be determined by the formulas

$$D = dD_1 + (1 - d)D_2, \qquad 0 \le d \le 1;$$

$$D_1 = (\rho_1 u_1 - \rho_2 u_2)/(\rho_1 - \rho_2); \qquad (2.4.19)$$

$$D_2 = u_2 + (p_2 - p_1)/[\rho_2(u_2 - u_1)].$$

proposed in [2.69]. The quantities with subscripts "1" and "2" entering the right-hand sides of formulas (2.4.19) determine the state of a medium behind and before the front of a finite-difference shock wave, respectively.

Once zones of smearing of the shock waves have been found, the position of the shock fronts in these zones is determined as a point of intersection of the solution graphs for different h or τ. The above-presented procedure for the shock fronts localization was applied in [2.58] for the numerical determination of the positions of shock waves while using seven first- and second-order difference schemes, including the difference schemes considered in Section 2.4. As a result it was found that the error in locating a discontinuity (on the basis of the finite-difference shock wave center notion) did not exceed $0.3h$ for the schemes considered. We have also carried out numerical experiments on the comparison of the abscissas of finite-difference shock wave centers obtained on the grids with steps h, $h/2$ and on the grids with steps $h/2$, $h/4$. It turned out that these abscissas differ slightly from each other (by a value less than $h/4$); thereat the solution component values at the wave centers differed insignificantly. Thus, the finite-difference shock wave center determined by the numerical solution is a point at which the values of the components of the solution vector depend "most weakly" on the grid parameters h, τ.

The above method of strong discontinuity localization in the finite-difference solution needs for its realization an execution of two calculations of the same problem. Such a method is not economical when solving multidimensional problems with shock waves. Therefore, the questions of the development of algorithms for an approximate determination of the finite-difference shock wave center (with an error acceptable in practice) are of present interest. One such algorithm, based on determining the point of maximum of the artificial viscosity q, has been investigated in Section 2.2. However, in the actually used difference schemes the scheme viscosity is also present along with the artificial viscosity. Scheme viscosity generally has a very complicated structure as was shown in the foregoing examples. There arises the question as to whether it is possible to use the determination of maximum of some scheme viscosity norm in the differential analyzer algorithms similar to the case of the scalar artificial viscosity q. The first theoretical investigation in this direction had been carried out in [2.58] (see also Section 2.3). Previously, in [2.106], we had presented the results of numerical experiments confirming the possibility of direct application of the information on the structure of scheme viscosity obtained from the f.d.a. to the construction of differential analyzers of shock waves.

A numerical verification of the preliminary results of Section 2.3, showing that the point of extremum of the quantity (2.3.27) propagates along a shock wave trajectory, has been carried out in [2.58]. Three first-order schemes were used therein: the GODUNOV scheme [2.88], the FLIC scheme [2.100], and the LAX scheme [2.98]. Taking into account (2.3.32) the sought for shock front abscissa was determined by the inversion of linear interpolation of the grid values of the function $u(x, t) + c(x, t)$ at the point where $u + c = D$. In the case of the Godunov scheme and the FLIC scheme the error in shock front localization on the basis of the property (2.3.32) did not exceed $0.5h$, according to the data of [2.58]; in the case of the Lax scheme the above error reached $1.22h$; however, it should be noted that in the case of the Lax scheme the width of the zone of "smearing" of a shock wave exceeded $10h$ (see the graph in Figure 2.4 marked by the symbols h_1, τ_1). It follows, from the results presented, that it is possible to apply the property expressed by formula (2.3.32) for the development of some new algorithms for shock wave localization by shock-capturing computational results while using the first-order schemes.

Since in the general case the shock wave speed D in (2.3.32) is unknown, it is possible to determine the shock front point as a point of maximum of the quantity

$$\| B(\mathbf{w}, h, \tau)\, \partial \mathbf{w}/\partial x \|, \qquad\qquad (2.4.20)$$

taking into account the considerations of Section 2.3. This was realized for the first time by computations in [2.106].

2.5. The Application of Differential Analyzers in Problems of Shock Wave Formation

It has been shown in [2.41], in the examples of the simplest flows, that, as a rule, the solutions of gas dynamics equations remain continuous during a bounded time, and then discontinuities arise in the solution. At present there are a number of continuum mechanics problems which are important for applications in which the shock arise in a continuous flow. As examples, we mention the problems of shock formation in various shock tubes [2.107], the problem of the computation of the buffeting phenomenon on an airplane wing at transonic flight speeds [2.108], and a number of other applied problems (see the vast bibliography in [2.109]). In one of the well-known approaches to the determination of shock formation—the method of characteristics [2.41]–[2.44]—the shock wave formation is determined by the intersection of the characteristics of the same family. The shock-fitting method [2.110]–[2.113] also enables one to determine the shock waves formation. There have been developed versions of the shock-fitting method in which the shock formation can be determined without using the characteristics intersection [2.111]. However, it was pointed out in [2.41]–[2.44] that the method of shock fitting is especially difficult in programming, if the code is to include an algorithm for the determination of the shock waves formation.

2.5.1. An Analysis of the Shock Wave Formation Problem While Using a Uniform Grid

Finite-difference shock-capturing schemes have an advantage over the method of characteristics and the shock-fitting method, in that they are universal and simple in computer implementation. It is shown below that it is possible to solve numerically a problem on shock formation by using homogeneous difference schemes. In the process of the temporal evolution of a compression wave into a shock wave, the solution gradients in the compression zone are increasing and at the time of discontinuity formation they become infinite. In this connection it was reasonable to use a method for the determination of the shock formation moment of time based on a comparison (at each time level t_n) of the maximum gradient $|u_{xh}|_{max}$ of a difference solution in the compression wave with some theoretical value $|u_x|_{max}$ characterizing the maximum value of the difference solution gradients in a "smeared" shock wave. Further consideration will be restricted to a model equation

$$\partial u/\partial t + \partial\varphi(u)/\partial x = 0, \qquad (2.5.1)$$

where

$$\varphi(u) = 0.5u^2. \qquad (2.5.2)$$

Let us approximate equation (2.5.1) on a uniform mesh along the x-axis using the simplest explicit scheme with one-sided differences, assuming that $\varphi'(u) \geq 0$ in the computational domain

$$(u_i^{n+1} - u_i^n)/\tau + [\varphi(u_i^n) - \varphi(u_{i-1}^n)]/h = 0, \qquad (2.5.3)$$

where $u_i^n = u(ih, n\tau)$. Write the P-form of the f.d.a. of scheme (2.5.3)

$$\partial w/\partial t + \partial\varphi(w)/\partial x = (\partial/\partial x)[B(w, h, \tau) \, \partial w/\partial x], \qquad (2.5.4)$$

where

$$B(w, h, \tau) = (h/2)w - (\tau/2)w^2. \qquad (2.5.5)$$

As was shown in [2.91], [2.58], [2.114], in the case of a stationary shock wave, that is, when $D(t) = $ const, the progressive wave-type solutions of the f.d.a. equation (2.5.5) are applicable to the description of the difference solution u_h behavior in a smeared shock wave. Let $D = $ const. As in Section 1.4, we shall search among the solutions of the equation

$$-D \, dw/d\xi + d\varphi/d\xi = (d/d\xi)[B(w, h, \tau) \, dw/d\xi] \qquad (2.5.6)$$

for a solution $w(\xi)$ which satisfies the conditions

$$\lim_{\xi \to -\infty} w(\xi) = u_1, \qquad \lim_{\xi \to +\infty} w(\xi) = u_2, \qquad (2.5.7)$$

where u_1, u_2 are constant values equal to the solution value behind and before the shock wave front, respectively; $u_1 > u_2$ and, besides, we shall also take in the following, $u \geq 0$. The values u_1, u_2, D are related by the Rankine–Hugoniot condition

$$\varphi(u_1) - \varphi(u_2) = D(u_1 - u_2). \qquad (2.5.8)$$

Integrating (2.5.6) we obtain (with (2.5.7), (2.5.8) in view) an ordinary differential equation

$$dw/d\xi = g(w), \qquad (2.5.9)$$

where

$$g(w) = \{\varphi(w) - Dw - [\varphi(u_2) - Du_2]\}/B(w, h, \tau). \qquad (2.5.10)$$

Let us find the width after Prandtl X of a smeared shock wave zone determined by the solution of problem (2.5.9), (2.5.7). By the definition,

$$X = \min_\xi \frac{u_1 - u_2}{|dw/d\xi|} = \frac{u_1 - u_2}{\max_\xi |dw/d\xi|}. \qquad (2.5.11)$$

Employing formulas (2.5.10), (2.5.11), (2.5.2), it is easy to find that

$$X = h(1 + \sqrt{\chi})/(1 - \sqrt{\chi}), \qquad (2.5.12)$$

where

$$\chi = u_2(1 - \kappa u_1)/[u_1(1 - \kappa u_2)]; \qquad \kappa = \tau/h. \qquad (2.5.13)$$

Note that if the time step τ is taken from the stability domain of the difference scheme (2.5.3), that is, $\kappa u \leq 1$, then $\chi \geq 0$ as can easily be seen in formula (2.5.13). Consider now, for equations (2.5.1), (2.5.2), an initial function $u(x, 0)$ of the form

$$u(x, 0) = u_0(x) = \begin{cases} u_1, & x \leq x_1; \\ u_1 + k_1(x - x_1), & x_1 \leq x \leq x_2; \\ u_2, & x > x_2; \end{cases} \qquad (2.5.14)$$

where u_1, u_2, x_1, x_2, k_1 are constants and $u_1 > u_2 \geq 0$, $x_1 < x_2$, $k_1 = (u_2 - u_1)/(x_2 - x_1)$. The exact solution of the test problem (2.5.1), (2.5.2), (2.5.14) is easily found by means of a well-known functional relationship [2.41]

$$u(x, t) = u_0(x - u(x, t)t),$$

and has the form;

(a) $t < t_*$,

$$u(x, t) = \begin{cases} u_1, & x \leq x_1 + u_1 t; \\ [u_1 + k_1(x - x_1)]/(1 + k_1 t), & \\ & x_1 + u_1 t \leq x \leq x_2 + u_2 t; \\ u_2, & x > x_2 + u_2 t; \end{cases} \qquad (2.5.15)$$

(b) $t \geq t_*$,

$$u(x, t) = \begin{cases} u_1, & x \leq x_* + D_0(t - t_*); \\ u_2, & x > x_* + D_0(t - t_*); \end{cases} \qquad (2.5.16)$$

where (x_*, t_*) are coordinates of the point of shock formation,

$$x_* = x_1 - u_1/k_1, \qquad t_* = -1/k_1, \qquad D_0 = (u_1 + u_2)/2. \quad (2.5.17)$$

In Figure 2.9 we present a graph of the relation X_d/X_t as a function of

Figure 2.9. The graph of the
relation X_d/X_T as a function of
time

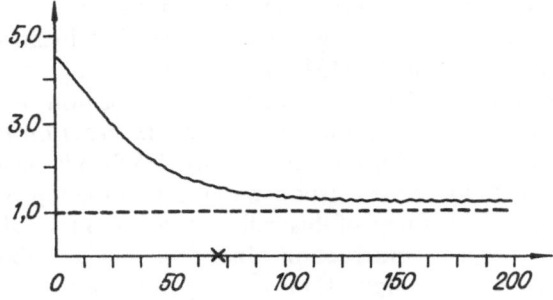

dimensionless time t/τ where X_d is a width after Prandtl of the finite-difference
shock wave zone which was computed on the basis of the difference solution
u^n obtained by the scheme (2.5.3), and

$$X_d = \min_i \, (u_1 - u_2)/|(u_{i+1}^n - u_{i-1}^n)/(2h)|,$$

X_t is the width of a smeared shock wave zone obtained above on the basis of
f.d.a. and computed by formulas (2.5.12), (2.5.13). In Figure 2.9 the moment of
time $t = t_*$ of the shock formation is marked by a cross on the t/τ-axis. It is
seen in Figure 2.9 that the formation of a stationary profile of a stationary
shock wave occurs in the difference solution with a large lag with respect to
the true moment of time of the shock formation.

2.5.2. Nonuniform Moving Grid Adapting to the Flow

It is clear that the absolute error in determining the values x_*, t_* from the
difference solution can be substantially reduced by using a sufficiently fine
grid on the x-axis. However, in the case of a uniform grid a very large number
of nodes would be required. On the other hand, it is known that the compression zone size decreases in comparison with the overall computational domain
as the time t increases, and since a shock wave is formed in the difference
solution within the limits of a compression zone, it is reasonable to use a
fine mesh only inside the compression zone. Since the compression zone
moves along the x-axis with time, a mesh subdomain having small steps
should generally also change its position, that is, the mesh should be solution-
adaptive. A nonuniform mesh of such kind was proposed in [2.115] in
connection with the numerical solution of the Burgers equation (2.1.1) and
the modified Burgers equation. Subsequently, this mesh was successfully used
by the authors of [2.116]. A similar idea has been realized in [2.117] in the
computations of shock wave propagation in reacting flows. The details of
the corresponding algorithm are not presented in [2.117]. In [2.115], in the
process of constructing an adaptive grid, a "wave center" notion has been used

which in its geometric sense coincides with the definition of a finite-difference shock wave center given in [2.47]. The location of the wave center was determined in [2.115] by $\max|\partial u/\partial x|$. Taking into account the results of Section 2.3 this technique gives the position of the wave center with some error; more accurate is the use of the criterion $\max|0.5(hu - \tau u^2)\,\partial u/\partial x|$ in accordance with formulas (2.5.4), (2.5.5). We have realized the algorithm of [2.115] for the construction of a solution-adaptive grid; with some modifications the features of this will be clear from the subsequent short presentation of the grid construction algorithm. Let us subdivide all the necessary sequences of computations into separate stages.

Stage 1. Find the number I_n of a cell such that in this cell $|\Delta_{1x}(u_i^n)|$ achieves its maximum. Here

$$0 = x_0^n < x_1^n < \cdots < x_{N_n}^n = l, \qquad (2.5.18)$$

Δ_{1x} is an operator of differencing on a nonuniform grid which approximates the operator $\partial/\partial x$;

$$\Delta_{1x}(u_i^n) = (u_i^n - u_{i-1}^n)/h_i,$$

where $h_i = x_i^n - x_{i-1}^n$, $i = 1, \ldots, N_n$. Taking the above remarks into account, the abscissa x_{I_n} approximately determines a compression wave center which at $t \geq t_*$ passes to the finite-difference shock wave center. As in [2.113], we check whether the wave center has moved by a distance necessary for the rezoning of a mesh on the x-axis.

Stage 2. If it is already necessary to rezone a mesh, we begin with the computations of the new abscissas of the grid nodes. For this purpose, we compute the value

$$\varepsilon = 1/|\Delta_{1x}(u_{I_n}^n)|. \qquad (2.5.19)$$

Let $h^{(1)}$ be a user-specified reference value of the step size outside the compression zone. Then we introduce (in the vicinity of the node I_n) a uniform grid with a step

$$h^{(2)} = \max(h_{\min}, \alpha\varepsilon h^{(1)}), \qquad (2.5.20)$$

where $\alpha > 0$, for example, $\alpha = 0.4$ (see [2.115]); $h_{\min}$ is a user-specified minimal allowed value of the step $h^{(2)}$, for example, $h_{\min} = 10^{-4}$. This value is introduced in order that there be no computation with very small steps along the x-axis at $t \geq t_*$ (if one computes $h^{(2)}$ simply by a formula $h^{(2)} = \alpha\varepsilon h^{(1)}$, then at $t \cong t_*$ we will obtain the values $h^{(2)} < 10^{-7}$). Then we compute the $M_l^{(n)}$ and $M_r^{(n)}$ abscissas of equally spaced nodes with a step $h^{(2)}$ to the left and to the right of the node x_{I_n}, respectively. As a result we obtain a subdomain $G_r^{(n)}$ with a regular grid on the x-axis.

Stage 3. Let X_l^n be the abscissa of the left boundary of the domain $G_r^{(n)}$. Moving from right to left from X_l^n we compute M_2 steps h_i by the formula

$$h_i = h^{(2)}(1 + \beta h^{(1)})^i, \qquad i = 1, \ldots, M_2. \qquad (2.5.21)$$

In formula (2.5.21), $\beta = \text{const} > 0$, for example, $\beta = 2.5$, according to [2.115]. The number M_2 is determined here in the process of computation as the least number for which $h_{M_2+1} > h^{(1)}, h_{M_2} < h^{(1)}$. After that we compute the abscissa X_{l1}^n of the left boundary of the cell of the size h_{M_2}, and the interval $[0, X_{l1}^n]$ is subdivided into M_1 equal cells where $M_1 = \max(1, X_{l1}^n/h^{(1)})$. Now let X_r^n be the abscissa of the right boundary of the domain $G_r^{(n)}$. Moving from left to right from X_r^n we compute M_3 steps h_i by formula (2.5.21). The number M_3 is determined by analogy with M_2. After that we compute the abscissa X_{r1}^n of the right boundary of the cell of the size h_{M_3}. The interval $[X_{r1}^n, l]$ is subdivided into M_4 equal cells where

$$M_4 = \max(1, X_{r1}^n/h^{(1)}).$$

Stage 4. Compute the solution u^n on a new grid by a cubic spline interpolation [2.118] using the values u^n on an old grid.

It has been shown in [2.115] that the computation of grid steps by formula (2.5.21) (that is, h_i is a member of geometric progression) provides a uniform accuracy of a difference solution throughout the computational domain. Note that in [2.119] there have been obtained meshes from some optimality criterion which include, as a particular case, the law of the grid steps variation (2.5.21).

While using the above nonuniform grid the time t_* of the shock formation was determined as the least value of t_n,

$$t_n = \sum_{k=1}^{n} \tau_k,$$

such that for this the inequality

$$X_d < \delta_1 \cdot h^{(2)} \tag{2.5.22}$$

is satisfied, where δ_1 is a user-specified constant equal to a mean number of cells on the x-axis, which is typical of the difference scheme employed and within which there occurs (in accordance with the definition of a width after Prandtl) the sharpest change of the solution u_h in the zone of a smeared shock wave. For a preliminary estimation of the value δ_1 we have used formula (2.5.12), which in this case is applicable, because the mesh is uniform in the vicinity of the compression wave center by virtue of its construction; one should set $h = h^{(2)}$ in (2.5.12). Of course, formula (2.5.12) is inapplicable in the general case of the nonstationary shock wave. Figure 2.10 illustrates the influence of the choice of a constant δ_1 in the inequality (2.5.22) on the accuracy of determination of the moment t_* of the shock formation. The value $\delta t = (t_* - t_{*h})/t_*$ is measured along the ordinate axis in Figure 2.10, where t_* is the time of the shock formation in accordance with the exact solution of the test problem (2.5.1), (2.5.2), (2.5.14), and t_{*h} is the time of the shock formation as determined with the aid of the inequality (2.5.22). Curve 1 in Figure 2.10 is obtained at $M_l^{(n)} = M_r^{(n)} = 30$, $h_{\min} = 10^{-4}$, in formula (2.5.20); Curve 2 is

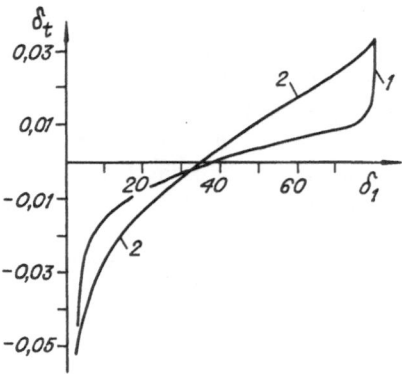

Figure 2.10. The dependence of δt on δ_1.

obtained at $M_l^{(n)} = M_r^{(n)} = 20$, $h_{\min} = 2 \cdot 10^{-4}$. We have taken in (2.5.14) the values $u_1 = 3$, $u_2 = 1$, $x_1 = 0.175$, $x_2 = 0.525$; $h^{(1)} = 0.05$, $\alpha = 0.5$, and $\beta = 2.0$ in formula (2.5.21). It follows from Figure 2.10 that at $3.5 \le \delta_1 \le 80$ the value of the error $|\delta t|$ does not exceed 4.5%. In Figure 2.11 we present (by solid lines for different moments of time t) the difference solution graphs obtained by the scheme

$$(u_i^{n+1} - u_i^n)/\tau_n + [\varphi(u_i^n) - \varphi(u_{i-1}^n)]/(x_i^n - x_{i-1}^n) = 0$$

when solving the problem (2.5.1), (2.5.2), (2.5.14); the initial data are the same as for curve "1" in Figure 2.10. The exact solution is plotted in Figure 2.11 by dashed lines. In the computational example of Figure 2.11 the number of computing cells on the x-axis reached 186 by the time $t \cong t_*$. Note that one would have to take 10,000 cells to reach a comparable difference solution accuracy in the neighborhood of a shock wave in the case of using a uniform grid.

Thus, the information on the location of subdomains with large gradients

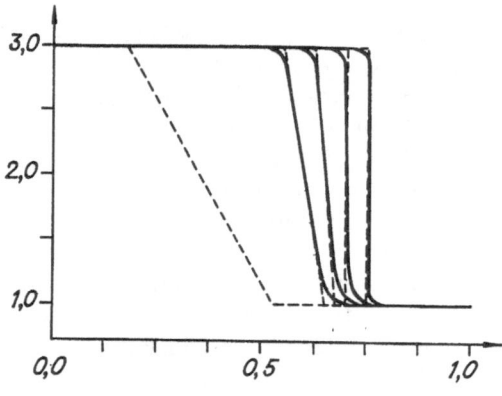

Figure 2.11. Results of computation of the problem on shock formation.

(which can be obtained with the aid of a differential analyzer) may be used to construct a dynamic grid concentrating on the above subdomains, and which enables one to increase substantially the difference solution accuracy at a moderate number of grid nodes. As a result the accuracy of determining the location and time of the discontinuities formation in the process of numerical studies of those problems where these phenomena may take place increases substantially.

Note that there may take place, in some fluid mechanics problems, the phenomena of shock wave decay when the shock wave decays and transforms to a continuous flow. The above method of computation on a solution-adaptive grid can be used after introducing some insignificant alterations for the numerical investigation of such problems.

Differential Analyzers of Shock Waves in Two-Dimensional Gas Dynamic Computations

Two- and three-dimensional gas dynamic problems are at present the most important in practice. An increase by one in the number of spatial variables leads to a significant increase in the amount of numerical data by means of which the computer-aided solution of a problem is represented. This takes place, in particular, in cases when the finite-difference shock-capturing schemes are used for the numerical solution of multidimensional problems of gas dynamics. As in the one-dimensional case, the strong discontinuities in the difference solutions of two- and three-dimensional problems are "smeared". This smearing generally takes place in the direction of each spatial variable. This impedes substantially the efficient use of numerical information on problem solution obtained with the aid of a computer, in particular, the extraction of information on the position and on the type of discontinuity surfaces. In this connection the questions of the development of the algorithms for the singularities localization in numerical solutions of multidimensional gas dynamics problems, and of the methods for investigating the accuracy of these algorithms are of present interest.

An extension of the above techniques, for the investigation of differential analyzers of one-dimensional shock waves for the multidimensional case, faces the necessity of taking into account a number of specific features which are inherent in multidimensional computations on a fixed rectangular Eulerian grid. First, the increase in the number of spatial variables gives rise to a problem such as the noninvariance of a difference scheme under the transformations, with respect to which the original Euler equation system governing the flow of an inviscid compressible gas is invariant [3.1]. This noninvariance leads, in the general case, to an anisotropy of the difference solution properties in various directions in the plane of spatial variables (x, y). For example, it was pointed out in [3.2] that in the case of a shock located at an angle to the computing mesh lines the Rankine–Hugoniot conditions are not necessarily exactly satisfied, since by virtue of the shock smearing a nonzero gradient of the normal momentum flux can arise in the direction tangential to the shock. Therefore one cannot automatically apply the one-dimensional technique presented in Chapter 2 to the considerations of the difference solution properties along the beams orthogonal to the shock front, even in the case when, according to the exact solution, the tangential velocity component behind and before the front of the shock is equal to zero at chosen points.

The above circumstances have led to the fact that until now there is no detailed technique for the investigation of differential analyzers of shock waves even at a "differential" level, that is, within the framework of the first differential approximations (f.d.a.s) of difference schemes. Below we present a number of results of the investigation of differential analyzers of shock waves in two-dimensional gas dynamic flows.

At first we consider in Section 3.1 a question on the criteria by means of which the zones of "smearing" of shock fronts may be found in the difference solutions of two-dimensional problems. The implementation of such criteria makes it possible to eliminate from further consideration those subdomains which surely do not contain the shock waves (for example, the rarefaction domains and the constant flow domains) while searching for shock waves in a two-dimensional domain.

After the zones of shock wave front smearing in a two-dimensional flow are localized, a question arises on the dependence of the width of these zones upon the orientation of the discontinuity zone with respect to the lines of a rectangular computational grid in the (x, y)-plane. Information of this kind is of importance, for example, in the cases when different strong discontinuities (in the process of their propagation within the computational domain) draw so close together that their zones of smearing begin to overlap. This causes an increase in the solution errors in the overlapping domains and, as a consequence of this, the deterioration of the accuracy of the algorithms for the localization of strong discontinuities by shock-capturing computational results. In this connection we present in Section 3.1.2 a method for investigating the properties of "smearing" of two-dimensional shock waves with curved fronts. The presentation is carried out in the example of an asymptotic difference scheme of the Harlow particle-in-cell method [3.3] that coincides with the scheme of the "fluid-in-cell" method [3.4].

In Sections 2.2–2.4 we have used, in particular, the considerations of the theory of similarity and dimensions when analyzing the differential equations with two small parameters h and τ arising in the study of differential analyzers of shock waves for one-dimensional flows. In Section 3.2 an analogous analysis is carried out in the two-dimensional case for certain classes fo the f.d.a.s of the difference schemes containing three small parameters h_x, h_y, and τ, where h_x, h_y are the dimensions of a cell of a rectangular computational grid in the (x, y)-plane. It is shown that in the case when the grid parameters h_x, h_y, and τ enter the approximation and the artificial viscosity in the f.d.a. equations with regard to some similarity and dimensional considerations, then it is possible to formulate some necessary conditions for the smeared shock wave center existence along a beam orthogonal to the shock front. After that, we present in Section 3.2 a number of differential analyzers of shock waves which are based on the use of scheme viscosity norms.

At the beginning of Chapter 2 we have mentioned a number of well-known techniques for shock waves localization in two-dimensional flows, and among them a widespread technique such as the determination of discontinuity location by the coalescence of isolines. In recent years there has taken place an intense development of an approach to solving fluid mechanics problems with big gradients, in which the mesh is assumed to be dynamic and adapting to the flow in such a way that the lines of this mesh concentrate in the domains of large solution gradients. Therefore, one can determine in this case the position of shock lines by the concentration of grid lines. In this sense the application of moving solution-adaptive grids may be considered as one of the techniques of the singularities localization in the solution. In Section 3.3 we consider a number of questions related to the realization of the idea of adaptive grids: the form of the flow equations in moving coordinates, the grid control, etc. Localization of shock surfaces by means of moving grids can be considered as part of a vast problem of the localization of singular zones and singularities in the numerical solutions of fluid mechanics problems. Here by singular zones are meant the boundary layers, flame fronts, viscous mixing layers, displacement fronts in filtration problems, shock transitions in a viscous, heat-conducting gas, etc. [3.5]–[3.7]. It was pointed out in [3.8] that, for example, such singularities as the contact discontinuities, flame fronts, detonation waves, ionization fronts, and thermal waves driven by radiation transfer can be treated in much the same way as the shock waves.

3.1. Method for Investigating the Properties of Curvilinear Shock Front "Smearing"

3.1.1. Inequalities in the Zone of Smearing of a Two-Dimensional Shock Wave

As was pointed out above, the localization of the zones of shock wave "smearing" should precede the calculation of the coordinates of the shock front points. In the one-dimensional case the inequality $\partial u/\partial x < 0$ and the Rankine–Hugoniot conditions were checked, where u is the gas velocity. Let us derive an analogue of the inequality $\partial u/\partial x < 0$ for the two-dimensional case when the shock wave is some curve in the plane of the (x, y) coordinates. According to Zemplén's theorem (see Section 1.1.2) $[u_n] < 0$ on a stable shock wave. Suppose that the values of the velocity components u, v have been calculated in the nodes of a computational grid in the (x, y)-plane at some moment of time $t = t^m$. Let us determine the u, v solution for the rest of the (x, y) points of the computational domain by means of spline functions of sufficient smoothness (for example, by means of polynomial cubic splines). Denote the corresponding splines by $S_u(x, y, t^m)$ and $S_v(x, y, t^m)$. Using these functions we can find the velocity component u_n normal to the shock front:

$u_n = u_n(x, y, t^m)$. Let us draw a normal to the shock line at a point on this line where the tangent to the shock line exists and is unique. If a transition from the state behind the front to the state before the front occurs monotonously along the above normal in the zone of smearing of a shock wave, then, by virtue of the inequality $[u_n] < 0$, the inequality

$$\partial u_n / \partial n < 0 \qquad (3.1.1)$$

takes place in this zone if the coordinate n increases in the direction of the propagation of a shock wave.

Consider now an arbitrary curved shock wave. Let $F_0(x_0, y_0)$ be a fixed point of the shock wave front at $t = t_0$. Let us place the origin of a Cartesian rectangular system (τ, n) into the (x_0, y_0) point, and let us draw the $F_0 n$ axis in the direction of a normal to the shock wave front at the point F_0. Let the n coordinate increase in the direction of the shock front propagation. We further draw the $F_0 \tau$ axis along the tangent to the front at the F_0 point in such a way that the coordinate system (n, τ) is a right one. Let $\mathbf{n} = \{n_1, n_2\}$ be the unit vector of the $F_0 n$ axis and let n_1, n_2 be its coordinates in the Cartesian system (x, y). Then

$$u_n = n_1 u + n_2 v, \qquad u_\tau = n_2 u - n_1 v, \qquad (3.1.2)$$

where u, v are the projections of the velocity vector on the x- and y-axis, respectively; they are calculated at the (x, y) point by means of the above-introduced splines S_u, S_v. Taking into account the inequality (3.1.1) we obtain the formula

$$n_1 (\partial/\partial x)(n_1 u + n_2 v) + n_2 (\partial/\partial y)(n_1 u + n_2 v) < 0. \qquad (3.1.3)$$

Unknown coordinates n_1, n_2 enter the inequality (3.1.3). Let us show that in a particular case, when in the zone of a smeared shock wave in the neighborhood of the (x_0, y_0) point the gas flow is normal to the front, the inequality (3.1.3) turns into

$$\operatorname{div} \mathbf{u} < 0, \qquad (3.1.4)$$

where $\mathbf{u}$ is the velocity vector. In fact, let $y = \varphi(x, t)$ be the equation of a shock wave front curve. We shall assume that the function $\varphi(x, t)$ has a continuous derivative $\partial \varphi / \partial x$. Then it is easy to calculate the coordinates of the $\mathbf{n}$ vector at the (x_0, y_0) point

$$n_1 = -\dot{\varphi}_0 (1 + \dot{\varphi}_0^2)^{-0.5}, \qquad n_2 = (1 + \dot{\varphi}_0^2)^{-0.5}, \qquad (3.1.5)$$

where $\dot{\varphi}_0 = (\partial \varphi / \partial x)|_{x=x_0, t=t_0}$. Taking into account (3.1.5), formula (3.1.3) may be rewritten as

$$(1 + \dot{\varphi}_0^2)^{-1} \left\{ \dot{\varphi}_0^2 \frac{\partial u}{\partial x} + \frac{\partial v}{\partial y} - \dot{\varphi}_0 \left(\frac{\partial v}{\partial x} + \frac{\partial u}{\partial y} \right) \right\} < 0. \qquad (3.1.6)$$

Suppose that the gas flow is normal to the shock wave front, that is, $u_\tau = 0$, at the point (x_0, y_0) and in some neighborhood of this point. Then, taking (3.1.2), (3.1.5) into account, we obtain at the point (x_0, y_0) and in its neighborhood the equation

$$u + v\dot\phi_0 = 0. \tag{3.1.7}$$

Let us differentiate both sides of equation (3.1.7) with respect to x and substitute into formula (3.1.6) the expression $-\partial u/\partial x$ instead of $\dot\phi_0 \, \partial v/\partial x$. Then replace in (3.1.6) $\partial u/\partial y$ by $-\dot\phi_0 \, \partial v/\partial y$. As a result we obtain

$$\partial u/\partial x + \partial v/\partial y = \text{div } \mathbf{u} < 0.$$

Lemma 3.1. *If in the zone of a smeared stationary shock wave with a plane front moving into a gas at rest with an angle β to the positive direction of the x-axis the gas flow is normal to the front, then the following relationships* [3.9] *are valid*

$$u_x < 0, \qquad v_y < 0, \qquad \text{div } \mathbf{u} < 0;$$

$$u_y = v_x; \qquad u_y < 0 \quad \text{at} \quad 0 < \beta < \pi/2, \quad \pi < \beta < 3\pi/2;$$

$$u_y > 0 \quad \text{at} \quad \pi/2 < \beta < \pi, \quad 3\pi/2 < \beta < 2\pi.$$

PROOF. In the coordinate system (x^*, y^*), the x^*-axis of which is directed in parallel with the wave propagation the flow is one-dimensional, and is parallel to the x^*-axis. Let $u^* = u^*(x^*, t)$ be the velocity of the one-dimensional flow. Let us make use of the formulas

$$x^* = x \cos \beta + y \sin \beta, \qquad y^* = -x \sin \beta + y \cos \beta, \tag{3.1.8}$$

$$u = u^* \cos \beta, \qquad v = u^* \sin \beta. \tag{3.1.9}$$

Differentiating both sides of equations (3.1.9) with respect to x we get

$$u_x = u^*_{x^*} \cos^2 \beta, \qquad u_y = v_x = u^*_{x^*} \sin \beta \cos \beta,$$
$$v_y = u^*_{x^*} \sin^2 \beta, \qquad u_x + v_y = u^*_{x^*}, \tag{3.1.10}$$

from where the assertion of the lemma follows with regard to inequality (3.1.1) which in the case under consideration may be written in the form $\partial u^*/\partial x^* < 0$. □

3.1.2. An Analysis on the Basis of First Differential Approximation

Below we present a method for investigating the properties of curvilinear shock wave smearing that was proposed in [3.10]. The aim of this investigation is to elucidate the specific behavior of the dependence of a smeared curved shock wave zone width on the angle between the normal to the shock front and the lines of rectangular computing mesh in the plane of the Eulerian Cartesian coordinates (x, y). Information of this kind, as was pointed out in the foregoing, can be useful when studying the accuracy of algorithms for the localization of shock waves interacting with other shocks or free surfaces and rigid walls. In addition, the method of transforming the f.d.a. by means of simple trigonometric formulas presented below may be useful in the study of the invariance of the f.d.a. of a difference scheme under the rotation transformation.

The presentation is carried out in the example of the HARLOW particle-in-cell method [3.3], [3.11]–[3.13]. Despite the fact that over 30 years have passed since the development of this method it is still applied in numerical calculations [3.14]–[3.22]. This is explained by the efficiency of the particle-in-cell method in the computations of problems with large deformations of a medium and with hydrodynamically unstable interfaces.

Assuming an infinite number of particles used in the computation by the Harlow method let us replace the discrete mass, momentum, and energy fluxes by difference approximations introduced for the first time in [3.23]. Then the mass, momentum, and energy transport accomplished by the particles is described in the following finite difference equations [3.4], [3.10]

$$h_x h_y [(\rho w)_{jk}^{m+1} - \rho_{jk}^m \tilde{w}_{jk}] = [d_{1jk} \tilde{w}_{j-1k} + (1 - d_{1jk}) \tilde{w}_{jk}] \Delta M_{j-1/2,k}^m$$

$$- [d_{3jk} \tilde{w}_{j+1,k} + (1 - d_{3jk}) \tilde{w}_{jk}] \Delta M_{j+1/2,k}^m$$

$$+ [d_{2jk} \tilde{w}_{jk-1} + (1 - d_{2jk}) \tilde{w}_{jk}] \Delta M_{j,k-1/2}^m$$

$$- [d_{4jk} \tilde{w}_{j,k+1} + (1 - d_{4jk}) \tilde{w}_{jk}] \Delta M_{j,k+1/2}^m, \quad (3.1.11)$$

where h_x, h_y are the step sizes in the directions x, y, respectively, m is the number of a time step, δt is the size of this step, indices (j, k) refer to the cell center, w is the vector, $w = \{1, u, v, E\}$ where $E = \varepsilon + 0.5(u^2 + v^2)$, and ε is the specific internal energy.

$$d_{1jk} = \begin{cases} 1, & \tilde{u}_{j-1/2,k} > 0, \\ 0, & \tilde{u}_{j-1/2,k} \le 0; \end{cases} \qquad d_{3jk} = \begin{cases} 1, & \tilde{u}_{j+1/2,k} < 0, \\ 0, & \tilde{u}_{j+1/2,k} \ge 0; \end{cases}$$

$$d_{2jk} = \begin{cases} 1, & \tilde{v}_{j,k-1/2} > 0, \\ 0, & \tilde{v}_{j,k-1/2} \le 0; \end{cases} \qquad d_{4jk} = \begin{cases} 1, & \tilde{v}_{j,k+1/2} < 0, \\ 0, & \tilde{v}_{j,k+1/2} \ge 0, \end{cases}$$

$$\Delta M_{j+1/2,k}^m = \begin{cases} h_y \rho_{jk}^m \tilde{u}_{j+1/2,k} \, \delta t, & \tilde{u}_{j+1/2,k} > 0, \\ h_y \rho_{j+1k}^m \tilde{u}_{j+1/2,k} \, \delta t, & \tilde{u}_{j+1/2,k} \le 0; \end{cases} \tag{3.1.12}$$

$$\Delta M_{j,k+1/2}^m = \begin{cases} h_x \rho_{jk}^m \tilde{v}_{j,k+1/2} \, \delta t, & \tilde{v}_{j,k+1/2} > 0, \\ h_x \rho_{j,k+1}^m \tilde{v}_{j,k+1/2} \, \delta t, & \tilde{v}_{j,k+1/2} \le 0. \end{cases}$$

The quantities appearing as a result of computations at the Eulerian stage of the computation at the same time step [3.3] are denoted by the symbol $\sim$. Let us consider first the case when $\tilde{u}_{j\pm1/2,k} > 0$, $\tilde{v}_{j,k\pm1/2} > 0$. Then formula (3.1.11) can be written, in view of (3.1.12), as follows:

$$[(\rho w)_{jk}^{m+1} - \rho_{jk}^m \tilde{w}_{jk}]/\delta t = (\tilde{u}_{j-1/2,k} \rho_{j-1k}^m \tilde{w}_{j-1k} - \tilde{u}_{j+1/2,k} \rho_{jk}^m \tilde{w}_{jk})/h_x$$

$$+ (\tilde{v}_{j,k-1/2} \rho_{jk-1}^m \tilde{w}_{jk-1} - \tilde{v}_{jk+1/2} \rho_{jk}^m \tilde{w}_{jk})/h_y. \quad (3.1.13)$$

Employing formula (3.1.13), we shall obtain the f.d.a. of an integral step scheme of the PIC method when using the Harlow scheme [3.3] at the Eulerian stage of computation. Let us introduce the following notations to shorten the amount of further calculations

$$\tilde{U}_{l-1/2}^{\alpha} = \begin{cases} \tilde{u}_{l-1/2,k}, & \alpha = 1, \\ \tilde{v}_{j,l-1/2}, & \alpha = 2; \end{cases} \qquad \tilde{W}_l^{\alpha} = \begin{cases} \tilde{w}_{l,k}, & \alpha = 1, \\ \tilde{w}_{j,l}, & \alpha = 2; \end{cases}$$

$$l = \begin{cases} j, & \alpha = 1, \\ k, & \alpha = 2; \end{cases} \qquad u_1 \equiv u, \quad u_2 \equiv v, \quad h_1 = h_x, \quad h_2 = h_y. \quad (3.1.14)$$

Then equation (3.1.13) may be rewritten in the form

$$[(\rho w)_{jk}^{m+1} - \rho_{jk}^m \tilde{w}_{jk}]/\delta t = \sum_{\alpha=1}^{2} (\tilde{U}_{l-1/2}^{\alpha} \rho_{l-1}^{\alpha} \tilde{W}_{l-1}^{\alpha} - \tilde{U}_{l+1/2}^{\alpha} \rho_l^{\alpha} \tilde{W}_l^{\alpha})/h_{\alpha}. \quad (3.1.15)$$

Let us write the difference equations of the Eulerian stage as

$$\tilde{u}_{\alpha jk} = u_{\alpha jk}^m - (\delta t/\rho_{jk}^m)F_{1jk}^{\alpha}, \qquad \alpha = 1, 2;$$
$$\tilde{\varepsilon}_{jk} = \varepsilon_{jk}^m - (\delta t/\rho_{jk}^m)(F_{2jk}^1 + F_{2jk}^2). \quad (3.1.16)$$

Then the following equation for $\tilde{E}_{jk}$ may be obtained from (3.1.16) by the use of algebraic operations

$$\tilde{E}_{jk} = E_{jk}^m - (\delta t/\rho_{jk}^m)F_{3jk}, \quad (3.1.17)$$

where

$$F_{3jk} = \sum_{\alpha=1}^{2} [F_{2jk}^{\alpha} + u_{\alpha jk}F_{1jk}^{\alpha} - (\delta t/2)(F_{1jk}^{\alpha})^2/\rho_{jk}^m].$$

Thus with regard to (3.1.14), (3.1.16), (3.1.17) we have

$$\tilde{U}_l^{\alpha} = U_l^{\alpha} - (\delta t/\rho_{jk}^m)F_{1jk}^{\alpha}, \qquad \tilde{E}_{jk} = E_{jk}^m - (\delta t/\rho_{jk}^m)F_{3jk}. \quad (3.1.18)$$

Let us set $F_l = \{0, F_{1jk}^1, F_{1jk}^2, F_{3jk}\}^T$. Then the Eulerian stage scheme written in vector form is

$$\tilde{W}_l^{\alpha} = W_l^{\alpha} - (\delta t/\rho_l)F_l. \quad (3.1.19)$$

When accounting for the terms of the first order of smallness with respect to δt, h_1, h_2, the expansions of the components of the vector F_l into Taylor series at the point $(x_j, y_k, m\,\delta t)$, where (x_j, y_k) are the coordinates of a cell center, have the following form in a case of the Harlow scheme:

$$F_l = \begin{pmatrix} 0 \\ p_{x_1} \\ p_{x_2} \\ \sum_{\alpha=1}^{2} [pu_{\alpha} - (\delta t/2)pp_{x_{\alpha}}/\rho]_{x_{\alpha}} \end{pmatrix}, \quad (3.1.20)$$

where the notations $x_1 \equiv x$, $x_2 \equiv y$, $p_{x_{\alpha}} \equiv \partial p/\partial x_{\alpha}$, etc. are introduced.

Substituting formulas (3.1.18), (3.1.19) into (3.1.15) and expanding the quantities entering (3.1.15) into Taylor series with respect to the point $(x_j, y_k, m\,\delta t)$, one obtains the f.d.a. of an integral step scheme of the form

$$(\rho w)_t + (\delta t/2)(\rho w)_{tt} + F = \sum_{\alpha=1}^{2} \{-(\rho u_\alpha w)_{x_\alpha} + (h_\alpha/2)(u_\alpha \rho w_{x_\alpha})_{x_\alpha}$$
$$+ (h_\alpha/2)(u_\alpha \rho_{x_\alpha} w)_{x_\alpha} + \delta t[(F_1^\alpha w)_{x_\alpha} + (u_\alpha F)_{x_\alpha}]\}(3.1.21)$$

Substituting into formula (3.1.21) the components of the w-vector and the corresponding components of the vector F (see (3.1.20)) and returning to the initial notations u, v, x, y, we obtain the f.d.a. of the Harlow particle-in-cell method in the form

$$\partial\rho/\partial t + \partial\rho u/\partial x + \partial\rho v/\partial y = (\partial/\partial x)[(h_x/2)u\,\partial\rho/\partial x + \delta t\,\partial p/\partial x]$$
$$+ (\partial/\partial y)[(h_y/2)v\,\partial\rho/\partial y + \delta t\,\partial p/\partial y] - (\delta t/2)\,\partial^2\rho/\partial t^2;$$

$$\partial\rho u/\partial t + \partial(p + \rho u^2)/\partial x + \partial\rho uv/\partial y = (\partial/\partial x)[(h_x/2)u\,\partial\rho u/\partial x + 2\delta tu\,\partial p/\partial x]$$
$$+ (\partial/\partial y)[(h_y/2)v\,\partial\rho u/\partial y + \delta t(u\,\partial p/\partial y + v\,\partial p/\partial x)] - (\delta t/2)\,\partial^2\rho u/\partial t^2;$$
$$(3.1.22)$$

$$\partial\rho v/\partial t + \partial\rho uv/\partial x + \partial(p + \rho v^2)/\partial y$$
$$= (\partial/\partial x)[(h_x/2)u\,\partial\rho v/\partial x + \delta t(v\,\partial p/\partial x + u\,\partial p/\partial y)]$$
$$+ (\partial/\partial y)[(h_y/2)v\,\partial\rho v/\partial y + 2\delta tv\,\partial p/\partial y] - (\delta t/2)\,\partial^2\rho v/\partial t^2;$$

$$\partial\rho E/\partial t + \partial(pu + \rho uE)/\partial x + \partial(pv + \rho vE)/\partial y$$
$$= (\partial/\partial x)[(h_x/2)u\,\partial\rho E/\partial x + \delta t((E + p/(2\rho))\,\partial p/\partial x + u\,\partial pu\,\partial x)]$$
$$+ (\partial/\partial y)[(h_y/2)v\,\partial\rho E/\partial y + \delta t((E + p/(2\rho))\,\partial p/\partial y + v\,\partial pv/\partial y)]$$
$$- (\delta t/2)\,\partial^2\rho E/\delta t^2.$$

Let us turn to the Cartesian coordinate system (n, τ) whose axis On has an angle β with the positive direction of the axis Ox, and the positive direction of the $O\tau$ axis is chosen in such a way that the coordinate system (n, τ) was the right one. Let us introduce some notations. Let

$$w = w(x, y, t) = \{\rho, u, v, E\}^T.$$

Let us set

$$w^* = w^*(n, \tau, t) = w(x(n, \tau), y(n, \tau), t) = \{\rho^*, u^*, v^*, E^*\}^T.$$

Denote by u_n, u_τ the velocity vector components along the On- and $O\tau$-axes. Then

$$u^* = u_n \cos\beta - u_\tau \sin\beta, \qquad v^* = u_\tau \cos\beta + u_n \sin\beta, \qquad (3.1.23)$$

$$\frac{\partial w}{\partial x} = -\sin\beta\frac{\partial w^*}{\partial \tau} + \cos\beta\frac{\partial w^*}{\partial n}, \qquad \frac{\partial w}{\partial y} = \cos\beta\frac{\partial w^*}{\partial \tau} + \sin\beta\frac{\partial w^*}{\partial n}. \qquad (3.1.24)$$

Employing formulas (3.1.22)–(3.1.24), it is easy to obtain the f.d.a. of the continuity equation with respect to the variables n, τ, t:

$$\partial\rho^*/\partial t + \partial\rho^* u_\tau/\partial\tau + \partial\rho^* u_n/\partial n$$

$$= (\partial/\partial\tau)\{-\sin\beta[(h_x/2)(-(\sin\beta)u_\tau + (\cos\beta)u_n)$$

$$\times ((\partial\rho^*/\partial\tau)(-\sin\beta) + (\partial\rho^*/\partial n)\cos\beta)$$

$$+ \delta t((-\sin\beta)\,\partial p^*/\partial\tau + (\partial p^*/\partial n)\cos\beta)]$$

$$+ \cos\beta[(h_y/2)(u_\tau\cos\beta + u_n\sin\beta)((\partial\rho^*/\partial\tau)\cos\beta + (\partial\rho^*/\partial n)\sin\beta)$$

$$+ \delta t((\partial p^*/\partial\tau)\cos\beta + (\partial p^*/\partial n)\sin\beta)]\} + (\partial/\partial n)\{\cos\beta[(h_x/2)$$

$$\times (-u_\tau\sin\beta + u_n\cos\beta)(-(\partial\rho^*/\partial\tau)\sin\beta + (\partial\rho^*/\partial n)\cos\beta)$$

$$+ \delta t(-(\partial p^*/\partial\tau)\sin\beta + (\partial p^*/\partial n)\cos\beta)]$$

$$+ \sin\beta[(h_y/2)(u_\tau\cos\beta + u_n\sin\beta)((\partial\rho^*/\partial\tau)\cos\beta + (\partial\rho^*/\partial n)\sin\beta)$$

$$+ \delta t((\partial p^*/\partial\tau)\cos\beta + (\partial p^*/\partial n)\sin\beta)]\} - (\delta t/2)\,\partial^2\rho^*/\partial t^2. \qquad (3.1.25)$$

Let us show that the right-hand side of (3.1.25) depends explicitly on β. For this purpose let us collect the terms of similar structures on the right-hand side of (3.1.25). As a result equation (3.1.25) may be written in the form

$$\partial\rho^*/\partial t + \partial\rho^* u_\tau/\partial\tau + \partial\rho^* u_n/\partial n$$

$$= (h_x/2)\{\partial/\partial\tau[a_1(\beta,\lambda)u_\tau\,\partial\rho^*/\partial\tau + a_2(\beta,\lambda)(u_\tau\,\partial\rho^*\,\partial n + u_n\,\partial\rho^*/\partial\tau)$$

$$+ a_3(\beta,\lambda)u_n\,\partial\rho^*/\partial n + \delta t\,\partial p^*/\partial\tau] + \partial/\partial n[a_2(\beta,\lambda)u_\tau\,\partial\rho^*/\partial\tau + a_3(\beta,\lambda)$$

$$\times (u_\tau\,\partial\rho^*/\partial n + u_n\,\partial\rho^*/\partial\tau) + a_4(\beta,\lambda)u_n\,\partial\rho^*/\partial n + \delta t\,\partial p^*/\partial n]\}$$

$$- (\delta t/2)\,\partial^2\rho^*/\partial t^2, \qquad (3.1.26)$$

where

$$\lambda = h_y/h_x; \qquad a_1(\beta,\lambda) = \lambda\cos^3\beta - \sin^3\beta;$$

$$a_2(\beta,\lambda) = \sin^2\beta\cos\beta + \lambda\cos^2\beta\sin\beta;$$

$$a_3(\beta,\lambda) = \lambda\sin^2\beta\cos\beta - \cos^2\beta\sin\beta;$$

$$a_4(\beta,\lambda) = \cos^3\beta + \lambda\sin^3\beta.$$

It is easy to show that the right-hand side of (3.1.26) depends explicitly on β. It is sufficient to note for this that $da_j(\beta,\lambda)/d\beta \not\equiv 0$ at $\beta \in [0, 2\pi]$, $\lambda > 0$. This means that the f.d.a. of a difference approximation of the continuity equation by the scheme (3.1.11) is not invariant under the rotation transformation. We shall derive similar conclusion when writing the f.d.a. of the momentum and energy equations in the variables n, τ, t. These formulas are not presented here because of their bulky form. Previously, the noninvariance of the particle-in-cell method approximation viscosity under the rotation transformation was indicated in [3.13]. This noninvariance may lead to large errors while calculating the spherical converging flows [3.13], [3.3]. Note that the above method of the f.d.a. analysis, with respect to the invariance under the rotation transformation, is slightly simpler than the more general approach presented in [3.1], and applied in [3.24] when constructing the invariant schemes with the splitting up of the Harlow type.

Consider the cylindrical shock problem. Initial data for $t = 0$ were set in [3.10] as follows: the computational domain in the plane (x, y) (having the form of a rectangle) was subdivided into two subdomains I and II. The subdomain I was a circle sector of a finite radius equal to several mesh sizes in the direction of the x-axis, with the top in the coordinate origin and with a top angle of $90°$; the subdomain II comprised the remainder of the computational domain. The initial velocity of a gas with the equation of state (1.8) was set to zero in the computational domain, and the flow at $t > 0$ arose as a result of the pressure and density jump on both sides of a discontinuity. At the boundaries $x = 0$ and $y = 0$ the rigid wall conditions were set, and the conditions of the flux continuity (matter outflow boundaries) were set at two other boundaries of a rectangular computational domain.

Let us take, at fixed $t > 0$, a point $F_0(x_0, y_0)$ at the front of a cylindrical shock wave lying at a finite distance from the coordinate origin O. Take the coordinate system (n, τ) with the origin at the point F_0 and with the axis $F_0 n$ normal to the front at the point F_0. The axis $F_0 \tau$ is directed normal to the axis $F_0 n$ so that the coordinate system (n, τ) is the right one. Denote by β $(0 \le \beta \le \pi/2)$ an inclination angle of the axis $F_0 n$ with the axis Ox. Suppose that the distance of the point F_0 from the origin "O" is sufficiently large, and that the main contribution to the smearing of the shock wave front in the direction of a normal to the shock surface is introduced only by those terms on the right-hand sides of the f.d.a. equations of the type (3.1.26), which contain the normal velocity component u_n and the derivatives $\partial^2 w^*/\partial n^2$. Neglecting, in the right-hand sides of the f.d.a equations, the terms containing u_τ, $\partial w^*/\partial \tau$ and the derivatives of these terms with respect to n and τ, we obtain from formula (3.1.26), as well as from the f.d.a. of the momentum and energy equations, the f.d.a. of a PIC method integral step scheme in the coordinate system (n, t)

$$\partial \rho^*/\partial t + \partial \rho^* u_n/\partial n = (\partial/\partial n)[a(\beta, h_x, h_y) u_n \, \partial \rho^*/\partial n + \delta t \, \partial p^*/\partial n]$$

$$- (\delta t/2) \, \partial^2 \rho^*/\partial t^2;$$

$$\partial \rho^* u_n/\partial t + \partial(p^* + \rho^* u_n^2)/\partial n = (\partial/\partial n)[a(\beta, h_x, h_y) u_n \, \partial \rho^* u_n/\partial n$$

$$+ 2\delta t \, u_n \, \partial p^*/\partial n] - (\delta t/2) \, \partial^2 \rho^* u_n/\partial t^2;$$

$$(3.1.27)$$

$$\partial \rho^* E^*/\partial t + \partial(\rho^* E^* u_n + p^* u_n)/\partial n$$

$$= (\partial/\partial n)[a(\beta, h_x, h_y) u_n \, \partial \rho^* E^*/\partial n + \delta t((E^* + p^*/(2\rho)) \, \partial p^*/\partial n$$

$$+ u_n \, \partial p^* u_n/\partial n \, b(\beta))] - (\delta t/2) \, \partial^2 \rho^* E^*/\partial t^2,$$

where

$$a(\beta, h_x, h_y) = (h_x/2) \cos^3 \beta + (h_y/2) \sin^3 \beta; \qquad (3.1.28)$$

$$b(\beta) = \cos^4 \beta + \sin^4 \beta. \qquad (3.1.29)$$

Following Section 1.4 we shall search for the particular solutions of the system (3.1.27) of the form $w^* = w^*(\xi)$ where $\xi = n - Dt$, D is the shock wave speed which is assumed to be locally constant. Then the system (3.1.27) is integrated once and converts to a system of ordinary differential equations which may be transformed to the form

$$\delta t \, dp/d\xi + (au - \delta t D^2/2) \, d\rho/d\xi = \rho(u - D) - A_1; \qquad (3.1.30)$$

$$(au\rho - \delta t D^2 \rho/2) \, du/d\xi + \delta tu \, dp/d\xi = p - A_2 + uA_1; \qquad (3.1.31)$$

$$au\rho \, dE/d\xi + \delta tp/(2\rho) \, d\rho/d\xi + \delta tbu \, dpu/d\xi - (\delta t/2)D^2\rho \, dE/d\xi$$
$$= pu - A_3 + A_1 E. \qquad (3.1.32)$$

Since we use the equation of state (1.8), equation (3.1.32) may be rewritten as

$$(a\rho u^2 + \delta tbup - \delta t D^2 \rho u/2) \, du/d\xi$$
$$+ [au/(\gamma - 1) + \delta t(0.5p/\rho + bu^2) - (\delta t/2)D^2/(\gamma - 1)] \, dp/d\xi$$
$$+ [(\delta t/2)D^2 p/(\rho(\gamma - 1)) - au p/(\rho(\gamma - 1))] \, d\rho/d\xi$$
$$= pu - A_3 + A_1 E(u, p, \rho). \qquad (3.1.33)$$

In equations (3.1.30)–(3.1.33) the upper index $*$ by the quantities p, ρ, E is omitted for brevity of notation, as well as the lower index n by u. The quantities A_1, A_2, A_3 are the integration constants.

Let us denote the values determining the state behind and before the shock front by the lower indices "1" and "2", respectively, and by the upper index "0". Let $\mathbf{f} = (u, p, \rho)^T$ and let $X(u_1^0, u_2^0, \mathbf{f}, h_x, h_y, \beta)$ be the width after Prandtl of a smeared shock wave zone in the velocity $u(\xi)$ profile, that is,

$$X = |u_1^0 - u_2^0| \Big/ \max_\xi |du/d\xi|. \qquad (3.1.34)$$

To estimate in some way the width X, we solve the equation system (3.1.30), (3.1.31), (3.1.33) with respect to the derivatives $du/d\xi$, $dp/d\xi$, $d\rho/d\xi$. The determinant of this system has the form

$$d = [au\rho - (\delta t/2)D^2\rho](b_{01}a^2 + b_{02}a\delta t + b_{03}\delta t^2), \qquad (3.1.35)$$

where

$$b_{01} = u^2/(\gamma - 1);$$

$$b_{02} = u\{b(\beta)p/(\rho(\gamma - 1)) + (3/2)[b(\beta) - 1]u^2 - D^2/(\gamma - 1)\};$$

$$b_{03} = D^4/[4(\gamma - 1)] - b(\beta)[3D^2u^2/4 + D^2p/(2\rho(\gamma - 1)) + u^2p/\rho]$$
$$+ 3D^2u^2/4. \qquad (3.1.36)$$

In the cylindrical shock problem under consideration $u > 0$ in the domain of the most significant change of the solution in the zone of smearing of a shock wave. It follows from formula (3.1.35) and from the expressions (3.1.36) for b_{0k} that $d \neq 0$ in the domain of large gradients at sufficiently small δt. Solving the system (3.1.30), (3.1.31), (3.1.33) with respect to $du/d\xi$, $dp/d\xi$, $d\rho/d\xi$, we obtain for $du/d\xi$ the expression

$$du/d\xi = (1/d)(b_{11}a^2 + b_{12}a\delta t + b_{13}\delta t^2), \qquad (3.1.37)$$

where the determinant d is computed by (3.1.35), (3.1.36),

$$b_{11} = (\gamma - 1)^{-1}u^2(p - A_2 + uA_1);$$

$$b_{12} = -\{u^2p/[\rho(\gamma - 1)]\}[\rho(u - D) - A_1]$$
$$-u^2(pu - A_3 + A_1E) + u(p - A_2 + uA_1)$$
$$\times [(3b(\beta) - 1)u^2/2 + (b(\beta) - 1)p/(\rho(\gamma - 1)) - 3D^2/(2(\gamma - 1))]; \qquad (3.1.38)$$

$$b_{13} = (D^2/2)up[\rho(\gamma - 1)]^{-1}[\rho(u - D) - A_1]$$
$$+ (uD^2/2)(pu - A_3 + A_1E) - (D^2/2)(p - A_2 + uA_1)$$
$$\times \{(3b(\beta) - 1)u^2/2 + [b(\beta) - 1]p/[\rho(\gamma - 1)] - D^2/(2(\gamma - 1))\}.$$

If the quantities u, p, ρ are bounded in the zone of shock wave smearing, then we obtain at sufficiently small δt with regard to (3.1.35)–(3.1.38)

$$X(u_1^0, u_2^0, \mathbf{f}, h_x, h_y, \beta) = |u_1^0 - u_2^0| \min_\xi (|au^3\rho/((\gamma - 1)b_{11})|) + O(\delta t). \quad (3.1.39)$$

Taking into account (3.1.39) let us introduce the quantity

$$X_0(u_1^0, u_2^0, \mathbf{f}, h_x, h_y, \beta) = |u_1^0 - u_2^0| \min_\xi |au^3\rho/((\gamma - 1)b_{11})|. \quad (3.1.40)$$

It follows from formula (3.1.39) that at sufficiently small δt the width X will differ little from the value X_0 determined by (3.1.40). Let us fix the point ξ and compute the relation

$$\frac{X_0(u_1^0, u_2^0, \mathbf{f}, h_x, h_y, \beta)}{X_0(u_1^0, u_2^0, \mathbf{f}, h_x, h_y, 0)} = \tilde{a}(\lambda, \beta), \qquad (3.1.41)$$

where

$$\lambda = h_y/h_x, \qquad \tilde{a}(\lambda, \beta) = \cos^3 \beta + \lambda \sin^3 \beta. \qquad (3.1.42)$$

It is easy to find the minimum of the function $\tilde{a}(\lambda, \beta)$ at fixed λ in the interval $0 \le \beta \le \pi/2$. This minimum is achieved at $\beta = \arctan(1/\lambda)$, and

$$\min_{0 \le \beta \le \pi/2} \tilde{a}(\lambda, \beta) = \lambda/\sqrt{\lambda^2 + 1} = h_y/\sqrt{h_x^2 + h_y^2}. \qquad (3.1.43)$$

By analogy with formula (3.1.34) we can define the widths after Prandtl

$$X(p_1^0, p_2^0, \mathbf{f}, h_x, h_y, \beta), \qquad X(\rho_1^0, \rho_2^0, \mathbf{f}, h_x, h_y, \beta),$$

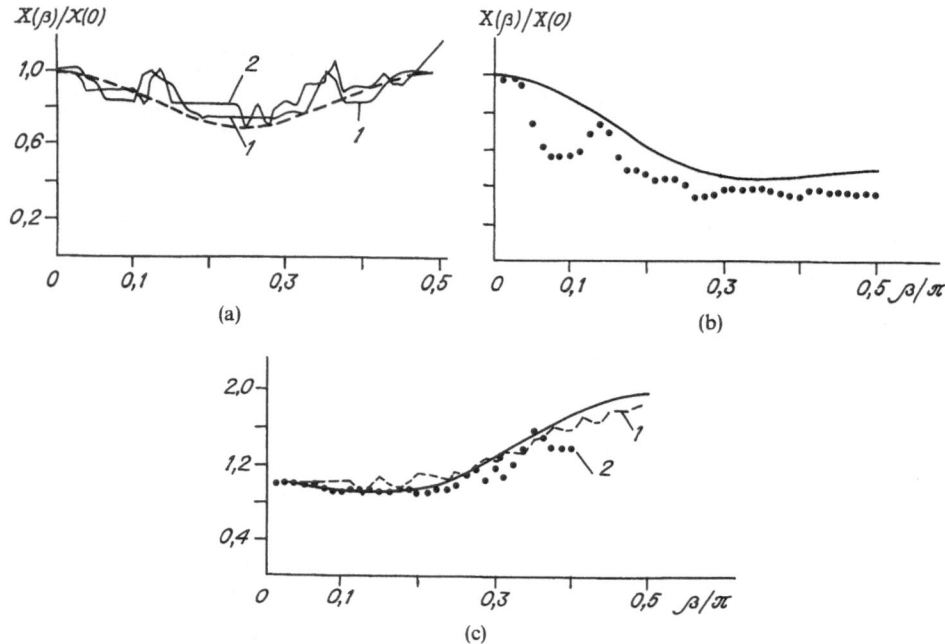

Figure 3.1. The graph of the relation $X(\beta)/X(0)$: (a) $\lambda = 1$; (b) 0.5; (c) 2.

of the zones of smearing of a shock wave in the profiles of functions $p(\xi)$, $\rho(\xi)$. Introducing, analogously to the foregoing, the quantities

$$X_0(p_1^0, p_2^0, \mathbf{f}, h_x, h_y, \beta), \quad \text{and} \quad X_0(\rho_1^0, \rho_2^0, \mathbf{f}, h_x, h_y, \beta),$$

we can find for these quantities the dependence on the angle β which coincides with that presented by formulas (3.1.41), (3.1.42). The graph of the function $\tilde{a}(\lambda, \beta)$ for the values of λ being equal to 1, 0.5, and 2 is plotted by a dashed line in Figure 3.1(a) and by solid lines in Figure 3.1(b), (c). Solid lines in Figure 3.1(a) and dotted and dashed lines in Figure 3.1(b), (c) represent the graphs of the relation $X(\beta)/X(0)$ obtained in the computations of a cylindrical Riemann problem by the particle-in-cell method [3.10]. The figures "1" and "2" in Figure 3.1(a), (c) refer to the numerical solution curves for the moments of time $t = 20\delta t$ and $t = 40\delta t$, respectively. Statistical fluctuations of the density inherent in the Harlow method result in the oscillations of the width $X(\beta)$; nevertheless, it is seen from Figure 3.1(a) that the quantity $X(\beta)$ achieves its minimum at $\beta \cong \pi/4$. To reduce the effects due to the discrete behavior of density we have also carried out the computations by the modified Harlow method, in which the pressure was computed by the equation of state (1.8) with a density ρ obtained by a recalculation by a difference scheme approximating the continuity equation [3.13], [3.25]. Corresponding computational results for $h_y/h_x = 2$ are presented in Figure 3.1(c). In comparison with Figure 3.1(b) the width $X(\beta)$ behavior in Figure 3.1(c) is more monotone owing to the density recalculation. Thus, the results of the numerical computations agree completely with conclusions drawn from an analysis of the f.d.a. of an asymptotic difference scheme of the integral step scheme of the Harlow particle-in-cell method. Note that the results of the above consideration are also applicable to the FLIC method [3.4], since the difference scheme used above in the analysis coincides with the scheme of the method [3.4].

3.2. Localization of the Smeared Shock Wave Center in the Case of a Straight Front

3.2.1. Analysis on the Basis of Progressive Wave-Type Solutions of the First Differential Approximation Equations

Consider in the (x, y)-plane a stationary shock wave with an infinite straight front propagating under a constant nonzero angle to the lines of the Eulerian rectangular computational grid. Consider a class of first-order difference schemes whose f.d.a. P-form is representable in the form [3.1]

$$\partial w/\partial t + \partial F(w)/\partial x + \partial G(w)/\partial y$$

$$= (\partial/\partial x)(C_{11}\, \partial w/\partial x + C_{12}\, \partial w/\partial y) + (\partial/\partial y)(C_{21}\, \partial w/\partial x + C_{22}\, \partial w/\partial y), \quad (3.2.1)$$

where the column vectors w, F, G are determined by (1.30b), and C_{ij}, $i, j = 1, 2$, are the square matrices of dimension 4×4, such that

$$C_{ij} = C_{ij}(w, h_x, h_y, \tau), \qquad i, j = 1, 2. \tag{3.2.2}$$

Let $\mathbf{D}$ be the velocity vector of a shock wave, and D_1, D_2 are the projections of this vector on the x-and y-axes, respectively. By analogy with [3.26] we shall consider the progressive wave-type solution of the f.d.a. (3.2.1), that is, the solutions depending only on two self-similar variables

$$\xi = x - D_1 t, \qquad \eta = y - D_2 t. \tag{3.2.3}$$

Lemma 3.2. *If the matrices C_{ij} in (3.2.1) are representable in the form*

$$C_{ij} = h_x \tilde{C}_{ij}(w, \lambda, \kappa), \qquad i, j = 1, 2, \tag{3.2.4}$$

where

$$\lambda = h_y/h_x, \qquad \kappa = \tau/h_x, \tag{3.2.5}$$

then the equations $\lambda = $ const, $\kappa = $ const, represent necessary conditions for the existence of a smeared shock wave center in the profile of the solution $w(\xi, \eta)$ along any beam orthogonal to the shock front.

PROOF. Let us rewrite the system (3.2.1) in a more compact form

$$\partial w/\partial t + \partial F/\partial x + \partial G/\partial y = R, \tag{3.2.6}$$

where R is a column vector,

$$R = (R_\rho, R_u, R_v, R_E)^{\mathrm{T}},$$

the superscript T denotes the transposition operation. Introduce the notations

$$U = u - D_1, \qquad V = v - D_2. \tag{3.2.7}$$

Substituting $w = w(\xi, \eta)$ into (3.2.6) we obtain, with a view to notations (3.2.7), the f.d.a. of the continuity equation as

$$\partial \rho\, U/\partial \xi + \partial \rho\, V/\partial \eta = R_\rho. \tag{3.2.8}$$

Consider now the second equation of the system (3.2.6). Multiply both sides of equation (3.2.8) by D_1 and subtract from both sides of the second equation in the system (3.2.6). As a result we obtain

$$\partial(p + \rho U^2)/\partial \xi + \partial \rho\, UV/\partial \eta = R_u - D_1 R_\rho. \tag{3.2.9}$$

Similarly,

$$\partial \rho\, UV/\partial \xi + \partial(p + \rho V^2)/\partial \eta = R_v - D_2 R_\rho. \tag{3.2.10}$$

Multiply both sides of equation (3.2.8) by $0.5\,(D_1^2 + D_2^2)$ and subtract from both sides of the fourth equation in the system (3.2.6). Then multiply both sides of equation (3.2.9) by D_1 and subtract from both sides of the fourth equation in the system (3.2.6). Multiply both sides of (3.2.10) by D_2 and subtract from both sides of the f.d.a. of the energy equation in (3.2.6). As a result we get the equation

$$(\partial/\partial \xi)(pU + \rho U\mathscr{E}) + (\partial/\partial \eta)(pV + \rho V\mathscr{E})$$

$$= R_E + (1/2)(D_1^2 + D_2^2)R_\rho - D_1 R_u - D_2 R_v, \tag{3.2.11}$$

where $\mathscr{E} = \varepsilon + 0.5(U^2 + V^2)$. The right-hand sides of equations (3.2.8)–(3.2.11) represent linear combinations (with constant coefficients) of the quantities R_ρ, R_u, R_v, R_E. Therefore, taking into account (3.2.4), we can rewrite the system (3.2.6) in the form

$$\partial F(W)/\partial \xi + \partial G(W)/\partial \eta = (\partial/\partial \xi)[Q_1(W, h_x\, \partial W/\partial \xi, \lambda h_x\, \partial W/\partial \eta, \kappa, \lambda, D_1, D_2)]$$

$$+ (\partial/\partial \eta)[Q_2(W, h_x\, \partial W/\partial \xi, \lambda h_x\, \partial W/\partial \eta, \kappa, \lambda, D_1, D_2)], \tag{3.2.12}$$

where $W = (\rho, \rho U, \rho V, \rho \mathscr{E})^{\mathrm{T}}$. Let us make, in the system (3.2.12), a change of variables

$$\xi' = (\xi - \xi_0)/h_x, \qquad \eta' = (\eta - \eta_0)/h_y,$$

where ξ_0, η_0 are arbitrary constants. Then the system (3.2.12) takes the form

$$\partial F(W)/\partial \xi' + \partial G(W)/\partial \eta' = (\partial/\partial \xi')[Q_1(W, \partial W/\partial \xi', \lambda\, \partial W/\partial \eta', \kappa, \lambda, D_1, D_2)]$$

$$+ (\partial/\partial \eta')[Q_2(W, \partial W/\partial \xi', \lambda\, \partial W/\partial \eta', \kappa, \lambda, D_1, D_2)]. \tag{3.2.13}$$

It follows from (3.2.13) that the solution W depends on $\xi', \eta', \lambda, \kappa, D_1, D_2$. Repeating the same argument as in Section 2.4 we arrive at the conclusion that the equations $\lambda = \mathrm{const}$, $\kappa = \mathrm{const}$, are necessary for the existence of a smeared shock wave center in the solution $w(\xi, \eta)$. ☐

Corollary 1. *The assertion formulated above in the form of Lemma 3.2 is also valid for difference schemes whose f.d.a. H-form is representable in the form*

$$\partial w/\partial t + \partial F(w)/\partial x + \partial G(w)/\partial y$$
$$= -(\tau/2)\, \partial^2 w/\partial t^2 + (\partial/\partial x)(C_{11}\, \partial w/\partial x + C_{12}\, \partial w/\partial y)$$
$$+ (\partial/\partial y)(C_{21}\, \partial w/\partial x + C_{22}\, \partial w/\partial y), \qquad (3.2.14)$$

where the matrices C_{ij} possess the property (3.2.4).

Indeed, the operator $\partial^2/\partial t^2$ in (3.2.14), acting on the progressive wave-type solutions $w = w(\xi, \eta)$, is replaced by a linear operator $\partial^2/\partial t^2 = (D_1\, \partial/\partial\xi + D_2\, \partial/\partial\eta)^2$.

Corollary 2. *Employing formulas (3.2.8)–(3.2.11) and performing the same calculations as in the case of obtaining the Bernoulli integral, we can derive the equation*

$$U(\partial/\partial\xi)[\gamma p/(\rho(\gamma - 1)) + (U^2 + V^2)/2]$$
$$+ V(\partial/\partial\eta)[\gamma p/(\rho(\gamma - 1)) + (U^2 + V^2)/2]$$
$$= (1/\rho)[\bar{R}_E - R_\rho(\gamma p/(\rho(\gamma - 1)) + (U^2 + V^2)/2)], \quad (3.2.15)$$

where γ is the constant in the equation of state (1.8),

$$\bar{R}_E = R_E + (1/2)(D_1^2 + D_2^2)R_\rho - D_1 R_u - D_2 R_v.$$

Formula (3.2.15) can be used as follows. Let the equation

$$U(\partial/\partial\xi)[\gamma p/(\rho(\gamma - 1))$$
$$+ (U^2 + V^2)/2] + V(\partial/\partial\eta)[\gamma p/(\rho(\gamma - 1)) + (U^2 + V^2)/2] = 0 \quad (3.2.16)$$

take place in some gas dynamic flow. Consider the stream line $d\eta/d\xi = V/U$. If the property

$$\gamma p/[\rho(\gamma - 1)] + (U^2 + V^2)/2 = \text{const} \qquad (3.2.17)$$

takes place in some section $\xi = \xi_*$, then this property by virtue of (3.2.16) also takes place in the region $\xi \geq \xi_*$, and the relationship (3.2.17) is called the Bernoulli integral. Then in the process of constructing difference schemes for the system (1.30) it is reasonable to require that the f.d.a. (3.2.6) possesses the property

$$\bar{R}_E - R_\rho[\gamma p/(\rho(\gamma - 1)) + (U^2 + V^2)/2] = 0. \qquad (3.2.18)$$

EXAMPLE. Consider the f.d.a. *H*-form of the Lax scheme [3.2], [3.27]

$$\partial w/\partial t + \partial F/\partial x + \partial G/\partial y$$
$$= -(\tau/2)\, \partial^2 w/\partial t^2 + [h_x^2/(4\tau)]\, \partial^2 w/\partial x^2 + [h_y^2/(4\tau)]\, \partial^2 w/\partial y^2. \quad (3.2.19)$$

In this case

$$C_{12} = C_{21} = 0; \qquad C_{11} = (h_x^2/(4\tau))I; \qquad C_{22} = (h_y^2/(4\tau))I; \quad (3.2.20)$$

in equation (3.2.14), where I is the identity matrix. Employing the notation (3.2.5) we rewrite formulas (3.2.20) as follows

$$C_{12} = C_{21} = 0; \qquad C_{11} = (h_x/(4\kappa))I; \qquad C_{22} = h_x(\lambda^2/(4\kappa))I.$$

It can be shown that in the case of the Lax scheme equation (3.2.18) does not take place at any relations between τ, h_x, h_y, D_1, D_2.

3.2.2. Application of the Scheme Viscosity Norm in the Algorithms of Shock Wave Analyzers

A direct application of the finite-difference shock wave center notion for the purpose of shock wave localization in two-dimensional flows requires the execution of two runs of the same problem, which is generally not economical. It has been shown in Section 2.3 that in the case when the matrix $B(w, h, \tau)$ of the scheme viscosity coefficients in the f.d.a. P-form (1.59) satisfies a number of conditions, then the determination of the points of maximum of the quantity (2.4.19) can be used in the differential analyzer algorithms. The proof of an analogue of this property of scheme viscosity in the two-dimensional case is difficult. However, it is possible to construct in the two-dimensional case the differential analyzers which, in the particular cases of propagation of a part of the shock front in the direction of the x- or y-axis, coincide with the one-dimensional case expressed by formula (2.4.19). Algorithms of this kind were for the first time tested in [3.9]. Let us enumerate some of these differential analyzers. In [3.9] a search for the shock front points was carried out along a certain array of the beams intersecting the computational domain at various angles to the x- and y-axes, in such a way that these beams were approximately orthogonal to the rising shock waves. Available information on the qualitative behavior of shock propagation in the two-dimensional problem under consideration was used in [3.9]. Let Q be some norm of the scheme viscosity at a chosen point (x, y). The values of Q on the beams were computed by interpolation. Let λ be the number of the beam under consideration. Mention some of the forms of the quantity Q considered in [3.9].

Algorithm 1. Let

$$
\begin{aligned}
Q_1 &= |C_{11}(\partial w/\partial x) + C_{12}(\partial w/\partial y)|, & \operatorname{div} \mathbf{u} < 0; \\
Q_2 &= |C_{21}(\partial w/\partial x) + C_{22}(\partial w/\partial y)|, & \operatorname{div} \mathbf{u} < 0.
\end{aligned}
\qquad (3.2.21)
$$

The notation (3.2.21) means that the quantities Q_1, Q_2 were computed only at those points where $\operatorname{div} \mathbf{u} < 0$.

Let $(x_\lambda, y_\lambda)_1$, $(x_\lambda, y_\lambda)_2$ be the coordinates of the maximum points of the quantities Q_1 and Q_2, respectively, at a beam with number λ. Then the sought for shock front point on this beam was computed by the formula

$$(x_\lambda, y_\lambda) = 0.5[(x_\lambda, y_\lambda)_1 + (x_\lambda, y_\lambda)_2]. \tag{3.2.22}$$

The nonnegative value $|C_{j1}\, \partial w/\partial x + C_{j2}\, \partial w/\partial y|$ in (3.2.21) represents a conventional Euclidean norm of the vector $C_{j1}\, \partial w/\partial x + C_{j2}\, \partial w/\partial y$ entering the f.d.a. P-form (3.2.1).

Algorithm 2.

$$Q = Q_{u,v} * Q_{u,v} = \sum_{i,j=1}^{2} q_{ij}^2; \qquad \text{div } \mathbf{u} < 0; \tag{3.2.23}$$

that is, in the case considered Q represents a convolution of the matrix $Q_{u,v}$ with itself, where $Q_{u,v}$ was computed on the basis of (3.2.1) by the derivation from (3.2.1) of an f.d.a. P-form of the momentum equations transformed to a nondivergence form

$$\rho\, d\mathbf{u}/dt + \nabla p = \text{div } Q_{u,v},$$
$$d/dt = \partial/\partial t + u\, \partial/\partial x + v\, \partial/\partial y, \tag{3.2.24}$$

$$Q_{u,v} = \begin{pmatrix} q_{11} & q_{12} \\ q_{21} & q_{22} \end{pmatrix}. \tag{3.2.25}$$

As an example in [3.9] the expressions for the functions q_{ij}, $i, j = 1, 2$, were written (see formula (3.2.25)) for the case of the HARLOW particle-in-cell method [3.3]. These formulas can easily be obtained from the f.d.a. H-form (3.1.23) by turning to the P-form and executing a number of algebraic transformations of the f.d.a. P-form equations. Note that the expressions for the scheme viscosity coefficients in a nondivergence f.d.a. (3.2.24) have a simpler structure than in the divergence system (3.2.1). An analysis of the behavior of the maximum of the quantity (3.2.23), as a function of an inclination angle β to the x-axis, was carried out in [3.9] under the assumption of a one-dimensional flow in a shock wave with a straight front having a nonzero inclination angle to the y-axis. This analysis, carried out for the case of the Harlow method scheme and for the case of square cells of a mesh in the (x, y)-plane, showed that the value $Q_{u,v} * Q_{u,v}$ achieved its maximum at $\beta = \pi/4$.

Algorithm 3. Direct application of the f.d.a. (3.2.1) leads to rather bulky formulas for the computation of Q. Therefore, it was of interest to elucidate the possibility of using simpler expressions for Q which are not related to the use of f.d.a. As was shown in Section 2.2, the point of $\max|\partial u/\partial x|$, where u is the gas speed along a normal to the shock front, coincides with the center of a smeared shock wave center at a certain structure of the artificial viscosity. In the two-dimensional case, as was shown in Section 3.1, the formula

$$\partial u_n/\partial n = \text{div } \mathbf{u} \tag{3.2.26}$$

takes place at $u_\tau = 0$ where u_τ is the velocity component tangential to the shock front surface. The following algorithm of the differential analyzer based on (3.2.26) was considered in [3.9]:

$$Q = |\text{div } \mathbf{u}|, \qquad \text{div } \mathbf{u} < 0. \tag{3.2.27}$$

3.3. Shock Localization by Moving Grids

The algorithms for the localization of shock waves by the shock-capturing computing results considered in Section 3.2 were based on the assumption of a fixed rectangular computing mesh in the plane of the spatial coordinates x, y. The existence of the problem of shock localization in the two-dimensional case is related to the fact that the width of the zone of "smearing" of a discontinuity on a uniform rectangular grid can be significant, especially in cases when one has to use a relatively small number of mesh cells in the computation. At the same time, it is possible to achieve an actual reduction of the width of the smeared shock zones on a moderate number of nodes, by applying moving grids which automatically adapt to the solution in such a way that the grid lines concentrate in the domains of large solution gradients [3.28]–[3.35]. In [3.36] the local refinement of mesh cells in the direction of the largest gradients of the flow parameters was proposed as a form of extraction of flow singularities.

We can distinguish between two approaches to the development of moving solution-adaptive grids, depending on the manner of coupling the mesh computation algorithm with the algorithm of solving a system of differential equations governing the fluid flow [3.37]. In the first of these approaches the solution of the original differential equations is determined simultaneously with the necessary distribution of the computing mesh nodes. In the second case the computations are carried out sequentially. At first a solution is computed on a given mesh, and then the new position of the mesh is computed on the basis of this solution. We have presented in Section 2.5 an algorithm of this type. It should be noted that there is a substantial difference in the realization of the above approaches. In the first of these approaches one has to use the gas dynamics equations written in general moving coordinates. To complete these equations one has to add, in the general case, two more equations for the computation of the coordinates $x_{ij}(t)$, $y_{ij}(t)$ of the nodes of the moving grid. The second of the above approaches makes use of a system of equations describing the flow in the original Cartesian system of the Eulerian coordinates x, y, for example, the system (1.30). However, an interpolation for the computation of the solution components on an "old" grid should be used in the second approach. In Section 2.5 we have used cubic splines for this purpose. In the general case the interpolation of the gas dynamic parameters from one grid to another should be carried out with account being taken of the laws of mass, momentum, and energy conservation. This problem of the conservative transfer of information from one grid to another is also a classic problem in the Lagrangian computational fluid dynamics [3.38]. One of the basic requirements for the algorithms of conservative mesh rezoning is efficiency. For example, the incorporation of such an algorithm into a computer code for solving a hydrodynamic problem should generally impose no additional restriction on a time step. Let us briefly enumerate the algorithms [3.38]–[3.43] of conservative mesh rezoning. Such a rezoning can be carried out relatively simply in the case when the density and other components of the vector w in (1.30b) are constant within the boundaries of a quadrilateral computing cell of a two-dimensional mesh. Then the rezoning reduces to a computation of contributions of the volumes Δv_{kl} of old cells to each new (i, j) cell. After the values Δv_{kl} are computed, the material density in the new cell is computed by the formula

$$\bar{\rho}_{ij} = \sum_{k,l} \Delta v_{kl} \rho_{kl} / \bar{v}_{ij}, \qquad (3.3.1)$$

where $\bar{v}_{ij}$ is the volume of the new cell and ρ_{kl} is the density fluid in the old cell. It was proposed in [3.39] to carry out a partial rezoning on the basis of the formulas of type (3.3.1), only in the domains of large distortions of a computing mesh. Such an approach enabled one to reduce the computing time by a factor of 9 while solving certain two-dimensional problems. In the ALE method [3.40] the rezoning imposes an additional restriction on a time step. In addition, it has been noted in [3.38] that the method of [3.40] introduces a strong numerical diffusion. A method of conservative rezoning, based on a reduction of the volume integrals to surface integrals by means of the divegence theorem (the Green formula), was proposed in [3.38]. A variant of integration by the Monte Carlo method (that is, the method of counting particles) was proposed in [3.41] for the conservative rezoning. This method alleviates the difficulties with logic by means of some loss in accuracy, and by substantial additional computational expenses and memory. A similar method of conservative rezoning was proposed in [3.42] where the discrete particles were packed into the cells of a distorted mesh. Each particle then transferred (into the cells of a new mesh) its mass, momentum, and energy— similar to the HARLOW method [3.3]. The idea of the local organization of computation, similar to that of [3.39] and which was realized in [3.43], has resulted in an economical algorithm of conservative mesh rezoning.

3.3.1. Equations of Inviscid Gas Flow in Moving Coordinates

Let us consider in more detail the first approach to the construction of adaptive grids when the coupled equations of gas flow and mesh coordinates are solved. The derivation of these equations may be found in [3.30], [3.35]. In the case of a time-dependent inviscid gas flow these equations (without equations for the computation of mesh node coordinates) have the form [3.35]

$$(\partial/\partial t)(\rho\Delta) + (\partial/\partial q^\alpha)[\rho\Delta(v^\alpha - \omega^\alpha)] = 0,$$

$$(\partial/\partial t)[\rho\Delta(v^i - \omega^i)] + (\partial/\partial q^\alpha)[\rho\Delta(v^i - \omega^i)(v^\alpha - \omega^\alpha) + \Delta \cdot p^{i\alpha}]$$

$$+ \Delta\Gamma^i_{\alpha\beta}[p^{\alpha\beta} + \rho(v^\alpha - \omega^\alpha)(v^\beta - \omega^\beta)] = 0, \quad (3.3.2)$$

$$(\partial/\partial t)(\Delta\rho E) + (\partial/\partial q^\alpha)[\rho E\Delta(v^\alpha - \omega^\alpha) + p^{k\alpha}g_{k\beta}v^\beta] = 0,$$

$$\Delta = \det\|\partial x^\alpha/\partial q^\beta\|, \qquad \omega^\alpha = \partial x^\alpha/\partial t, \qquad \alpha, \beta = 1, 2. \quad (3.3.3)$$

In the system (3.3.2) ρ is the density, q^α are curvilinear coordinates, and the summation is carried out over repeating indices; $p^{i\alpha} = pg^{i\alpha}$ and v^α is the projection of the velocity vector onto the x^α-axis,

$$E = \varepsilon + (1/2)[(v^1)^2 + (v^2)^2], \qquad g^{\alpha\beta} = g^{\beta\alpha} = \begin{cases} 0, & \alpha \neq \beta, \\ 1, & \alpha = \beta. \end{cases}$$

The Christoffel symbols $\Gamma^i_{\alpha\beta}$ are computed by the formulas

$$\Gamma^i_{\alpha\beta} = A^i_s \, \partial a^s_\beta/\partial q^\alpha \qquad (\alpha, \beta, i, s = 1, 2),$$

where

$$a^s_\beta = \partial x^s/\partial q^\beta, \qquad A^i_s = \partial q^i/\partial x^s, \quad i, s = 1, 2, \qquad a^\alpha_\beta A^\beta_\gamma = \delta^\alpha_\gamma.$$

To complete the system (3.3.2) we need an equation of state of the form (1.5) as well as the equation of mesh motion for the determination of the vector $\mathbf{w} = \{w^i\}$, being a contravariant vector of the velocity of coordinate mesh. The vector $\mathbf{w}$ may be considered as a vector of mesh control. Unlike the widespread geometric approach in which the mesh is determined by geometry of the computational domain, but not by the solution itself, in the dynamic approach under consideration a problem is solved on the construction of a continual differential mapping $\mathbf{x} \rightarrow \mathbf{y}$ which corresponds to the state of the overall flow. After that a discretization of this mapping is carried out and an efficient difference scheme adapting to the flow is constructed on the basis of a discrete mesh.

3.3.2. Equations of Grid Motion

Consider the question of grid control by solution gradients. The simplest technique for computing the coordinates $x(q_i)$ of the mesh nodes in the one-dimensional case, with account being taken of the gradients of the sought for function $f(x, t)$, is based on the use of the relationship

$$[x(q_{i+1}) - x(q_i)](|\partial f/\partial x| + \delta) = \text{const},$$

where δ is a user-specified positive constant. For some two-dimensional gas dynamics problems, for example, in the case of a gas flow around convex bodies, it is sufficient to restrict oneself to the grid condensation in one coordinate direction (along a normal to the body surface). Assume, as in [3.35], that the "natural" coordinates $x^1 = s$, $x^2 = n$, where the s-axis is directed along the tangent to the body surface and the n-axis is normal to the body surface, are employed as the curvilinear coordinates x^1, x^2 in the computations of a flow around a body. In the problem of a flow around a blunt body (see Figure 3.7) the zones of the large solution gradients are located in a region of the boundary layer and a shock transition whose position is not known *a priori*. This corresponds in the curvilinear orthogonal coordinates to the maximum gradients of functions in the direction x^2. To construct a difference grid automatically concentrating in the domains of large gradients, a coordinate transformation

$$q^1 = q^1(x^1), \qquad q^2 = q^2(x^1, x^2),$$

was introduced in [3.35] which mapped the computational domain onto a unit square $R\{0 \le q_i \le 1\}$, $i = 1, 2$. A uniform mesh with steps h_1, h_2 was taken in the standard domain R, and the corresponding mesh in the physical domain was assumed to be nonuniform and was given by an inverse transformation

$$x^1 = x^1(q^1), \qquad x^2 = x^2(q^1, q^2). \tag{3.3.4}$$

Along the coordinate direction x^1 the mesh steps were chosen to be uniform. In the x^2 direction the coordinate transformation (3.3.4) was specified by the equation

$$(|\partial f/\partial x^2|^\alpha + \delta)(\partial x^2/\partial q^2) = \text{const}, \tag{3.3.5}$$

where α, δ are user-specified positive constants and $\partial f/\partial x^2$ is the gradient of the function f in the direction x^2. Any of the functions ρ, v^1, v^2, ε or a combination of them may be chosen as f. The coefficient α prescribes a necessary coalescence of the nodes of a difference grid (the coordinate lines coalescence) in the x^2 direction, and the parameter δ is chosen to be nonzero in order to avoid mesh singularities at the nodes where $\partial f/\partial x^2 = 0$. The condition (3.3.5) was obtained in [3.35] on the basis of a variational principle whose satisfaction ensures small values of the flow gradients $\partial v/\partial q^2$. In the time-relaxation computations the value of the coordinate x^2 was determined from the solution of the equation

$$\partial x^2/\partial t = (\partial/\partial q^2)[(|\partial f/\partial x^2|^\alpha + \delta) \, \partial x^2/\partial q^2] \tag{3.3.6}$$

with stationary boundary conditions. In the process of relaxation the solution of (3.3.6) converges to the solution of equation (3.3.5).

A variational functional of a rather general form for the mesh control was proposed in [3.31] (see also [3.34], [3.35]). The functions $x^1(q^1, q^2, t)$ and $x^2(q^1, q^2, t)$ minimizing this functional simultaneously meet several requirements for the mesh where x^1, x^2 are moving Cartesian coordinates and q^1 and

q^2 are curvilinear coordinates. Following [3.31], [3.34], let us set

$$h_1 = \sum_{i=1}^{2} (v^i - \omega^i)^2;$$

$$h_2 = (\partial q^1/\partial x^1 - \partial q^2/\partial x^2)^2 + (\partial q^1/\partial x^2 - \partial q^2/\partial x^1)^2;$$

$$h_3 = \Delta = \det \|\partial x^\alpha/\partial q^i\|, \qquad i, \alpha = 1, 2.$$

In [3.31] the quantity h_1 was considered as a measure of deviation of the mesh from the Lagrangian mesh (the measure of "non-Lagrangianness"), h_2 as a measure of deformation characterizing the deviation from the conformal mapping, and h_3 as a measure of grid coalescence. The variational principle of grid construction was formulated in [3.31] as a problem of finding a minimum of the functional

$$\Phi(q^i) = \int_\Omega \varphi(h_1, h_2, h_3) \, dx^1 \, dx^2 \, dt, \tag{3.3.7}$$

where φ is an arbitrary nonnegative function of three arguments h_1, h_2, h_3 that additionally depends on the gas dynamic parameters and their gradients. The extremal surface satisfies (in the Cartesian coordinates) the Euler equation

$$(\partial/\partial t)[\partial\varphi/\partial(\partial q^i/\partial t)] + (\partial/\partial x^j)[\partial\varphi/\partial(\partial q^i/\partial x^j)] = 0$$

which in the new coordinates t, q^i takes the form

$$(\partial/\partial t)[\Delta \, \partial\varphi/\partial(\partial q^i/\partial t)]$$
$$+ (\partial/\partial q^j)\{\Delta[\partial\varphi/\partial(\partial q^i/\partial x^k) \, \partial q^j/\partial x^k - \omega \, \partial\varphi/\partial(\partial q^i/\partial t)]\} = 0.$$

Set

$$\varphi = \varepsilon_1 h_1 \Delta^\alpha + \varepsilon_2 h_2 + \varepsilon_3 \Delta^\beta, \tag{3.3.8}$$

where ε_i are the functions depending on the gas dynamic quantities and their gradients. In this specific case the Euler equations in the q^i coordinates have the form

$$(\partial/\partial t)[\varepsilon_1 \Delta^{\alpha+1}(v^i - \omega^i)] + (\partial/\partial q^\alpha)[(v^i - \omega^i)(v^\alpha - \omega^\alpha)\Delta^{\alpha+1}$$
$$- \delta^{ij}\varepsilon_1 \Delta^{\alpha+1} \sum_{\beta=1}^{2} (v^\beta - \omega^\beta)$$
$$+ \varepsilon_2((\Delta g^{i\alpha} - \delta^{i\alpha}) - \delta^{i\alpha}\beta\varepsilon_2 \Delta^{\beta+1}] = 0, \tag{3.3.9}$$

where

$$g_{ij} = (\partial q^i/\partial x^k)(\partial q^j/\partial x^k), \qquad \delta^{ij} = \begin{cases} 0, & i \neq j, \\ 1, & i = j. \end{cases}$$

Augmenting the system of gas flow equations in the moving coordinates (3.3.2), by the equation of state (1.5) and by the mesh equations (3.3.9), we obtain a closed equation system for the composite vector $\{\rho, v^i, E, \omega^i\}$ characterizing the state of the material and information medium [3.35]. Until now the equations of the mesh motion (3.3.8) in their complete form, that is, at $\varepsilon_1 \neq 0$, $\varepsilon_2 \neq 0$, $\varepsilon_3 \neq 0$, have not been used in gas dynamic computations because of the complexity of these equations. It was already mentioned above that for a number of gas dynamics problems it is sufficient to condensate the mesh in only one coordinate direction, say, q^2. In this case one must find a one-dimensional transformation $x^2 = x^2(q^2, t)$ such that the gradients $\partial f_i/\partial q^2$ of the gas dynamic quantities are not large with respect to the new coordinate q^2. These transformations may be determined by the result of minimization of the following functionals [3.44], [3.34]:

$$\Phi_1(q^2, t) = \int_{x^2} [\sum a_i |\partial f_i/\partial q^2|^{\alpha_i} + b \ln \Delta] \, dx^2;$$

$$\Phi_2(q^2, t) = \int_{x^2} [\sum a_i |\partial f_i/\partial q^2|^{\alpha_i} + b\Delta^\beta + c\Delta^{-\eta}] \, dx^2;$$

where $\alpha_i, b, c, \eta, \beta$ are real parameters and the a_i are the functions of gradients of the gas dynamic quantities. The transformation $x^2(q^2, t)$ satisfies the Euler equations

$$(\partial/\partial q^2)[(\sum a_i \alpha_i |\partial f_i/\partial q^2|^{\alpha_i} + b) \, \partial x^2/\partial q^2] = 0;$$

$$(\partial/\partial q^2)[\sum a_i \alpha_i |\partial f_i/\partial q^2|^{\alpha_i} \, \partial x^2/\partial q^2 + b\beta(\partial x^2/\partial q^2)^{\beta+1} - c\eta(\partial x^2/\partial q^2)^{-\eta+1} = 0.$$

Good reviews of the methods for the construction of solution-adaptive grids were made by Thompson [3.45]–[3.47]. In particular, in [3.48], [3.49], a functional for the grid construction has been proposed which is similar to the functional (3.3.7), (3.3.8)

$$I = I_s + \lambda_V I_V + \lambda_0' I_0',$$

where I_s, I_V, I_0' are functionals characterizing various grid properties. Here

$$I_s = \int_D [(\nabla \xi)^2 + (\nabla \eta)^2] \, dV; \qquad I_V = \int_{\mathscr{D}} wJ \, dV;$$

$$I_0' = \int_D (\nabla \xi \nabla \eta)^2 J^3 \, dV,$$

where $\xi = \xi(x, y)$, $\eta = \eta(x, y)$, and J is the Jacobian of the transformation $(x, y) \to (\xi, \eta)$. The integral I_s characterizes the mesh smoothness and I_0' the degree of mesh orthogonality; the integral I_V is included to take into account the solution singularities by a proper choice of the weight function w.

Another approach to the mesh construction minimizing the local error has been proposed in [3.34], [3.50]. Let $\mathscr{L}(u) = 0$ be a system of differential equations with respect to the vector function u, and let $\Lambda(u_i)$ be an approximation to $\mathscr{L}$ on a moving grid. Denote by ε the leading term of the error in u and by R the leading term in the approximation error for $\mathscr{L}$. As is shown in

[3.51], ε and R are related by the equation $\mathscr{L}_1(\varepsilon) = R$, where the left-hand side $\mathscr{L}_1$ is the variation equation for the system $\mathscr{L}(u) = 0$. From the equation $\mathscr{L}_1(\varepsilon) = R$ at $R = 0$ one gets $\varepsilon = 0$. It is obvious that the mesh for which $R = 0$ would be optimum. Therefore, the optimum mapping is found if the equation $R = 0$ is soluble with respect to the mesh point coordinates x^i. But the equation $R = 0$ is insoluble for an arbitrary scheme $\Lambda(u_i) = 0$ approximating the system $\mathscr{L}(u) = 0$. In this case the problem of the optimum mesh construction is solved simultaneously with the problem of the construction of a scheme such as $\Lambda(u_i) = 0$ for which the equation $R = 0$ is soluble with respect to the mesh point coordinates x_i. The examples of such a construction of the mesh for the equation $du/dx = f(u, x)$ have been presented in [3.50]. Note that the mesh constructed in this way depends on solution gradients, and is shown in the examples in [3.50].

It has been proposed in [3.52] to use the differential approximations of difference schemes while constructing adaptive grids. An algorithm of the adaptive mesh construction for the computation of potential transonic flow around airfoils has been constructed in [3.53]. In this algorithm, only the solution values on the airfoil surface are used, and the mesh nodes are also rearranged only on the airfoil surface. Thus, the problem of mesh rezoning is by one dimension less than the original problem of flow calculation. It has been shown that the application of the proposed algorithm of the grid construction substantially increases the numerical solution accuracy. The algorithm of [3.53] was realized in the following sequence. At first a conventional grid was generated around an airfoil by the solution of the Poisson equation. The full-potential equation was solved on this mesh. Information on solution gradients obtained from this solution was used to redistribute the grid nodes on the airfoil surfaces. After that a new grid was generated in the computational domain (by the same program as for the Poisson equations) at a given new distribution of nodes on the airfoil. Then the numerical solution of the full-potential equation was carried out on this new grid.

In the cases when there are (in the problem under study) flow singularities (shock waves, contact surfaces) that have a complicated geometric configuration and that rapidly change with time, the use of moving grids adapting to the flow, with regard to the solution gradients, may cause certain difficulties. It appears that the HARLOW particle-in-cell method [3.3], [3.12], [3.13] is still the most powerful tool for numerical simulation in these cases. Mention must also be made of the recently developed method of concentrations [3.54], by which it was possible to calculate the instability of the tangential discontinuity in the velocity in compressible gases (the Kelvin–Helmholtz instability) [3.55]. One can also mention a number of complicated problems, a typical feature of which is the presence of many interacting singularities and, as a consequence of this, it is difficult to realize in a numerical study of such problems the idea of moving grids whose lines could be conveniently mapped onto a rectangular grid in a square: these are, for example, the problems with complex types of the Mach shock reflections [3.56], [3.57] and supersonic flows in inlets [3.58]. In connection with the above it is reasonable to take into account the properties of various types of grids at an initial stage of the choice of a computational technique for the numerical investigation of a specific class of problems. A number of useful recommendations on the choice of grids may be found in [3.2], [3.46].

3.4. Computational Examples

When using the differential analyzers of shock waves based on formulas (3.2.21), (3.2.23), one must at first find the P-form of the f.d.a. of difference scheme employed. The f.d.a. P-forms for 23 difference schemes approximating the Euler equation system (1.30) have been written in [3.59]. In [3.9] the f.d.a. P-forms of the following difference schemes were used:

(a) the invariant scheme of [3.60];
(b) the scheme of the particle-in-cell method with an explicit scheme of the Eulerian computing stage from [3.3]; and
(c) the scheme of the particle-in-cell method with an implicit predictor–corrector scheme for the Eulerian computing stage that was constructed on the basis of a general methodology of the predictor–corrector schemes presented in [3.61]:

the predictor stage

$$(u_{j+1/2,k}^{n+\beta/2} - u_{j+1/2,k}^n)/(\beta\tau) = -(p_{j+1,k}^{n+\beta/2} - p_{jk}^{n+\beta/2})/(\rho_{j+1/2,k}^n h_x);$$

$$v_{jk}^{n+\beta/2} = v_{jk}^n, \qquad \rho_{jk}^{n+\beta/2} = \rho_{jk}^n;$$

$$[\partial f(p_{jk}^n, \rho_{jk}^n)/\partial p](p_{jk}^{n+\beta/2} - p_{jk}^n)/(\beta\tau) = -(p_{jk}^n/\rho_{jk}^n)(u_{j+1/2,k}^{n+\beta/2} - u_{j-1/2,k}^{n+\beta/2})/h_x;$$

$$u_{jk}^{n+\beta} = u_{jk}^{n+\beta/2}, \qquad \rho_{jk}^{n+\beta} = \rho_{jk}^{n+\beta/2};$$

$$(v_{j,k+1/2}^{n+\beta} - v_{j,k+1/2}^{n+\beta/2})/(\beta\tau) = -(p_{j,k+1}^{n+\beta} - p_{jk}^{n+\beta})/h_y; \qquad (3.4.1)$$

$$[\partial f(p_{jk}^{n+\beta/2}, \rho_{jk}^{n+\beta/2})/\partial p](p_{jk}^{n+\beta} - p_{jk}^{n+\beta/2})/(\beta\tau)$$
$$= -(p_{jk}^{n+\beta/2}/\rho_{jk}^{n+\beta/2})(v_{j,k+1/2}^{n+\beta} - v_{j,k-1/2}^{n+\beta})/h_y;$$

the corrector stage

$$\tilde{\rho}_{jk} = \rho_{jk}^n;$$

$$(\tilde{u}_{jk} - u_{jk}^n)/\tau = -(p_{j+1/2k}^{n+\beta} - p_{j-1/2,k}^{n+\beta})/h_x;$$

$$(\tilde{v}_{jk} - v_{jk}^n)/\tau = -(p_{jk+1/2}^{n+\beta} - p_{jk-1/2}^{n+\beta})/h_y;$$

$$(\tilde{E}_{jk} - E_{jk}^n)/\tau = -\{[(pu)_{j+1/2,k}^{n+\beta} - (pu)_{j-1/2,k}^{n+\beta}]/h_x \qquad (3.4.2)$$
$$+ [(pv)_{j,k+1/2}^{n+\beta} - (pv)_{j,k-1/2}^{n+\beta}]/h_y\}/\rho_{jk}^n.$$

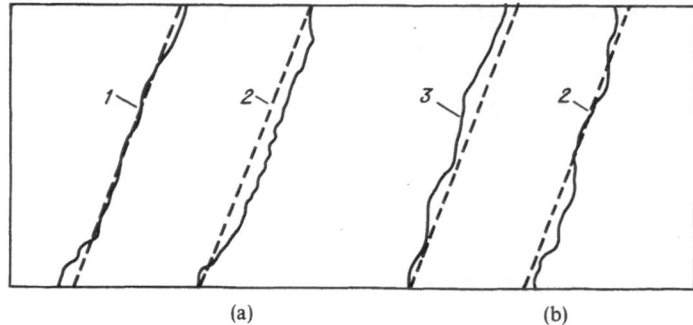

(a)   (b)

Figure 3.2. Results of shock wave propagation computation: (a) by the Yanenko–Shokin scheme [3.60]; (b) by the particle-in-cell method with an explicit Eulerian stage scheme from [3.3]. 1—$t = 0.016$; 2—0.036; 3—0.018.

The notations in (3.4.1), (3.4.2) are similar to those applied in Section 3.1. In addition, the function $f(p, \rho)$ is the function entering the equation of state (1.6). In equations (3.4.1) $u_{j+1/2,k}^n = 0.5(u_{jk}^n + u_{j+1,k}^n)$, $\rho_{j+1/2,k}^n = 0.5(\rho_{jk}^n + \rho_{j+1,k}^n)$, $v_{j,k+1/2}^{n+\beta/2} = 0.5(v_{jk}^{n+\beta/2} + v_{j,k+1}^{n+\beta/2})$; in equations (3.4.2) $p_{j+1/2,k}^{n+\beta} = 0.5(p_{jk}^{n+\beta} + p_{j+1k}^{n+\beta})$, $p_{j,k+1/2}^{n+\beta} = 0.5(p_{jk}^{n+\beta} + p_{j,k+1}^{n+\beta})$. The factor $\beta \in [0.5, 1]$ is introduced into formulas (3.4.1) with the purpose of a further increase in the stability of the Eulerian stage scheme (3.4.1), (3.4.2).

The computations of a model problem on the propagation of a one-dimensional shock wave at an angle to the lines of the rectangular computing mesh in the (x, y)-plane (which were carried out in [3.9] by the above three schemes) showed that the absolute error in shock front localization on the basis of differential analyzers (3.2.21), (3.2.23), (3.2.27) did not exceed the value $\max(h_x, h_y)$. To illustrate this we present in Figure 3.2 the results of the computation of propagation of a shock whose front is inclined at an angle of $70°$ to the x-axis. In these runs the differential analyzer (3.2.27) was used in the case of Figure 3.2(a), and the analyzer (3.2.23) was used in the case of Figure 3.2(b) for shock front localization.

The particle-in-cell method is one of the most efficient methods for the numerical modeling of high-velocity impact problems [3.62]–[3.66]. Let us present an algorithm of the shock wave differential analyzer for one of the problems of the class under consideration that was realized in [3.67] within the framework of the Harlow method. Consider the problem of the high-velocity impact of a cylinder onto a plate of finite dimensions (see Figure 3.3). Suppose that the projectile touches the target at $t = 0$. From an interface between the two bodies the shock waves propagate at $t > 0$ in opposite directions. Free surfaces initiate the rarefaction waves in the process of their interaction with shock waves. As the process proceeds the rarefaction waves emanating from the lateral free surfaces result in an ejection of the projectile and target material in the direction opposite to the impact direction. When

Figure 3.3. The scheme of the beams disposition in the projectile and in the obstacle.

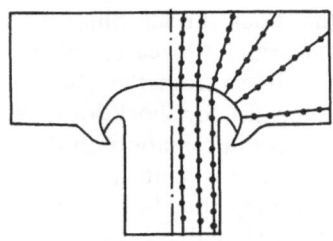

the longitudinal shock wave in a target arrives at a rear free surface it is reflected in the opposite direction by the rarefaction wave and increases the bulging of the mass of the projectile and target material. This wave picture of interaction, especially at small impact velocities, results in the formation of cracks in the medium and in its fragmentation. The motion in a domain which is not influenced by rarefaction waves remains one-dimensional, and this circumstance can be useful for a check on the calculation accuracy.

In the process of development of the differential analyzer we have used the fact that the Rankine–Hugoniot conditions on a shock wave, written in coordinates along a normal and a tangent to the shock, coincide with the one-dimensional conditions. Therefore, the accuracy of the shock localization will be higher if the position of a finite-difference shock wave center is determined on each of the beams in the (x, y)-plane that are drawn in a direction to the shock front which differs as little as possible from the normal direction.

Indeed, let us take an arbitrary point (x, y) in the zone of shock smearing, and let us place at this point the origin of a Cartesian rectangular coordinate system (n, τ) whose n-axis is directed along the normal to the shock surface, and the τ-axis is directed along the tangent to the above surface. Let $\mathbf{l}$ be some unit vector having the direction cosines $\cos \alpha_n$, $\cos \alpha_\tau$, and let the smeared shock surface be described by the function $u(n, \tau)$. Then the derivative along the $\mathbf{l}$ direction is calculated by the formula

$$du/d\mathbf{l} = (\partial u/\partial n) \cos \alpha_n + (\partial u/\partial \tau) \cos \alpha_\tau. \qquad (3.4.3)$$

If the change in the function u along the shock front is small, that is, $\partial u/\partial \tau \cong 0$, then we have from (3.4.3) that

$$du/d\mathbf{l} \cong (\partial u/\partial n) \cos \alpha_n. \qquad (3.4.4)$$

Let us determine the region of the sharpest change of the solution u in the zone of smearing in the direction determined by the vector $\mathbf{l}$ as the width after Prandtl:

$$X = |U_1 - U_2|/\max |du/d\mathbf{l}|, \qquad (3.4.5)$$

where U_1, U_2 are the values characterizing the state of the medium on different sides of the original nonsmeared shock surface. Then from (3.4.5) we obtain, with regard to (3.4.4), that the width X will be minimum at $\alpha_n = 0$, that is, when the direction of the vector $\mathbf{l}$ coincides with the direction of a normal to the shock surface. Since the shock position is determined within the zone of smearing, it is reasonable to expect a higher localization accuracy at $\alpha_n = 0$. It follows from the above that the success of the application of such an algorithm for shock localization depends substantially on the availability of *a priori* information on the orientation of shock surfaces with respect to the spatial coordinate axes.

Taking the above considerations into account, we have developed a differential analyzer code with the following sequence of computations:

the beams are drawn perpendicular to the contact boundary across each "marker" particle of the contact boundary (see Figure 3.3), in order to determine the shock wave position on a target;

the point of intersection of a beam with the upper (or with the right) boundary of the computational domain is determined;

the beam segment between the contact boundary marker and the found end is divided into N parts;

at each point of this segment the value div **u** is determined by interpolation, and then among the points where div **u** < 0 a point is sought at which the quantity |div **u**| achieves its maximum.

The coordinates of this point are printed out. The choice of beams for shock front localization in the projectile is analogous to the foregoing and is depicted in Figure 3.3. In Figure 3.4 we present the pictures of penetration of a pro-

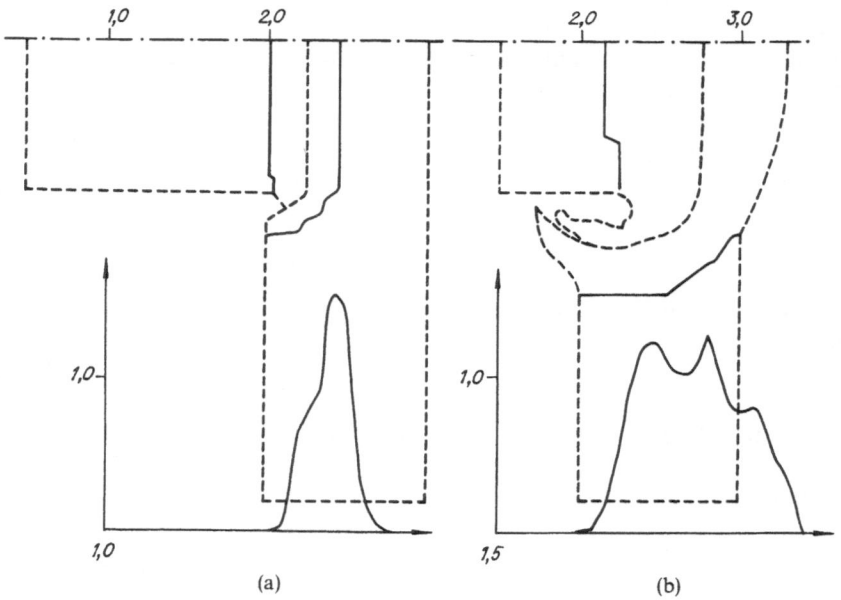

Figure 3.4. Computation by the Harlow method: (a) $t = 20\tau$; (b) $t = 60\tau$.

jectile into a target, which were obtained in the computation of a plane two-dimensional flow by the modified Harlow method. The particle-in-cell method modification consisted of the introduction of shape parameters of particles in the two-dimensional case. The particles had a circular shape in the case of a plane flow and had the shape of a rotation toroid with a circular cross section in the case of an axisymmetrical flow. The idea of introducing the shape parameters in the particles method was first proposed in [3.13], and was realized in the one-dimensional case of "layered particles" in [3.68]. A description of an algorithm for the circular particles transport in the modified Harlow method, as well as the results of studies on the influence of shape parameters of particles on the properties of the approximation viscosity of the method, have been presented in [3.69]. The impact velocity in the run of Figure 3.4 was equal to 10 km/s at $t = 0$. At the bottom of Figure 3.4 we present the graphs of the pressure p on the symmetry axis. The ideal gas equation of state was used in the computation. The shock fronts are depicted in Figure 3.4 by solid lines and the contact surfaces and free surfaces by dashed lines. In the pressure graph corresponding to the time $t = 60\tau$, two pressure peaks are formed which characterize the two shock waves propagating in opposite directions (see Figure 3.4). The shock wave front in the target is distorted on the right by the interaction of rarefaction waves emanating from the free surface.

In Figure 3.5 we present the result of a computation, of a high-velocity impact of two bodies onto an obstacle, by the method of circular particles [3.18]. The shock waves emanating from the surfaces of contact of the projectiles with the obstacle begin to interact with each other at some $t > 0$. The letters S and R in Figure 3.5 refer, respectively, to an incident shock and to a reflected shock; M is the Mach shock wave; the letters a, b refer to the position of the projectiles at the 50th time step; the numbers 40, 50, and 60 mark the

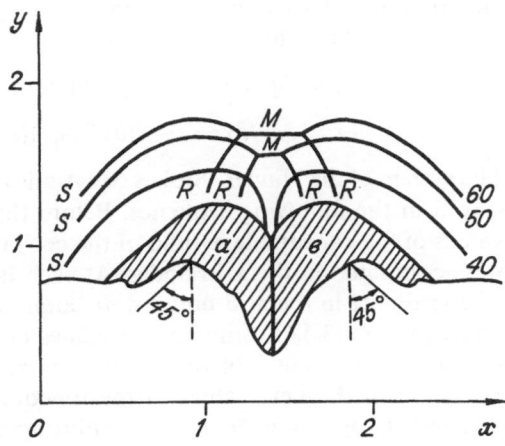

Figure 3.5. Computation of the impact of two bodies onto an obstacle.

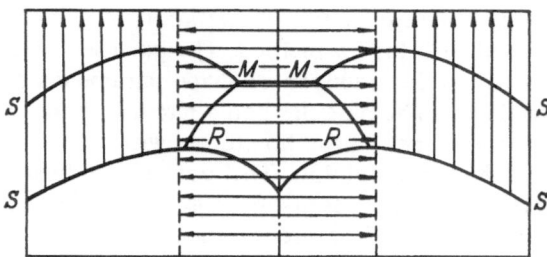

Figure 3.6. To the construction of the differential analyzer algorithm (the notation is the same as in Figure 3.5).

shock fronts at the times 40τ, 50τ, and 60τ, respectively. A triple shock wave configuration of shock waves (Mach configuration) can arise in the obstacle, depending on the values of the initial velocity impact of the bodies and on the angles α, β of the directions of the velocity vectors of the projectiles. In Figure 3.5 $\alpha = -\beta = 45°$, $|\mathbf{u}(x, y, 0)| = \sqrt{2 \cdot 9.659}$ km/s for the projectiles. The scheme of the beams disposition in the obstacle for shock front localization in the differential analyzer algorithm developed for this problem is shown in Figure 3.6. The shock front points were determined on the beams by $\max |\text{div } \mathbf{u}| < 0$, as in Figure 3.4.

In Figure 3.7 we present the examples of computations from [3.34], [3.35] of the hypersonic flow (a) around a convex body, and (b) around a nonconvex composite body of a viscous compressible heat-conductive gas with the use of a moving grid governed by equation (3.3.5). The equation of grid (3.3.6) was solved numerically by relaxation, and at each time level n an implicit difference equation [3.35]

$$[(x^2)^{n+1} - (x^2)^n]/\tau = \bar{\Lambda}_2 b^n \Lambda_2 (x^2)^{n+1} \qquad (3.4.6)$$

was solved where

$$b = |\partial f/\partial x^2|^\alpha + \delta = |z_1 \Lambda_2 f^n|^\alpha + \delta,$$

after that the values of the coefficients z and z_1 were computed to second-order accuracy by the formulas

$$\begin{aligned} z_1 &= \partial q^2/\partial x^2 = (\partial x^2/\partial q^2)^{-1}; \\ z &= \partial q^2/\partial x^1 = -[(\partial x^1/\partial q^1)(\partial x^2/\partial q^2)]^{-1}\, \partial x^2/\partial q^1. \end{aligned} \qquad (3.4.7)$$

The system of the Navier–Stokes equations and the grid equation (3.3.6) were solved in the following sequence. Before the beginning of computation the values of the coordinate x^1 and of the coefficient $z_0 = \partial q^1/\partial x^1 = 1/(\partial x^1/\partial q^1)$ were computed in the grid nodes. At each iteration level n the values of the x^2 coordinate in the grid nodes were computed by numerical solution of the grid equation (3.4.6) using known values of the components of the solution vector w^n. The values of the coefficients z_1, z were computed by formulas (3.4.7). Then the new values of the functions w^{n+1} at the level $n + 1$ were determined by using an implicit splitting-up scheme on the basis of the

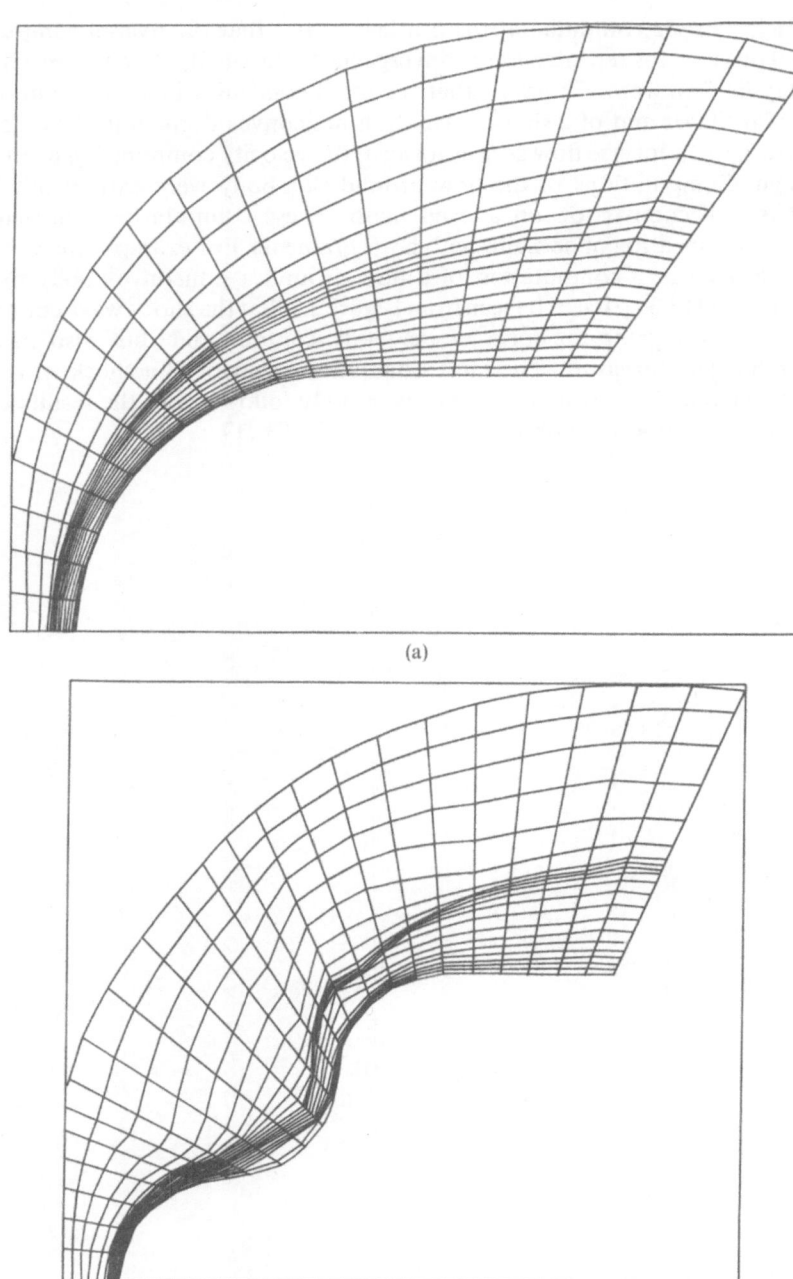

(a)

(b)

Figure 3.7. Examples of hypersonic flow computation: (a) the freestream Mach number $M_\infty = 25$, the Reynolds number $\mathrm{Re} = 3 \cdot 10^3$; (b) $M_\infty = 15$, $\mathrm{Re} = 3 \cdot 10^3$, the wall temperature $T_w = 15 T_\infty$.

coefficients z_1, z computed at the nth level. After that the overall computational process was repeated until convergence to the steady state was reached. It may be seen from Figure 3.7 that the mesh condenses in the domain of a boundary layer and of a shock wave. In a nonconvex domain in the case of Figure 3.7(b) a reverse flow zone is formed: the velocity component u changes its sign. Computations of the flow around this body were carried out for purposes of comparison on a fixed mesh. These computations showed a significant smearing of domains with large gradients. For example, the vortex did not arise in a computation of a flow around a nonconvex body on a uniform grid [3.35]. It can be seen from Figure 3.7 that the shock wave contour qualitatively repeats the form of the body contour and thus also has a nonmonotone curvature. The same qualitative behavior of a shock wave in the hypersonic flow around a nonconvex body follows from the results obtained by G. MORETTI with his λ-scheme [3.70], [3.71].

Differential Analyzers of Contact Discontinuities in One-Dimensional Gas Flows

By definition, the contact discontinuity in an ideal fluid is a discontinuity surface across which there is no fluid flow (see Section 1.1). A contact discontinuity may take place, for example, at the interface between two different ideal fluids or between the layers of the same fluid that move with different speeds (a tangential discontinuity). In some cases new contact boundaries can appear in the process of evolution of a gas dynamic flow; for example, after the collision of two shock waves of different intensity, after the coincidence of shock waves propagating in the same direction [4.1], after the transition of a regular shock wave reflecting into an irregular one (the Mach reflection) [4.2], [4.3], and at the interaction between the shock wave and the rarefaction wave [4.4], [4.1].

When using homogeneous difference schemes, singularities of the shock wave and the contact boundary type are approximated by continuous transition zones occupying several mesh intervals. The width of such a transition zone is approximately constant in the case of a shock wave. However, in the case of a contact boundary this width increases with increasing time t when calculating in Eulerian variables. As was indicated, for example, in [4.5], [4.6], the width X of a contact strip increases proportionally to $n^{1/2}$ when using the Lax scheme, where n is a number of time steps. In [4.7], [4.8] it is also communicated that the width X increases proportionally to $n^{1/2}$ when employing a series of other difference schemes of first-order accuracy. In [4.9]–[4.14] a more general asymptotic estimate of the width X having the form

$$X = \text{const} \cdot n^{1/(r+1)}$$

is given, where r is the order of accuracy in the finite-difference scheme.

This fact makes the interpretation of gas dynamic computing results difficult. It also poses two problems: the increase in the accuracy of the computation of problems with contact discontinuities, and the problem of the development and foundation of the algorithms for the localization of contact discontinuities in the computational region by the results of shock-capturing computations.

Let us enumerate the techniques for the localization of contact discontinuities in the computational region that were proposed previously for the cases when the flow is computed by a finite-difference scheme in Eulerian variables. One of the well-known techniques for the localization of contact boundaries on the basis of shock-smoothing computation consists of the visual determination of the position of such discontinuities by the coalescence of different isolines. As was noted in [4.15], by virtue of the fact that the pressure is continuous across the contact surface, the lines of constant pressure do not coalesce in the vicinity of such a discontinuity surface. Therefore, it is necessary to take as isolines only the lines of isovalues of those functions which undergo a discontinuity on the contact boundary (for example, density and temperature). In the cases when the initial contact boundary position at $t = 0$ is known, its motion at $t > 0$ may be followed with the help of the "marker" particles; at $t = 0$ these particles are placed on the contact boundary and then moved at each subsequent time step, using the velocity values obtained by the solution of the finite-difference equations (see, for example, [4.16]–[4.18] and Figure 2.1). As was noted in [4.18], such an algorithm for tracking contact boundaries provides especially good results, if in the process of problem solution one takes care of the distances between neighboring marker particles: when these distances become larger than some maximally allowed value it is very useful to introduce an additional marker particle with coordinates in the middle of a line segment connecting the two neighboring marker particles under consideration.

A similar technique was applied in [4.19], [4.20], with the difference that the discontinuity front was approximated by rectilinear segments whose ensemble forms some broken line. It is clear that the error in determining the contact boundary position by the marker particles increases with t, which is associated with the error accumulation when approximating the particle velocity integral by some quadrature formula. In [4.21] the technique for the contact boundary localization in the steady flow was based on the change in the Bernoulli integral constant which was different for the flows on both sides of the contact boundary. In [4.22] a method for the computation of gas dynamic flows was proposed which makes use of the notion of the mass concentration of a substance. This method adheres to the methods of the "particle-in-cell" type [4.23]. The localization of contact discontinuities is carried out in [4.22] on the basis of an analysis of the concentrations field. Thereby the contact boundaries are localized with an accuracy up to the size of one computing cell. Let us describe in more detail the algorithm of [4.22] (see also [4.24], [4.25]). In this method homogeneous and mixed cells of a spatial grid are considered: the homogeneous cells are occupied by a single fluid, and in the mixed cells there are a few fluids each of which has its own equation of state of the form (1.5). In addition to the vector **w** components in (1.30b) in the mixed cells, the mass concentrations of substances $\alpha_i = M_i/M$ and the specific internal energies ε_i are computed where M_i is the mass of the ith substance in the cell and M is the cell mass. The pressure and densities of the components in a mixed cell are computed in [4.22] with the aid of a procedure which is similar to the one used in the HARLOW method [4.23],

[4.26]–[4.27]. Let us consider in more detail the computation of the mass and energy fluxes and the other quantities in a mixed cell. To prevent numerical smearing, the following simple idea was realized in [4.22]: at first the substance which is contained in a homogeneous cell moves from the neighboring mixed cell into this homogeneous cell; thereby the numerical smearing is limited. In the general case an analysis of the concentrations field is performed; on the basis of the concentrations field the shape of the contact boundary is determined and then the fluxes from the mixed cells are determined. In the case when a mixture of substances α and β flows out of a mixed cell it is assumed (depending on the contact boundary shape) that the velocities u_α and u_β of the substances are different: these velocities are determined on the basis of the velocities of the nearest mesh nodes belonging to the corresponding homogeneous cells; that is, a slip is introduced by which the tangential discontinuities can be taken into account. In this way the mass and energy fluxes are determined for each component. In Figure 4.1 we show the temporal evolution of an initially sinusoidal interface between two fluids (the Helmholtz–Kelvin instability). The sinusoid has an amplitude $a_0 = 0.1$ and a wavelength $\lambda = 1$.

The front tracking algorithm [4.28], [4.29], which we have already presented at the beginning of Chapter 2, also enables us to solve problems with strongly distorting interfaces (as, for example, in the Rayleigh–Taylor problem [4.29]). However, this algorithm looks to be logically more complicated than the algorithm of the concentrations method [4.22]. In addition, there exist, in the method of [4.28], the difficulties of taking into account the disappearance of certain discontinuity surfaces or the formation of such new surfaces. It is promised in [4.28] to take these effects into account in future.

For the localization of the contact discontinuities in the two-dimensional unsteady problems on the interaction of jets with obstacles in [4.30], a differential analyzer, based on the determination of the points of maxima of

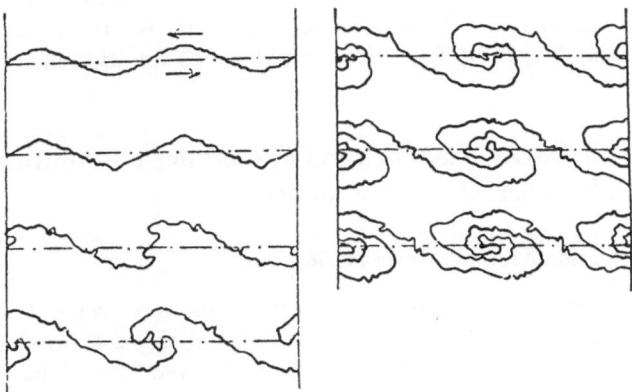

Figure 4.1. Interfaces at $t = 0$; 0.3; 0.7; 1.1; 1.5; 2.1; 2.5.

the quantity $|\text{div } \rho \mathbf{u}|$, was applied where ρ is the density and $\mathbf{u}$ is the velocity vector. In the computations of a problem on the Rayleigh–Taylor instability by the "coarse particles" method, the contact interface was localized in [4.31] as an ensemble of points at which the density $\rho = 0.5(\rho_1 + \rho_2)$, where ρ_2 is the density of a heavy fluid and ρ_1 is the density of a light fluid. The same technique was previously used in [4.6].

The idea of the differential analyzer method was also used in the numerical modeling of plane and axisymmetric flows of nonhomogeneous incompressible fluids in the gravity field. A problem on the evolution of a "spot" of homogeneous fluid in a stratified medium was considered in [4.32]. It was of interest to construct for this problem an algorithm for the determination of a "spot" boundary. In [4.32] such an algorithm was constructed in the following way. For the description of the flow, the Navier–Stokes equations were augmented by an equation of conservation of a nondiffusive passive admixture whose concentration at the initial moment of time was set constant, and nonzero in the "spot" and zero outside it. The numerical modeling of the problem was carried out on a sequence of "pairs" of meshes; the spot boundary was taken as a line along which the difference of concentrations $c_1^n - c_2^n$, for a given pair of meshes $(\Delta t, h_1, h_2)$ and $(\Delta t, h_1/2, h_2/2)$ where the subscripts 1 and 2 refer to the above two meshes, vanished. Thus, the obtained temporal change of the horizontal "spot" size in the planar problem proved to be in sufficiently good agreement with known experimental data.

Below we present a technique for the investigation of the differential analyzers of contact discontinuities which was previously proposed by the present authors in [4.33]–[4.37] for one- and two-dimensional problems of gas dynamics computed in the Eulerian variables. It is shown that within the framework of this technique it is possible to introduce the notion of a contact strip center, which in its geometric sense is analogous to the notion of a smeared shock wave center introduced in Chapter 2. We also present some methods, proposed in [4.36], [4.38]–[4.43], for increasing the accuracy of numerical solutions in the vicinity of a contact discontinuity for the cases when the equation of state used is different from an ideal gas equation of state.

4.1. Methods for the Localization of Contact Discontinuities in the Presence of K-Consistence

4.1.1. Investigation Method. Basic Definitions

When considering the differential analyzers of shock waves in Chapter 2, a problem on the propagation of a stationary shock wave was taken as a problem on a one-dimensional flow with a shock wave. The relative simplicity of this model problem enabled us to apply the apparatus of the progressive wave-type solutions, for the analysis of the approximation viscosity properties

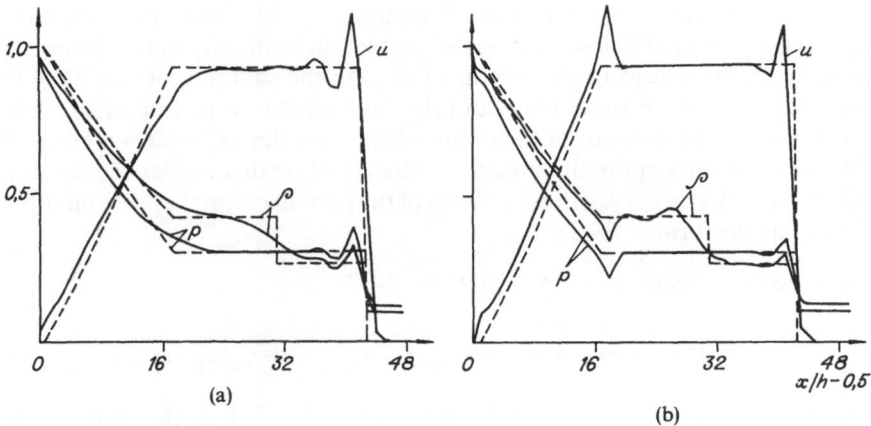

Figure 4.2. Numerical solutions of the breakdown of the discontinuity problem: (a) by the FLIC method [4.44]; (b) by the Lax–Wendroff scheme [4.45].

of a rather wide class of difference schemes, from the point of view of the constructed theory of differential analyzers. When studying the numerical "smearing" of a contact discontinuity, it was natural to address the case of the Riemann problem which is characterized by the presence of the contact boundary in the flow. Alongside the contact boundary in the flow arising at $t > 0$, shock and rarefaction waves may also be present (see Section 1.1 and Figure 4.2). By the broken lines in Figure 4.2 the exact solution which contains a rarefaction wave is depicted—a contact discontinuity and a shock wave in this example. Generally, the intersection of the zones of smearing of the contact discontinuity and the shock wave is possible in the numerical solution of the Riemann problem (as in Figure 4.2(a))—which creates additional diffi-culties for the mathematical investigation of the "smeared" solution. Note from Figure 4.2 the existence of a region in the vicinity of the contact dis-continuity in the numerical solution where the pressure $p(x, t)$ and velocity $u(x, t)$ functions are constant.

Let us consider, for the Euler equation system (1.11), (1.9) augmented by the equation of state (1.6), the Riemann problem with the following initial data:

$$t = 0, \quad u(x, t) = u_0, \quad p(x, t) = p_0, \quad \rho(x, t) = \begin{cases} \rho_1, & x < x_0, \\ \rho_2, & x > x_0, \end{cases} \quad (4.1.1)$$

where $u_0, p_0, \rho_1, \rho_2, x_0$ are constants and $\rho_1 \neq \rho_2$. The exact solution of this problem has the form

$$t \geq 0, \quad u(x, t) = u_0, \quad p(x, t) = p_0,$$

$$\rho(x, t) = \begin{cases} \rho_1, & x < x_0 + u_0 t, \\ \rho_2, & x > x_0 + u_0 t, \end{cases} \quad (4.1.2)$$

and thus contains only the contact discontinuity. For definiteness, the case $u_0 > 0$ is considered below; that is, the contact discontinuity moves from left to right. In the following the function $f(p, \rho)$ in the equation of state (1.6) is assumed to be three times continuously differentiable with respect to both arguments in the domain of definition of this function. Consider a class of difference schemes approximating the system (1.11) with an order of accuracy $O(h^r) + O(\tau^r)$, $1 \leq r \leq 2$, whose H-form of the first differential approximation (f.d.a.) has the form

$$\Gamma \mathbf{w} \equiv \partial \mathbf{w}/\partial t + \partial \boldsymbol{\varphi}(\mathbf{w})/\partial x - \sum_{\substack{i,j \\ i+j=r}} h^i \tau^j \mathbf{F}_{ij}(\mathbf{w}, \partial \mathbf{w}/\partial x, \partial \mathbf{w}/\partial t, \dots,$$

$$\partial^{r+1} \mathbf{w}/\partial x^{r+1}, \partial^{r+1} \mathbf{w}/\partial x^r \, \partial t, \dots, \partial^{r+1} \mathbf{w}/\partial t^{r+1}) = 0, \quad (4.1.3)$$

where $\mathbf{w} = \{w_1, w_2, w_3\}^T$, $r \geq 1$, $\mathbf{F}_{ij} = \{F_{ij1}, F_{ij2}, F_{ij3}\}^T$, h is the step of the uniform computing mesh on the x-axis, and τ is the time step. Let us express the derivatives $\partial^m \mathbf{w}/\partial x^k \, \partial t^{m-k}$, $m - k > 0$, which enter $\mathbf{F}_{ij}$ in terms of the derivatives with respect to x making use of formulas (2.3.2). As a result we obtain from (4.1.3) the P-form of the f.d.a.

$$\Pi \mathbf{w} \equiv \partial \mathbf{w}/\partial t + \partial \boldsymbol{\varphi}(\mathbf{w})/\partial x$$

$$- \sum_{\substack{i,j \\ i+j=r}} h^i \tau^j \mathscr{F}_{ij}(\mathbf{w}, \partial \mathbf{w}/\partial x, \dots, \partial^{r+1} \mathbf{w}/\partial x^{r+1}) = 0, \quad (4.1.4)$$

where

$$\mathscr{F}_{ij}(\mathbf{w}, \partial \mathbf{w}/\partial x, \dots, \partial^{r+1} \mathbf{w}/\partial x^{r+1})$$

$$= \mathbf{F}_{ij}(\mathbf{w}, \partial \mathbf{w}/\partial x, \mathbf{f}_{1,0}(\mathbf{w}, \partial \mathbf{w}/\partial x), \dots, \mathbf{f}_{r+1,0}(\mathbf{w}, \dots, \partial^{r+1} \mathbf{w}/\partial x^{r+1})). \quad (4.1.5)$$

In accordance with (1.9) let us set in (4.1.3), (4.1.4)

$$\mathbf{w} = \{w_1, w_2, w_3\}^T, \qquad w_1 \equiv \rho, \qquad w_j = f_j(\rho, u, p, \varepsilon(p, \rho)), \qquad j = 2, 3.$$

Definition 4.1. Let us call the quasi-linearization operation L of the first differential approximation (f.d.a.) (4.1.3) or (4.1.4) the equation

$$\Gamma L \mathbf{w} = 0, \qquad \Pi L \mathbf{w} = 0,$$

where

$$L\mathbf{w} = \{w_1, Lw_2, Lw_3\}^T; \qquad Lw_j = f_j(\rho, u_0, p_0, \varepsilon(p_0, \rho)), \qquad j = 2, 3,$$

where u_0, p_0 are constant velocity and pressure values entering (4.1.1).

Introduce the column vector

$$U = \{1, u_0, 0.5u_0^2 + f(p_0, \rho) + \rho \, \partial f(p_0, \rho)/\partial \rho\}^T. \quad (4.1.6)$$

Definition 4.2. The first differential approximation (f.d.a.) (4.1.3) will be called K-consistent if

$$\Gamma Lw = U\left[\partial\rho/\partial t + u_0\, \partial\rho/\partial x \right.$$
$$\left. - \sum_{\substack{i,j \\ i+j=r}} h^i \tau^j F_{ij1}(Lw, \partial Lw/\partial x, \partial Lw/\partial t, \ldots, \partial^{r+1} Lw/\partial t^{r+1}) \right] = 0,$$

that is, the left-hand sides of the equations of a system $\Gamma Lw = 0$ differ from each other only by a scalar multiplier. K-consistence of the f.d.a. P-form (4.1.4) is defined analogously. Note that the initial equation system (1.1)–(1.3) possesses the K-consistence property. Indeed, applying the quasi-linearization operation L to (1.11), (1.9) we obtain

$$L(\partial \mathbf{u}/\partial t + \partial \boldsymbol{\varphi}(\mathbf{u})/\partial x) = (\partial \rho/\partial t + u_0\, \partial \rho/\partial x)U = 0. \qquad (4.1.7)$$

We see from (4.1.7) that under the initial data (4.1.1) the solution of the Cauchy problem (1.11), (1.9), (1.6), (4.1.1) reduces to the solution of a linear equation

$$\partial \rho/\partial t + u_0\, \partial \rho/\partial x = 0. \qquad (4.1.8)$$

The discontinuities in the solutions of equation (4.1.8) propagate along the characteristics having the equation $dx/dt = u_0$. In accordance with the general definitions given in [4.46] (see also [4.47]), the discontinuities in the solutions of equation (4.1.8) are contact, and the characteristic $dx/dt = u_0$ is called contact [4.1], [4.6]. This conclusion agrees with the well-known fact for the system (1.11), (1.9) that this system is not "genuinely nonlinear" in the sense of the definition given in [4.46] (see also [4.1], [4.47]).

Lemma 4.1. *From the K-consistence of the H-form of the first differential approximation (f.d.a.) (4.1.3), the K-consistence of the f.d.a. P-form (4.1.4) follows.*

The proof makes use of Definition 4.1 and formula (4.1.5) and is omitted in view of its triviality.

In the following we shall consider the f.d.a. P-form. From Definition 4.2 it follows that the investigation of the K-consistent f.d.a. (4.1.4), subject to the initial condition (4.1.1), is reduced to the investigation of a single equation

$$\partial \rho/\partial t + u_0\, \partial \rho/\partial x = \sum_{\substack{i,j \\ i+j=r}} h^i \tau^j \mathscr{F}_{ij1}(Lw, \partial Lw/\partial x, \ldots, \partial^{r+1} Lw/\partial x^{r+1}). \qquad (4.1.9)$$

Equation (4.1.9) is studied below in the strip $-\infty < x < \infty, 0 \le t \le T$, where $0 < T < \infty$, with the initial condition

$$\rho(x, 0) = \begin{cases} \rho_1, & x < x_0, \\ \rho_2, & x > x_0, \end{cases} \qquad (4.1.10)$$

and boundary conditions

$$\lim_{x \to -\infty} \rho(x, t) = \rho_1; \qquad \lim_{x \to +\infty} \rho(x, t) = \rho_2, \qquad 0 < t \le T;$$

$$\lim_{x \to -\infty} \partial\rho/\partial x = 0, \quad r = 2; \quad \lim_{|x| \to \infty} \partial\rho/\partial x = 0, \quad r = 3; \qquad (4.1.11)$$

$$\lim_{x \to -\infty} \partial^2\rho/\partial x^2 = \lim_{|x| \to \infty} \partial\rho/\partial x = 0, \quad r = 4; \qquad 0 < t \le T.$$

In the case when the solution $\rho(x, t)$ of the problem (4.1.9)–(4.1.11) is monotone, the contact strip G_{KD} may be defined as an open simply connected domain in the (x, t)-plane, such that for $t > 0$ at any point $(x, t) \in G_{KD}$ the following conditions are satisfied:

(a) $\qquad\qquad\qquad\qquad$ sign $\partial\rho(x, t)/\partial x = \text{sign}(\rho_2 - \rho_1);$

(b) $\qquad\qquad\qquad\qquad \rho_1(1 - \delta) > \rho > \rho_2(1 + \delta), \qquad \rho_1 > \rho_2;$

$\qquad\qquad\qquad\qquad\qquad \rho_1(1 + \delta) < \rho < \rho_2(1 - \delta), \qquad \rho_1 < \rho_2;$

where δ is a small positive number;
(c) if $(x, t) \notin G_{KD}$, then at least one of the relationships (a), (b) is violated.

Note that this definition is close, in its geometric interpretation, to that of a shock wave zone in a viscous, heat-conducting gas given by Taylor and Maccoll [4.48].

Definition 4.3. Let us call the central line of the contact strip G_{KD} the line $L_{KD} \subset G_{KD}$ on which the value of the function $\rho(x, t)$, where ρ is the solution of the problem (4.1.9)–(4.1.11), does not depend on the values of the grid parameters h, τ.

Denote by $w_h(x, t, h, \tau)$ a vector-function which coincides at the grid points in the (x, t)-plane with the solution of the finite-difference equations approximating the system (1.11), (1.9) under initial conditions (4.1.1), and is determined for the rest of the points (x, t) by the use of interpolation. In the $\rho_h(x, t, h, \tau)$ profile obtained for fixed $t > 0$ the contact discontinuity is smeared over some interval whose width depends on t, as was indicated above. Find for $t = t^n = n\tau$ the boundaries (x_l^n, t^n), (x_r^n, t^n) on the x-axis of the zone of "smearing" of the contact discontinuity in the difference solution ρ_h. Performing this operation for increasing $n = 1, \ldots, N$ we obtain N points (x_l^n, t^n) and N points (x_r^n, t^n). Connecting the neighboring points (x_l^n, t^n), (x_l^{n+1}, t^{n+1}) and then the neighboring points (x_r^n, t^n), (x_r^{n+1}, t^{n+1}), $n = 0, 1, \ldots, N - 1$ by straight line segments we obtain in the (x, t)-plane two lines. These lines, as well as the line $t = N\tau$, yield the boundaries of a domain which we shall call the contact strip G_K. If in the problem under consideration the values ρ_1 and ρ_2 of the density on both sides of the contact discontinuity are known, then for the determination of the boundary points (x_l^n, t^n) and (x_r^n, t^n) one can, for example, check the satisfaction of the inequality

$$\text{sign}[\rho_h(x + h, t, h, \tau) - \rho_h(x - h, t, h, \tau)] = \text{sign}(\rho_2 - \rho_1).$$

Let us now find, in the contact strip G_K at fixed $t = t^n = n\tau$, the point (x_k^n, t^n) of the intersection of the profiles $\rho_h(x, t, h_1, \tau_1)$ and $\rho_h(x, t, h_2, \tau_2)$ where $|h_1 - h_2| + |\tau_1 - \tau_2| \neq 0$. Performing this operation for increasing $n = 1, 2, \ldots, N$ we obtain N points (x_k^n, t^n). Connecting the neighboring points (x_k^n, t^n), (x_k^{n+1}, t^{n+1}), $n = 0, \ldots, N - 1$, by straight line segments we obtain line L_K which we shall call the central line of the contact strip G_K.

Remark. The notion of a central line L_{KD} introduced above has a simple geometric interpretation. Indeed, if there exists the unique line L_{KD}, then the density (ρ) graphs obtained, for example, for $0 < h_1 < h_2 < h_3$, will intersect at the same point C (see Figure 4.3), for each fixed $t > 0$.

Lemma 4.2. *If the equation*

$$\sum_{\substack{i,j \\ i+j=r}} h^i \tau^j \mathscr{F}_{ij1}(\partial w, \partial Lw/\partial x, \ldots, \partial^{r+1} Lw/\partial x^{r+1}) = 0$$

takes place on the line L_{KD} and the line L_{KD} is unique, then L_{KD} coincides with the trajectory of the contact discontinuity.

Taking into account (4.1.9), the equality $\partial \rho/\partial t + u_0\, \partial \rho/\partial x = 0$ signifies that the function $\rho(x, t)$ is constant along the line $dx/dt = u_0$. This follows from the assertion of the lemma in view of the uniqueness of the L_{KD} line.

Definition 4.4. Let us call the quantity

$$X(t, h, \tau, u_0, p_0) = |\rho_1 - \rho_2| \Big/ \max_{x, x \in G_{KD}} |\partial \rho/\partial x| \qquad (4.1.12)$$

the width, after Prandtl, of the contact strip where $\rho(x, t)$ is the solution of the problem (4.1.9)–(4.1.11).

It was shown in [4.38] (see also [4.35], [4.41], [4.42]) that a definition of the K-consistence, being similar to Definition 4.2, may be given directly for

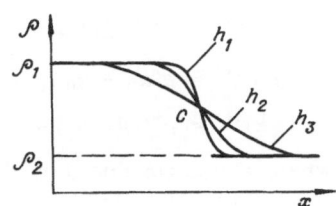

Figure 4.3. Density profiles for different h, $h_1 < h_2 < h_3$.

the finite-difference schemes approximating the Euler equation system (1.1)–(1.3).

Suppose, following [4.38], [4.42], that an explicit (one- or multistep) difference scheme approximating the system (1.1)–(1.3), (1.6) can be represented in the form

$$\rho^{n+1} = H_1(u^n, p^n, \rho^n, \varepsilon^n, T_1, h, \tau); \tag{4.1.13}$$

$$(\rho u)^{n+1} = H_2(u^n, p^n, \rho^n, \varepsilon^n, T_1, h, \tau); \tag{4.1.14}$$

$$(\rho E)^{n+1} = H_3(u^n, p^n, \rho^n, \varepsilon^n, T_1, h, \tau); \tag{4.1.15}$$

after elimination of the intermediate steps where $u^n = u(x, t^n)$, etc., $t^n = n\tau$, τ is the time step, n is an integer, $n = 0, 1, 2, \ldots$, T_1 is the shift operator along the x-axis, that is, $T_1 u(x, t) = u(x + h, t)$, $T_1^m u(x, t) = u(x + hm, t)$, and h is the step of a uniform grid on the x-axis.

Definition 4.5. Let us call the quasi-linearization operation $\tilde{L}$ of the difference expression $H_j(u^n, p^n, \rho^n, \varepsilon^n, T_1, h, \tau)$, $1 \le j \le 3$, the following operation

$$\tilde{L}H_j \equiv H_j(u_0, p_0, \rho^n, f(p_0, \rho^n), T_1, h, \tau),$$

where the function $f(p, \rho)$ enters the equation of state (1.6) and u_0, p_0 are constant values of the velocity and pressure entering (4.1.1).

Definition 4.6. We shall say that the difference momentum equation (4.1.14) is K-consistent with the difference continuity equation (4.1.13), if

$$\tilde{L}H_2 = u_0 \tilde{L}H_1. \tag{4.1.16}$$

Definition 4.7. We call the difference scheme (4.1.13)–(4.1.15), approximating the system of equations (1.1)–(1.3), (1.6), K-consistent if the momentum equation (4.1.14) is K-consistent with the difference continuity equation (4.1.13) and

$$F(\tilde{L}H_1, (\tilde{L}H_3 - 0.5(\tilde{L}H_2)^2/\tilde{L}H_1)/\tilde{L}H_1) \equiv p_0, \tag{4.1.17}$$

where the function $F(\rho, \varepsilon)$ enters the equation of state (1.5) and satisfies the identity (1.7).

EXAMPLE. Applying the quasi-linearization operation $\tilde{L}$ to the difference equations of the FLIC method [4.44] we obtain in the case $u_0 > 0$ (see [4.38]):

$$\rho^{n+1} = \rho^n - \kappa(I - T_{-1})\rho^n,$$

$$(\rho u)^{n+1} = \rho^n u_0 - u_0 \kappa(I - T_{-1})\rho^n, \tag{4.1.18}$$

$$(\rho E)^{n+1} = \rho^n f(p_0, \rho^n) + 0.5\rho^n u_0^2 - \kappa(I - T_{-1})[\rho^n f(p_0, \rho^n) + 0.5u_0^2\rho^n],$$

where $\kappa = u_0\tau/h$ and I is the identity operator. Let, for definiteness, the function $F(\rho, \varepsilon)$ in (1.5) have the form

$$F(\rho, \varepsilon) = a\rho\varepsilon + B_1(\rho - \rho_0), \tag{4.1.19}$$

where a, B_1, ρ_0 are some positive constants. Then the function $f(p, \rho)$ entering (1.6) obviously has the form

$$f(p, \rho) = [p - B_1(\rho - \rho_0)]/(a\rho) \tag{4.1.20}$$

and it is easy to be convinced that the substitution of formulas (4.1.18), (4.1.20) into (4.1.19) yields the identity (4.1.17).

4.1.2. Analysis of Schemes in the Case of the First Differential Approximation K-Consistence

Theorem 4.1. *If*

(i) *the f.d.a. of the difference scheme of the rth order of accuracy is K-consistent ($1 \leq r \leq 2$);*
(ii) *equation (4.1.9) has the form*

$$\partial\rho/\partial t + u_0\,\partial\rho/\partial x = (-1)^{r+1}\mu(h, \tau, u_0, p_0)\,\partial^{r+1}\rho/\partial x^{r+1}; \tag{4.1.21}$$

(iii) $$\mu(h, \tau, u_0, p_0) > 0,$$

then there exists a unique central line L_{KD} of the contact strip and this line coincides with the trajectory of the contact discontinuity.

The proof of this theorem is reduced to obtaining the exact solution of the problem (4.1.21), (4.1.10), (4.1.11) and the subsequent study of its properties. At first let us prove the theorem for the case $r = 1$. The exact solution of equation (4.1.21) at the initial condition (4.1.10) has the form [4.49]

$$\rho(x, t) = 0.5(\rho_1 + \rho_2) + 0.5(\rho_2 - \rho_1)\,\mathrm{erf}(\xi(x, t)), \tag{4.1.22}$$

where

$$\xi(x, t) = (x - x_0 - u_0 t)/([2(\mu t)^{1/2}]; \qquad \mathrm{erf}(\xi) = \frac{2}{\sqrt{\pi}}\int_0^\xi e^{-\lambda^2}\,d\lambda; \tag{4.1.23}$$

hence $\mathrm{erf}(\xi)$ is the probability integral. From formulas (4.1.22), (4.1.23) it follows that only at $\xi = 0$ the value of the density ρ does not depend on h, τ, it is equal to $0.5(\rho_1 + \rho_2)$ at $t > 0$. According to (4.1.23) the equation $\xi = 0$ coincides with the equation for the contact discontinuity trajectory in the (x, t)-plane, that was to be proved. $\qquad\square$

Consider now the case $r = 2$. We search for the solution of equation (4.1.21) having the form $\rho(x, t) = \tilde{\rho}(\tilde{\xi})$ where

$$\tilde{\xi} = (x - x_0 - u_0 t)/(\beta(\mu t)^{1/3}), \qquad \beta = \mathrm{const}.$$

For $\tilde{\rho}(\tilde{\xi})$ one obtains from (4.1.21) an ordinary differential equation

$$3\beta^{-3} \, d^3\tilde{\rho}/d\tilde{\xi}^3 - \tilde{\xi} \, d\tilde{\rho}/d\tilde{\xi} = 0. \qquad (4.1.24)$$

In (4.1.24) let us make a change

$$d\tilde{\rho}/d\tilde{\xi} = v(\xi), \qquad \tilde{\xi} = \alpha\xi, \qquad \alpha = \text{const.} \qquad (4.1.25)$$

As a result we obtain from (4.1.24)

$$d^2v/d\xi^2 - (\alpha^3\beta^3/3)\xi v = 0. \qquad (4.1.26)$$

Equation (4.1.26) coincides with the Airy equation [4.50] $d^2v/d\xi^2 = \xi v$ if one takes $\alpha^3 = 3/\beta^3$. The general solution of the Airy equation has the form

$$v(\xi) = aAi(\xi) + bBi(\xi),$$

where $Ai(\xi)$, $Bi(\xi)$ are the Airy functions and a, b are arbitrary constants. The variable $\xi = \tilde{\xi}/\alpha$ already does not contain an arbitrary constant β. With (4.1.25) in view one obtains the general solution of equation (4.1.24) in the form

$$\tilde{\rho}(\xi) = \int_0^\xi (aAi(\lambda) + bBi(\lambda)) \, d\lambda + c, \qquad (4.1.27)$$

where c is an integration constant. Let us find the constants a, b, c making use of the initial conditions (4.1.10), the boundary conditions (4.1.11), and the properties of the Airy functions. According to [4.50], [4.51]

$$\lim_{|\xi| \to \infty} Ai(\xi) = 0; \qquad \lim_{\xi \to -\infty} Bi(\xi) = \infty;$$

therefore we should set $b = 0$ in (4.1.27) to satisfy the boundary conditions (4.1.11). According to [4.51],

$$\int_0^\infty Ai(\lambda) \, d\lambda = 1/3, \qquad \int_0^{-\infty} Ai(\lambda) \, d\lambda = -2/3. \qquad (4.1.28)$$

Taking formulas (4.1.28), (4.1.27), (4.1.10) into account we obtain the solution $\rho(x, t)$ as

$$\rho(x, t) = \tilde{\rho}(\xi) = (2\rho_2 + \rho_1)/3 + (\rho_2 - \rho_1) \int_0^\xi Ai(\lambda) \, d\lambda, \qquad (4.1.29)$$

where

$$\xi = (x - x_0 - u_0 t)/[(3\mu t)^{1/3}]. \qquad (4.1.30)$$

It follows from formulas (4.1.29), (4.1.30) that on the line $x = x_0 + u_0 t$, the contact discontinuity trajectory, the density value does not depend on h, τ and is equal to $(2\rho_2 + \rho_1)/3$. Let us show that the value of the density ρ does not depend on h, τ only on the line $x = x_0 + u_0 t$. Suppose that there exists one more line $x = X(t)$ belonging to G_{KD}, and on this line the density value also does not depend on h and τ. From (4.1.29) we have:

$$\rho(X(t), t) = (2\rho_2 + \rho_1)/3 + (\rho_2 - \rho_1) \int_0^{\xi(X(t), t)} Ai(\lambda) \, d\lambda, \quad (4.1.31)$$

where

$$\xi(X(t), t) = (X(t) - x_0 - u_0 t)(3\mu t)^{-1/3}. \quad (4.1.32)$$

In accordance with the definition of the line $x = X(t)$

$$\partial \rho(X(t), t)/\partial h = \partial \rho(X(t), t)/\partial \tau = 0.$$

Let us make use of the formula

$$\partial \rho(X(t), t)/\partial h = [\partial \tilde{\rho}(\xi(X(t), t))/\partial \xi] [\partial \xi(X(t), t)/\partial \mu](\partial \mu/\partial h). \quad (4.1.33)$$

Since μ depends on h, see (4.1.21), $\partial \mu/\partial h \neq 0$. We have from (4.1.29) that

$$\partial \tilde{\rho}(\xi(X(t), t))/\partial \xi = (\rho_2 - \rho_1) Ai(\xi(X(t), t)). \quad (4.1.34)$$

Since the line $x = X(t)$ belongs to the contact strip G_{KD}, sign $\partial \rho(X(t), t)/\partial x = $ sign$(\rho_2 - \rho_1)$, hence $Ai(\xi(X(t), t)) > 0$ by virtue of (4.1.34) and condition (iii). Therefore, we obtain from the requirement $\partial \rho(X(t), t)/\partial h = 0$ and taking (4.1.33) into account that

$$\partial \xi(X(t), t)/\partial \mu = -(1/3)\mu^{-4/3}(3t)^{-1/3}[X(t) - x_0 - u_0 t] = 0,$$

from where we find that $X(t) = x_0 + u_0 t$.

Corollary 1. *At $r = 1$ along the contact discontinuity trajectory the equation $\partial^2 \rho/\partial x^2 = 0$ takes place. In view of the function erf(ξ) properties it is easy to obtain that the line L_{KD} is the line of the extremum of the function $\partial \rho(x, t)/\partial x$.*

Corollary 2. *At $r = 2$ the abscissa of the quantity $|\partial \rho/\partial x|$ maximum, where $\rho(x, t)$ is defined by (4.1.29), (4.1.30), lies to the left of the contact discontinuity abscissa.*

In fact, let us consider the formula

$$\partial^2 \rho/\partial x^2 = -(dAi(\xi)/d\xi)(\rho_1 - \rho_2)(3\mu t)^{-2/3}.$$

It is well-known that the equation $Ai'(\xi) = 0$ has roots only for $\xi < 0$. Denote by ξ_1 the abscissa of the first local minimum of the function

$$a(\xi) = \int_0^\xi Ai(\lambda) \, d\lambda \quad (4.1.35)$$

when one moves along the ξ-axis to the left of the coordinate origin (see Figure 4.4). Let ξ_0 be the least in modulus root of the equation $dAi(\xi)/d\xi = 0$, according to [4.52] $\xi_0 \cong -1.018793$, $\xi_1 = -2.338107$. Thus $\xi_0 > \xi_1$, and

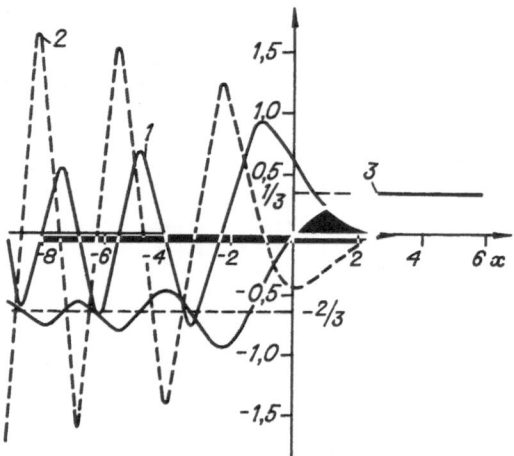

Figure 4.4. The graphs of the functions. $1-\sqrt{\pi}Ai$; $2-\sqrt{\pi}Ai'$; and $3-\int_0^x Ai(x)\,dx$.

consequently the maximum value of the derivative $|\partial\rho/\partial x|$ in the "smeared" contact discontinuity zone, is achieved at $\xi = \xi_0$. Let x_* be the abscissa of the point where $\max|\partial\rho/\partial x|$ takes place in the contact strip. Then with (4.1.22) in view

$$x_* = x_0 + u_0 t + \xi_0(3\mu t)^{1/3} \tag{4.1.36}$$

it follows that the distance between the abscissa of the point where $\max_x|\partial\rho/\partial x|$ takes place and the contact discontinuity abscissa increases proportionally to $t^{1/3}$.

Corollary 3. *At $r = 1$ (schemes of first-order accuracy)*

$$X = 2(\pi\mu t)^{1/2}. \tag{4.1.37}$$

Corollary 4. *At $r = 2$ (schemes of second-order accuracy)*

$$X = (3\mu t)^{1/3}/Ai(\xi_0), \tag{4.1.38}$$

where according to Corollary 2, $\xi_0 \cong -1.018793$ and $Ai(\xi_0) \cong 0.535657$.

Corollary 5. *At $r = 2$ the abscissa of the local extremum of the function $\partial^2\rho/\partial x^2$ in the contact strip coincides with the exact position of the contact discontinuity.*

According to formula (4.1.21) and Lemma 4.2, $\rho_{xxx} = 0$ along the contact discontinuity trajectory. From (4.1.29) we get

$$\partial^3\rho/\partial x^3 = \xi(\rho_2 - \rho_1)Ai(\xi)/(3\mu t);$$
$$\partial^4\rho/\partial x^4|_{\xi=0} = (\rho_2 - \rho_1)/[3^{2/3}\Gamma(2/3)(3\mu t)^{4/3}]. \tag{4.1.39}$$

The relations (4.1.39) mean that along the line $\xi = 0$ and along the line $\xi = \xi_1$

the derivative $\partial^2 \rho / \partial x^2$ achieves its local extrema. However, $\partial \rho / \partial x|_{\xi=\xi_1} = 0$, thus there is only one extremum of the derivative $\partial^2 \rho / \partial x^2$ in a zone of smeared contact discontinuity at $\xi = 0$.

The change of variables of the form

$$\xi(x, t) = (x - x_0 - u_0 t)/(2\sqrt{\mu t})$$

is sometimes called the Boltzmann change (Boltzmann was the first to apply it [4.53]).

Consider two examples of the application of Theorem 4.1.

EXAMPLE 1 (The Lax Scheme [4.5]). According to Sections 2.3 and 2.4 the f.d.a. P-form of this scheme has the form

$$\partial w/\partial t + \partial \varphi(w)/\partial x = (h^2/(2\tau)) \, \partial^2 w/\partial x^2 - (\tau/2)(\partial/\partial x)(A^2 \, \partial w/\partial x), \quad (4.1.40)$$

where $A = \partial \varphi / \partial w$. Let us apply the quasi-linearization operation L to both sides of (4.1.40). As a result we obtain the system of equations

$$U(\partial \rho / \partial t + u_0 \, \partial \rho / \partial x) = \mu (\partial / \partial x)(U \, \partial \rho / \partial x), \quad (4.1.41)$$

where

$$\mu(h, \tau, u_0, p_0) = 0.5(h^2/\tau - \tau u_0^2). \quad (4.1.42)$$

From (4.1.41) it follows that the f.d.a. of the Lax scheme is not K-consistent in the general case. However, it is easy to see that if the function $f(p, \rho)$ in the equation of state (1.6) satisfies the relation

$$(\partial^2 / \partial \rho^2)[\rho f(p_0, \rho)] = 0, \quad (4.1.43)$$

then the K-consistence of the f.d.a. (4.1.40) takes place, and the system (4.1.41) is reduced to a single equation (4.1.21) where $r = 1$. Note that it is not difficult to identify the class of the equations of state which satisfy the condition (4.1.43). For this purpose let us integrate equation (4.1.43) as an ordinary differential equation at fixed $p = p_0$. The solution results in the form

$$\varepsilon = f(p, \rho) = f_1(p) + f_2(p)/\rho. \quad (4.1.44)$$

With regard to the known stability condition of the Lax scheme from (4.1.42) the inequality $\mu > 0$ follows. Thus, if equation (4.1.43) is satisfied, then there exists the unique central line L_{KD} of the contact strip for the Lax scheme. Note that, for example, the function

$$\varepsilon = f(p, \rho) = p/[\rho(\gamma - 1)], \quad (4.1.45)$$

where $\gamma = \text{const} > 1$ (ideal gas), satisfies the condition (4.1.43).

The importance of the notion (4.1.12) of the width of the contact strip for the construction of differential analyzers of the contact discontinuities follows from the fact that, in accordance with formulas (4.1.37), (4.1.42), the density graphs for fixed t will intersect on the L_{KD} line not only at different h steps, but also at different τ. Furthermore, if one introduces the coordinate $x' = x - x_0 - u_0 t$, then the density graphs with respect to the x'-axis will intersect at the same point $x' = 0$ for different t in accordance with formulas (4.1.22), (4.1.23), (4.1.37), differing from each other only by the "steepness" of the profiles.

EXAMPLE 2 (The One-Step Lax–Wendroff Scheme [4.45]). The f.d.a. H-form of this scheme has the form [4.54]

$$\partial \mathbf{w}/\partial t + \partial \boldsymbol{\varphi}(\mathbf{w})/\partial x = -(\tau^2/6)\, \partial^3 \mathbf{w}/\partial t^3 - (h^2/6)\, \partial^3 \boldsymbol{\varphi}/\partial x^3. \qquad (4.1.46)$$

Applying the quasi-linearization operation L to both sides of equation (4.1.46) we obtain the system

$$U(\partial \rho/\partial t + u_0\, \partial \rho/\partial x) = -(\tau^2/6)(\partial^2/\partial t^2)(U\, \partial \rho/\partial t)$$
$$- (h^2/6)u_0(\partial^2/\partial x^2)(U\, \partial \rho/\partial x). \qquad (4.1.47)$$

It is easy to see from (4.1.47) that the K-consistence of the f.d.a. (4.1.46) takes place if equation (4.1.43) is satisfied. In this case it is easy to obtain the f.d.a. P-form from the f.d.a. (4.1.47)

$$U(\partial \rho/\partial t + u_0\, \partial \rho/\partial x) = -\mu U\, \partial^3 \rho/\partial x^3, \qquad (4.1.48)$$

where

$$\mu(h, \tau, u_0) = (h^2/6)u_0 - (\tau^2/6)u_0^3, \qquad (4.1.49)$$

Taking into account the Lax–Wendroff scheme stability condition $u_0 \tau/h < 1$ always, therefore, it follows from (4.1.49) that μ is positive.

Consider now the f.d.a. (2.4.12), (2.4.13) of the "breakdown-of-discontinuity" scheme [4.55]. Applying the quasi-linearization operation L to both sides of equations (2.4.12), (2.4.13) and passing then to the f.d.a. P-form with the use of the algorithm described in Section 1.2, we obtain a system of the form (4.1.41) where one should now set

$$\mu(h, \tau, u_0) = 0.5|u_0|(h - \tau|u_0|). \qquad (4.1.50)$$

Consider now the FLIC method [4.44]. The f.d.a. H-form of the computational scheme of this method has the form (2.4.4) in the absence of an artificial viscosity q introduced additively into the pressure in [4.44]. In the case when, at the Lagrangian stage of the "coarse particles" method [4.56], the first-order accuracy formulas are used the f.d.a. H-form of the integral step scheme of the method [4.56] coincides with the system (2.4.4), if one omits the term $(\tau/2)(pp_x/\rho)_x$ in the energy equation of the system (2.4.4). It is clear that this term does not affect the result of the application of the quasi-lineariztion operation L to the system (2.4.4). Thus, in the case of the FLIC method and of the "coarse particles" method we obtain (after execution of the operation L) the system (4.1.41) where μ is again computed by formula (4.1.50), and the condition (4.1.43) is assumed to be satisfied. Thus, for the FLIC method, for the "breakdown-of-discontinuity" scheme, and for the "coarse particles" method there exists the unique central line L_{KD} of a contact strip, and along this line the properties of the solution (4.1.22), (4.1.23) formulated in the corollaries of Theorem 4.1 take place. Denote by X_{Lax} the width after Prandtl of a contact strip for the Lax scheme and by X_{FLIC} the width for the FLIC method. Then

$$X_{Lax}/X_{FLIC} = \sqrt{1 + h/(\tau|u_0|)}. \qquad (4.1.51)$$

It follows from (4.1.51) that at $0 < |u_0| < \infty$ the width of "smearing" of a contact discontinuity by the Lax scheme is larger than in the case of the FLIC scheme. It follows from (4.1.50) that the width X for the FLIC method increases with the velocity u_0 of the contact discontinuity propagation. A similar feature of the FLIC method was also found in the computations of a flow with shock wave [4.44].

The Rusanov Scheme [4.7]. The f.d.a. H-form of this scheme has the form (2.4.2) where ω is a constant coefficient, $\omega = O(1)$. It is easy to be convinced of the fact that the f.d.a. (2.4.2) is K-consistent if the condition (4.1.43) is satisfied, and equation (4.1.9) takes the form

$$\partial\rho/\partial t + u_0\, \partial\rho/\partial x = [\omega(h/2)|u_0| - (\tau/2)u_0^2]\, \partial^2\rho/\partial x^2$$
$$+ (\partial/\partial x)[\omega(h/2)(\gamma p_0/\rho)^{0.5}\, \partial\rho/\partial x] \qquad (4.1.52)$$

in the case when the equation of state (4.1.45) is used. Rewrite equation (4.1.52) in the form

$$\partial\rho/\partial t + u_0\, \partial\rho/\partial x = (\partial/\partial x)[b(\rho)\, \partial\rho/\partial x], \qquad (4.1.53)$$

where

$$b(\rho) = \mu + \beta\rho^{-1/2},$$
$$\mu = \omega h|u_0|/2 - \tau u_0^2/2, \qquad \beta = 0.5\omega h(\gamma p_0)^{1/2}. \qquad (4.1.54)$$

It follows from (4.1.54) that at $\beta \neq 0$ equation (4.1.53) is nonlinear. This justifies the term "quasi-linearization" which we have introduced in Section 4.1.1. We search for the solution of the problem (4.1.53), (4.1.10) in the form $\rho = \tilde{\rho}(\xi)$ where ξ is determined by (4.1.23). For $\tilde{\rho}(\xi)$ we get the equation

$$-2\xi\, d\tilde{\rho}/d\xi = (1/\mu)(d/d\xi)[b(\tilde{\rho})\, d\tilde{\rho}/d\xi]. \qquad (4.1.55)$$

Let us make a change

$$\tilde{\rho}(\xi) = (\beta^2/\mu^2)\hat{\rho}(\xi), \qquad (4.1.56)$$

assuming that $\beta > 0$, $\mu > 0$. In addition, we shall assume that the steps τ and h are related by formula (2.4.5). In this case we have with (4.1.54) in view that

$$\alpha \equiv \beta^2/\mu^2 = \omega^2\gamma p_0/[(\omega|u_0| - \kappa u_0^2)^2]. \qquad (4.1.57)$$

It follows from (4.1.57) that at $\tau/h = \kappa = $ const the relation β^2/μ^2 remains constant when τ, h vary. Substituting the right-hand side of equation (4.1.56) instead of $\tilde{\rho}$ into equation (4.1.55) we obtain for the determination of $\hat{\rho}(\xi)$ the equation

$$-2\xi \, d\hat{\rho}/d\xi = (d/d\xi)[(1 + \hat{\rho}^{-1/2}) \, d\hat{\rho}/d\xi]. \tag{4.1.58}$$

Taking into account formulas (4.1.11), (4.1.56), (4.1.57) we shall solve equation (4.1.58) under following boundary conditions:

$$\lim_{\xi \to -\infty} \hat{\rho}(\xi) = \rho_1^0; \qquad \lim_{\xi \to +\infty} \hat{\rho}(\xi) = \rho_2^0; \tag{4.1.59}$$

where $\rho_\nu^0 = (1/\alpha)\rho_\nu$, $\nu = 1, 2$. In the following we shall assume the presence of the inequalities $0 < \rho_\nu^0 < \infty$, $\nu = 1, 2$. Let $\hat{b}(\hat{\rho}) = 1 + \hat{\rho}^{-1/2}$. Making use of the change

$$\tilde{v} = \int \tilde{b}(\hat{\rho}) \, d\hat{\rho} = \hat{\rho} + 2\sqrt{\hat{\rho}}$$

we can reduce equation (4.1.58) to the form

$$d^2v/d\xi^2 = -\xi G(v) \, dv/d\xi, \tag{4.1.60}$$

where

$$v(\xi) = \tilde{v}(\hat{\rho}(\xi)), \qquad G(v) = 2/\tilde{b}(\hat{\rho}(v)),$$

$$\hat{\rho}(v) = v + 2 + [(v + 2)^2 - v^2]^{1/2}.$$

The boundary conditions for $v(\xi)$ may be written, with (4.1.59) in view, as

$$\lim_{\xi \to -\infty} v(\xi) = v_1^0, \qquad \lim_{\xi \to +\infty} v(\xi) = v_2^0, \tag{4.1.61}$$

where $v_\nu^0 = \tilde{v}(\rho_\nu^0)$, $\nu = 1, 2$. It appears that on the basis of the results of [4.57], [4.58] one should be able to prove the unique solvability of the problem (4.1.60), (4.1.61). However, the present authors did not aim at the investigation of these questions, and all the subsequent considerations are based on the assumption that a solution of the problem (4.1.60), (4.1.61) exists.

Consider the question of existence and uniqueness of the central line L_{KD} of a contact strip in the solution of the problem (4.1.52), (4.1.10), (4.1.11), assuming that there exists a unique bounded and monotone solution of the problem (4.1.60), (4.1.61). Let us take some $\delta \neq 0$. Consider in the (x, t)-plane a point with coordinates $x_* = x_0 + u_0 t - \delta$, $t = t$. Then we have in accordance with (4.1.23) that $\xi(x_*, t) = \tilde{\xi}(h, t)$ where

$$\tilde{\xi}(h, t) = -\delta/(2\sqrt{\mu t}) = -\delta h^{-1/2}[2t(\omega|u_0| - \kappa u_0^2)]^{-1/2}.$$

Then

$$\partial \rho(x_*, t)/\partial h = (\beta^2/\mu^2) \, d\hat{\rho}(\tilde{\xi}(h, t))/d\xi \, 0.5\delta[2t(\omega|u_0| - \kappa u_0^2)]^{-1/2}h^{-3/2}. \tag{4.1.62}$$

In formula (4.1.62) $d\hat{\rho}/d\xi \neq 0$ by virtue of the assumption on the monotonicity of the solution of the problem (4.1.60), (4.1.61). Therefore, the equation $\partial\rho(x_*, t)/\partial h = 0$ takes place only at $\delta = 0$ with regard to (4.1.53). Thus, under the above-stated assumptions on the $v(\xi)$ solution there exists the unique central line L_{KD} of a contact strip in the solution of the problem (4.1.52), (4.1.10), (4.1.11), and the equation $x = x_0 + u_0 t$ of this line coincides with the exact contact discontinuity trajectory in the solution of a model Riemann problem (1.1)–(1.3), (1.5), (4.1.1).

Consider now a question on the computation of the width after Prandtl X of a contact strip in the solution of the problem (4.1.52), (4.1.10), (4.1.11). In a particular case $\beta = 0$ we have from (4.1.54) that $b(\rho) = \mu$, therefore, $X = 2(\pi\mu t)^{0.5}$. Consider the case $\beta > 0$. Let ξ_1 be a point at which $d\rho/d\xi = A_0$ where $0 < |A_0| < \infty$. Such a point exists in the case of a bounded solution $\rho(\xi) \in C^2(-\infty, \infty)$. Let $\xi = \xi_0$ be such a value of the ξ variable at which $\rho_{\xi\xi} = 0$. Such a value of ξ exists. In fact, if one assumes that sign $\rho_{\xi\xi} = $ const $\neq 0$, then from the formula

$$\rho(\xi) = \rho(\xi_1) + \int_{\xi_1}^{\xi}\left[A_0 + \int_{\xi_1}^{\zeta}(d^2\rho(\lambda)/d\lambda^2)\, d\lambda\right]d\zeta$$

one gets that $\rho(\xi) \to \infty$ at $\xi \to \infty$, which is in contradiction with the boundedness of the solution $v(\xi)$. We have from (4.1.55) that

$$-2\xi_0(d\tilde\rho/d\xi)|_{\xi=\xi_0} = [(1/\mu)(db/d\tilde\rho)(d\tilde\rho/d\xi)^2]_{\xi=\xi_0}. \tag{4.1.63}$$

Let us find the maximum value of the derivative $|d\tilde\rho/d\xi|$ in a contact strip by using the relationship (4.1.63)

$$d\tilde\rho/d\xi = -2\xi_0\mu/(db/d\tilde\rho) = 4\xi_0\beta^{-1}\mu\tilde\rho(\xi_0)^{3/2}. \tag{4.1.64}$$

Making use of formulas (4.1.12), (4.1.37), (4.1.64) we can find the following expression for the width X of a contact strip when using the Rusanov scheme

$$X = \delta_{p_0}^0 2(\pi\mu t)^{1/2} + (1 - \delta_{p_0}^0)|\rho_2 - \rho_1|0.5\beta(t/\mu)^{0.5}|\xi_0|^{-1}\tilde\rho(\xi_0)^{-3/2}, \tag{4.1.65}$$

where $\delta_{p_0}^0$ is the Kronecker symbol. Let us show that sgn $\xi_0 = \text{sgn}(\rho_2 - \rho_1)$ in formula (4.1.64). In accordance with (4.1.55), the equation

$$(db/d\tilde\rho)(d\tilde\rho/d\xi)^2 + b(\tilde\rho)\, d^2\tilde\rho/d\xi^2 = 0$$

takes place on the line $\xi = 0$. Therefore, at $\xi = 0$ we have, with (4.1.54) in view, that $d^2\tilde\rho/d\xi^2 > 0$. Assume that the density profile $\tilde\rho(\xi)$ is monotone in a contact strip. Consider first the case $\rho_1 > \rho_2$. Then $d^2\tilde\rho/d\xi^2 > 0$ at $\xi > \xi_0$, from where it follows that $0 > \xi_0$. The case $\xi < \xi_0$ is considered similarly.

It follows from formula (4.1.65) that the RUSANOV scheme [4.7] has the following shortcoming: the stagnant contact discontinuity ($u_0 = 0$) will be smeared in computations by this scheme if $p_0 \neq 0$. The same shortcoming is inherent in the modifications of the Rusanov scheme proposed in [4.59]–[4.62] as well as in the second-order scheme of BALAKIN and BULANOV [4.63]–[4.65].

At the same time, such schemes as the GODUNOV scheme [4.55], the FLIC scheme [4.44], and the Lax–Wendroff scheme [4.45] do not have the above shortcoming (see below Table 4.1 and the comments on it).

Taking into account the foregoing, as well as Lemma 4.2, we obtain that the abscissa of the point of maximum of the quantity

$$|[(\omega h/2)(|u| + c) - (\tau/2)u^2](\partial\rho/\partial x)|, \qquad x \in G_{KD}, \qquad (4.1.66)$$

where c is the local sound speed, will coincide with the exact position of a contact discontinuity. It is interesting to note that the well-known technique of the visual determination of the contact discontinuity position (by the coalescence of isochors) is generally inapplicable in this case, by virtue of the fact that $\zeta_0 \neq 0$, as was found above. Thus, in the process of the development of algorithms of differential analyzers for contact discontinuities it is necessary to take into account both the dissipative mechanism, which is implicitly present in a scheme, and the artificial dissipators introduced into a scheme.

Table 4.1 summarizes the investigation results for a number of difference schemes for first- and second-order accuracy. These investigations (see above) were carried out under the assumption that the equation of state meets the requirement (4.1.43) and thus the f.d.a. K-consistence takes place. A specific form of the function $\mu(h, \tau, u_0, p_0)$ entering (4.1.21) is indicated in Table 4.1 for each scheme that enables one to evaluate the width of a sharp change region in density within the contact strip.

In addition to Table 4.1 it is possible to introduce, by analogy with [4.66], [4.67], the notion of the grid Péclet number

$$P_\Delta = u_0 h/\mu \qquad (4.1.67)$$

for equation (4.1.21) at $r = 1$ (first-order schemes). This is a dimensionless parameter characterizing the difference scheme from the point of view of the size of the numerical diffusion in a contact strip. If $\mu = 0$ (no numerical diffusion of a contact boundary), then $P_\Delta = \infty$. Let us evaluate in a number of examples the grid Péclet number (4.1.67) when using some first-order schemes.

(a) The Lax scheme. Let us introduce the notation $\kappa = u_0\tau/h$. Then it is easy to compute with the aid of Table 4.1 that $P_\Delta = 2\kappa/(1 - \kappa^2)$. Write the Courant–Friedrichs–Lewy stability condition for the Lax scheme in the form $(\kappa_{max} + \kappa_c) \leq 1$ where $\kappa_{max} = \max_x |u|\tau/h$, $\kappa_c = \max_x c\tau/h$. It is clear that $0 \leq \kappa \leq \kappa_{max} \leq 1 - \kappa_c$, therefore in the case of the Lax scheme the grid Péclet number varies in the limits

$$0 \leq P_\Delta \leq 2(1 - \kappa_c)/[1 - (1 - \kappa_c)^2] \leq 2.$$

(b) In the case of the FLIC method [4.44] we have $P_\Delta = 2/(1 - \kappa)$, from where the following limits of the P_Δ number variation are obtained

$$2 \leq P_\Delta \leq 2/[1 - (1 - \kappa_c)^2].$$

Table 4.1

Method	Order of approximation	μ	X	Practical criterion for finding the line L_K
Lax [4.5]	$O(\tau) + O(h^2/2\tau)$	$h^2/(2\tau) - (\tau/2)u_0^2$	$2(\pi\mu t)^{1/2}$	By $\max\lvert\partial\rho/\partial x\rvert$ at $\operatorname{sign}(\partial\rho/\partial x) = \operatorname{sign}(\rho_2 - \rho_1)$
Godunov [4.55]	$O(\tau) + O(h)$	$\lvert u_0\rvert(h/2 - (\tau/2)\lvert u_0\rvert)$	$2(\pi\mu t)^{1/2}$	
FLIC [4.44]	$O(\tau) + O(h)$	$\lvert u_0\rvert(h/2 - (\tau/2)\lvert u_0\rvert)$	$2(\pi\mu t)^{1/2}$	
"coarse particles" [4.56]	$O(\tau) + O(h)$	$\lvert u_0\rvert(h/2 - (\tau/2)\lvert u_0\rvert)$	$2(\pi\mu t)^{1/2}$	
Rusanov [4.7]	$O(\tau) + O(h)$	$(\omega h\lvert u_0\rvert - \tau u_0^2)/2$	See (4.1.65)	By $\max\left\lvert[\omega h(\lvert u\rvert/2 + c/2) - \tau u^2/2]\partial\rho/\partial x\right\rvert$ at $\operatorname{sgn}\rho_x = \operatorname{sgn}(\rho_2 - \rho_1)$
Lax–Wendroff one-step [4.45]	$O(\tau^2) + O(h^2)$	$(\lvert u_0\rvert/6)(h^2 - \tau^2 u_0^2)$	$(3\mu t)^{1/3}/Ai(\xi_0),$ $Ai(\xi_0) \cong 0.535657$	By $\max\lvert\partial^2\rho/\partial x^2\rvert$ at $\operatorname{sign}(\partial\rho/\partial x) = \operatorname{sign}(\rho_2 - \rho_1)$
Lax–Wendroff two-step [4.45]	$O(\tau^2) + O(h^2)$	$(\lvert u_0\rvert/6)(h^2 - \tau^2 u_0^2)$		
MacCormack [4.82]	$O(\tau^2) + O(h^2)$	$(\lvert u_0\rvert/6)(h^2 - \tau^2 u_0^2)$		

It is clear from the definition (4.1.67) that the P_Δ number is inversely proportional, at $u \neq 0$, to the value of the coefficient μ of the numerical viscosity. Therefore, the larger P_Δ is for a specific scheme, the more advantageous is this scheme for problems with contact discontinuities. Compare, for example, the grid Péclet numbers $(P_\Delta)_{\text{Lax}}$ and $(P_\Delta)_{\text{FLIC}}$ for the Lax and FLIC methods, respectively:

$$(P_\Delta)_{\text{Lax}}/(P_\Delta)_{\text{FLIC}} = \kappa/(1 + \kappa) < 1,$$

that is, the numerical diffusion across the contact boundary is larger in the Lax scheme than in the FLIC scheme.

From the above consideration of the properties of the solution of equation (4.1.21) one can deduce (for practical computations using the first- and second-order schemes satisfying the conditions of Theorem 4.1) the following three algorithms of the differential analyzer of the contact discontinuity.

Algorithm I. Localization of the line L_K directly on the basis of its definition.

Algorithm II. Determining at each $t > 0$ the point of the line L_K as a point where $\max|\partial^r \rho/\partial x^r|$ takes place at $x \in G_K$.

Algorithm III. On the line L_{KD} the property $\rho = 0.5(\rho_1 + \rho_2)$ at $r = 1$ and $\rho = (2\rho_2 + \rho_1)/3$ at $r = 2$ is satisfied. Therefore, at known values for ρ_1, ρ_2 one can find approximately the abscissa of the point of the line L_K by inverse interpolation of the function $\rho_h(x, t, h, \tau)$.

It is easy to see that the well-known technique for determining the position of contact boundaries by the coalescence of isochors corresponds to Algorithm II at $r = 1$, since it is clear that the maximum coalescence of isochors takes place in the region of maximum density gradients. On the other hand, at $r = 2$, in view of formula (4.1.36), this technique for the localization of contact boundaries is inapplicable, because (by the choice of time t in (4.1.36)) it is easy to obtain any prescribed deviation of the place of the most intense coalescence of isochors from the true contact discontinuity position.

In the case of the RUSANOV scheme [4.7] the position of domains of maximum coalescence of isochors does not coincide with the true position of the contact boundaries. It is reasonable to use here not the lines of constant density, but the level curves of a function which can easily be obtained by turning to formulas (4.1.53), (4.1.54) and to Table 4.1:

$$R(p, \rho) = \int b(\rho)\, d\rho = [\omega(h/2)|\mathbf{u}| - (\tau/2)|\mathbf{u}|^2]\rho + \omega h(\gamma p \rho)^{0.5}. \quad (4.1.68)$$

We now make a few remarks following on from an analysis of Table 4.1. Except for the LAX scheme [4.5] and the RUSANOV scheme [4.7] all the first-

and second-order schemes listed in Table 4.1 possess the following positive property: they do not smear out stagnant contact discontinuities. This property for the "breakdown-of-discontinuity" scheme was previously noted in [4.68]. The relationship $\tau|u_0|/h = 1$ cannot be realized in practice for schemes from Table 4.1, in view of the Courant–Friedrichs–Lewy stability condition $\tau(|u| + c)/h \leq 1$.

The absence of smearing of a stagnant contact discontinuity noted above for some schemes is related to the K-property formulated in [4.69], [4.70]. In fact, a relation between the K-property and the vanishing of one of the eigenvalues of the matrix $A = \partial\varphi/\partial w$ was established in [4.70]. As is known, in the case of the Euler equations (1.11), (1.9), the eigenvalues λ_i of the A matrix are expressed by the formula

$$\lambda_1 = u - c, \qquad \lambda_2 = u, \qquad \lambda_3 = u + c.$$

The contact discontinuity moves along the characteristic $dx/dt = u$, therefore, in the case of a stagnant contact discontinuity, $\lambda_2 \equiv 0$. Thus, the presence of the K-property in a difference scheme for the system (1.11), (1.9) means that a stagnant contact discontinuity will not be smeared when using such a difference scheme in computations. It was shown in [4.70] that the Lax–Wendroff scheme possesses the K-property, however, the Lax scheme does not possess this property. These conclusions are in agreement with Table 4.1. It appears that the requirement of the absence of smearing of a stagnant contact discontinuity was for the first time consistently realized in [4.10], in the process of the practical construction of third-order difference schemes. Classes of difference schemes for the system (1.11), (1.9), (4.1.44) that possess the "K-property" have been constructed in [4.71]–[4.72]. Some of these schemes also possess the invariance properties in the sense defined in [4.71], [4.72], in particular, invariance with respect to the Galilean transformation. It follows from the expressions for μ, given in Table 4.1, that all the schemes listed in the table are noninvariant with respect to the Galilean transformation. This conclusion coincides with the results of studies carried out in [4.44], [4.71]–[4.74].

A boundary-value problem for the wave equation

$$\partial^2 u/\partial t^2 = c^2\, \partial^2 u/\partial x^2 \tag{4.1.69}$$

was considered in [4.75]. Equation (4.1.69) was solved in [4.75] by the method of lines in which the x-derivatives were approximated by central differences. The system of ordinary differential equations obtained in [4.75] by the method of lines was solved exactly. The exact solution of these equations was compared to the solution of the equation of the f.d.a.

$$\partial^2 w/\partial t^2 = c^2[\partial^2 w/\partial x^2 + (h^2/12)\, \partial^4 w/\partial x^4] \tag{4.1.70}$$

obtained by expansion of the quantities $u(x_k \pm h, t)$ entering the equations of the lines method with respect to the point (x_k, t) where $x_k = kh$. The solution of equation (4.1.70) at the function $w(0, t)$ of the form $w(0, t) = H(t)(H(t)$ is the Heavyside function), obtained in [4.75] by the method of matched asymptotic expansions, differed from the exact solution of ordinary differential equations of the method of lines only in the phase of error propagation. The relative phase error, due to neglecting the terms of order higher than $O(h^2)$ made, in the differential approximation, in accordance with [4.75], an amount which did not exceed 4%. This was the basis for the conclusion that the dispersive character of the numerical solution, giving rise to the Gibbs phenomenon, is due mainly to the first term of the approximation error obtained by the Taylor series expansion of a difference quotient, which in this case has the form $(c^2 h^2/12)\, \partial^4 w/\partial x^4$ in accordance with (4.1.70). Let us find a relationship between equations (4.1.70) and (4.1.21) at $r = 2$. For this purpose let us differentiate both sides of equation (4.1.21) with respect to t, and replace the derivatives ρ_{xt}, ρ_{xxxt} appearing in the equation thus obtained by the x-derivatives, making use of the formulas

$$\rho_{tx} = -u_0 \rho_{xx} - \mu \rho_x^{(4)}; \qquad (\rho_{tx})_{xx} = -(u_0 \rho_{xx})_{xx} - \mu \rho_x^{(6)}.$$

As a result we obtain the equation

$$\partial^2 \rho/\partial t^2 = u_0^2\, \partial^2 \rho/\partial x^2 + 2u_0\mu\, \partial^4 \rho/\partial x^4 + \mu^2\, \partial^6 \rho/\partial x^6. \qquad (4.1.71)$$

Neglecting in (4.1.71) the term of the order $O(\mu^2) = O(h^4)$, and making a substitution $t = \alpha t'$, $x = \alpha x'$, where

$$\alpha = [24\mu/(u_0 h^2)]^{0.5},$$

we obtain the equation coinciding with (4.1.70) at $c \equiv u_0$. The established relationship between equation (4.1.70) and equation (4.1.21) may serve as the foundation for the use of the f.d.a. of second-order difference schemes in the theoretical studies of the difference schemes properties in the neighborhood of a contact discontinuity. Thus, there exists a close relationship between dispersive properties and the Gibbs phenomenon stated in [4.75], for a semidiscretized approximation of the wave equation, and the properties of the solution (4.1.29).

Let us now show how one can investigate the entropy behavior in a contact strip by using Theorem 4.1. According to the second law of thermodynamics the entropy change is described by the formula

$$T\,dS = (\partial f/\partial p)_\rho\, dp + (\partial f/\partial \rho)_p\, d\rho - \rho^{-2} p\, d\rho, \qquad (4.1.72)$$

where the function $f(p, \rho)$ enters the equation of state (1.6). As is well known from thermodynamics, from the three functions T, p, ρ two are independent, therefore there exists a dependence $T = T(p, \rho)$. For the isobaric process $dp = 0$, so from (4.1.72) we obtain the formula for the entropy $S(x, t)$

$$S - S_0 = \int [\partial f(p_0, \rho)/\partial \rho - \rho^{-2} p_0] T^{-1}(p_0, \rho)\, d\rho, \qquad (4.1.73)$$

where S_0 is an arbitrary constant. Let the difference scheme under consideration satisfy the conditions of Theorem 4.1. Then the function $\rho(x, t)$ is continuous at $t > 0$ according to (4.1.22), (4.1.29), and from formula (4.1.73) follows the continuity of the entropy $S(x, t)$ in the contact strip at $t > 0$. Since the values h, τ do not explicitly enter the expression (4.1.73), we conclude that along the central line of the contact strip the value of the entropy S does not depend on the values of the grid parameters h, τ. Analogously to (4.1.12) one can introduce a notion of the width after Prandtl X_S of the transition zone in the entropy profile. From formula (4.1.73) with regard to Theorem 4.1 we obtain that $X_S = C(h, \tau, u_0, p_0)t^{1/(r+1)}$, where the form of the function $C(h, \tau, u_0, p_0)$ may be determined by the use of (4.1.73) at given functions $f(p, \rho), T(p, \rho)$.

In concluding this section consider a question on whether it is possible to investigate and to substantiate the contact discontinuity localization algorithms completely at a discrete level, without using the f.d.a. of a difference scheme. Suppose that a specific difference scheme (4.1.13)–(4.1.15) is K-consistent in the sense of Definition 4.7. Then scheme (4.1.13) approximates in a contact strip arising in the difference solution of the problem (1.1)–(1.3), (1.6), (4.1.1) at $t > 0$ the equation (4.1.8), that is, the equation with constant coefficients. The asymptotics at $n \to \infty$ and at the fixed values of τ, h of the solution $\rho^n(x)$ of an explicit difference equation

$$\rho^{n+1}(x) = \sum_{\nu=\nu_1}^{\nu_2} a_\nu \rho^n(x + \nu h), \qquad -\infty < x < +\infty, \qquad (4.1.74)$$

approximating equation (4.1.8) at discontinuous initial data for $\rho(x, 0)$ of the "step" form (see (4.1.1)), was investigated in [4.11], [4.76], [4.77]. In (4.1.74) $a_\nu, \nu = \nu_1, \ldots, \nu_2$, are some constants. In [4.11], [4.76] a theorem was proved which asserted that at any fixed ξ and $n \to \infty$ the functions

$$\rho^n(n u_0 \tau + \xi(b_{r+1} n)^{1/(r+1)} h), \qquad (4.1.75)$$

where b_{r+1} is a positive constant and r is the order of approximation of the scheme (4.1.74), weakly converge to the function

$$\tilde{\rho}_r(\xi) = \rho_2 + (\rho_2 - \rho_1)F_r(\xi) \qquad (4.1.76)$$

with

$$F_r(\xi) = 1 - \int_{-\infty}^{\xi} f_r(z)\, dz, \qquad F_r(+\infty) = 0; \qquad (4.1.77)$$

$$f_r(z) = \begin{cases} \dfrac{1}{\pi} \displaystyle\int_0^\infty \exp\left(-\dfrac{s^{r+1}}{r+1}\right) \cos(zs)\, ds, & r \text{ is odd}, \\[4mm] \dfrac{1}{\pi} \displaystyle\int_0^\infty \cos\left(\dfrac{s^{r+1}}{r+1} + zs\right) ds, & r \text{ is even}. \end{cases} \qquad (4.1.78)$$

It was proved in [4.77] that the sequence of functions (4.1.75) converges to $\tilde{\rho}_r(\xi)$ in the metrics of $C(-\infty, \infty)$.

Thus, at sufficiently large n, the functions $\rho^n(x)$ will be approximated to any desired degree of accuracy by the function

$$\rho^n(x) \approx \tilde{\rho}_r((x - x_0 - u_0 \tau n)/(b_{r+1} n)^{1/(r+1)}h). \tag{4.1.79}$$

The formulas (4.1.76)–(4.1.78) at $r = 1$ and $r = 2$ coincide with formulas (4.1.22), (4.1.23) and (4.1.29), (4.1.30), respectively, if one takes in (4.1.79) $b_{r+1} = 4h^{-2}\mu\tau$ at $r = 1$ and $b_{r+1} = 3h^{-3}\mu\tau$ at $r = 2$.

It should be noted that in cases when the linear equation (4.1.8) is approximated by a nonlinear difference scheme, for example, by the RUSANOV scheme [4.7] (see an analysis of this scheme within the f.d.a. framework above, in this section), the convergence results being analogous to the ones obtained in [4.11], [4.76], [4.77] for the linear case, are absent in the literature. The f.d.a. method also enables one to draw definite conclusions on the structure of the contact discontinuity localization algorithms in the nonlinear case, as was shown above in the example of the Rusanov scheme [4.7] (see formulas (4.1.65), (4.1.66), (4.1.68)). In addition, a consideration within the f.d.a. framework enables one to find, for a specific difference scheme, an explicit expression for the constant b_{r+1} in (4.1.79), which is important in the determination of the actual size of the width of the zone of smearing of a contact discontinuity (see Table 4.1).

It is possible to deduce from formulas (4.1.74), (4.1.76)–(4.1.78), (4.1.79) certain conclusions on the structure of the procedures which are suitable for the localization of contact discontinuities, on the basis of difference solutions obtained by the schemes of the third and higher orders of approximation. Suppose that, as the result of execution of the quasi-linearization operation $\tilde{L}$ (see Definition 4.5), the difference equation (4.1.13) goes over into the equation of the form (4.1.74). Then the above-cited results of [4.11], [4.76], [4.77] are applicable and, consequently, a contact discontinuity can be located as a point of intersection of the density profiles $\rho^n(x)$ in a contact strip obtained at different h or τ. This localization algorithm follows from formula (4.1.79). From the consideration of the f.d.a. one can derive one more algorithm for the localization of a contact discontinuity. Indeed, the solution (4.1.76)–(4.1.78) at a proper choice of the constant b_{r+1} satisfies equation (4.1.21) where $r \geq 3$. Then along the central line L_{KD} of a contact strip the quantity $|\partial^r\rho/\partial x^r|$ achieves its maximum by virtue of Lemma 4.2. This fact can also be established by direct use of the asymptotic formulas (4.1.76)–(4.1.78). Thus, if as the result of execution of the quasi-linearization operation $\tilde{L}$, the difference equations (4.1.13), (4.1.14) go over into a difference equation of the form (4.1.74) with constant coefficients, then one can assert that at sufficiently large n the use of Algorithms I–II of the differential analyzers of the contact discontinuities is justified.

4.2. *K*-Consistence Property of the First Differential Approximation in the Two-Dimensional Case

The technique of Section 4.1 proved to be a convenient tool in investigating the difference scheme properties in the neighborhood of a contact discontinuity of the form (4.1.2) in the one-dimensional flow. Therefore, it appears reasonable to consider the possibility of extending the above technique to the case of two-dimensional flows. It should be noted that the investigation of difference scheme properties in the vicinity of contact discontinuities in two-dimensional flows is substantially complicated compared with the one-dimensional case, by virtue of the fact that the fluid velocity component tangential to the contact discontinuity surface may undergo a discontinuity; then one has to deal with tangential discontinuity, the instability of which was investigated in [4.78]. As far as we know, at present there is no technique for investigating the scheme viscosity influence on the behavior of difference solutions in the neighborhood of a tangential discontinuity. A generalization, presented below, of the technique of Section 4.1 refers to a simpler way of investigating the difference solution properties in the neighborhood of purely contact discontinuities in two-dimensional flows [4.37], [4.79].

Suppose that in (1.9), (1.11), (4.1.1), (4.1.2) the x-coordinate is replaced by some other spatial coordinate x^* in such a way that there takes place a one-dimensional flow along the x^*-axis. Consider now a Cartesian coordinate system (x_1, x_2) such that the x^*-axis makes some angle β with the positive direction of the x_1-axis, and $\beta \neq (\pi/2)k$, $k = 0, 1, 2, 3$. Then the fluid velocity vector in the (x_1, x_2) system will have two components u_1, u_2, and consequently it is necessary to make use of the Euler equations governing the two-dimensional flow to describe this essentially one-dimensional flow along the x^*-axis:

$$\partial u/\partial t + \partial F_1(u)/\partial x_1 + \partial F_2(u)/\partial x_2 = 0 \tag{4.2.1}$$

where

$$u = \begin{pmatrix} \rho \\ \rho u_1 \\ \rho u_2 \\ \rho E \end{pmatrix}, \quad F_1 = \begin{pmatrix} \rho u_1 \\ p + \rho u_1^2 \\ \rho u_1 u_2 \\ p u_1 + \rho u_1 E \end{pmatrix}, \quad F_2 = \begin{pmatrix} \rho u_2 \\ \rho u_1 u_2 \\ p + \rho u_2^2 \\ p u_2 + \rho u_2 E \end{pmatrix}, \tag{4.2.2}$$

$$E = \varepsilon + (u_1^2 + u_2^2)/2.$$

Let h_1, h_2 be the steps of a uniform rectangular computing mesh in the (x_1, x_2)-plane, and let τ be time step. Consider, by analogy with Section 4.1.1, a class of difference schemes approximating the system (4.2.1), (4.2.2) with an rth order of accuracy, $1 \leq r \leq 2$, and such that their f.d.a. H-form has the form

$$\Gamma w \equiv \partial w/\partial t + \sum_{j=1}^{2} \partial F_j(w)/\partial x_j - \sum_{\substack{i,j,k \\ i+j+k=r}} h_1^i h_2^j \tau^k$$

$$\times F_{ijk}(w, \partial w/\partial x_1, \partial w/\partial x_2, \partial w/\partial t, \ldots, \partial^{r+1} w/\partial x_1^{r+1},$$

$$\partial^{r+1} w/\partial x_1^r \, \partial x_2, \ldots, \partial^{r+1} w/\partial x_2 \, \partial t^r, \partial^{r+1} w/\partial t^{r+1}) = 0, \quad (4.2.3)$$

where in accordance with (4.2.2)

$$w = \{w_1, w_2, w_3, w_4\}^{\mathrm{T}}, \qquad F_{ijk} = \{F_{ijk}^{(1)}, F_{ijk}^{(2)}, F_{ijk}^{(3)}, F_{ijk}^{(4)}\}^{\mathrm{T}},$$

$$w_1 \equiv \rho, \qquad w_j = f_j(\rho, u_1, u_2, p, \varepsilon(p, \rho)), \qquad j = 2, 3, 4.$$

Let us express the derivatives

$$\partial^m w/\partial x_1^{k_1} \, \partial x_2^{k_2} \, \partial t^{m-k_1-k_2}, \qquad m - k_1 - k_2 > 0,$$

entering F_{ijk} in terms of the derivatives with respect to x_1, x_2 and making use of the algorithm described in Section 1.3

$$\partial^m w/\partial x_1^{k_1} \, \partial x_2^{k_2} \, \partial t^{m-k_1-k_2}$$

$$= f_{m,k_1,k_2}(w, \partial w/\partial x_1, \ldots, \partial w/\partial x_2, \ldots, \partial^m w/\partial x_1^m, \partial^m w/\partial x_1^{m-1} \, \partial x_2, \ldots, \partial^m w/\partial x_2^m).$$

As a result we obtain from (4.2.3) the f.d.a. P-form

$$\Pi w \equiv \partial w/\partial t + \sum_{j=1}^{2} \partial F_j(w)/\partial x_j$$

$$- \sum_{\substack{i,j,k \\ i+j+k=r}} h_1^i h_2^j \tau^k \mathscr{F}_{ijk}(w, \partial w/\partial x_1, \partial w/\partial x_2, \ldots,$$

$$\partial^{r+1} w/\partial x_1^{r+1}, \partial^{r+1} w/\partial x_1^r \, \partial x_2, \ldots, \partial^{r+1} w/\partial x_2^{r+1}) = 0, \quad (4.2.4)$$

where

$$\mathscr{F}_{ijk} = F_{ijk}(w, \partial w/\partial x_1, \partial w/\partial x_2, f_{1,0,0}(w, \partial w/\partial x_1, \partial w/\partial x_2), \ldots,$$

$$\partial^{r+1} w/\partial x_1^{r+1}, \partial^{r+1} w/\partial x_1^r \, \partial x_2, \ldots, f_{r+1,0,0}(w, \partial w/\partial x_1,$$

$$\partial w/\partial x_2, \ldots, \partial^{r+1} w/\partial x_1^{r+1}, \partial^{r+1} w/\partial x_1^r \, \partial x_2, \ldots, \partial^{r+1} w/\partial x_2^{r+1})).$$

Definition 4.8. Let us call the quasi-linearization operation L of the f.d.a. (4.2.3) or (4.2.4) the operation

$$\Gamma L w = 0, \qquad \Pi L w = 0,$$

where

$$L w = \{w_1, L w_2, L w_3, L w_4\}^{\mathrm{T}},$$

$$L w_j = f_j(\rho, u_{10}, u_{20}, p_0, \varepsilon(p_0, p)), \qquad j = 2, 3, 4,$$

where u_{10}, u_{20} are the components of the constant fluid velocity vector,

$u_{10} = u_0 \cos \beta$, $u_{20} = u_0 \sin \beta$, and u_0, p_0 are constant velocity and pressure values entering (4.1.1).

Let us introduce the column vector

$$U = \{1, u_{10}, u_{20}, 0.5(u_{10}^2 + u_{20}^2) + f(p_0, \rho) + \rho \, \partial f(p_0, \rho)/\partial \rho\}^{\mathrm{T}}, \quad (4.2.5)$$

where $f(p, \rho)$ is the function entering the equation of state (1.6).

Definition 4.9. The f.d.a. (4.2.3) will be called *K*-consistent if

$$\Gamma Lw = U\left[\partial \rho/\partial t + \sum_{j=1}^{2} u_{j0} \, \partial \rho/\partial x_j \right.$$

$$- \sum_{\substack{i,j,k \\ i+j+k=r}} h_1^i h_2^j \tau^k F_{ijk}^{(1)}(Lw, \partial Lw/\partial x_1, \partial Lw/\partial x_2, \partial Lw/\partial t, \dots,$$

$$\partial^{r+1}Lw/\partial x_1^{r+1}, \partial^{r+1}Lw/\partial x_1^r \, \partial x_2, \dots,$$

$$\left. \partial^{r+1}Lw/\partial x_2 \, \partial t^r, \partial^{r+1}Lw/\partial t^{r+1}) \right] = 0,$$

that is, the left-hand sides of the equations of a system $\Gamma Lw = 0$ differ from each other only by a scalar multiplier.

K-consistence of the f.d.a. *P*-form (4.2.4) is defined analogously. Note that the initial equation system (4.2.1)–(4.2.2) possesses the *K*-consistence property. Indeed, applying the quasi-linearization operation L to (4.2.1), (4.2.2) we obtain

$$L\left(\frac{\partial u}{\partial t} + \sum_{j=1}^{2} \frac{\partial F_j(u)}{\partial x_j}\right) = \left(\frac{\partial \rho}{\partial t} + \sum_{j=1}^{2} u_{j0} \frac{\partial \rho}{\partial x_j}\right) U = 0. \quad (4.2.6)$$

We see from (4.2.6) that under initial conditions of the form (4.1.1) the solution of the Cauchy problem (4.2.1), (4.2.2), (1.5), (4.1.1) reduces to the solution of a linear equation

$$\partial \rho/\partial t + \partial u_{10}\rho/\partial x_1 + \partial u_{20}\rho/\partial x_2 = 0. \quad (4.2.7)$$

Let

$$\rho(x_1, x_2, 0) = R(x_1, x_2), \quad (4.2.8)$$

where R is a given function. Then, as is known, the exact solution of the problem (4.2.7), (4.2.8) has the form [4.80], [4.81]

$$\rho(x_1, x_2, t) = R(x_1 - u_{10} \cdot t, x_2 - u_{20} \cdot t),$$

from which it follows that the quantity ρ remains constant along the lines $x_1 - u_{10} \cdot t = c_1$, $x_2 - u_{20} \cdot t = c_2$, where c_1, c_2 are arbitrary constants. Note that the two-dimensional Euler equations (4.2.1), (4.2.2) have been studied under the assumption of constant pressure p in [4.81].

Below we shall consider the f.d.a. P-form (4.2.4) following Section 4.1. From Definition 4.9 it follows that the investigation of the K-consistent f.d.a. (4.2.4), subject to the initial condition (4.1.1), is reduced to the investigation of a single equation

$$\partial\rho/\partial t + \partial u_{10}\rho/\partial x_1 + \partial u_{20}\rho/\partial x_2$$

$$= \sum_{\substack{i,j,k \\ i+j+k=r}} h_1^i h_2^j \tau^k \mathscr{F}_{ijk}^{(1)}(Lw, \partial Lw/\partial x_1, \partial Lw/\partial x_2, \dots,$$

$$\partial^{r+1}Lw/\partial x_1^{r+1}, \partial^{r+1}Lw/\partial x_1^r \partial x_2, \dots, \partial^{r+1}Lw/\partial x_2^{r+1}). \quad (4.2.9)$$

EXAMPLE 1. The f.d.a. H-form of the integral step scheme of the FLIC method [4.44] may be written in the form (3.1.22). Let us apply the quasi-linearization operation L to both sides of (3.1.22) assuming that the function $f(p, \rho)$ in the equation of state (1.6) satisfies the identity (4.1.43). After that let us express the derivative ρ_{tt} in the terms of the x_1, x_2 derivatives with the purpose of obtaining the f.d.a. P-form. As a result we obtain a system

$$U \cdot \left\{ \rho_t + \sum_{\alpha=1}^{2} \left[(\rho u_{\alpha 0})_{x_\alpha} - (h_\alpha/2)(|u_{\alpha 0}| \rho_{x_\alpha})_{x_\alpha} \right. \right.$$

$$\left. \left. - (\tau/2)u_{\alpha 0}(u_{10}\rho_{x_1} + u_{20}\rho_{x_2})_{x_\alpha} \right] \right\} = 0, \quad (4.2.10)$$

where the vector U is defined by formula (4.2.5). From (4.2.10) it follows that the f.d.a. (3.1.22) is K-consistent under the condition that (4.1.43) is satisfied.

EXAMPLE 2. Let us approximate the system (4.2.1), (4.2.2) by the MacCormack difference scheme [4.82]

$$\tilde{u}_{ij}^{n+1} = u_{ij}^n - (\tau/h_1)(F_{1i+1,j}^n - F_{1ij}^n) - (\tau/h_2)(F_{2ij+1}^n - F_{2ij}^n); \quad (4.2.11)$$

$$u_{ij}^{n+1} = 0.5[u_{ij}^n + \tilde{u}_{ij}^{n+1} - (\tau/h_1)(\tilde{F}_{1ij}^{n+1} - \tilde{F}_{1i-1j}^{n+1}) - (\tau/h_2)(\tilde{F}_{2ij}^{n+1} - \tilde{F}_{1ij-1}^{n+1})], \quad (4.2.12)$$

where $\tilde{F}_{\alpha ij}^{n+1} = F_\alpha(\tilde{u}_{ij}^{n+1})$. Performing calculations similar to those presented in [4.39], [4.43] we obtain the f.d.a. H-form of the scheme (4.2.11), (4.2.12) as follows:

$$w_t + \sum_{\alpha=1}^{2} F_{\alpha x_\alpha} = -(\tau^2/6)w_{ttt} + (\tau/4) \sum_{\alpha=1}^{2} (\partial/\partial x_\alpha)\left[\sum_{j=1}^{4} (\partial F_\alpha/\partial w_j) \cdot \left(\sum_{\beta=1}^{2} h_\beta F_{\beta x_\beta x_\beta}^{(j)} \right) \right]$$

$$- (1/6) \sum_{\alpha=1}^{2} h_\alpha^2 F_{\alpha x_\alpha x_\alpha x_\alpha} - (\tau/4) \sum_{\alpha=1}^{2} h_\alpha(\partial^2/\partial x_\alpha^2)$$

$$\times \left[\sum_{j=1}^{4} (\partial F_\alpha/\partial w_j)\left(\sum_{\beta=1}^{2} F_{\beta x_\beta}^{(j)} \right) \right]. \quad (4.2.13)$$

In (4.2.13)
$$F_\alpha = \{F_\alpha^{(1)}, F_\alpha^{(2)}, F_\alpha^{(3)}, F_\alpha^{(4)}\}^{\mathrm{T}}, \qquad \alpha = 1, 2.$$

The application of the quasi-linearization operation L in the case when the identity (4.1.43) is satisfied yields

$$U\left[\rho_t + \sum_{\alpha=1}^{2} (\rho u_{\alpha 0})_{x_\alpha} + (\tau^2/6)\rho_{ttt} + (1/6) \sum_{\alpha=1}^{2} h_\alpha^2 (\rho u_{\alpha 0})_{x_\alpha x_\alpha x_\alpha}\right] = 0. \quad (4.2.14)$$

The f.d.a. P-form obtained from (4.2.14) may be written as follows:

$$\rho_t + \sum_{\alpha=1}^{2} (\rho u_{\alpha 0})_{x_\alpha} = -\sum_{\alpha=1}^{2} \mu_\alpha \rho_{x_\alpha x_\alpha x_\alpha}$$

$$+ (\tau^2/2) u_{10} u_{20} (\partial^2/\partial x_1\, \partial x_2) \sum_{\alpha=1}^{2} u_{\alpha 0} \rho_{x_\alpha}, \quad (4.2.15)$$

where by analogy with [4.39], [4.43] we have set

$$\mu_\alpha = h_\alpha^2 u_{\alpha 0}/6 - \tau^2 u_{\alpha 0}^3/6, \qquad \alpha = 1, 2.$$

Equation (4.2.15) coincides with the f.d.a. obtained in [4.83] for the case when the MacCormack scheme (4.2.11), (4.2.12) approximates equation (4.2.7). From (4.2.14) it follows that the f.d.a. of the difference scheme under consideration is K-consistent under the condition that (4.1.43) is satisfied.

Suppose that the function $R(x_1, x_2)$ in (4.2.8) is piecewise continuous and let $\mathcal{L}$ be one of the discontinuity lines of the function R in the (x_1, x_2)-plane. Let us take some point $(x_{10}, x_{20}) \in \mathcal{L}$ and at $t > 0$ consider in the (x_1, x_2)-plane the line described parametrically by the equations

$$x_1 = X_1(x_{10}, x_{20}, t), \qquad x_2 = X_2(x_{10}, x_{20}, t),$$

and
$$\lim_{t \to 0} X_j(x_{10}, x_{20}, t) = x_{j0}, \qquad j = 1,2; \quad (4.2.16)$$

$$\sum_{\substack{i,j,k \\ i+j+k=r}} h_1^i h_2^j \tau^k \mathcal{F}_{ijk}^{(1)}(Lw, \ldots, \partial^{r+1} Lw/\partial x_2^{r+1})\big|_{x_1=X_1, x_2=X_2} = 0. \quad (4.2.17)$$

Suppose that the line $\mathcal{L}$ possessing the properties (4.2.16), (4.2.17) is unique. Then at arbitrary $t > 0$ at the point (X_1, X_2) the value of the function ρ being the solution of equation (4.2.9) does not depend on h_1, h_2, τ. In fact, by virtue of (4.2.17) the equality (4.2.7) is satisfied along $\mathcal{L}$ from which it follows that ρ is constant along the line $x_1 - u_{10} \cdot t = c_1$, $x_2 - u_{20} \cdot t = c_2$. Since, by assumption, the line $\mathcal{L}$ satisfying (4.2.16), (4.2.17) is unique,

$$X_j(x_{10}, x_{20}, t) = u_{j0} \cdot t + x_{j0}, \qquad j = 1, 2,$$

from which it follows that the point (X_1, X_2) moves with increasing t at an accurate speed of contact discontinuity. From this consideration it follows that the determination of the locus of points at which the right-hand side of (4.2.9) vanishes may be used in an algorithm of the differential analyzer of contact discontinuity for which it is known *a priori* that this discontinuity is not tangential.

4.3. Methods of K-Inconsistence Suppression

4.3.1. Preliminary Discussion

As was shown in Section 4.1, equation (4.1.43) is a necessary condition for the f.d.a. K-consistence of a wide class of difference schemes for gas dynamics problems. Let us now assume that equation (4.1.43) is not satisfied. Consider the model problem (1.11), (1.9), (1.6), (4.1.1). In the exact solution (4.1.2) of this problem the pressure and the velocity are constant independently of the form of the equation of state employed. Applying the quasi-linearization operation L to the Euler equation system (1.11), (1.9) we have obtained in Section 4.1 the system (4.1.7). This system is compatible at the arbitrary equations of state (1.5) or (1.6), because the solution of equation (4.1.8) is also the solution of the second and third equations of the system

$$(\partial \rho / \partial t + u_0 \, \partial \rho / \partial x) U = 0,$$

where the column vector U is determined by the formula (4.1.6). Let us now see whether the f.d.a. compatibility of a difference scheme will be conserved in the case when the equation of state used is such that equation (4.1.43) is not satisfied. Take, for example, the Lax scheme. Let us write out all three equations of the system (4.1.41). With regard to the definition (4.1.6) we have

$$\partial \rho / \partial t + u_0 \, \partial \rho / \partial x = \mu \, \partial^2 \rho / \partial x^2; \tag{4.3.1}$$

$$u_0(\partial \rho / \partial t + u_0 \, \partial \rho / \partial x) = \mu u_0 \, \partial^2 \rho / \partial x^2; \tag{4.3.2}$$

$$[0.5u_0^2 + f(p_0, \rho) + \rho \, \partial f(p_0, \rho)/\partial \rho](\partial \rho / \partial t + u_0 \, \partial \rho / \partial x)$$
$$= \mu(\partial / \partial x)\{[0.5u_0^2 + f(p_0, \rho) + \rho \, \partial f(p_0, \rho)/\partial \rho](\partial \rho / \partial x)\}. \tag{4.3.3}$$

The solution of (4.3.1) at the initial condition (4.1.10) has the form (4.1.22), (4.1.23). Let us now multiply both sides of equation (4.3.1) by a quantity $-[0.5u_0^2 + f(p_0, \rho) + \rho \, \partial f(p_0, \rho)/\partial \rho]$ and add the result to both sides of equation (4.3.3). We obtain:

$$\mu\{\partial^2 / \partial \rho^2 [\rho f(p_0, \rho)]\}(\partial \rho / \partial x)^2 = 0,$$

from where we find the solution $\rho(x, t) = a(t)$ where $a(t)$ is an arbitrary function. Thus, in the case when equation (4.1.43) does not take place the system (4.3.1)–(4.3.3) proves to be incompatible. In this connection in [4.40] an assumption was made that in the absence of the f.d.a. K-consistence the difference solutions for the pressure and velocity, obtained by the numerical solution of the model problem (1.11), (1.19), (1.6), (4.1.1), would not be constant in a contact strip but would be certain functions of x, t. Computations of the Riemann problem (1.1)–(1.3), (4.1.1) using a number of equations of state, different from the ideal gas equation of state and not satisfying the condition (4.1.43), were carried out in [4.40] with the purpose of checking the above assumption. In these computations a number of well-known first- and second-order difference schemes were used. In [4.40] there were errors obtained in the pressure and the velocity which ranged from several percent up to several dozen percent depending on the specific form of the equation of state used and on the magnitude of the density jump across the contact discontinuity. It was shown in [4.43], in the example of the Lax–Wendroff scheme and the MacCormack scheme (see also Section 4.3.3 below), that in cases when the equation of state employed was not too complicated, then one was able to get an explicit expression for the error in the difference solution for the pressure at the first two–three steps (analytic calculations become very cumbersome for subsequent steps).

The above errors in pressure and velocity in a contact strip arising in the numerical solution of the problem (1.1)–(1.3), (1.6), (4.1.1) are not relevant to physical properties of the flow under consideration, they have a purely computational nature. In this connection we propose in Sections 4.3.2 and 4.3.3 a number of techniques for the reduction or suppression of the numerical solution errors in a contact strip when using the equations of state which do not satisfy the condition (4.1.43). As will be shown in Section 4.4, the realization of the corresponding algorithms for K-inconsistence suppression substantially improves the accuracy of the contact discontinuities localization by means of the differential analyzers considered in Section 4.1.

With regard to existing construction techniques for finite-difference schemes approximating the system (1.1)–(1.3) the following reasons can be found which may cause the f.d.a. K-inconsistence [4.40].

(1) Some of the three difference equations approximating the system (1.1)–(1.3) use the grid stencil different from that for the rest of the difference equations.

(2) Some of the three difference equations approximating the system (1.1)–(1.3) have a structure of approximation viscosity which is different from the approximation viscosity entering the rest of the equations, and, particularly, contain "viscous" terms proportional to the derivatives $\partial^k \rho / \partial x^k$, $k = 1, 2, \ldots$.

This list of possible reasons for K-inconsistence does not pretend to be exhaustively complete. In Sections 4.3.2 and 4.3.3 we restrict ourselves only to the study of the influence of the equation of state form on the K-consistence. In fact, many equations of state now in use in the numerical solution of continuum mechanics problems do not satisfy equation (4.1.43).

The function
$$B(p_0, \rho) \equiv (\partial^2/\partial\rho^2)[\rho f(p_0, \rho)] \tag{4.3.4}$$

plays an important role in the construction of K-inconsistence suppression algorithms for the difference schemes that are presented in Sections 4.3.2 and 4.3.3. From (4.1.43) follows the necessary condition of K-consistence which has the form $B(p_0, \rho) = 0$. Since the equation (or equations) of state is used in gas dynamic computations in various forms, for example, in the form (1.5) or (1.6), or in the form

$$p = G(V, S), \qquad T = T(V, S), \tag{4.3.5}$$

where S is the entropy, $V = 1/\rho$, and T is the temperature, the formulas for $B(p_0, \rho)$ are needed in each of these cases. Let us derive an expression for $B(p_0, \rho)$ in the case when the equation of state (1.5) is employed. Differentiate both sides of the identity (1.7) with respect to ρ assuming that $p = \text{const}$. We obtain

$$\partial f/\partial\rho = -\frac{\partial F}{\partial\rho}\bigg/\frac{\partial F}{\partial\varepsilon}. \tag{4.3.6}$$

With regard to (4.3.6) we can rewrite (4.3.4) in the form
$$B(p_0, \rho) = -(\partial F/\partial\rho)/(\partial F/\partial\varepsilon) - (\partial/\partial\rho)[\rho(\partial F/\partial\rho)/(\partial F/\partial\varepsilon)]. \tag{4.3.7}$$

Assuming that the argument ε in (1.5) depends in turn on ρ: $\varepsilon = f(p_0, \rho)$ let us calculate the second derivatives entering (4.3.7)

$$\begin{aligned}
(d/d\rho)(\partial F/\partial\rho) &= \partial^2 F/\partial\rho^2 - (\partial^2 F/\partial\rho\,\partial\varepsilon)(\partial F/\partial\rho)/(\partial F/\partial\varepsilon), \\
(d/d\rho)(\partial F/\partial\varepsilon) &= \partial^2 F/\partial\varepsilon\,\partial\rho - (\partial^2 F/\partial\varepsilon^2)(\partial F/\partial\rho)/(\partial F/\partial\varepsilon).
\end{aligned} \tag{4.3.8}$$

Substituting formulas (4.3.8) into (4.3.7) we obtain finally:
$$B(p_0, \rho) = -2F_\rho/F_\varepsilon - \rho(F_\varepsilon F_{\rho\rho} - F_{\rho\varepsilon}F_\rho - F_{\varepsilon\rho}F_\rho + F_\rho^2 F_{\varepsilon\varepsilon}/F_\varepsilon)/F_\varepsilon^2, \tag{4.3.9}$$

where
$$F_\varepsilon \equiv \partial F(\rho, \varepsilon)/\partial\varepsilon, \qquad F_{\rho\rho} \equiv \partial^2 F(\rho, \varepsilon)/\partial\rho^2, \qquad \text{etc.}$$

Note that in the case of sufficiently complicated equations of state of the form (1.5) we can use, for the computation of the derivatives entering (4.3.9), the specialized programs performing analytic calculations on a computer, including the differentiation of analytical functions (see, for example, [4.84]). Let us present a technique for the calculation of the quantity $B(p_0, \rho)$ without using formula (4.3.9), when the equation of state may be written in the form

$$\sum_{k=0}^{N} f_k(p, \rho)\varphi_k(\rho\varepsilon) = 0, \tag{4.3.10}$$

where f_k, φ_k are twice differentiable functions, $N \geq 1$:

$$B(p_0, \rho) = (2S_1 S_2 S_3 - S_5 S_2^2 - S_4 S_1^2)/(S_2^3), \tag{4.3.11}$$

where

$$S_1 = \sum_{k=0}^{N} f_k' \varphi_k, \qquad S_2 = \sum_{k=0}^{N} f_k \varphi_k', \qquad S_3 = \sum_{k=0}^{N} f_k' \varphi_k',$$

$$S_4 = \sum_{k=0}^{N} f_k' \varphi_k'', \qquad S_5 = \sum_{k=0}^{N} f_k'' \varphi_k;$$

$$\varphi_k' \equiv (\partial/\partial \rho \varepsilon) \varphi_k(\rho \varepsilon), \qquad f_k' \equiv (\partial/\partial \rho) f_k(p_0, \rho), \qquad \text{etc.}$$

To obtain formula (4.3.11) we differentiate both sides of equation (4.3.10) with respect to ρ assuming that p is fixed. As a result we obtain

$$(\partial/\partial \rho)[\rho \varepsilon(p_0, \rho)] = -\left(\sum_{k=0}^{N} f_k' \varphi_k \right) \Big/ \left(\sum_{k=0}^{N} f_k \varphi_k' \right). \qquad (4.3.12)$$

Differentiating both sides of (4.3.12) with respect to ρ and expressing the derivative $\partial/\partial \rho [\rho \varepsilon(p_0, \rho)]$ in terms of the right-hand side of (4.3.12) we finally derive formula (4.3.11).

Sometimes the quantities V, S, where V is the specific volume and S is the entropy, are used as independent variables in the equation of state. Let us derive the expression for $B(p_0, \rho)$ for the case when the relations (4.3.5) are known. We shall make use of the relationship

$$F(1/V, \varepsilon(V, S)) \equiv G(V, S), \qquad (4.3.13)$$

where the function $F(\rho, \varepsilon)$ enters the equation of state (1.5). Differentiating both sides of equation (4.3.13) with respect to V and S we obtain a system of algebraic equations for computation of the derivatives $\partial F/\partial \rho, \partial F/\partial \varepsilon$:

$$-(1/V^2)F_\rho + F_\varepsilon \, \partial \varepsilon(V, S)/\partial V = G_V;$$
$$F_\varepsilon \, \partial \varepsilon(V, S)/\partial S = G_S; \qquad (4.3.14)$$

where

$$G_V \equiv \partial G(V, S)/\partial V, \qquad G_S \equiv \partial G(V, S)/\partial S.$$

Making use of the well-known thermodynamic relations [4.1]

$$\partial \varepsilon(V, S)/\partial V = -p(V, S); \qquad \partial \varepsilon(V, S)/\partial S = T(V, S), \qquad (4.3.15)$$

we rewrite the system (4.3.14) as follows

$$-(1/V^2)F_\rho - pF_\varepsilon = G_V;$$
$$TF_\varepsilon = G_S. \qquad (4.3.16)$$

From (4.3.16) we find

$$F_\varepsilon = G_S/T, \qquad F_\rho = -V^2(G_V + GG_S/T). \qquad (4.3.17)$$

Differentiating both sides of the first of the equations of system (4.3.14) with respect to V we obtain

$$F_{\rho\rho}(1/V^4) + F_{\rho\varepsilon}2G/(V^2) + F_{\varepsilon\varepsilon}G^2 = G_{VV} + F_\varepsilon G_V - 2V^{-3}F_\rho, \qquad (4.3.18)$$

similarly, differentiating both sides of the second equation in (4.3.16) with respect to V we obtain

$$(F_{\varepsilon\rho} \, d\rho/dV + F_{\varepsilon\varepsilon} \, \partial\varepsilon/\partial V)T(V, S) + F_{\varepsilon}T_V = G_{SV}. \tag{4.3.19}$$

Let us use one of the four well-known Maxwell equations in thermodynamics [4.1], [4.85].

$$\partial T(V, S)/\partial V + \partial p(V, S)/\partial S = 0. \tag{4.3.20}$$

Taking into account (4.3.20), (4.3.15) let us rewrite equation (4.3.19) in the form

$$[-V^{-2}F_{\varepsilon\rho} - GF_{\varepsilon\varepsilon}]T(V, S) - F_{\varepsilon}G_S = G_{SV}. \tag{4.3.21}$$

Differentiating both sides of the second of the equations of the system (4.3.16) with respect to S and using the relationships (4.3.15) we obtain the equation

$$F_{\varepsilon\varepsilon}T^2(V, S) + F_{\varepsilon} \, \partial T(V, S)/\partial S = G_{SS}. \tag{4.3.22}$$

Thus, we have obtained a system of three algebraic equations (4.3.18), (4.3.21), (4.3.22) for the calculation of $F_{\rho\rho}$, $F_{\rho\varepsilon}$, $F_{\varepsilon\varepsilon}$ in terms of the derivatives of the functions $G(V, S)$, $T(V, S)$. When collecting the terms of similar structure we assume here and in what follows that $F_{\rho\varepsilon} = F_{\varepsilon\rho}$. Solving the system (4.3.18), (4.3.21), (4.3.22) we get the expressions sought for

$$F_{\varepsilon\varepsilon} = T^{-2}(G_{SS} - T_S G_S/T);$$

$$F_{\varepsilon\rho} = -(V^2/T)(GG_S/T + G_V)_S; \tag{4.3.23}$$

$$F_{\rho\rho} = V^4[2GT^{-1}(GG_S/T + G_V)_S - (G/T)^2$$

$$\times (G_{SS} - T_S G_S/T) + G_{VV} + T^{-1}G_V G_S + 2V^{-1}(G_V + GG_S/T)].$$

Substituting (4.3.17), (4.3.23) and the relation $\rho = 1/V$ into (4.3.9) yields a formula for $B(p_0, \rho)$ sought for.

EXAMPLE 1. Let the function $F(\rho, \varepsilon)$ in equation (1.5) have the form (1.8) where $\gamma = c_p/c_V = \text{const}$. Then the functions $G(V, S)$, $T(V, S)$ in equations (4.3.5) are [4.1]

$$G(V, S) = (a^2/\gamma)V^{-\gamma} \exp(S/c_V);$$

$$T(V, S) = \beta V^{1-\gamma} \exp(S/c_V); \tag{4.3.24}$$

where $a^2 = \text{const}$, $\beta = a^2/(R\gamma)$, R is the universal gas constant, and $R = c_p - c_V$. Substituting expressions (4.3.24) for the function G and T into (4.3.23), (4.3.17) and then using formula (4.3.9) it is very easy to find that $B(p_0, \rho) \equiv 0$. This result coincides with the one indicated previously in Section 4.1 for equation (4.1.45).

EXAMPLE 2. Consider a binomial equation of state

$$p = \beta\rho\varepsilon + \rho_0 c_0^2(\rho/\rho_0 - 1)^k, \tag{4.3.25}$$

where β, ρ_0, c_0, k are constants. At $k = 1$ equation (4.3.25) was used, for example, in [4.68], [4.86]–[4.88]. The term $(\rho/\rho_0 - 1)^k$ at $k \neq 1$ was used, for example, in [4.89], [4.90]. Employing formula (4.3.9) it is easy to compute

$$B(p_0, \rho) = -(kc_0^2/(\beta\rho_0))(k - 1)(\rho/\rho_0 - 1)^{k-2}. \qquad (4.3.26)$$

From (4.3.26) it follows that the condition (4.1.43) is satisfied, for instance, at $c_0 = 0$ (ideal gas); or at $k = 1$, $c_0 \neq 0$ (binomial equation of state (4.3.25)).

EXAMPLE 3. The equation of state [4.90]

$$p = [a + b/(\varepsilon/(I_0\eta^2) + 1)]\varepsilon\rho + A\zeta + B\zeta^2, \qquad (4.3.27)$$

where $\eta = \rho/\rho_0 = \zeta + 1$ and $a, b, A, B, I_0, I_S, \rho_0$ are constants for the particular material. Equation (4.3.27) is part of the modified Tillotson equation of state from [4.90] and is valid for material in the condensed state $(\rho/\rho_0 > 1)$ or in any cold state $(\varepsilon < I_S)$. The complete formula for $B(p_0, \rho)$ has a very bulky form in this case. In practice, the quantity $B(p_0, \rho)$ was computed by a special subroutine in which, at first, the derivatives F_ρ, $F_{\rho\varepsilon}$, etc. were found by using analytic expressions, and then the value of the function $B(p_0, \rho)$ at given ρ, ε was computed by formula (4.3.9). In Figure 4.5 we present diagrams of the surfaces $p = p(x, t)$ obtained by the solution of a Riemann problem (1.1)–(1.3),

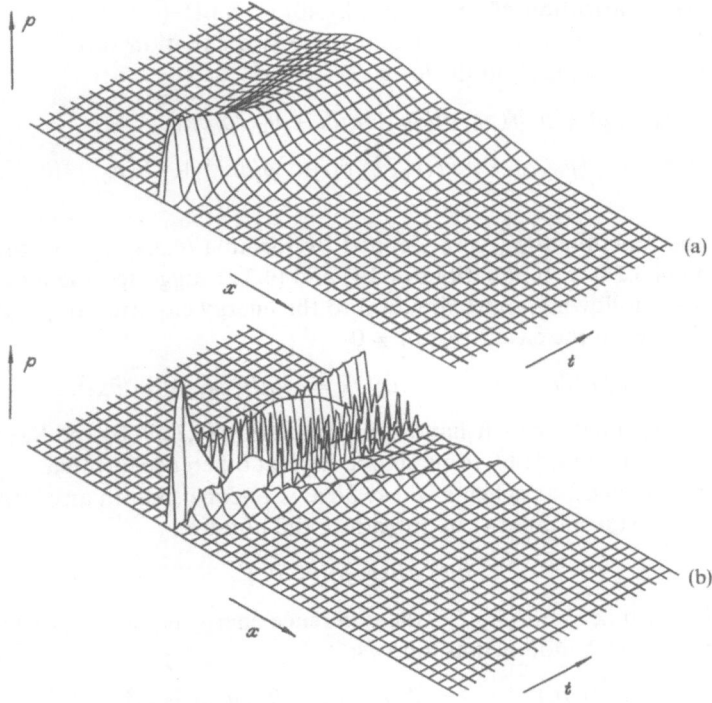

Figure 4.5. The surfaces $p = p(x, t)$.

(4.3.27), (4.1.1) with the LAX scheme [4.5], Figure 4.5(a), and with the one-step Lax–Wendroff scheme [4.45], Figure 4.5(b). It can be seen from Figure 4.5 that the difference solution for the pressure in the absence of K-consistence differs substantially from the exact solution which in this case represents a plateau of constant pressure. In this computational example the errors in pressure reached 18% for the Lax scheme and 50% for the Lax–Wendroff scheme.

4.3.2. Construction of K-Inconsistence Suppression Algorithms in the First-Order Schemes

Let us illustrate the basic idea of the algorithms in a number of examples. In [4.40] the construction of an algorithm for K-inconsistence suppression was performed in the example of the LAX scheme [4.5]. Note that this scheme is used as an intermediate scheme in such well-known difference methods as the two-step Lax–Wendroff method [4.45] and the third-order RUSANOV scheme [4.10]. As will be shown below, the K-inconsistence suppression algorithms may be constructed similarly for a number of other well-known first-order schemes.

Let us consider the f.d.a. P-form (4.1.40) of the Lax scheme. The result of the quasi-linearization of equations (4.1.40) is (4.3.1)–(4.3.3). Consider equations (4.3.1), (4.3.2). It is evident that these two equations are K-consistent. Rewrite equation (4.3.3) in the form

$$[0.5u_0^2 + f(p_0, \rho) + \rho \, \partial f(p_0, \rho)/\partial \rho](\partial \rho/\partial t + u_0 \, \partial \rho/\partial x)$$
$$= \mu[0.5u_0^2 + f(p_0, \rho) + \rho \, \partial f(p_0, \rho)/\partial \rho](\partial^2 \rho/\partial x^2) + \mu B(p_0, \rho)(\partial \rho/\partial x)^2,$$
$$(4.3.28)$$

where the function $B(p_0, \rho)$ is defined by formula (4.3.4). A comparison of equation (4.3.28) with equations (4.3.1) and (4.3.2) suggested the idea of considering the following approximation to the energy equation to get the f.d.a. K-consistence in the case $B(p_0, \rho) \neq 0$

$$\partial \rho E/\partial t + \partial(pu + \rho uE)/\partial x = -\mu B(p_0, \rho)(\partial \rho/\partial x)^2. \qquad (4.3.29)$$

Let us approximate the left-hand side of equation (4.3.29) by the Lax scheme and the right-hand side by some simple explicit difference formula. As a result we get a modified Lax scheme whose f.d.a. is K-consistent in any form of the equations of state (1.5) or (1.6). Introduce the notation

$$q = \mu B(p, \rho)(\partial \rho/\partial x)^2. \qquad (4.3.30)$$

Taking into account (4.3.29), the difference energy equation (4.1.15) of the original scheme is modified as follows:

$$(\rho E)^{n+1} = H_3(u^n, p^n, \rho^n, \varepsilon^n, T_1, h, \tau) - \tau q^n, \qquad (4.3.31)$$

where q^n is an approximation of the corrective term (4.3.30) at $t = t^n$. In the case of the Lax scheme in [4.42] the following approximation for q^n was considered:

$$q^n = \mu^n B(p^n, \rho^n)[((T_1 - T_{-1})/(2h))\rho^n]^2, \qquad (4.3.32)$$

where

$$\mu^n = h^2/(2\tau) - (\tau/2)(u^n)^2. \qquad (4.3.33)$$

The quantity B in (4.3.32) is computed by formula (4.3.9) when the equation of state (1.5) is used. It was proposed in [4.40] to use instead of (4.3.31) the difference approximation of the form

$$(\rho E)^{n+1} = H_3(u^n, p^n, \rho^n, \varepsilon^n, T_1, h, \tau) - \tau(T_{1/2} - T_{-1/2})R^n/h, \qquad (4.3.34)$$

where

$$R = \int_0^x q(\xi, t) \, d\xi, \qquad (4.3.35)$$

$0 \leq x \leq l$ is the integration interval on the x-axis. Employing various quadrature formulas for evaluation of the integral (4.3.35) we obtain, generally speaking different approximations to the function q. In particular, it was proposed in [4.40] to utilize the trapezoid rule for the computation of R.

Consider now the f.d.a. H-forms of the FLIC method (2.4.4) and of the "breakdown-of-discontinuity" scheme (2.4.12), (2.4.13). Applying the quasi-linearization operation L to both sides of the equations of these systems and passing then to the f.d.a. P-form, as was done in Section 4.1, we again obtain the system of the form (4.1.41) where μ is now computed by formula (4.1.49). Thus, to obtain the f.d.a. K-consistence of the difference schemes under consideration in the case $B(p_0, \rho) \neq 0$, one should modify the difference energy equation by formulas (4.3.31) or (4.3.34), (4.3.35), where the value μ entering the approximation of the corrective term q is computed at $t = t^n$ in accordance with (4.1.50) by the formula

$$\mu^n = 0.5|u^n|(h - \tau|u^n|), \qquad (4.3.36)$$

where u^n is the velocity value at $t = t^n$. Taking into account the results of Section 4.1 the corrective term (4.3.30), where μ is computed by (4.1.50), is also applicable in the case of the "coarse particles" method [4.56] when difference formulas of the first order of accuracy are used at the Lagrangian stage of the computation in this method.

An analysis of the stability of the modified Lax scheme (4.3.31)–(4.3.33) was carried out in [4.42] in the approximation of linearized difference equations: therein also the value $\beta = \mu B \, \partial\rho/\partial x$ was "frozen". From this analysis a conclusion was drawn on the insignificant influence of the introduced term on the computational stability. In the case when the inequality $a_{23}\beta < 0$ was satisfied, where $a_{23} = (1/\rho) \, \partial F/\partial\varepsilon$, the stability analysis of [4.42] yielded an increase in stability compared with the case $\beta = 0$. Similar conclusions were derived in [4.42] in the result of the stability analysis of the GODUNOV scheme [4.55] in the case $u > c > 0$ (supersonic flow).

4.3.3. K-Inconsistence Suppression in Second-Order Schemes

At present finite-difference second-order schemes are widely used in numerical simulations of compressible fluid flows [4.1], [4.80]. For example, the MacCormack scheme [4.82] and its numerous modifications are widely used in aerodynamic computations. It follows from the results of Section 4.1 that the width of a contact strip increases substantially slower with time t in the case of second-order schemes than in the case of first-order schemes. The presence of parasitic oscillations of the difference solution in the region behind a strong discontinuity is a short-coming of second-order schemes. In the case of a contact discontinuity an analysis of the f.d.a. in Section 4.1 has led to the solution (4.1.29), (4.1.30) which contains an integral of the Airy function $Ai(\xi)$ which is, as is known [4.52], an oscillatory function for $\xi < 0$ (see Figure 4.4). As was shown in [4.40], in the presence of K-inconsistence of second-order difference schemes, there takes place in the one-dimensional contact strip the oscillations of not only the density, but also of such continuous functions as pressure and velocity (see above Figure 4.5(b)). The amplitude of the oscillations caused by the K-inconsistence proved to be very significant (up to 50% of the exact pressure values in the contact strip) when using such second-order schemes as the one-step Lax–Wendroff scheme [4.92], [4.45], [4.1], [4.80] and the MacCormack scheme [4.82], [4.93], [4.94].

In what follows we show, in an example of these two schemes, how to generalize the basic idea of the K-inconsistence suppression algorithms presented in the foregoing section for the case of second-order schemes.

As was shown in [4.40], the amplitude of the pressure and velocity oscillations decreases in the presence of K-inconsistence as the time t increases. Therefore, it is important for the estimation of the size of this amplitude to determine its value at $t = \tau$, that is, at the first time step. Since we consider for the system (1.1)–(1.3) the initial data (4.1.1), we can utilize these data for the computation of the grid solution at $t = \tau$. Let us show that in the case when the equations of state (1.5) or (1.6), which do not satisfy condition (4.1.43), are not too complicated then one can obtain explicit expressions for the errors in difference solutions for the pressure and velocity in the contact strip at $t = \tau$. For this purpose let us take the equation of state

$$p = a\rho\varepsilon + \sum_{k=1}^{m} B_k(\rho - \rho_0)^k, \qquad (4.3.37)$$

where a, B_k, ρ_0 are constant and $a > 0, B_k \geq 0, m$ is an integer, $m \geq 1$. Consider for definiteness the case $m = 2$. Let the abscissa of the right boundary of the jth cell on the x-axis coincide with the initial abscissa x_0 of a contact discontinuity. Let us take, in accordance with (4.1.1), the following initial grid distribution of the quantities ρ, p, u, ε:

$$\rho_i^0 = \begin{cases} \rho_1, & i \leq j, \\ \rho_2, & i > j, \end{cases} \qquad p_i^0 = p_0, \quad u_i^0 = u_0, \quad \varepsilon_i^0 = f(p_0, \rho_i^0), \qquad (4.3.38)$$

where according to (1.6), (4.3.37)

$$f(p, \rho) = \left[p - \sum_{k=1}^{2} B_k (\rho - \rho_0)^k \right] / (a\rho). \qquad (4.3.39)$$

Then at $t = \tau$, and after some algebraic calculations, we obtain in the case of the one-step Lax–Wendroff scheme the following exact result:

$$\rho_i^1 = \rho_1, \quad i \leq j - 1, \qquad \rho_i^1 = \rho_2, \quad i \geq j + 2,$$

$$\rho_j^1 = \rho_1 + (1/2)(\rho_2 - \rho_1)(\kappa^2 - \kappa),$$

$$\rho_{j+1}^1 = \rho_2 - (1/2)(\rho_2 - \rho_1)(\kappa^2 + \kappa),$$

$$u_i^1 = u_0, \qquad \forall i, \qquad (4.3.40)$$

$$p_i^1 = \begin{cases} p_0, & i \leq j - 1, i \geq j + 2; \\ p_0 + (1/4)B_2(\rho_2 - \rho_1)^2\kappa(\kappa^2 - 1)(\kappa - 2), & i = j; \\ p_0 + (1/4)B_2(\rho_2 - \rho_1)^2\kappa(\kappa^2 - 1)(\kappa + 2), & i = j + 1; \end{cases}$$

where $\kappa = u_0 \tau / h$ is the dimensionless speed in the contact strip. Note that by virtue of the stability condition of the Lax–Wendroff scheme $|\kappa|$ is always less than 1.

Let use write the MacCormack scheme for the system (1.11) as follows:

$$\tilde{w}_i^{n+1} = w_i^n - (\tau/h)(\varphi_{i+1}^n - \varphi_i^n); \qquad (4.3.41)$$

$$w_i^{n+1} = (1/2)(w_i^n + \tilde{w}_i^{n+1}) - (1/2)(\tau/h)(\tilde{\varphi}_i^{n+1} - \tilde{\varphi}_{i-1}^{n+1}); \qquad (4.3.42)$$

where $\varphi_i^n \equiv \varphi(w_i^n)$, $\tilde{\varphi}_i^{n+1} \equiv \varphi(\tilde{w}_i^{n+1})$, etc. From equation (4.3.41) we have at $t = \tau$ (that is, at $n = 0$ in (4.3.41))

$$\tilde{\rho}_i^1 = \begin{cases} \rho_1, & i \leq j - 1, \\ \rho_2, & i \geq j + 1, \\ \rho_1 - \kappa(\rho_2 - \rho_1), & i = j; \end{cases} \qquad \tilde{u}_i^1 = u_0, \quad \forall i; \qquad (4.3.43)$$

$$\tilde{p}_i^1 = \begin{cases} p_0, & i \leq j - 1, i \geq j + 1; \\ p_0 + B_2(\rho_1 - \rho_2)^2\kappa(\kappa + 1), & i = j. \end{cases} \qquad (4.3.44)$$

With regard to (4.3.43), (4.3.44), (4.3.38) we obtain from (4.3.42) the following grid solution at $t = \tau$:

$$\rho_i^1 = \begin{cases} \rho_1, & i \leq j - 1; \\ \rho_2, & i \geq j + 2; \\ \rho_1 + (1/2)(\rho_2 - \rho_1)(\kappa^2 - \kappa), & i = j; \\ \rho_2 - (1/2)(\rho_2 - \rho_1)(\kappa^2 + \kappa), & i = j + 1; \end{cases}$$

$$u_i^1 = u_0, \qquad \forall i; \qquad (4.3.45)$$

$$p_i^1 = \begin{cases} p_0, & i \leq j - 1, i \geq j + 2; \\ p_0 + (1/4)B_2(\rho_1 - \rho_2)^2\kappa(\kappa + 1)[(\kappa - 1)(\kappa - 2) - 2\kappa a], & i = j; \\ p_0 + (1/4)B_2(\rho_1 - \rho_2)^2\kappa(\kappa + 1)[(\kappa - 1)(\kappa + 2) + 2\kappa a], & i = j + 1. \end{cases}$$

Let $\Delta p^1 = \max_i|p_i^1 - p_0|$. Employing formulas (4.3.40) it is easy to verify that the following relations are valid in the case of the one-step Lax–Wendroff scheme and the MacCormack scheme

$$\Delta p^1 = \begin{cases} |p_j^1 - p_0|, & \kappa < 0; \\ |p_{j+1}^1 - p_0|, & \kappa > 0. \end{cases} \tag{4.3.46}$$

Denote by $(\Delta p^1)_{\text{LW}}$, $(\Delta p^1)_{\text{M}}$ the values of the quantity Δp^1 for the Lax–Wendroff and MacCormack schemes, respectively. Let $\delta = (\Delta p^1)_{\text{M}}/(\Delta p^1)_{\text{LW}}$. Employing formulas (4.3.40), (4.3.45) it is easy to show that at $-1 < \kappa < 0$ the inequality $\delta > 1$ takes place. If κ is positive and $\kappa < 1$, then at

$$\kappa > \min\{[((3 + a)/2) - (((3 + a)/2)^2 - 2)^{1/2}],$$
$$[-((1 + a)/2) + ((1 + a)^2/4 + 2)^{1/2}]\} \tag{4.3.47}$$

also $\delta > 1$. Note that at $a > 0$ the right-hand side of inequality (4.3.47) is a positive number less than unity. Thus, if κ satisfies any of the inequalities $-1 < \kappa < 0$ and (4.3.47), the MacCormack scheme (4.3.41), (4.3.42) generates more significant pressure oscillations in the contact strip than does the one-step Lax–Wendroff scheme. In the process of deriving formula (4.3.45) we can easily see that the appearance of the item $2\kappa a$ in (4.3.45) is caused by the use of the quantities with tilda in the corrector (4.3.42) (see formulas (4.3.43), (4.3.44)).

Consider the question of the construction of the K-inconsistence suppression algorithm in the case when the one-step Lax–Wendroff scheme is used in the computations. Following [4.43] let us derive from (4.1.46) the f.d.a. P-form using the standard algorithm presented, for example, in Section 1.3:

$$U(\partial\rho/\partial t + u_0 \, \partial\rho/\partial x) = -\mu(u_0, h, \tau) \, \partial^2/\partial x^2(U \, \partial\rho/\partial x), \tag{4.3.48}$$

where the coefficient μ is determined by (4.1.48). Rewrite the system (4.3.48) as follows:

$$U(\partial\rho/\partial t + u_0 \, \partial\rho/\partial x) = -\mu U \, \partial^3\rho/\partial x^3 - (\partial/\partial x)[\mu B(\partial\rho/\partial x)^2]$$
$$- \mu B(\partial\rho/\partial x)(\partial^2\rho/\partial x^2).$$

For the purpose of restoring the K-consistence of the f.d.a. of the difference energy equation in the Lax–Wendroff scheme consider, by analogy with Section 4.3.2, the following approximation of the energy equation:

$$\partial\rho E/\partial t + \partial(pu + \rho uE)/\partial x = q, \tag{4.3.49}$$

where

$$q = (\partial/\partial x)[\mu B \cdot (\partial\rho/\partial x)^2] + \mu B(\partial\rho/\partial x)(\partial^2\rho/\partial x^2). \tag{4.3.50}$$

Employing Theorem 4.1 from Section 4.1 it is easy to unite formula (4.3.50) for q and the formulas for q (derived in Section 4.3.2 for first-order schemes) into a more general formula

$$q = (-1)^r \mu \sum_{k=0}^{r-1} C_r^k \left\{ \frac{\partial^{r-k-1}}{\partial x^{r-k-1}} \left[B(p_0, \rho) \frac{\partial \rho}{\partial x} \right] \right\} \left(\frac{\partial^{k+1} \rho}{\partial x^{k+1}} \right), \qquad (4.3.51)$$

where r is the order of approximation of a difference scheme,

$$\partial B(p_0, \rho)/\partial x = [dB(p_0, \rho)/d\rho] \, \partial \rho/\partial x;$$

$$\partial^0 g(x, t)/\partial x^0 \equiv g(x, t); \qquad 0! = 1; \qquad C_r^k = r!/(k!(r-k)!).$$

Then instead of the initial energy equation in the system (1.11), (1.9) one should approximate in the contact strip an equation (4.3.49) where q is determined by (4.3.51).

Consider now a question on the construction of K-inconsistence suppression algorithms in the MacCormack scheme (4.3.41), (4.3.42). Two different approaches to constructing such algorithms in the case of predictor–corrector schemes were realized in [4.38], [4.39], [4.43]. The first of these approaches assumes a consequent K-inconsistence suppression at each of the intermediate stages. The second approach is based on the K-inconsistence suppression only in the process of computing a final solution vector at the corrector stage. In the following we show that these approaches lead to different computational algorithms (even if only the formulas of the corrector stage are compared).

Algorithm 1. To suppress the oscillations in the intermediate pressure and velocity profiles let us apply the general procedure presented in Section 4.1 to the predictor scheme (4.3.41). To this end, let us first write the f.d.a. H-form of the scheme (4.3.41)

$$\mathbf{w}_t + \boldsymbol{\varphi}_x = -(\tau/2)\mathbf{w}_{tt} - (h/2)\boldsymbol{\varphi}_{xx}. \qquad (4.3.52)$$

Applying the quasi-linearization operation L (which we have defined previously in Section 4.1) to both sides of (4.3.52) we obtain the system

$$U(\rho_t + u_0 \rho_x) = -(\tau/2)(U\rho_t)_t - (hu_0/2)(U\rho_x)_x, \qquad (4.3.53)$$

where the vector U is determined by formula (4.1.6) and u_0 is a constant velocity value from (4.1.1). Employing the general procedure for obtaining the f.d.a. P-form from the f.d.a. H-form, it is easy to obtain the f.d.a. P-form from (4.3.53)

$$U(\rho_t + u_0 \rho_x) = -\mu_0 U\rho_{xx} - \mu_0 (dU/d\rho)\rho_x^2, \qquad (4.3.54)$$

where

$$\mu_0 = 0.5h(1 + \kappa), \qquad \kappa = u_0 \tau/h, \qquad (4.3.55)$$

$$dU/d\rho = \{0, 0, B(p_0, \rho)\}^{\mathrm{T}}, \qquad (4.3.56)$$

where the function $B(p_0, \rho)$ is defined by formula (4.3.4). It follows from (4.3.54) that for restoration of the f.d.a. K-consistence for the energy equation it is necessary, with (4.3.56) in view, to consider the following approximation to the energy equation at the predictor stage:

$$(\rho E)_t + (pu + \rho uE)_x = \mu_0 B(p_0, \rho)\rho_x^2. \tag{4.3.57}$$

Suppose that we have

$$\tilde{p}_i^{n+1} = p_0, \qquad \tilde{u}_i^{n+1} = u_0 \tag{4.3.58}$$

for all i as a result of the application of the approximation (4.3.57). Then it can easily be found from the difference momentum equation of the scheme (4.3.42) that $u_i^{n+1} = u_0$ for all i. Rewrite the difference energy equation of the corrector scheme (4.3.42) with regard to the relations (4.3.58) and $u_i^{n+1} = u_0$

$$(\rho\varepsilon)_i^{n+1} + \rho_i^{n+1}u_0^2/2 = (1/2)[(\rho\varepsilon)_i^n + (u_0^2/2)(\rho_i^n + \tilde{\rho}_i^{n+1}) + (\tilde{\rho}\tilde{\varepsilon})_i^{n+1}]$$
$$- (\kappa/2)[(\tilde{\rho}\tilde{\varepsilon})_i^{n+1} - (\tilde{\rho}\tilde{\varepsilon})_{i-1}^{n+1} + (u_0^2/2)(\tilde{\rho}_i^{n+1} - \tilde{\rho}_{i-1}^{n+1})]. \tag{4.3.59}$$

Let us now make use of the difference equations of the predictor scheme (4.3.41). Let $u_i^n = u_0$, $p_i^n = p_0$, for all i at $t = t^n$. Then we obtain from the difference energy equation of the predictor scheme

$$(\tilde{\rho}\tilde{\varepsilon})_i^{n+1} + \tilde{\rho}_i^{n+1}u_0^2/2 = (\rho E)_i^n - \kappa[(\rho E)_{i+1}^n - (\rho E)_i^n]. \tag{4.3.60}$$

Employing (4.3.60) we obtain from (4.3.59) a difference equation

$$(\rho\varepsilon)_i^{n+1} + \rho_i^{n+1}u_0^2/2 = (\rho E)_i^n - 0.5\kappa[(\rho E)_{i+1}^n - (\rho E)_{i-1}^n]$$
$$+ 0.5\kappa^2[(\rho E)_{i+1}^n - 2(\rho E)_i^n + (\rho E)_{i-1}^n] \tag{4.3.61}$$

after some computations. Applying the quasi-linearization operation L to the difference continuity equation in the scheme (4.3.42), we obtain the following difference equation:

$$\rho_i^{n+1} = 0.5[2\rho_i^n - \kappa(\rho_{i+1}^n - \rho_i^n)]$$
$$- 0.5\kappa[(\rho_i^n - \rho_{i-1}^n) - \kappa(\rho_{i+1}^n - 2\rho_i^n + \rho_{i-1}^n)]. \tag{4.3.62}$$

Substituting now the expression (4.3.62) instead of ρ_i^{n+1} into (4.3.61) we obtain the following difference equation

$$(\rho\varepsilon)_i^{n+1} = (\rho\varepsilon)_i^n - 0.5\kappa[(\rho\varepsilon)_{i+1}^n - (\rho\varepsilon)_{i-1}^n]$$
$$+ 0.5\kappa^2[(\rho\varepsilon)_{i+1}^n - 2(\rho\varepsilon)_i^n + (\rho\varepsilon)_{i-1}^n], \tag{4.3.63}$$

The f.d.a. P-form of equation (4.3.63) is as follows:

$$(\rho\varepsilon)_t + u_0(\rho\varepsilon)_x = -\mu_1(\rho\varepsilon)_{xxx}, \tag{4.3.64}$$

where
$$\mu_1 = (h^2/6)u_0 - (\tau^2/6)u_0^3. \tag{4.3.65}$$

Applying the quasi-linearization operation L to the differential equation (4.3.64) it is easy to find that for the restoration of the f.d.a. K-consistence of the energy equation in the scheme (4.3.42) it is necessary to introduce a corrective term q into the energy equation by formulas (4.3.49), (4.3.50) where one should set $\mu = \mu_1$. Thus, in the case when the corrective term is introduced at the predictor stage in accordance with (4.3.57) the function form of the corrective term q for the corrector stage coincides with the expression for q (4.3.50) found above for the one-step Lax–Wendroff scheme.

Algorithm 2. Setting in (2.4.15) $\alpha = 1$, $\beta = 0$, in accordance with the MacCormack scheme (4.3.41), (4.3.42) let us write the f.d.a. H-form of this scheme as follows:
$$\mathbf{w}_t + \varphi(\mathbf{w})_x = (h^2/6)\{-\sigma^2 \mathbf{w}_{ttt} - \varphi(\mathbf{w})_{xxx} + \tfrac{3}{2}[\mathscr{B}(\varphi_x^0, \varphi_x^1)]_x\}, \tag{4.3.66}$$
where
$$\sigma = \tau/h, \qquad \varphi_0 = \mathbf{w} + \sigma\varphi(\mathbf{w}), \qquad \varphi_1 = -\sigma\varphi(\mathbf{w}),$$

and $\mathscr{B}$ is the bilinear symmetric transformation, $\mathscr{B} = \varphi''(\mathbf{w})$. We must find the result of the application of the quasi-linearization operation L to equation (4.3.66) in order to find the structural form of the corrective term q from the requirement of the f.d.a. (4.3.66) K-consistence. At first let us obtain a more detailed expression for (4.3.66). Let us substitute the right-hand side of equation (4.3.41) into (4.3.42) instead of $\mathbf{w}_i^{n+1}$ and expand the quantities $\mathbf{w}_i^n$, φ_i^n, φ_{i+1}^n into Taylor series with respect to the point with indices (i, n) in the (x, t)-plane. The quantities $\tilde{\varphi}_i^{n+1}$, $\tilde{\varphi}_{i-1}^{n+1}$ will be expanded with respect to the point $(i, n + 1)$ so far. Retaining the terms having only the first and second order of smallness with respect to h, τ we obtain from (4.3.42) the equation

$$\mathbf{w}_t + (\tau/2)\mathbf{w}_{tt} + (\tau^2/6)\mathbf{w}_{ttt} + (1/2)\varphi_x + (1/2)\tilde{\varphi}_x^{n+1}$$
$$= (h/12)(3\tilde{\varphi}_{xx}^{n+1} - h\tilde{\varphi}_{xxx}^{n+1} - 3\varphi_{xx} - h\varphi_{xxx}). \tag{4.3.67}$$

In (4.3.67) $\tilde{\varphi}^{n+1} = \varphi(\tilde{\mathbf{w}}^{n+1})$. Let us find expressions for the derivatives $\tilde{\varphi}_{xx}^{n+1}$, $\tilde{\varphi}_{xxx}^{n+1}$ in terms of the derivatives of the functions $\mathbf{w}(x, t^n)$, $\varphi(\mathbf{w}(x, t^n))$. Taking (4.3.41) into account let us introduce the vector
$$\mathbf{v} = \mathbf{w} - \tau(\varphi_x + 0.5h\varphi_{xx}). \tag{4.3.68}$$

It is clear that $\mathbf{v} = \mathbf{v}(w_1, w_2, w_3)$. Let $\mathbf{v} = \{v_1, v_2, v_3\}^T$. Then
$$v_i = v_i(w_1, w_2, w_3); \qquad i = 1, 2, 3, \tag{4.3.69}$$

where we have according to (1.9) that
$$w_1 = \rho, \qquad w_2 = \rho u, \qquad w_3 = \rho E, \qquad \mathbf{w} = \{w_1, w_2, w_3\}^T.$$

By virtue of the Taylor formula we have

$$\varphi(v_1, v_2, v_3) = \varphi(v_1^0, v_2^0, v_3^0) + \sum_{j=1}^{3} (\partial\varphi/\partial v_j)|_{v^0}(v_j - v_j^0)$$

$$+ (1/2) \sum_{i=1}^{3} \sum_{j=1}^{3} (\partial^2\varphi/\partial v_i\, \partial v_j)|_{v^0}(v_i - v_i^0)(v_j - v_j^0) + \cdots.$$

$$(4.3.70)$$

Let us set in (4.3.70), with the definition (4.3.68) in view,

$$\mathbf{v} = \{v_1, v_2, v_3\}^{\mathrm{T}} = \mathbf{w} - \tau(\varphi_x + 0.5h\varphi_{xx}),$$

$$\mathbf{v}^0 = \mathbf{w}(x, t^n) \qquad\qquad (4.3.71)$$

Then differentiating both sides of equation (4.3.70) a needed number of times we obtain approximate expressions for the derivatives $\tilde{\varphi}_{xx}^{n+1}$, $\tilde{\varphi}_{xxx}^{n+1}$ in terms of the derivatives of the functions $\mathbf{w}(x, t^n)$, $\varphi(\mathbf{w}(x, t^n))$. Let us introduce the following conventional notation:

$$A = \partial\varphi/\partial\mathbf{w} = \|a_{ik}\| = \|\partial\varphi_i/\partial w_k\|.$$

The expressions for a_{ik} in the case when the equation of state (1.5) is employed are given by the formulas (1.14). The matrix $H_3 = \|\partial^2\varphi_3/\partial w_i\, \partial w_j\|$ is, by definition, the Hesse matrix corresponding to the function $\varphi_3(w_1, w_2, w_3)$ (see, for instance, [4.94]). Let us present the expressions for the elements of the H_3 matrix which we shall subsequently make use of.

$$\partial^2\varphi_3/\partial w_1^2 = \partial a_{31}/\partial w_1 = (u/\rho)[2E + 2p/\rho - 2\,\partial F/\partial\rho$$

$$- \rho^{-1}(\partial F/\partial\varepsilon)(5u^2 - 4E)] + u[\partial^2 F/\partial\rho^2$$

$$+ 2(\partial^2 F/\partial\rho\, \partial\varepsilon)(u^2 - E)/\rho + F_{\varepsilon\varepsilon}(u^2/\rho - E/\rho)^2];$$

$$\partial^2\varphi_3/\partial w_1\, \partial w_2 = \partial a_{31}/\partial w_2 = \partial a_{32}/\partial w_1$$

$$= -\rho^{-1}(E + p/\rho - \partial F/\partial\rho + (\partial F/\partial\varepsilon)E/\rho)$$

$$- \rho^{-3}u^2[-4\rho F_\varepsilon + \rho^2 F_{\rho\varepsilon} + \rho F_{\varepsilon\varepsilon}(u^2 - E)];$$

$$\partial^2\varphi_3/\partial w_1\, \partial w_3 = \partial a_{31}/\partial w_3 = \partial a_{33}/\partial w_1$$

$$(4.3.72)$$

$$= -\rho^{-1}u[1 + (2/\rho)\,\partial F/\partial\varepsilon - \partial^2 F/\partial\rho\, \partial\varepsilon$$

$$- \rho^{-1}(\partial^2 F/\partial\varepsilon^2)(u^2 - E)];$$

$$\partial^2\varphi_3/\partial w_3^2 = \partial a_{33}/\partial w_3 = \rho^{-2}u\, \partial^2 F/\partial\varepsilon^2;$$

$$\partial^2\varphi_3/\partial w_2^2 = \partial a_{32}/\partial w_2 = -3\rho^{-2}u\, \partial F/\partial\varepsilon + \rho^{-2}u^3\, \partial^2 F/\partial\varepsilon^2;$$

$$\partial^2\varphi_3/\partial w_3\, \partial w_2 = \partial a_{33}/\partial w_2 = \partial a_{32}/\partial w_3$$

$$= (1/\rho)(1 + (\partial F/\partial\varepsilon)/\rho) - (u/\rho)^2\, \partial^2 F/\partial\varepsilon^2.$$

Employing the definition (4.3.71) of the vector-functions $\mathbf{v}$, $\mathbf{v}^0$ it is easy to show that

$$L(v_1 - v_1^0) = -\tau(u_0\rho + 0.5hu_0\rho_x)_x;$$

$$L(v_2 - v_2^0) = -\tau(u_0^2\rho + 0.5hu_0^2\rho_x)_x; \tag{4.3.73}$$

$$L(v_3 - v_3^0) = -\tau\{u_0(\varepsilon + 0.5u_0^2) + 0.5h[u_0\rho(\varepsilon + 0.5u_0^2)]_x\}_x,$$

where L is the quasi-linearization operation. Using formula (4.3.73) we get

$$\sum_{j=1}^{3} La_{3j}L(v_j - v_j^0) = -\tau u_0^2[\rho\varepsilon_x + (\varepsilon + u_0^2/2)\rho_x]$$

$$- (\tau u_0 h/2)[(F_\rho + 0.5u_0^2)\rho_{xx} + (\rho\varepsilon)_{xx}$$

$$+ (F_\varepsilon/\rho)(2\rho_x\varepsilon_x + \rho\varepsilon_{xx})]. \tag{4.3.74}$$

Making use of formulas (4.3.72) (4.3.73) we obtain, after some calculations,

$$\sum_{i=1}^{3}\sum_{j=1}^{3} L(\partial^2\varphi_3/\partial w_i\,\partial w_j)L(v_i - v_i^0)L(v_j - v_j^0)$$

$$= \tau^2 u_0^3(F_{\rho\rho}\rho_x^2 + 2F_{\rho\varepsilon}\rho_x\varepsilon_x - (2/\rho)F_\varepsilon\rho_x\varepsilon_x + F_{\varepsilon\varepsilon}\varepsilon_x^2). \tag{4.3.75}$$

Applying the quasi-linearization operation L to the rest of the terms in (4.3.67), (4.3.70), we obtain the quasi-linearization result of the f.d.a. of the energy equation of the MacCormack scheme (4.3.41), (4.3.42) in the form

$$[\rho(\varepsilon + 0.5u_0^2)]_t + u_0[\rho(\varepsilon + 0.5u_0^2)]_x$$

$$= -(\tau^2/6)[\rho(\varepsilon + 0.5u_0^2)]_{ttt} - (h^2/6)u_0$$

$$\times [\rho(\varepsilon + 0.5u_0^2)]_{xxx} + \mu_2(F_\rho\rho_{xx} + F_\varepsilon\varepsilon_{xx} + (2/\rho)F_\varepsilon\rho_x\varepsilon_x)_x, \tag{4.3.76}$$

where

$$\mu_2 = (u_0 h^2/4)(\kappa + \kappa^2), \qquad \kappa = u_0\tau/h. \tag{4.3.77}$$

Let us transform the last term in the right-hand side of (4.3.76) employing the assumption that at $t = t^n$, $u(x, t^n) = u_0$, $p(x, t^n) = p_0$. Let us differentiate twice with respect to x both sides of (1.5) assuming that in the left-hand side of (1.5) $p = $ const. After some computations we find

$$F_\rho\rho_{xx} + F_\varepsilon\varepsilon_{xx} + (2/\rho)F_\varepsilon\rho_x\varepsilon_x = (F_\varepsilon/\rho)B(p_0, \rho)\rho_x^2, \tag{4.3.78}$$

where the function $B(p_0, \rho)$ is determined by formula (4.3.4) and it can be calculated for example, by formula (4.3.9) when the equation of state (1.5) is employed. With regard to (4.3.78) formula (4.3.76) may be rewritten as follows:

$$\mathscr{E}(\rho_t + u_0\rho_x) = -\mu_1\mathscr{E}\rho_{xxx} - \mu_1[(dB(p_0, \rho)/d\rho)\rho_x^3 + 3B\rho_x\rho_{xx}]$$

$$+ \mu_2[(F_\varepsilon/\rho)B(p_0, \rho)\rho_x^2]_x, \tag{4.3.79}$$

where $\mathscr{E} = \varepsilon + \rho\,\partial f(p_0, \rho)/\partial\rho + 0.5u_0^2$ and μ_1 is determined by (4.3.65). In the process of derivation of formula (4.3.79) we at first have passed from the f.d.a. H-form (4.3.76) to the f.d.a. P-form employing the standard procedure described in Section 1.3. From (4.3.79) we obtain the following formula for the term q in equation (4.3.49):

$$q = q_{cor} = (\partial/\partial x)[(\mu_1 - \mu_2 F_\varepsilon/\rho)B(p_0, \rho)\rho_x^2] + \mu_1 B\rho_x\rho_{xx}. \quad (4.3.80)$$

We emphasize that the quantity (4.3.80) is introduced only at the corrector stage of the MacCormack scheme.

4.4. The Contact Residual Subtraction Method

Definitions 4.5–4.7 of the K-consistence of difference schemes approximating the Euler equation system (1.1)–(1.3) were given in Section 4.1. By using these definitions it is possible to construct accurate difference algorithms for the elimination of the K-inconsistence of difference schemes. The inclusion of these algorithms in the computer codes realizing homogeneous shock-capturing difference schemes reverts these schemes back to being inhomogeneous in the sense of the definition given in Section 1.3.1. This property of the algorithms presented in this section constitutes their substantial difference from the homogeneous algorithms presented in the foregoing section. Indeed, the computational formulas of the algorithms of Section 4.3 do not reduce the order of approximation of the difference scheme employed (see, for example, formulas (4.3.49), (4.3.51)), therefore, the modified energy equation (4.3.49), (4.3.51)) may also be used in principle outside the contact strips.

The algorithms presented below possess a higher efficiency than the approximate algorithms of Section 4.3 which were obtained from an analysis of the f.d.a. Indeed, they enable one to reduce the errors in pressure and velocity in a contact strip to the level of machine rounding-off errors. As a result of this, there takes place a substantial increase in the accuracy of contact discontinuity localization with the aid of Algorithms I–III of the differential analyzers presented in Section 4.1, as was shown in [4.41].

Two different algorithms for the K-inconsistence elimination of difference schemes were proposed in [4.41]. The first of these algorithms is based on the following definition of K-consistence [4.41].

Definition 4.10. Let us call the difference scheme (4.1.13)–(4.1.15) (approximating the system of equations (1.1)–(1.3), (1.6)) K-consistent if the momentum equation (4.1.14) is K-consistent with the difference continuity equation (4.1.13) and

$$\tilde{L}H_1 \cdot [f(p_0, \tilde{L}H_1) + u_0^2/2] - \tilde{L}H_3 \equiv 0, \quad (4.4.1)$$

where the function $f(p, \rho)$ enters the equation of state (1.6) and $\tilde{L}$ is the quasi-linearization operation given by Definition 4.5.

The following two theorems from [4.41] establish the conditions for the equivalence of Definitions 4.7 and 4.10.

Theorem 4.2. *If the difference scheme* (4.1.13)–(4.1.15) *is K-consistent in the sense of Definition* 4.10 *and, in addition, the identity* (1.7) *holds, then the difference scheme* (4.1.13)–(4.1.15) *is K-consistent in the sense of Definition* 4.7.

Theorem 4.3. *If the following conditions are satisfied:*

(a) *the difference scheme* (4.1.13)–(4.1.15) *is K-consistent in the sense of Definition* 4.7;
(b) *the function* $F(\rho, \varepsilon)$ *in the equation of state* (1.5) *is such that the only value of p corresponds to each pair of values of* (ρ, ε);
(c) *at* $p \geq 0$, $\rho \geq 0$, *the identity* (1.7) *takes place;*

then the difference scheme is K-consistent in the sense of Definition 4.10.

The proofs of these assertions may be found in [4.41]. Definition 4.10 is more constructive than Definition 4.7, since it enables one to prove a number of theorems on K-consistence of difference schemes of the form (1.1)–(1.3), (1.6).

Theorem 4.4. *If the equation of state is such that the equality* (4.1.43) *takes place for any fixed* $p \geq 0$, *then the Lax scheme* [4.5] *approximating the system* (1.1)–(1.3) *is K-consistent.*

The proof of this theorem, which has been presented in [4.41], proceeds as follows. At first the difference expressions H_1, H_2, and H_3 entering (4.1.13)–(4.1.15) are written down. Then the value of the pressure p is fixed in the general solution (4.1.44) of equation (4.1.43). After that the right-hand side of equation (4.1.44) is substituted (instead of $f(p, \rho)$) into the expression for H_3, and the quasi-linearization operation $\tilde{L}$ is applied to H_1, H_2, and H_3. The resulting expressions for $\tilde{L}H_1$, $\tilde{L}H_2$, and $\tilde{L}H_3$ are substituted into formula (4.4.1) and it is directly verified that the identity (4.4.1) takes place. $\qquad\qquad\square$

A theorem analogous to Theorem 4.4 is also valid for the FLIC method [4.44], one needs only to replace the words "the Lax scheme [4.5]" in the formulation of Theorem 4.4 by "the integral step scheme of the FLIC method".

Theorem 4.5. *To assure the K-consistence of the one-step Lax–Wendroff scheme approximating the system* (1.1)–(1.3), *one should satisfy the following conditions:*

(a)
$$\tilde{L}\pi_{i+1/2}^n = 0, \qquad \forall i,$$

where

$$
\pi_{i+1/2}^n \equiv \frac{\partial F(\rho_{i+1/2}^n, \varepsilon_{i+1/2}^n)}{\partial \rho}(T_1 - I)\rho_i^n
$$

$$
+ \frac{\partial F(\rho_{i+1/2}^n, \varepsilon_{i+1/2}^n)}{\partial \varepsilon}(T_1 - I)\varepsilon_i^n, \tag{4.4.2}
$$

$$
\rho_{i+1/2}^n = (I + T_1)\rho_i^n/2, \qquad \varepsilon_{i+1/2}^n = (I + T_1)\varepsilon_i^n/2,
$$

where T_1 is the shift operator, that is, $T_1 \rho(x, t) = \rho(x + h, t)$, I is the identity operator, $\tilde{L}$ is the quasi-linearization operation, and $F(\rho, \varepsilon)$ is the function entering the equation of state (1.5);

(b) *the equation of state (1.6) should satisfy condition (4.1.43).*

The proof of this theorem is carried out analogously to the proof of Theorem 4.4 (see [4.41]).

Suppose now that the equation of state (1.6) is such that condition (4.1.43) is not satisfied. Then in accordance with Definition 4.10

$$\tilde{L}H_1 \cdot [f(p_0, \tilde{L}H_1) + u_0^2/2] - \tilde{L}H_3 = \tilde{L}R, \qquad (4.4.3)$$

where $\tilde{L}R \neq 0$. Let equation (4.1.14) be K-consistent with the difference continuity equation (4.1.13). Let us modify the operator H_3 in (4.1.15) in such a way that the K-consistence of the difference scheme (4.1.13)–(4.1.15) takes place. For this purpose it is sufficient to take the expression

$$\tilde{H}_3 = H_3 + R \qquad (4.4.4)$$

instead of H_3 in (4.1.15). Then

$$\tilde{L}H_1 \cdot [f(p_0, \tilde{L}H_1) + u_0^2/2] - \tilde{L}\tilde{H}_3 = 0$$

by virtue of (4.4.3), (4.4.4). In [4.41] the quantity R in (4.4.4) was called a contact residual, because in the case when $\tilde{L}R \neq 0$ within the contact strip, the K-consistence property of the difference scheme (4.1.13)–(4.1.15) does not take place.

EXAMPLE. Consider the Lax scheme [4.5]. In this case the expression for the contact residual R may easily be found with the aid of (4.4.3):

$$R_i = \rho_i^{n+1} \cdot f(p_i^n, \rho_i^{n+1}) - 0.5[(\rho\varepsilon)_{i-1}^n + (\rho\varepsilon)_{i+1}^n]$$
$$+ 0.5(\tau/h)[(\rho u\varepsilon)_{i+1}^n - (\rho u\varepsilon)_{i-1}^n], \qquad (4.4.5)$$

where the subscript "i" by the quantities ρ, u, p, ε signifies that the grid value of each of these functions is computed in the center x_i of the ith cell of a uniform grid with the step h on the x-axis, and the superscript "n" indicates the number of the time level $t^n = n\tau$. Let the relations

$$u_i^n = u_0, \qquad p_i^n = p_0, \qquad (x_i, t^n) \in G_K, \qquad (4.4.6)$$

be satisfied in the contact strip G_K at $t = t^n$. Under the assumptions (4.4.6) it is easy to prove, with the aid of the Taylor series expansions of the difference quantities entering the right-hand side of (4.4.5), that

$$R_i = (1/2)B(p_0, \rho)h^2(\partial\rho/\partial x)^2(\kappa^2 - 1) + O(\tau h^2) + O(\tau^3), \qquad (4.4.7)$$

where $B(p_0, \rho)$ is the function determined by (4.3.4) and $\kappa = u_0\tau/h$. Comparing formula (4.4.7) with formula (4.3.30), for the corrective term q obtained in

Section 4.3.2 from the requirement of the K-consistency of the f.d.a. of the Lax scheme, we can see that the f.d.a.s of both modifications of the Lax scheme coincide and are K-consistent. But unlike (4.4.5), the explicit approximations (4.3.32)–(4.3.35) of the corrective term q were used in Section 4.3.2. The application of the method for the subtraction of the contact residual R assures the exact satisfaction of the K-consistency property of the difference scheme (4.1.13)–(4.1.15), unlike the approximate methods for the K-inconsistence suppression of difference schemes that were described in Section 4.3. At the same time, the present method has the following shortcoming: in the flow subdomains in which some of the derivatives u_x, p_x have an order of smallness $O(1)$ the inclusion of a contact residual R in the difference approximation (4.1.15) of the energy equation leads to the violation of an approximation of the original energy equation (1.3) (see [4.41]). To eliminate this negative effect two techniques were proposed in [4.41]. The first one consists of the introduction of two functions g_1 and g_2 such that the relationship $\tilde{L}(g_1 R + g_2) = \tilde{L}R$ is satisfied. In addition, the functions g_1 and g_2 should satisfy the requirements:

(a) for $\partial u/\partial x = O(1)$, $\partial p/\partial x = O(1)$, they should have an order of smallness $O(\tau h^r) + O(\tau^{r+1})$ where r is the order of approximation of the original scheme;
(b) the functions g_1 and g_2 may use only the grid points that are located within the stencil of the difference scheme considered;
(c) the stability of the modified scheme should not become worse;
(d) the functions g_1 and g_2 should involve the least possible number of arithmetic operations.

In [4.41] the function g_1 was set with regard to the above requirements as one of the simplest continuous approximations of the Dirac delta function

$$g_1 = g_1(h, \tau, \zeta) = a \exp(-b\zeta^2) + 1 - a, \qquad (4.4.8)$$

where

$$\delta = K_1 \cdot \min(\tau^{r+1}, \tau h^r), \qquad \zeta = K_2|u_x| + K_3|p_x|, \qquad (4.4.9)$$

$$a = (1 - K_4\delta)/[1 - \exp(-b\delta^2)]. \qquad (4.4.10)$$

In (4.4.8)–(4.4.10) r is the approximation order of the scheme, K_1, K_2, K_3, K_4 are positive constants, and $K_4 > 1$; b is a great positive constant, such that

$$0 < 1 - a \le \delta(K_4 - 1).$$

It is easy to show that the function g_1 (4.4.8) satisfies the inequality

$$|g_1 - \delta| \le (K_4 - 1)\delta \qquad \text{at} \quad \zeta \ge \delta. \qquad (4.4.11)$$

The function g_2 in [4.41] was taken in the form $g_2 \equiv 0$. The use of the contact residual

$$\tilde{R} = g_1 R \qquad (4.4.12)$$

provides the same order of approximation that the difference scheme had before the incorporation of the contact residual. This fact follows from the choice of the constant δ in the form (4.4.9) and from the inequality (4.4.11).

Another technique, which was proposed and realized on a computer in [4.41], also assures the conservation of the approximation order and is determined by the formula

$$(\rho E)_i^{n+1} = H_{3i} + \tilde{\tilde{R}}_i, \qquad (4.4.13)$$

where

$$\tilde{\tilde{R}}_i = \begin{cases} R_i & (|\rho_{i+1}^n - \rho_{i-1}^n| > \delta_1 \rho_i^n) \wedge (|p_{i+1}^n - p_{i-1}^n| < \delta_2 |p_i^n|) \\ & \wedge (|u_{i+1}^n - u_{i-1}^n| < \delta_3 |u_i^n|), \\ 0, & \text{otherwise.} \end{cases} \qquad (4.4.14)$$

In (4.4.14) $\delta_1, \delta_2, \delta_3$ are dimensionless constants and $\wedge$ is the conjunction sign. The constants δ_2 and δ_3 should satisfy the inequality

$$\delta_j \le O(h^{r+1}) + O(\tau^r h), \qquad j = 2, 3, \qquad (4.4.15)$$

where r is the approximation order of the original scheme. The check-up of the inequality $|\rho_{i+1}^n - \rho_{i-1}^n| > \delta_1 \rho_i^n$ prevents the computation of R in the domains of constant flow, and that saves machine time.

The expressions for the contact residual R were also obtained in [4.41] for the FLIC method [4.44] and for the one-step Lax–Wendroff scheme [4.45]. In the case of this second-order scheme it turns out that a contact residual r_i should be introduced into equation (4.1.14) to assure the K-consistence of the difference momentum equation:

$$(\rho u)_i^{n+1} = H_{2i} - r_i,$$

where

$$r_i = 0.5(\tau/h)^2 (u_{i+1/2}^n \pi_{i+1/2}^n - u_{i-1/2}^n \pi_{i-1/2}^n)$$

and the quantities $\pi_{i\pm1/2}^n$ are determined by formula (4.4.2).

One more method was proposed in [4.41] for the elimination of the K-inconsistence of difference schemes approximating the system of equations of one-dimensional flow of an inviscid gas. This method is based on the use of the well-known equation [4.45]

$$\partial p/\partial t + u \, \partial p/\partial x + \rho c^2 \, \partial u/\partial x = 0 \qquad (4.4.16)$$

instead of the energy equation (1.3) in the contact strip. In (4.4.16) c^2 is the square of the isentropic sound speed, which can be calculated with the aid of the formula

$$c^2 = (p/\rho^2 - \partial f/\partial \rho)/(\partial f/\partial p),$$

where the function $f(p, \rho)$ enters the equation of state (1.6). To reduce the influence of the nondivergence form of equation (4.4.16) on the conservation of the total energy in the computational domain, the difference approximation of equation (4.4.16) was used in [4.41] only within the limits of contact strips. Thus, from the point of view of the definition given in [4.1] (see also Section 1.3.1), the difference scheme obtained is inhomogeneous, as in the case of the application of the contact residual subtraction method.

The method of increasing the accuracy of difference solutions in the neighborhood of contact discontinuities in cases when condition (4.1.43) is not satisfied, which is based on the use of equation (4.4.16), is also applicable in the cases of nonzero pressure and velocity gradients in a one-dimensional contact strip.

4.5. Computational Examples

In this section we present some results of the application of the contact discontinuity localization techniques, presented in Section 4.1, to the computation of both a relatively simple flow containing only one discontinuity, the contact one, and the more complicated configurations containing not only the contact discontinuity but also shock waves and rarefaction fans.

Let us first present some results of computations of a Riemann problem (1.11), (1.9), (1.8), (4.1.1) by some first- and second-order difference schemes. Let K be the Courant number, $K = \max(|u| + c)\tau/h$. In Figure 4.6 we show the results of a Lax scheme computation at $K = 0.5$ (solid lines) and at $K = 0.9$ (dashed lines) for the moment of time when the contact discontinuity has propagated a distance of $\approx 70h_1$ [4.34]. In this computation $h_1 = 1/126$, $p_0 = u_0 = 1, \rho_1 = 2, \rho_2 = 1, \gamma = 2$. Within the diagram accuracy all five curves of the numerical solution intersect in the cell where the true contact discontinuity locates. Note that the contact strip width on the mesh $h = h_1, K = 0.5$, is equal, according to Figure 4.6, to $\approx 75h_1$, that is, this value is comparable with the integration interval size. Thus, the accuracy of the contact discontinuity localization on the basis of a definition of the contact strip central line (see Section 4.1), has proved to be very high. It follows from (4.1.37), (4.1.42) that, at fixed h, u_0, the width of a contact strip increases as the step τ decreases,

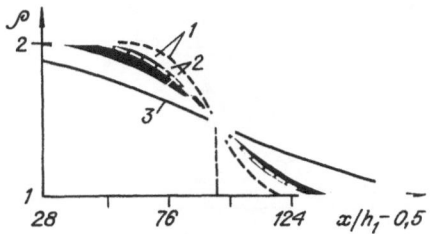

Figure 4.6. A computation by the Lax scheme at different h: 1—h_1; 2—$3h_1$; 3—$9h_1$.

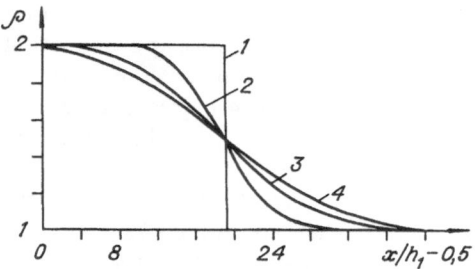

that is, as the Courant number decreases. Figure 4.6 clearly confirms this effect. The effect of smearing of a stagnant contact discontinuity ($u_0 = 0$) in computations by the Lax scheme is shown in Figure 4.7. At the same time, similar computations with $u_0 = 0$ by the "breakdown-of-discontinuity" scheme and by the FLIC method confirmed completely the conclusion on the absence of stagnant contact discontinuity smearing in computations by both the above schemes. The region of significant change in the density in a contact strip, in computations by the Lax scheme with Courant number $K = 0.5$, is shaded in Figure 4.8 by the lines parallel to the x-axis. The width X of the shaded region at fixed t was computed by formulas (4.1.37), (4.1.42) at $h = 1/126$. The middle of each interval of length X parallel to the x-axis was placed on the contact discontinuity line $dx/dt = u_0$, taking into account the properties of the solution (4.1.22), (4.1.23). In Figure 4.8 a similar region, obtained in the case of the one-step Lax–Wendroff scheme, is shaded by parallel lines making an angle with the x-axis—the values of K and h are the same as in the case of the Lax scheme. The width X has been computed in accordance with Table 4.1. It is seen from Figure 4.8 that at $t = 0.54$ the width after Prandtl of a contact strip X is in the case of the Lax–Wendroff scheme about three times less than the width X obtained in the case of the Lax scheme.

In Figure 4.9, similar to Figure 4.6, we present the graphs obtained by the GODUNOV method [4.55] and by the FLIC method [4.44]. Here $h_1 = 1/80$, $p_0 = u_0 = 1$, $\rho_1 = 2$, $\rho_2 = 1$, $K = 0.5$ (solid lines), $K = 0.9$ (dashed lines).

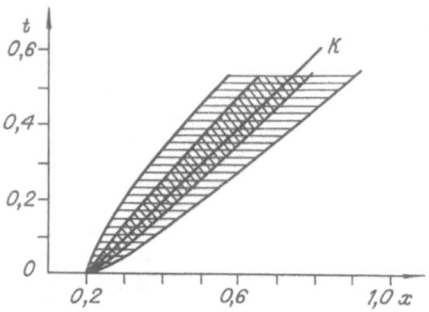

Figure 4.8. Contact strip in the (x, t)-plane. K—the line $dx/dt = u_0$.

Figure 4.9. Graph of the result of computation by the Godunov scheme.

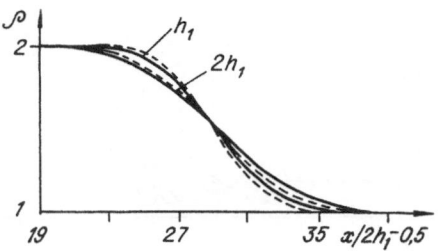

In Section 2.4 a necessary condition for the "smeared" shock wave center
existence is indicated when using the FLIC method [4.44] which has the form
$\tau/h = \text{const}$. On the other hand, from formula (4.1.23) for ξ the independence
of the position of the central line L_{KD} of the contact strip of the quantity τ/h
follows. Thus, we see an essential difference in the properties of the difference
scheme in the shock wave and in the contact strip.

When the one-step Lax–Wendroff scheme is used in computations we may
obtain from (4.1.29), (4.1.35) the condition for positiveness of the density $\rho(x, t)$:

$$\rho_1/\rho_2 > [a(\xi_1) + 2/3]/[a(\xi_1) - 1/3] \cong 0.21529. \qquad (4.5.1)$$

The inequality predicts quite well the appearance of negative densities in the
course of the numerical integration of the system (1.1)–(1.3) by the Lax–
Wendroff scheme. For example, at $\rho_1 = 1$, $\rho_2 = 5$ $(\rho_1/\rho_2 = 0.200)$ we have
obtained negative pressure and density values when using this scheme.

In Figure 4.10 the solution (4.1.29) for $\rho_1 = 1$, $\rho_2 = 4$, is represented by a
solid line, and the numerical solution via the Lax–Wendroff scheme obtained
at $K = 0.9$ is represented by circles. The exact solution at $t = 0$ and at $t = 0.54$
is shown by dashed lines (in this computational example $h = 1/40$, $u_0 = p_0 =
1$, $\gamma = 2$). In Figure 4.11 solid lines represent the solution of the same problem
as Figure 4.10 obtained by the Lax–Wendroff scheme at $h = 1/40$.

The above established oscillatory behavior of the solution for the density
ρ in a domain to the left of the contact discontinuity (see formula (4.1.29) and
Figures 4.4, 4.10, and 4.11), explains the oscillations in this domain which are
inherent in the second-order schemes and which arise in the numerical solution of a more general Riemann problem including, at $t > 0$, the rarefaction
wave, the contact discontinuity, and the shock wave [4.95]–[4.97].

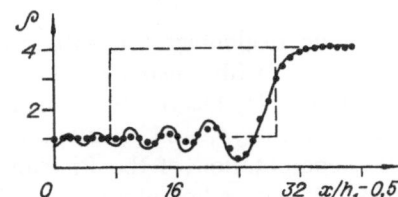

Figure 4.10. Computation of the contact discontinuity propagation by the Lax–Wendroff scheme.

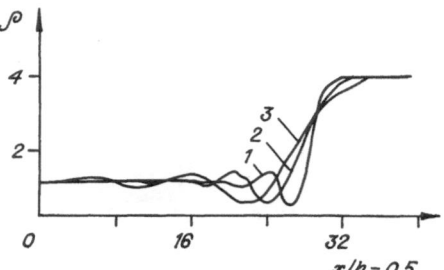

Figure 4.11. Computation by the Lax–Wendroff scheme at different h: 1—0.5h; 2—h; 3—2h.

The applicability of the above three Algorithms I–III for contact discontinuity localization in one-dimensional flows was shown in the computational examples in [4.34]. The error in localization did not exceed the step size h. On the central line L_K of the contact strip the value of the density ρ depends most weakly on the steps h and τ, similar to the case of the finite-difference shock-wave center localization (see Section 2.4).

Consider the behavior of numerical solutions obtained by the particle-in-cell method [4.26], [4.27], [4.28]–[4.103] in the neighborhood of a one-dimensional contact discontinuity. It was shown in [4.98] that in the limit of an infinite number of particles in each cell the computation of particles transport in the Harlow method is equivalent to a computation by a certain first-order difference scheme. This scheme coincides with the scheme of the FLIC method [4.44] which we have already considered above. In this connection it was of interest to compare the particle-in-cell method and the FLIC method properties when solving the problem (1.11), (1.9), (4.1.1) with a contact strip as $N \to \infty$, where N is the minimal number of particles in a cell. It follows from the formula for the width X of a contact strip presented in Table 4.1 for the FLIC method, that X increases proportionally to $\sqrt{t}$ in computations by the FLIC method and, in accordance with the above considerations, by the Harlow method. However, the available experience of the particle-in-cell method application to computing problems with contact discontinuities [4.27], [4.100] has shown that the Harlow method, unlike the known purely Eulerian schemes, preserves the contact strip width constant. This is undoubtedly an important positive property of the Harlow method. Consider the Riemann problem (1.11), (1.9), (4.1.45), (4.1.1) with the following initial data:

$$p_1 = p_2 = \rho_1 = u_1 = u_2 = 1, \qquad \rho_2 = 5. \tag{4.5.2}$$

The numerical integration with the data (4.5.2) was carried out in the interval $0 \le x \le 1.5$ with constant steps $h = 0.02$, $\tau = 0.005$, $\gamma - 1 = 1$, in the equation of state (4.1.45). Denote by N_1 and N_2 the number of particles in a cell, at $t = 0$, with densities ρ_1 and ρ_2, respectively. In Table 4.2 we present the results of a computation of the Riemann problem with the data (4.5.2) by two methods: the method of square particles [4.101] and the FLIC method [4.44].

Table 4.2

Particle-in-cell method, point particles				Particle-in-cell method, square particles				FLIC method		
t/τ \ N_1	4	20	40	t/τ \ N_1	1	2	8	t/τ	$h^{-1}X_d$	$h^{-1}X_T$
20	3.20	3.26	3.17	20	2.66	2.93	3.23	20	5.10	4.85
40	4.00	4.71	4.57	40	4.42	4.35	4.59	40	7.04	6.86
60	5.33	5.00	5.00	60	3.98	4.96	4.93	60	8.55	8.41
80	6.40	5.41	5.71	80	7.22	5.83	5.62	80	9.83	9.71
100	6.40	6.67	6.96	100	4.83	6.91	7.03	100	10.97	10.85
120	6.40	6.15	6.15	120	5.76	5.90	5.86	120	11.99	11.89
140	5.33	5.52	5.61	140	3.94	6.84	5.45	140	12.94	12.84
160	5.33	6.15	6.27	160	3.00	5.91	5.50	160	13.82	13.73
180	5.33	4.85	4.85	180	2.94	6.30	5.25	180	14.64	14.56
200	6.40	4.71	5.08	200	2.54	4.59	4.45	200	15.43	15.35

The width X_T was computed in accordance with a formula from Table 4.1, and the "difference" width X_d was computed by the formula $X_d = |\rho_1 - \rho_2| 2h/(\max_x|\rho_{i+1} - \rho_{i-1}|)$. Table 4.2 contains the numerical values of the quantity X_d (besides the extreme right column). In this series of computations the above number N_2 was taken to be dependent on N_1 in accordance with formula $\rho_1/\rho_2 = N_1/N_2$. This relation between the number of particles in cells with different initial densities is optimal for computations by the particle-in-cell method [4.23], [4.27]. The numerical solution profiles at time $t = 200\tau$ are shown in Figure 4.12; for the case $N_1 = 40$, $N_2 = 200$ by a solid line, for the exact solution by a dashed line, and for the FLIC solution by a line of open circles. Note the presence of strong pressure oscillations in the region behind the contact discontinuity when using the point particles (in the

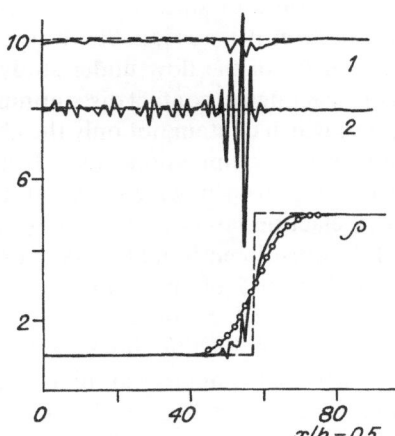

Figure 4.12. Computation of the contact discontinuity propagation by the PIC method and by the FLIC method. $1 — 10u$; $2 — 10p - 2$.

case of the FLIC method a difference solution for the pressure and velocity was obtained which coincided with the exact solution (4.1.2)). It follows from Table 4.2 that in the case of point particles, as well as in the case of square particles, the width X_d at first increases, and then it begins to oscillate around a certain constant value at the times $t \geq t_*$. This value of X_d is equal to $\approx 6h$. In the case of the FLIC method [4.44] formula (4.1.37) describes very well the actual behavior of the contact strip width. On the other hand, formula (4.1.37), as may be seen from Table 4.2, proves to be inapplicable in the case of the particle-in-cell method. We do not know of any rigorous theoretical explanation for this effect, however, it is clear that this effect is associated, first of all, with the use of Lagrangian particles in the Harlow method. Each particle moving at a contact discontinuity speed u_0 translates for itself information on that gas state which takes place on one of the sides of the contact discontinuity. The influence of geometric dimensions, and of the shape of particles on the properties of numerical solutions obtained by the particle-in-cell method in the vicinity of a contact discontinuity, has also been considered in [4.101]. In particular, it has been shown that in the case when the characteristic dimension of each particle is sufficiently large, then there takes place with time t an increase in the contact strip width. However, this increase proves to be slower than in the FLIC method.

In many problems the values ρ_1, ρ_2 on both sides of a contact discontinuity are not known in advance, therefore, Algorithm III of the contact discontinuity localization, which was presented in Section 4.1, has a limited domain of applicability. Let us consider in more detail the question of the computational realization of Algorithm II from Section 4.1. In the case of first-order schemes satisfying the conditions of Theorem 4.1 from Section 4.1 this algorithm reduces to the determination of the coordinates of the points of maximum of the gradient $|\partial \rho / \partial x|$ within the limits of contact strips. The value $|\partial \rho / \partial x|$ may be large not only in the zone of smearing of a contact discontinuity, but also in the zones of smearing of shock waves and compression waves. In this connection it is reasonable to use the available information on the general properties of the flow under study, as well as the experimental data, when constructing the contact discontinuity localization algorithms for complicated flows which contain not only the above type of discontinuities but also shock waves and compression waves. Consider, for example, a configuration of the Riemann problem which contains a shock wave, a contact discontinuity, and a rarefaction fan (see Figure 1.1(a)). It is known from the general properties of the one-dimensional Riemann problem solution [4.1], that within a certain neighborhood of the contact discontinuity the pressure and velocity are constant. Such a constancy of p and u in the numerical solutions of this problem, obtained by shock-capturing schemes generally, does not take place by virtue of the numerical solution errors. As the numerical experiments show (see, for example, Figure 4.2 above as well as numerous computational examples in [4.13], [4.14], [4.95], [4.96], [4.104]), in the presence of the f.d.a.

K-consistence of difference schemes the differences $|p_{i+1}^n - p_{i-1}^n|$, $|u_{i+1}^n - u_{i-1}^n|$ remain small in the neighborhood of a contact discontinuity. In this connection it was proposed in [4.42] to determine the position of contact strips in the numerical solutions of one-dimensional Riemann problems by checking the inequalities

$$|\rho_{i+1}^n - \rho_{i-1}^n| > \delta_1 \rho_i^n, \qquad |g_{i+1}^n - g_{i-1}^n| < \delta_2 |g_i^n|, \qquad (4.5.3)$$

where δ_1, δ_2 are nondimensional positive constants to be chosen empirically and g is some of the functions u, p. The check-up of the inequality $|\rho_{i+1}^n - \rho_{i-1}^n| > \delta_1 \rho_i^n$ serves the purpose of eliminating from further consideration the subdomains of a constant flow.

In the case of second-order schemes satisfying the conditions of Theorem 4.1 of Section 4.1 there are oscillations in the difference solution of the density. In terms of the ξ variable (see formula (4.1.30)), these oscillations take place at $\xi \leq \xi_0$ where $\xi_0 < -1$ (see Section 4.1). The practice of computations shows that in the subdomain $\xi > \xi_0$ the difference solution obtained (for example, by the Lax–Wendroff scheme [4.45]) does not contain oscillations, as does the solution (4.1.29). In this connection the satisfaction of the local monotonicity criterion of the form

$$\operatorname{sign}(\rho_{i+1}^n - \rho_i^n) = \operatorname{sign}(\rho_i^n - \rho_{i-1}^n), \qquad r = 2, \qquad (4.5.4)$$

was preliminarily checked alongside the inequalities (4.5.3) in the process of the determination of the coordinates of $|\partial^2 \rho / \partial x^2|$ maximum points within the limits of the contact strips. In Figure 4.13(a), (b) we show the results of locating a shock wave (triangles) and a contact discontinuity (circles) in the computation of an A-configuration in the Riemann problem. The values $\delta_1 = 0.01$, $\delta_2 = 0.05$, were used in the inequalities (4.5.3) (as in [4.42]); $g_i^n \equiv u_i^n$. The initial data for these computations were taken from [4.96] (see also Figure 4.2); the ideal gas equation of state (1.8) with the constant $\gamma = 1.4$ was used. The sequence of computations in the differential analyzer subroutine was as follows. At first the contact discontinuity location, at given $t = t^n$, was determined by $\max |\partial^r \rho / \partial x^r|$ with regard to the inequalities (4.5.3), (4.5.4). After that the shock wave front position was determined. For this purpose the mesh cells on the x-axis were "looked through" in a search for that cell, in which the quantity $|u_{i+1}^n - u_{i-1}^n|$ approximating the gradient $|2h \, \partial u / \partial x|$ on a uniform mesh achieved its maximum under the condition that the inequality $\partial u / \partial x < 0$ was satisfied (see also Section 3.4 for a description of a differential analyzer for the hypervelocity impact problems). This examination of cells was carried out in the case of an A-configuration in the direction from left to right departing from the contact discontinuity. In the case of a B-configuration (see Figure 4.13(b)), the cells were also "looked through" from right to left departing from the contact discontinuity for purpose of determining the location of a second shock wave in the flow. Note that by virtue of the smallness of the density jump across the contact discontinuity (in the given computational

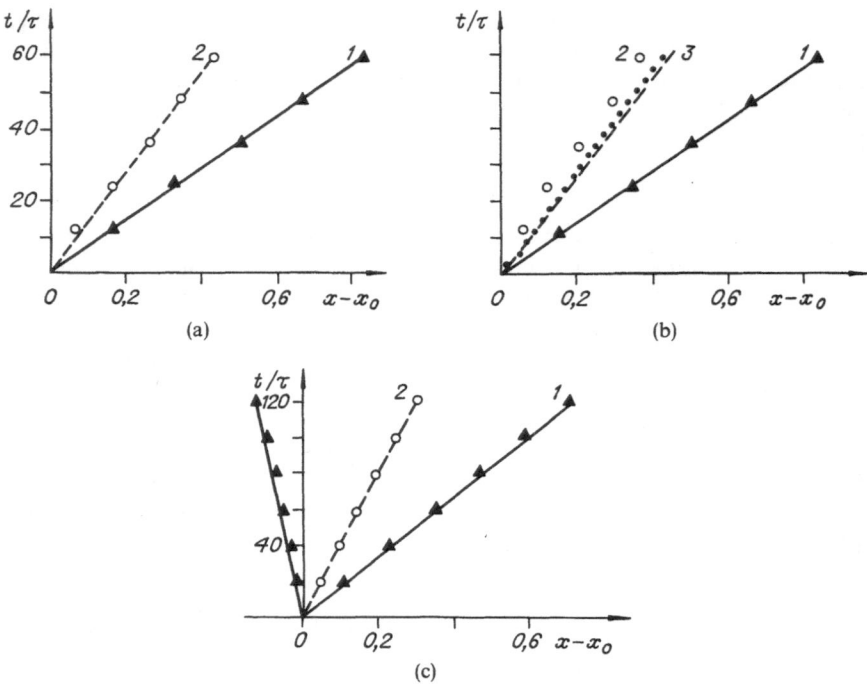

Figure 4.13. The calculation of the shock wave (1) and the contact discontinuity (2) trajectory: (a) by the FLIC method; (b) by the Lax–Wendroff method ((a) and (b) represent configuration *A*); (c) by the FLIC method (configuration *B*). 3—the contact discontinuity trajectory obtained on the basis of the function (5.6.7) minimization.

example $\rho_2 - \rho_1 \cong 0.016$, $\rho_1 = 4.7546$, see [4.42]; $h = 0.02$), and by virtue of the presence of oscillations in the numerical solution, a differential analyzer by $\max |\rho_x|$ proves to be inapplicable here. In this connection the x_k^{n+1} abscissa of a contact discontinuity (circles in Figure 4.13(b)) was determined by the formula $x_k^{n+1} = x_k^n + \tilde{u}_i \tau$, where "*i*" is the number of a cell on the *x*-axis in which the "marker" particle of a contact boundary was located at $t = t^n$, and $\tilde{u}_i$ is a velocity value obtained at the Eulerian stage of a FLIC method computation (see [4.44], [4.99] as well as Section 3.1). Solid and dashed lines in Figure 4.13 represent the trajectories of the shock waves and contact discontinuities, respectively, that were found numerically, to a high accuracy, by solving nonlinear algebraic relations which represent dynamic compatibility conditions on discontinuities.

    Let us now present some examples of computations to illustrate the efficiency of the *K*-inconsistence suppression algorithms which were described in Section 4.3. In Figure 4.14 (taken from [4.40]) we show the results of computation by the modified Lax scheme (4.3.32)–(4.3.35) of the problem (1.11), (1.9),

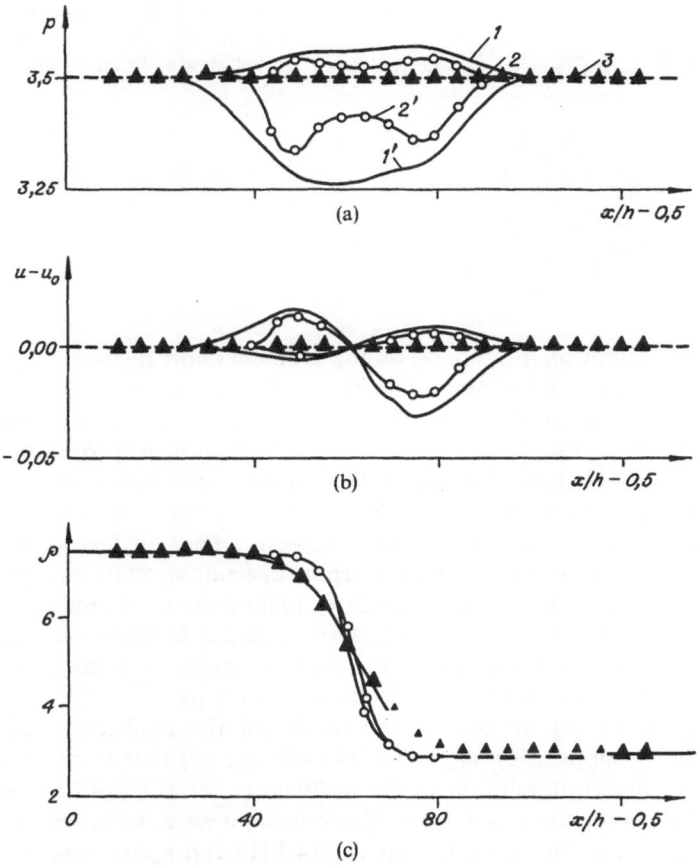

Figure 4.14. The graphs of numerical solutions in a contact strip: (a) the pressure; (b) the velocity; (c) the density. 1,1'—the Lax scheme; 2,2'—the Godunov scheme ($1,2-k = 0.5$; $1',2'-k = 2$); 3—the numerical solution by the Lax scheme at $k = 2$ in equation (4.3.25). The moment of time $t = 100\tau$; $\tau = 0.002$; $h = 0.02$.

(4.1.1) with the binomial equation of state (4.3.25)). It follows from Figure 4.14(b) that the density profile in a contact strip within the diagram accuracy coincides (in the case of an application of a K-inconsistence suppression algorithm) with the profile which was obtained in the presence of K-consistence, that is, when $k = 1$ in the equation of state (4.3.25). In Figure 4.15 (taken from [4.42]) are shown the results of a FLIC method computation (solid lines) of the problem (1.11), (1.9), (4.1.1) completed by the Tillotson equation of state (4.3.27). The numerical solution obtained with the K-consistence suppression algorithm (4.3.31), (4.3.32), (4.3.36) is depicted in Figure 4.15 by triangles. In Figure 4.16 we present the results of a FLIC

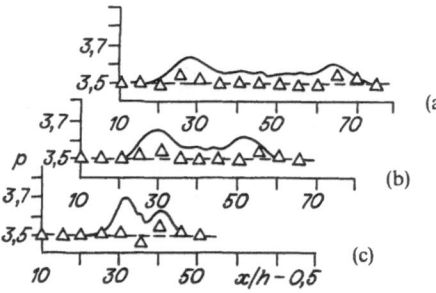

Figure 4.15. The calculation by the FLIC method with the use of the Tillotson equation of state: (a) $t = 60\tau$; (b) 40τ; (c) 20τ.

method computation (solid lines) on the problem of the high-velocity normal impact of a semi-infinite aluminum projectile onto a semi-infinite obstacle of the same material. At $t = 0$, we have set throughout the computational domain $p = 0$, $\rho = \rho_0$. $u_0 = 1$ cm/μs or $u_0 = 3$ cm/μs is the projectile velocity. The Tillotson equation of state (4.3.27) was used. In these computations the corrective term q (4.3.32), (4.3.36) was introduced into the scheme of the FLIC method only at those grid points on the x-axis where the inequalities (4.5.3) were satisfied. This was done for the purpose of economy of the computer time required for the computation of the $B(p_0, \rho)$ function by formula (4.3.9). The results of the shock waves and contact discontinuity localization in this flow, by means of the above presented differential analyzer subroutine, are shown in Figure 4.13(c) for the impact velocity $u_0 = 1$ cm/μs.

In Figure 4.17(a) we present the results of the application of the K-inconsistence suppression algorithm (4.3.49), (4.3.50) that were obtained in [4.43] by numerical solution of the problem (1.11), (1.9), (4.1.1) completed by the Tillotson equation of state. The scheme used was the one-step Lax–Wendroff scheme. The quantities entering (4.3.50) were approximated by using well-known differencing formulas (see, for example, [4.80])

$$\{(\partial/\partial x)[\mu B(\partial\rho/\partial x)^2]\}_i^n \cong [(\mu B)_{i+1/2}^n(\rho_{i+1}^n - \rho_i^n)^2 - (\mu B)_{i-1/2}^n(\rho_i^n - \rho_{i-1}^n)^2]/(h^3),$$

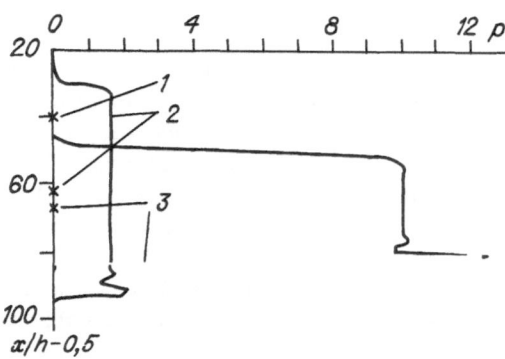

Figure 4.16. A high-velocity impact calculation by the FLIC method. 1—$t = 0$; 2—$u_0 = 1$, $t = 0.9$; 3—$u_0 = 3$, $t = 0.36$.

$$B_{i\pm1/2} = 0.5(B_i + B_{i\pm1}), \qquad (\partial\rho/\partial x)_i^n \cong (\rho_{i+1}^n - \rho_{i-1}^n)/(2h),$$

$$(\partial^2\rho/\partial x^2)_i^n \cong (\rho_{i+1}^n - 2\rho_i^n + \rho_{i-1}^n)/(h^2).$$

In Figure 4.17(b) are shown the results of solving the same Riemann problem by the MacCormack scheme (see also [4.43]).

Since the term q determined by formula (4.3.51) is not in divergence form, its use in computations by divergence schemes (considered above in Section 4.3) leads to the violation of the conservation of the total energy of the fluid contained within the computational domain. The practice of the computations shows (see [4.39], [4.42], [4.43]) that the size of the corresponding relative error in energy is not large: in all the computations in [4.39], [4.42], [4.43] it did not exceed 1% for the first-order schemes and 0.1% for the second-

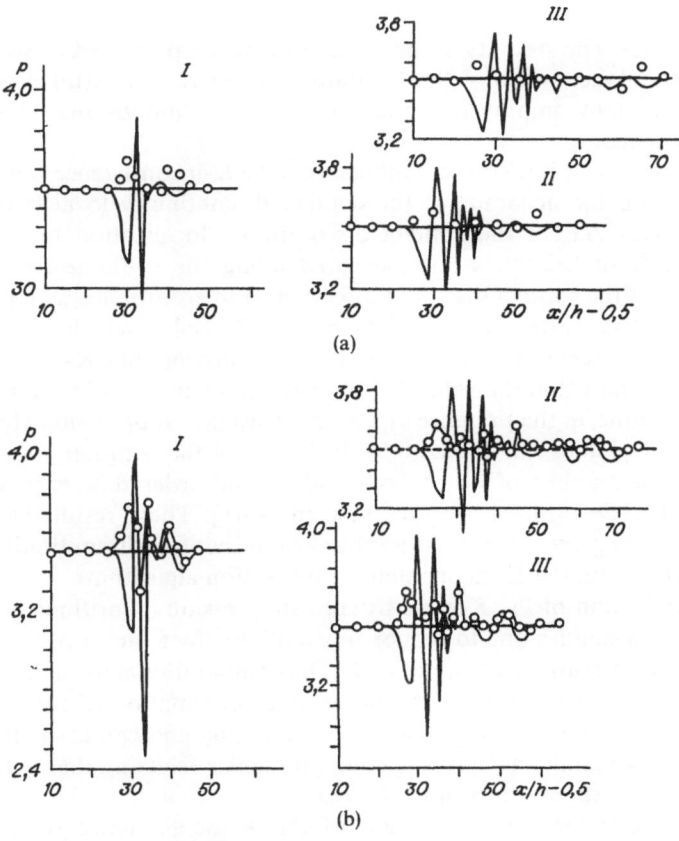

Figure 4.17. Calculation of the contact discontinuity propagation with the use of equation (4.3.27): (a) by the Lax–Wendroff scheme; (b) by the MacCormack scheme. I—$t = 20\tau$; II—40τ; III—60τ.

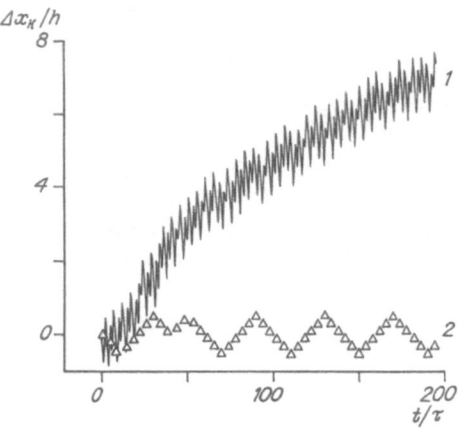

Figure 4.18. Error in the contact discontinuity localization as a function of time in the case of using Algorithm 2 in the Lax scheme. The equation of state (4.3.27). 1—the localization result in the absence of the K-inconsistence suppression algorithm; 2—the result of application of the algorithm (4.3.30), (4.3.33)–(4.3.35).

order schemes. The stability of the modified schemes presented in Section 4.3 took place at the values of the Courant number $K = \max(|u| + c)\tau/h \leq 1$ which are usually employed in the gas dynamic computations by original explicit schemes.

In Figure 4.18 we illustrate the influence of the K-inconsistence suppression algorithms on the accuracy of the contact discontinuity localization. The absolute error Δx_k of the contact discontinuity localization by means of Algorithm II of Section 4.1 is measured along the ordinate axis; $\Delta x_k = x_k - x_{da}$, x_k is the contact discontinuity abscissa in accordance with the exact solution, and x_{da} is the abscissa of the center of a cell in which $\max_x |\partial \rho / \partial x|$ is achieved. It is seen from Figure 4.18 that in the absence of a K-inconsistence suppression algorithm the value Δx_k increases, and at $t = 200\tau$ it reaches $8h$; at the same time, in the presence of a K-inconsistence suppression algorithm, the value $|\Delta x_k|$ does not exceed $0.5h$ during all of the computation. Similar graphs for a number of other first- and second-order difference schemes considered in Section 4.3 are presented in [4.41]. These results also show considerable improvement in the accuracy of the contact discontinuity localization when using the K-inconsistence suppression algorithms.

The application of the K-inconsistence suppression algorithms presented in Section 4.3 enables one to reduce substantially (by a factor of $5 \div 15$) the errors in the pressure in a contact strip. The computational formulas of these algorithms do not deteriorate the order of approximation of the difference scheme employed, they are simple in programming and can easily be incorporated into existing gas dynamic computer codes realizing the well-known finite-difference shock-capturing schemes.

The methods for the elimination of the K-inconsistence of difference schemes that were presented in [4.41] were tested on the problem (1.1)–(1.3), (4.1.1). For the numerical integration of the system (1.1)–(1.3) the method of Lax, the FLIC method, and the one-step Lax–Wendroff method were used.

Each of these methods was implemented with one of the following two equations of state: the Tillotson equation of state (4.3.27) and the Osborne equation of state [4.105]

$$p = \{\zeta(a_1 + a_2|\zeta|) + \rho_0 \varepsilon[b_0 + \zeta(b_1 + b_2\zeta) + \varepsilon(c_0 + c_1\zeta)]\}/(\rho_0\varepsilon + \varphi_0),$$

where $\zeta = \rho/\rho_0 - 1$, a_1, a_2, ρ_0, b_0, b_1, b_2, c_0, c_1 are the constants specifying the material. Denote by $\delta p(t)$, $\delta u(t)$ the relative errors in the pressure and velocity, respectively. Numerous computations by the above three schemes (the results of which have been presented in [4.41]) enable us to draw the following conclusions:

(1) the use of the contact residual subtraction method and of the method involving the difference approximation of the equation for the pressure (4.4.16) enable us to reduce the errors δp, δu up to the level of the machine rounding-off errors;
(2) the relative errors in the total energy are by one–two decimal orders smaller than in the case of the application of the approximate methods for the K-inconsistence suppression of difference schemes that were presented in Section 4.3.

The question of the influence of the algorithms for the K-inconsistence elimination (described in Section 4.4), on the accuracy of the contact discontinuity localization while using Algorithm II of the differential analyzer presented in Section 4.1 was also considered in [4.41]. The figures, analogous to Figure 4.18 and presented in [4.41], enable us to draw the conclusion that the use of the algorithms of Section 4.4 leads to a substantial increase in the accuracy of the contact discontinuity localization in the difference solutions. As was noted in [4.41], a specific form of the dependence of the localization error on t does not depend on the form of the equation of state (1.5) employed. This is easily explained by the fact that the construction of the difference schemes considered in [4.41] is such that the function H_1 in (4.1.13) does not depend on the specific internal energy ε; consequently, in the presence of the K-consistence the grid density ρ^{n+1} does not depend on the form of the equation of state (1.5) in the contact strip arising in the numerical solution of the problem (1.1)–(1.3), (1.5), (4.1.1).

Thus, the methods for the K-inconsistence elimination of difference schemes considered in Section 4.4 have a number of advantages over the approximate methods for the K-inconsistence suppression described in Section 4.3. This is not surprising, because the algorithms for the restoration of K-consistence have been obtained in Section 4.4 directly from the analysis of difference schemes, that is, without recourse to the first differential approximation. A generalization of the algorithms of Sections 4.3 and 4.4 for the case of two space variables was not carried out.

CHAPTER 5

Optimization Techniques of the Discontinuities Localization

At the beginning of Chapter 2 we have enumerated a number of well-known methods for shock wave localization by shock-capturing computational results, and among them we have mentioned a method proposed by Miranker and Pironneau [5.1]–[5.3]. This method, termed by the above authors the "global shock fitting method", reduces a problem on the localization of a strong discontinuity in the solution of the Cauchy problem for the Burgers equation to a problem on minimization of some integral functional. In the present chapter we analyze in detail the Miranker–Pironneau method with the purpose of elucidating a question on the accuracy of shock localization by this method. We also generalize the optimization approach of [5.1] [5.2] for the case of shock wave localization in the numerical solutions of one-dimensional problems of gas dynamics obtained by using finite-difference shock-capturing schemes. A modification of the Miranker–Pironneau method is proposed which uses, in the variational principle, information on the approximation viscosity of a difference scheme obtained from the first differential approximation (f.d.a.). It is shown by means of the classical variational calculus methods that the application of such a variational principle yields a trajectory which coincides with the true discontinuity trajectory for the case of a shock wave moving at a constant speed. On the basis of this modification of the Miranker–Pironneau method one more algorithm is proposed which reduces the shock localization problem to a problem of the minimization of a univariate function.

Investigations of modifications of the Miranker–Pironneau method carried out in Sections 5.2–5.5 enabled us to reveal a number of advantages of the optimization procedures for shock localization described in the present chapter, compared to the differential analyzers of shock waves which have been investigated in Chapters 2 and 3. These advantages are listed at the beginning of Section 5.6, which deals with questions of constructing the optimization techniques for the localization of contact discontinuities in one-dimensional gas flows. These techniques, unlike the differential analyzers considered in Section 4.1, are also applicable in cases of nonzero finite pressure and velocity gradients in a contact strip. Finally, in the concluding section of this chapter, Section 5.8, the Miranker–Pironneau method is extended for the case of shock localization in a two-dimensional flow computed by shock-capturing schemes. Here the polar coordinates have been used, and at fixed t the problem of determining the discontinuity line is reduced to a one-dimensional problem of unconstrained optimization.

Note that the methods considered in this chapter are applicable only to those problems in which the presence of shock waves in the flow to be modeled is known *a priori*. In the cases when a shock wave arises in the flow-in-process of a numerical computation, the optimization techniques of shock localization presented below should be combined with a method for determining the location and time of shock formation, for which purpose one can use the concept of the solution-adaptive grid which was presented in Section 2.5, which uses the notion of a compression wave center which at the time of shock formation goes over into a finite-difference shock wave center. The material of the present chapter is based on [5.1], [5.2], [5.4]–[5.7].

5.1. An Analysis of the Miranker–Pironneau Method

Following [5.1], [5.2], consider the equation

$$\partial u/\partial t + \partial \varphi(u)/\partial x = 0 \tag{5.1.1}$$

with the initial condition

$$u(x, 0) = u_0(x), \qquad x \in R, \tag{5.1.2}$$

$0 \le t \le T$. The function $\varphi(u)$ in (5.1.1) is assumed to be twice continuously differentiable, and $\varphi''(u) \ne 0$. The function $u_0(x)$ in condition (5.1.2) is assumed to have a discontinuity at a point x_0, $-\infty < x_0 < \infty$, and

$$u_0(x) = \begin{cases} U_1(x), & x < x_0, \\ U_2, & x \ge x_0, \end{cases} \tag{5.1.3}$$

where $U_2 = \text{const}$ and $U_1(x)$ is a continuous function, such that $U_1(x_0 - 0) - U_2 > 0$. In addition, the function $U_1(x)$ is assumed to be such that at $0 < t \le T$ no other shock waves arise in the solution $u(x, t)$. As in [5.1], [5.2], let us now introduce two auxiliary Cauchy problems P^+ and P^- for equation (5.1.1) with solutions u^+ and u^-, respectively. As initial values let us take the functions u_0^+ and u_0^- in $C^1(-\infty, \infty)$ such that

$$u_0^- \equiv u_0, \qquad x < x_0,$$

$$u_0^+ \equiv u_0, \qquad x > x_0.$$

The functions u_0^+ and u_0^- are chosen in such a way that the respective solutions u^+ and u^- do not contain shock waves in the interval $0 \le t \le T$. Further, as in [5.1], [5.2], let C^+ (resp. C^-) be a characteristic in the P^+ problem (resp. P^- problem) emanating from the point $(x_0, 0)$. Let the equations of these characteristics be $x(t) = C^{\pm}(t)$. Further let

$$\Omega = \{(x, t)|C^+(t) \le x \le C^-(t), 0 \le t \le T\}.$$

Note the following property of the characteristics C^+, C^- and the set Ω: they do not depend on the choice of u_0^+ and u_0^-, in addition, the values of u^+ and u^- in Ω also do not depend on this choice. Let $x = \xi(t)$ be the equation of a discontinuity trajectory in the solution of the problem (5.1.1), (5.1.2). Take some function $\zeta(t) \in C^1[0, T]$. Let us introduce the function [5.1]

$$J(\zeta(t), t) = \{\varphi(u^-(\zeta(t), t)) - \varphi(u^+(\zeta(t), t))\}/\{u^-(\zeta(t), t) - u^+(\zeta(t), t)\} \quad (5.1.4)$$

as well as the function

$$F(\zeta(t), \dot\zeta(t), t) = J(\zeta(t), t) - \dot\zeta(t), \quad (5.1.5)$$

where $\dot\zeta(t) \equiv d\zeta(t)/dt$. Now consider, as in [5.1], the nonnegative functional

$$I(\zeta) = \int_0^T F^2(\zeta(t), \dot\zeta(t), t)\, dt. \quad (5.1.6)$$

The Rankine–Hugoniot condition implies the equality $I(\xi) = 0$. Since at $t = 0$ the discontinuity position is known according to (5.1.3), we can set $\zeta(0) = x_0$. The position of the abscissa $x = \zeta(T)$ is generally unknown (it is to be determined). In this connection it is natural to impose at the right end $t = T$ the transversality condition [5.6], [5.8]

$$2FF_{\dot\zeta}|_{t=T} = 0, \quad (5.1.7)$$

where $F_{\dot\zeta} = (\partial/\partial\dot\zeta)F(\zeta, \dot\zeta, t)$. Thus, let us consider the following variational problem for the functional (5.1.6):

$$I(\zeta) \to \min, \qquad \zeta(0) = x_0, \qquad 2FF_{\dot\zeta}|_{t=T} = 0. \quad (5.1.8)$$

Regarding (5.1.3) let us set $u_0^+ = U_2$. Taking into account the properties of equation (5.1.1) we have that $u^+(x, t) = U_2$. The function $u^-(x, t)$, being the solution of the auxiliary problem P^-, may be determined similarly to [5.1] as the solution obtained by a shock-capturing scheme (in [5.1] the Lax–Wendroff scheme was used for this purpose). Thus, in the solution $u^-(x, t)$ the shock front is approximated by a smeared shock wave zone occupying several mesh intervals on the x-axis.

The Euler–Lagrange equation corresponding to the functional (5.1.6) has the form [5.6], [5.8]

$$d^2\zeta/dt^2 - (d\zeta/dt)(\partial F/\partial\zeta) - \partial F/\partial t - F\, \partial F/\partial\zeta = 0. \quad (5.1.9)$$

Employing formula (5.1.5) let us rewrite equation (5.1.9) as follows

$$\ddot\zeta - \partial J/\partial t - J\, \partial J/\partial\zeta = 0. \quad (5.1.10)$$

By virtue of the construction of the solution $u^-(x, t)$ the relationship

$$\partial u^-/\partial t = -\varphi'(u^-)\, \partial u^-/\partial x \quad (5.1.11)$$

is valid. Making use of formulas (5.1.4), (5.1.11) it is easy to find that

$$J_t + JJ_\zeta = -(u^- - U_2)^{-1}(\partial u^-/\partial x)(\zeta(t), t)$$
$$\times \langle \{ [\varphi(u^-) - \varphi(U_2)]/(u^- - U_2) \} - \varphi'(u^-) \rangle^2. \quad (5.1.12)$$

Let us now employ the formula

$$\varphi(U_2) = \varphi(u^-) + \varphi'(u^-)(U_2 - u^-) + 0.5\varphi''(u^*)(U_2 - u^-)^2, \quad (5.1.13)$$

where $u^* \in [U_2, u^-(\zeta(t), t)]$. With the formulas (5.1.12), (5.1.13) in view, let us rewrite (5.1.10) in the form

$$\ddot{\zeta} = -(u^- - U_2)[\partial u^-(\zeta(t), t)/\partial x][\varphi''(u^*)/2]^2. \quad (5.1.14)$$

When an approximate solution obtained by a shock-capturing scheme is used for $u^-(x, t)$ then, in the smeared shock wave zone, $\partial u^-/\partial x < 0$ (see [5.9]). In addition, we use a function $u^-(x, t)$ such that $u^- > U_2$ in the smeared shock wave zone. Therefore we obtained from (5.1.14) that, at the initial data (5.1.3), $\ddot{\zeta}$ is always greater than 0. This inequality means that shock wave found as a result of the minimization of functional (5.1.6) always accelerates independently of the actual behavior of the true discontinuity. Thus the solution of the variational problem (5.1.8) in the general case does not coincide with the true discontinuity trajectory in the solution of the problem (5.1.2).

Similar to [5.1], [5.2], the variational problem (5.1.8) for the functional (5.1.6) does not contain the entropy condition

$$\varphi'(u(\xi(t) + 0, t)) \le \dot{\xi}(t) \le \varphi'(u(\xi(t) - 0, t)) \quad (5.1.15)$$

in the form of constraints. Note that the inequality $u^- > U_2$, which was assumed satisfied in the proof of the inequality $\ddot{\zeta} > 0$, implies satisfaction of the entropy condition (5.1.15), at least in the case when $\varphi''(u) > 0$ [5.9]. In this connection we also note that the computational results presented in [5.1], [5.2], for the case of the Burgers equation (5.1.1) with $\varphi(u) = 0.5u^2$, confirm the above result $\ddot{\zeta} > 0$. On the other hand, the inclusion of conditions (5.1.15), in the formulation of an optimization problem for the functional (5.1.6) in the form of two constraints, makes the numerical solution of the arising nonlinear problem of the variational calculus substantially more difficult. In this connection we suggest and investigate, in Section 5.2, an alternative optimization problem using artificial or approximation viscosity in the basic functional. This suggestion was inspired by the well-known fact that the entropy condition can be enforced by introducing artificial viscosity terms (see, for example, [5.9]–[5.11]).

Now a natural question arises on the size of the difference $|\xi(t) - \zeta(t)|$. Let us show, following [5.6], that in the case when the Miranker–Pironneau functional (5.1.6) is used for shock localization in some model gas dynamic problem, that an exact formula can be obtained for the extremal $\zeta(t)$ which provides the solution of the problem (5.1.8).

Consider a problem on the motion of a stationary shock wave in a gas. Following Section 2.2, let us employ the one-dimensional gas dynamic equations in Eulerian variables in the presence of the artificial viscosity q (2.2.18)

$$q = ah^2\rho[\min(\partial u/\partial x, 0)]^2$$

introduced additively into the pressure. Consider the progressive wave-type solutions of the above-mentioned equation system, that is, the solutions depending only on the variable

$$y = x - Dt - x_0, \tag{5.1.16}$$

where $D = const$ is the speed of a steady shock wave and x_0 is an arbitrary constant. Then the equation system under consideration may be integrated once. As a result of this, the following system is obtained

$$\rho(u - D) = C_1 = m; \tag{5.1.17}$$

$$p + q + m(u - D) = C_2; \tag{5.1.18}$$

$$m[p/(\rho(\gamma - 1)) + (1/2)(u - D)^2] + (p + q)(u - D) = C_3. \tag{5.1.19}$$

In (5.1.17)–(5.1.19) C_1, C_2, C_3 are integration constants and γ is a constant entering the equation of state (1.8).

Let us ascribe to the arbitrary constant x_0 entering into (5.1.16) the following physical meaning: let x_0 be the abscissa of the smeared shock wave center in the solution of the system (2.2.18), (5.1.17)–(5.1.19) at $t = 0$, that is

$$\xi(0) = x_0. \tag{5.1.20}$$

Then the trajectory of the smeared shock wave center at $t \geq 0$ is described by the equation

$$x = \xi(t) = Dt + x_0, \tag{5.1.21}$$

and thus it coincides with the steady shock wave trajectory. The quantities determining the state behind and before the shock front will be marked by the subscripts "1" and "2", respectively. As in Section 2.2 these states are assumed to be constant. Let $V(y) = 1/\rho$. As in Section 2.2 let us find, from the system (5.1.17)–(5.1.19), ρ, u, p, q as functions of the specific volume V. Assuming that the value of m in (5.1.17) is given, the constants C_2, C_3 are determined from the conditions $q(V_1) = q(V_2) = 0$. Then ρ, u, p, q as functions of V are found as

$$\rho = 1/V, \qquad u = mV + D;$$

$$p = 0.5(\gamma - 1)m^2\{(\gamma + 1)V_1 V_2/[(\gamma - 1)V]$$
$$+ V - (\gamma + 1)(V_1 + V_2)/\gamma\}; \tag{5.1.22}$$

$$q = (\gamma + 1)m^2(V_2 - V)(V - V_1)/(2V).$$

Analogously to (5.1.4)–(5.1.6) let us introduce the functional $I(\zeta)$ by the formulas (5.1.23), (5.1.5), (5.1.6) where

$$J(\zeta(t),\,t) = [(p + \rho u^2)(\zeta(t),\,t) - (p_2 + \rho_2 u_2^2)]/[(\rho u)(\zeta(t),\,t) - \rho_2 u_2]. \quad (5.1.23)$$

It is assumed in (5.1.23) that $p,\,\rho,\,u$ is the solution of the system (5.1.22), (2.2.18) subject to the initial condition (see Section 2.2)

$$V(0) = 0.5(V_1 + V_2).$$

Then

$$V(y) = 0.5[V_1 + V_2 + (V_2 - V_1)\sin by], \quad (5.1.24)$$

where

$$b = (1/h)[(\gamma + 1)/(2a)]^{0.5}. \quad (5.1.25)$$

Since we consider solutions of the progressive-wave type, the Euler–Lagrange equation (5.1.10) may be rewritten in the form

$$\ddot{\zeta} - (J - D)(dJ/dy) = 0. \quad (5.1.26)$$

With (5.1.23), (5.1.16) in view let us introduce the function

$$v(t) = \zeta(t) - Dt - x_0.$$

Then equation (5.1.26) may be rewritten as follows:

$$\ddot{v} - [(J - D)(dJ/dy)]|_{y=v} = 0. \quad (5.1.27)$$

Employing the formulas (5.1.22), (5.1.23) it is easy to find that

$$J(v) = (1/D)[D^2 - (m^2/2)(\gamma + 1)V_2(V - V_1)], \quad (5.1.28)$$

where

$$V = V(v) = 0.5[V_1 + V_2 + (V_2 - V_1)\sin bv(t)]. \quad (5.1.29)$$

Making use of formulas (5.1.28), (5.1.29) let us rewrite equation (5.1.27) in the form

$$\ddot{v} - bc(1 + \sin bv)\cos bv = 0, \quad (5.1.30)$$

where

$$c = [m^2(\gamma + 1)V_2(V_2 - V_1)/(4D)]^2 \quad (5.1.31)$$

and the constant b is determined by (5.1.25). Let us search for the solution of equation (5.1.30) in the interval $[0,\,T]$ which satisfies the boundary conditions

$$v(0) = 0, \qquad J(v(T)) - \dot{v}(T) - D = 0. \quad (5.1.32)$$

Applying the substitution $\dot{v} = s(v)$ as well as substitutions

$$z = bv, \qquad \chi = \tan(z/2), \qquad t_1 = (\chi + 1)/(\chi - 1),$$
$$\sin^2\varphi = (a_1^2 + a_2^2)/(a_2^2 + t_1^2), \quad (5.1.33)$$

we can write the general solution of equation (5.1.30) as

$$t + A_2 = \sigma(A_1)F(\varphi\backslash\alpha), \quad (5.1.34)$$

where A_1, A_2 are integration constants,

$$\sigma(A_1) = (2/b)[A_1(1 + \lambda_1)(1 + \lambda_2)(a_1^2 + a_2^2)]^{-1/2};$$

$$\lambda_1 = v + \sqrt{v^2 - v}, \qquad \lambda_2 = v - \sqrt{v^2 - v}, \qquad v = c/A_1;$$

$$a_1^2 = (\lambda_1 - 1)/(\lambda_1 + 1), \qquad a_2^2 = (1 - \lambda_2)/(1 + \lambda_2);$$

$$\alpha = \arctan(a_2/a_1) = \arctan\left[\left(\frac{2\sqrt{v} - \sqrt{v - 1}}{2\sqrt{v} + \sqrt{v - 1}}\right)^{0.5}\right],$$

and $F(\varphi\backslash\alpha)$ is the incomplete elliptic integral of the first kind [5.12]. The solution (5.1.34) is valid for $v > 1$. Employing the boundary condition $v(0) = 0$, the chain of substitutions (5.1.33), and formula (5.1.34) it is easy to find the constant A_2 as a function of the constant A_1:

$$A_2 = \frac{2}{b}\{4A_1[v(v - 1)]^{0.5}\}^{-0.5} F\left(\arcsin\left[\frac{2(v^2 - v)^{0.5}}{1 + v + (v^2 - v)^{0.5}}\right]^{0.5} \alpha\right).$$

Let us find the integration constant A_1 employing the second condition in (5.1.32). Let us introduce the notation

$$r(t) = b\{A_1[v(v - 1)]^{0.5}\}^{0.5}(t + A_2). \qquad (5.1.35)$$

Then formula (5.1.34) may be rewritten, with (5.1.33) in view, as

$$v(t) = b^{-1}[2\arctan(\sqrt{a_1^2 + a_2^2}\, ds\, r(t)) - \pi/2], \qquad (5.1.36)$$

where by definition [5.12] $ds\, r = dn\, r/sn\, r$, dn and sn being elliptic Jacobi functions. Employing formulas (5.1.28), (5.1.29) let us write the transversality condition $J(v(T)) - \dot{v}(T) - D = 0$ in the form

$$\sqrt{c}\,[1 + \sin(bv(T))] + \dot{v}(T) = 0. \qquad (5.1.37)$$

Making use of formulas (5.1.33), (5.1.36) we find from condition (5.1.37) the following equation for the determination of the constant A_1:

$$cn\, r(T) = [(1 - \tilde{m})/\tilde{m}]^{0.5}, \qquad (5.1.38)$$

where $\tilde{m} = \sin^2\alpha = a_2^2/(a_1^2 + a_2^2)$. Considering the behavior of the left- and right-hand sides of equation (5.1.38) as functions of the constant A_1, it is not difficult to show that equation (5.1.38) has the single root $A_1 = c$. However, the solution (5.1.34) becomes invalid at such a value of A_1. Therefore the case $A_1 = c$ should be treated separately. As a result of this treatment the solution of equation (5.1.30) can easily be obtained in the form

$$v(t) = (2/b)\{-(\pi/4) - \arctan[1/(\pm b\sqrt{c}(t + A_2))]\} \qquad (5.1.39)$$

by using (5.1.33) where A_2 is an integration constant. From the condition $v(0) = 0$ we find that $A_2 = \mp 1/(b\sqrt{c})$, and from the transversality condition we easily find that the "$-$" sign should be taken in formula (5.1.39) before $\sqrt{c}$.

Thus, in the case $A_1 = c$, the solution of equation (5.1.30) under conditions (5.1.32) is given by the formula

$$v(t) = (2/b)[-(\pi/4) + \arctan(1/(b\sqrt{ct} + 1))]. \qquad (5.1.40)$$

In the case of a shock moving from left to right $m < 0$, $V_2 > V_1$, and then we get from (5.1.29) that $dV/dv > 0$ at $|v(t)| < \pi/(2b)$. Consequently, the use of the solution (5.1.29) in the functional (5.1.6) is valid only at values of $\zeta(t)$ such that

$$|v(t)| \leq \pi/(2b). \qquad (5.1.41)$$

Employing (5.1.40) it is easy to show that at any $a > 0$ in (2.2.18) the inequality (5.1.41) holds. It is easy to get an upper estimate for $|v(t)|$ from (5.1.40), with (5.1.25) in view,

$$|v(t)| \leq h[2a/(\gamma + 1)]^{0.5}(\pi/2), \qquad t \geq 0. \qquad (5.1.42)$$

If, in particular,

$$a \leq 2(\gamma + 1)/(\pi^2), \qquad (5.1.43)$$

then $|v(t)| \leq h$. Since, at $t > 0$, the inequality

$$\arctan(1/(1 + b\sqrt{ct})) < \pi/4$$

holds, it follows from (5.1.40) that $v(t) < 0$. This means that, at $t > 0$, the shock front calculated by the extremum of the functional (5.1.6), (5.1.28), (5.1.29) lags behind the true shock front. It is interesting to note that in the preceding example with the Burgers equation (5.1.1) the situation was the opposite. Substituting the exact solution (5.1.40) into the functional (5.1.6), (5.1.28), (5.1.29) it is easy to find that $I(\zeta) = 0$. Although the deviation $|v(t)|$ becomes greater as t increases, under a proper choice of the dimensionless coefficient a in (2.2.18) one can obtain that the inequality $|v(t)| \leq h$ will be valid for any $t \geq 0$. However, (5.1.43) yields values of the quantity a which are too small for the successful application of the pseudoviscosity (2.2.18) (cf. [5.10]). Thus, in practice, the value $|v(t)|$ can reach a magnitude of several intervals h. On the other hand, it follows from the estimate (5.1.42) that the difference $|v(t)|$ tends to zero with the mesh size. However, in practical computations we always use finite mesh sizes; furthermore, the computing mesh is generally more crude in multidimensional computations.

5.2. Incorporation of the Information on Approximation Viscosity into the Basic Functional

Below we consider a modification of the basic functional (5.1.4)–(5.1.6) for the purpose of increasing the discontinuity localization accuracy when an optimization procedure is used which is based on such a functional.

Let us write the Euler equation system for an inviscid, compressible, non-heat-conducting gas in the form (1.11), (1.9). This system is completed by the equation of state (1.8). Let us approximate the system (1.11) by a conservative difference scheme of the rth order of accuracy ($1 \leq r$). As was shown in Section 5.3, the first differential approximation (f.d.a.) of such a difference scheme is representable in the divergence form

$$\partial w/\partial t + \partial \varphi(w)/\partial x = \partial Q(x, t)/\partial x. \qquad (5.2.1)$$

In a more detailed form the vector $Q(x, t)$ may be written as

$$Q(x, t) = \tilde{Q}(w(x, t), h\, \partial w(x, t)/\partial x, \ldots, h^r \partial^r w(x, t)/\partial x^r, x, t, h, \tau), \qquad (5.2.2)$$

and

$$Q = O(h^r) + O(\tau^r). \qquad (5.2.3)$$

In (5.2.2), (5.2.3) τ is the time step and h is the step of the uniform computing mesh on the x-axis. In accordance with (1.9), (5.2.2) let us introduce the notations

$$w = \{w_1, w_2, w_3\}^T, \qquad \varphi = \{\varphi_1, \varphi_2, \varphi_3\}^T,$$

$$Q = \{Q_1, Q_2, Q_3\}^T,$$

where

$$w_1 \equiv \rho, \quad w_2 \equiv \rho u, \quad w_3 \equiv \rho E; \quad \varphi_1 = \rho u, \quad \varphi_2 = p + \rho u^2, \quad \varphi_3 = pu + \rho uE.$$

As in the foregoing section let us consider the progressive wave-type solutions of the system (5.2.1), that is, the solutions depending upon the variable y defined by (5.1.16). Then the system (5.2.1) may be integrated once

$$-Dw(y) + \varphi(w) + C = Q(y), \qquad (5.2.4)$$

where C is a constant vector. As in Section 2.3 we search for the solution of the system (5.2.4) which satisfies the conditions

$$w(y) = \begin{cases} W_1, & y \to -\infty, \\ W_2, & y \to +\infty, \end{cases} \qquad (5.2.5)$$

where W_1 and W_2 are constant vectors satisfying the Rankine–Hugoniot conditions. Then the vector C in (5.2.4) should satisfy the requirements

$$C = DW_1 - \varphi(W_1) = DW_2 - \varphi(W_2). \qquad (5.2.6)$$

As was shown in Section 1.4 the exact solution of the problem (5.2.4), (5.2.5) describes well the actual behavior of the difference solution in the zone of a smeared strong discontinuity. For example, the corresponding relative error obtained in Section 1.4 did not exceed 4% for shock waves of finite intensity. Let, in (5.2.5),

$$W_j = \{W_{j1}, W_{j2}, W_{j3}\}^T, \qquad j = 1, 2.$$

Employing formula (5.2.6) let us rewrite equation (5.2.4) for the kth component $(1 \leq k \leq 3)$ as

$$[\varphi_k(\mathbf{w}) - \varphi_k(\mathbf{W}_2) - Q_k(y)]/[w_k(y) - W_{2k}] - D = 0. \qquad (5.2.7)$$

Let us now consider, on the basis of (5.2.7), the functional

$$I_k(\zeta) = \int_0^T \left[\frac{\varphi_k(\mathbf{w}(\zeta(t), t)) - \varphi_k(\mathbf{W}_2) - Q_k(\zeta(t), t)}{w_k(\zeta(t), t) - W_{2k}} - \dot{\zeta}(t) \right]^2 dt. \qquad (5.2.8)$$

Consider, for the functional (5.2.8) by analogy with (5.1.9), the variational problem

$$I_k(\zeta) \to \min, \qquad \zeta(0) = x_0, \qquad (5.2.9)$$

$$[\varphi_k(\mathbf{w}(\zeta(T), T)) - \varphi_k(\mathbf{W}_2) - Q_k(\zeta(T), T)]/[w_k(\zeta(T), T) - W_{2k}] - \dot{\zeta}(T) = 0. \qquad (5.2.10)$$

Consider the functional (5.2.8) on the solutions $\mathbf{w} = \mathbf{w}(y)$ of the problem (5.2.4), (5.2.5). Suppose that the structure of the approximation viscosity $\mathbf{Q}$ in the system (5.2.1) is such that there takes place a smooth transition from state "1" to state "2" in a smeared shock wave described by the solution $\mathbf{w}(y)$ at fixed $h > 0, \tau > 0$. Then, with regard to formula (5.2.7), the condition $\zeta(0) = x_0$, and the transversality condition (5.2.10), one can easily find that, on the progressive wave-type solution $\mathbf{w} = \mathbf{w}(y)$, the functional (5.2.8) reaches its minimum at the function $\zeta(t) = x_0 + Dt$, which describes the exact trajectory of a stationary shock wave.

In the case of a nonstationary shock wave, propagating in a medium described by the nonconstant functions ρ, u, p, ε, let us consider the functional

$$I(\zeta) = \int_{t_0}^T \left\{ \sum_{k=1}^3 \alpha_k [J_k(\zeta(t), t) - \dot{\zeta}(t)]^2 \right\} dt, \qquad (5.2.11)$$

where the functions $J_k(\zeta(t), t)$ are determined by a formula

$$J_k(\zeta(t), t) = [\varphi_k(\mathbf{w}^-(\zeta(t), t)) - \varphi_k(\mathbf{w}^+(\zeta(t), t))$$
$$- Q_k(\zeta(t), t)]/[w_k^-(\zeta(t), t) - w_k^+(\zeta(t), t)], \qquad (5.2.12)$$

where $\mathbf{w}^-, \mathbf{w}^+$ are the solutions of the auxiliary Cauchy problems P^- and P^+ for the system (1.5), (1.9), (1.11), that is, $\mathbf{w}^-, \mathbf{w}^+$ are obtained at the initial data of the form (cf. Section 5.1)

$$\mathbf{w}^-(x, t_0) = \mathbf{u}(x, t_0), \qquad x < x_0,$$
$$\mathbf{w}^+(x, t_0) = \mathbf{u}(x, t_0), \qquad x > x_0, \qquad (5.2.13)$$

where x_0 is the abscissa of a discontinuity of the shock wave type in the initial profile $\mathbf{u}(x, t_0), t_0 < T$. Further, $\alpha_1, \alpha_2, \alpha_3$ in (5.2.11) are nonnegative constants, such that $\alpha_1 + \alpha_2 + \alpha_3 > 0$. Formula (5.2.8) is a particular case of the functional (5.2.11) at

$$\mathbf{w}^-(\zeta(t), t) = \mathbf{w}(\zeta(t), t), \qquad \mathbf{w}^+(\zeta(t), t) = \mathbf{W}_2, \qquad \alpha_j = \delta_k^j, \quad j = 1, 2, 3.$$

The above consideration of the functional (5.2.8), within the framework of the progressive wave-type solutions of the f.d.a. equations, gives some reasons for expecting a higher accuracy of locating the stationary shock waves on the basis of the functional (5.2.11) minimization, compared with the Miranker–Pironneau functional considered in Section 5.1.

EXAMPLE 1. Consider again the system (5.1.17)–(5.1.19). In this case $\mathbf{Q} = \{0, -q, -qu\}^\mathsf{T}$ where q, u can be uniquely determined as functions of the specific volume V by means of formulas (5.1.22). In its turn, formula (5.1.24) provides a smooth transition from state "1" behind the shock front to state "2" before the front, in the case of the quadratic artificial viscosity (2.2.18) (see also [5.10]). Earlier, in Section 2.2, we have shown that the accuracy of the shock front localization by max q depends substantially on the values of the dimensionless coefficients entering the expression for q. On the other hand, in the case of shock localization on the basis of the functional (5.2.8), it is only required (for the specific form of the viscosity q employed) to ensure a smooth transition from one state to another in a stationary shock wave. Thus, in the case when the functional (5.2.8) is used for the localization of a stationary shock wave, there takes place a relatively weaker dependence of the localization accuracy upon the specific form of q than in the case of a differential analyzer of Section 2.2, based on determining the points of maximum of the artificial viscosity q.

EXAMPLE 2. Consider the "breakdown-of-discontinuity" scheme of [5.13]. Below, in Section 5.5, we present some results of the practical application of the functional (5.2.8) with $k = 1$. In this case the expression for $Q_1(\zeta, t)$ can easily be found from the f.d.a. equation systems (2.4.12), (2.4.13)

$$Q_1(\zeta, t) = \begin{cases} -(\tau/2)(p + \rho u^2)_x + (h/2)[(1/c)(1 - u/c) \\ \qquad \times p_x + u\rho_x + (u/c)\rho u_x], \quad 0 < u < c, \quad (5.2.14) \\ -(\tau/2)(p + \rho u^2)_x + (h/2)(\rho u)_x, \quad u \geq c > 0. \end{cases}$$

Thus we have shown, with the aid of the classical variational calculus methods within the framework of the f.d.a. consideration, that the use (in the basic functional) of information on the artificial or approximation viscosity of a difference scheme obtained from the f.d.a. provides a result which coincides with the exact discontinuity trajectory; at least, in the case of a stationary shock wave. This constitutes the advantage of the functional (5.2.8) over the functional (5.1.4)–(5.1.6).

5.3. Shock Localization on the Basis of Function Minimization

It follows from the construction of the basic functional (5.1.4)–(5.1.6) or (5.2.8) that it is necessary to store in the computer memory the values of the quantities u, p, ρ, ε found as the solution of the finite difference equations approximating the Euler equation system (1.1)–(1.3) in a domain $\Omega(T)$ of the (x, t)-plane. To estimate the minimal size of this domain along the x-axis at $T = N\tau$ where N is a positive integer, let us make use of the Courant–Friedrichs–Lewy stability condition

$$(|u| + c)\tau/h \leq 1 \tag{5.3.1}$$

and Zemplén's theorem (1.29). Really, by virtue of the inequalities (5.3.1) and (1.29) the shock front cannot propagate over a distance exceeding h during time τ. Let x_0 be the discontinuity abscissa at $t = 0$. Then the domain $\Omega(T)$ includes all those (x, t) points for which the following inequalities are satisfied:

$$x_0 - nh \leq x \leq x_0 + nh, \qquad n = 1, \ldots, N.$$

We can get rid of the requirement for additional storage if, instead of the problem on minimization of the functional (5.2.11), the minimization problem for some function $F(\zeta(t))$ is considered for the required moments of time t. Let us consider the following function on the basis of (5.2.11):

$$F(\zeta(t^{n+1})) = \sum_{k=1}^{3} \alpha_k \{ [\varphi_k(\mathbf{w}^-(\zeta^{n+1}, t^{n+1})) - \varphi_k(\mathbf{w}^+(\zeta^{n+1}, t^{n+1}))$$

$$- \tilde{Q}_k(\mathbf{w}^-, h\partial \mathbf{w}^-/\partial x, \ldots, h^r \partial^r \mathbf{w}^-/\partial x^r, \zeta(t^{n+1}), t^{n+1}, h, \tau)]$$

$$/[w_k^-(\zeta^{n+1}, t^{n+1}) - w_k^+(\zeta^{n+1}, t^{n+1})] - \dot{\zeta}_{\tau_1}(t^{n+1}) \}^2, \tag{5.3.2}$$

where $\zeta^{n+1} = \zeta(t^{n+1})$, $t^{n+1} = (n + 1)\tau$, $\mathbf{w}^-, \mathbf{w}^+$ are the solutions of the auxiliary problems P^- and P^+, respectively (see Sections 5.1 and 5.2), $\dot{\zeta}_{\tau_1}(t^{n+1})$ is a difference approximation of the derivative $\dot{\zeta}(t^{n+1})$ employing the step $\tau_1 = \beta\tau$, $\beta \geq 1$. $\alpha_1, \alpha_2, \alpha_3$ are penalty constants, $\alpha_k \geq 0$, $\alpha_1 + \alpha_2 + \alpha_3 = 1$. The simplest approximation for $\dot{\zeta}_{\tau_1}(t^{n+1})$ is as follows:

$$\dot{\zeta}_{\tau_1}(t^{n+1}) = [\zeta(t^{n+1}) - \zeta(t^{n+1} - \tau_1)]/\tau_1. \tag{5.3.3}$$

Unlike the case of the functional (5.2.11) minimization in the process of the numerical solution of the problem

$$F(\zeta(t^{n+1})) \rightarrow \min, \qquad 0 \leq n \leq N - 1,$$
$$\zeta(0) = x_0, \tag{5.3.4}$$

there is no need to store the values of the quantities u, p, ρ, ε in the above domain $\Omega(T)$. In accordance with formulas (5.3.2), (5.3.3) it is sufficient for the computation of $\zeta(t^{n+1})$ to know the solutions $\mathbf{w}^-(x, t)$ and $\mathbf{w}^+(x, t)$ only at $t = t^{n+1}$, and the value of the abscissa $x = \zeta(t)$ at $t = t^{n+1} - \tau_1$. At $\mathbf{w}^-$ we can use, similar to Section 5.1, a finite-difference solution obtained by the difference scheme approximating equations (1.1)–(1.3); $\mathbf{w}^+(x, t)$ can be taken to be equal to $W_2 = $ const in the case when the shock wave propagates into a gas with constant parameters ρ, u, p, ε. Let $\mathbf{w}(y)$, where y is determined by formula (5.1.16), be the exact solution of the problem (5.2.4), (5.2.5). As was shown in Section 1.4, the difference solution $\mathbf{w}^-(x, t)$ is approximated well by the solution $\mathbf{w}(y)$ in the zone of a smeared stationary shock wave of finite intensity. Let us substitute into (5.3.2) $\mathbf{w}^- = \mathbf{w}(y)$, $\mathbf{w}^+ = W_2 = $ const. Then, with regard to (5.2.7) and (5.3.2), we obtain that

$$F(\zeta(t^{n+1})) = [D - \dot{\zeta}_{\tau_1}(t^{n+1})]^2. \tag{5.3.5}$$

At sufficiently small τ_1 in (5.3.3) we find that

$$[\zeta(t^{n+1}) - \zeta(t^{n+1} - \tau_1)]/\tau_1 = \partial\zeta(t^{n+1})/\partial t + O(\tau_1). \tag{5.3.6}$$

Let the value $\zeta(t^{n+1})$ be computed as the solution of a minimization problem (5.3.4). Then at sufficiently small τ_1 we obtain, with regard to (5.3.5), (5.3.6), that $d\zeta/dt \cong D$, thus the set of points $x = \zeta(t^{n+1})$ obtained for $n = 0, 1, \ldots,$ $N - 1$ (as a solution of the problem (5.3.4)) yields an approximation to the trajectory of a stationary shock wave.

Let us show that there exists a relationship between the procedure of the numerical solution of problem (5.3.4) and the numerical solution of some ordinary differential equation. Consider, for clarity, the case when $\alpha_{k_0} = 1$, $\alpha_k = 0$ at $k \neq k_0$ in formula (5.3.2), $1 \leq k_0 \leq 3$. Then in the sum on the right-hand side of equation (5.3.2) there is only one nonzero item. Consider an ordinary differential equation

$$d\zeta/dt = [\varphi_{k_0}(\mathbf{w}^-(\zeta(t), t)) - \varphi_{k_0}(\mathbf{w}^+(\zeta(t), t))$$
$$- \tilde{Q}_{k_0}(\mathbf{w}^-(\zeta(t), t), h \, \partial \mathbf{w}^-(\zeta(t), t)/\partial x, \ldots, h^r \, \partial^r \mathbf{w}^-(\zeta(t), t)/\partial x^r, \zeta(t), t, h, \tau)]$$
$$/[w_{k_0}^-(\zeta(t), t) - w_{k_0}^+(\zeta(t), t)]. \tag{5.3.7}$$

Let us approximate equation (5.3.7) by the Adams implicit difference scheme [5.14]

$$G(\zeta^{n+1}) = [\zeta(t^{n+1}) - \zeta(t^{n+1} - \tau_1)]/\tau_1$$
$$- [\varphi_{k_0}(\mathbf{w}^-(\zeta^{n+1}, t^{n+1})) - \varphi_{k_0}(\mathbf{w}^+(\zeta^{n+1}, t^{n+1}))$$
$$- \tilde{Q}_{k_0}(\mathbf{w}^-, h \, \partial\mathbf{w}^-/\partial x, \ldots, h^r \, \partial^r\mathbf{w}^-/\partial x^r,$$
$$\zeta(t^{n+1}), t^{n+1}, h, \tau)]/[w_{k_0}^-(\zeta^{n+1}, t^{n+1}) - w_{k_0}^+(\zeta^{n+1}, t^{n+1})] = 0, \tag{5.3.8}$$

where $\zeta^{n+1} = \zeta(t^{n+1})$. For equation (5.3.7) we pose an initial condition $\zeta(0) = x_0$. Then we can find $\zeta(t^{n+1})$ with the aid of the difference scheme (5.3.8). However, obtaining an exact solution $\zeta(t^{n+1})$ of (5.3.8) is difficult by virtue of the fact that $\zeta(t^{n+1})$ enters (5.3.8) nonlinearly because of the nonlinearity of the functions φ_{k_0}, Q_{k_0}. Therefore, one must use some approximate numerical method to compute $\zeta(t^{n+1})$. One of the well-known approaches to solving the nonlinear equation $G(\zeta^{n+1}) = 0$ (see, for example, [5.15]) consists of the solution of an optimization problem

$$(G(\zeta^{n+1}))^2 \to \min$$

which in the case of the Adams scheme (5.3.8), as may easily be seen, coincides with the minimization problem (5.3.4).

Thus, solving equation (5.3.7) under the initial condition $\zeta(0) = x_0$, with the aid of some stable difference scheme for the numerical integration of ordinary differential equations (see, on stability [5.16], for example), we obtain an approximation $\zeta(t)$ to the discontinuity trajectory. In the case of a stationary shock wave equation (5.3.7), as may easily be seen with (5.2.7) in view, goes over into the equation $d\zeta/dt = D$, whose solution under the initial condition $\zeta(0) = x_0$ obviously coincides with the exact shock wave trajectory $x = x_0 + Dt$.

5.4. Gradient Methods of Basic Functional Minimization

Let the initial data for the system (1.1)–(1.3) be given at some time $t = t_0$. For the numerical minimization of the basic functional (5.2.11), (5.2.12) it is possible to use some well-known methods; a description of a number of widespread methods may be found, for example, in [5.17]–[5.21]. In particular, gradient methods for functional minimization are widely used. Two different realizations of gradient methods, as applied to the functional of the form (5.2.11), (5.2.12), have been presented in [5.1], [5.2], [5.4].

In order to have the ability to use the derivations of [5.1], [5.2], we shall consider the function $\overline{\zeta}(t) = \zeta(t) - x_0$ instead of the approximation to the discontinuity trajectory. Thus $\overline{\zeta}(t_0) = 0$. In what follows the bar over ζ is omitted for brevity. Take the functions $\zeta(t)$ and $\zeta(t) + \delta\zeta(t)$ from $C^1[t_0, T]$, such that $\zeta(t_0) = \delta\zeta(t_0) = 0$. Making use of the formula

$$J_k(\zeta + \delta\zeta, t) = J_k(\zeta, t) + [\partial J_k(\zeta, t)/\partial x]\delta\zeta + o(\delta\zeta), \qquad (5.4.1)$$

we obtain (with the aid of (5.2.11)) an expression

$$I(\zeta + \delta\zeta) - I(\zeta) = 2 \int_{t_0}^{T} \left\{ \sum_{k=1}^{3} \alpha_k [(\dot{\zeta}(t) \right.$$
$$\left. - J_k(\zeta, t))(\delta\dot{\zeta} - (\partial J_k(\zeta, t)/\partial x)\delta\zeta)] \right\} dt + o(\delta\dot{\zeta}). \qquad (5.4.2)$$

Employing the formula

$$\int_{t_0}^{T} (\dot{\zeta} - J_k(\zeta, t))(\partial J_k(\zeta, t)/\partial x)\delta\zeta \, dt$$
$$= \int_{t_0}^{T} \delta\dot{\zeta}(t) \left\{ \int_{t}^{T} [\dot{\zeta}(\tau) - J_k(\zeta, \tau)](\partial J_k(\zeta, \tau)/\partial x) \, d\tau \right\} dt,$$

obtained by the integration by parts, let us rewrite (5.4.2) in the form

$$I(\zeta + \delta\zeta) - I(\zeta) = 2\int_{t_0}^{T}\left\{\sum_{k=1}^{3}\alpha_k[\dot{\zeta}(t) - J_k(\zeta, t)\right.$$
$$\left. - \int_{t}^{T}(\dot{\zeta}(\tau) - J_k(\zeta, \tau))(\partial J_k/\partial x)(\zeta, \tau)\,d\tau]\right\}\delta\dot{\zeta}(t)\,dt \quad (5.4.3)$$

which is obtained by neglecting the terms $o(\delta\dot{\zeta})$. Let the function $\zeta(t)$ be given along with $\zeta(t_0) = 0$ and with the corresponding value of $I(\zeta)$. Let us choose the following approximation $\zeta(t) + \delta\zeta(t)$ in such a way that $I(\zeta) > I(\zeta + \delta\zeta)$. Employing the steepest descent method, let us set

$$\delta\dot{\zeta}(t) = -\lambda\nabla I(\zeta)(t), \qquad (5.4.4)$$

where λ is a positive constant and where, with (5.4.3) in view,

$$\nabla I(\zeta)(t) = 2\left\{\sum_{k=1}^{3}\alpha_k[\dot{\zeta}(t) - J_k(\zeta(t), t)\right.$$
$$\left. - \int_{t}^{T}(\dot{\zeta}(\tau) - J_k(\zeta(\tau), \tau))(\partial J_k(\zeta(\tau), \tau)/\partial x)\,d\tau]\right\}. \qquad (5.4.5)$$

With such a choice of $\delta\zeta$ we have, taking formula (5.4.3) into account, that

$$I(\zeta + \delta\zeta) - I(\zeta) = -\lambda\int_{t_0}^{T}[\nabla I(\zeta)(t)]^2\,dt + o(\lambda\nabla I).$$

At sufficiently small $\lambda\nabla I$ the right-hand side of this equality is negative.

Let us describe an algorithm for the functional (5.2.11) minimization based on formulas (5.4.3)–(5.4.5). Denote by $\zeta_i(t)$ the ith approximation to the discontinuity trajectory $\xi(t)$, $i = 0, 1, 2, \ldots$.

Algorithm 1.

Step 1. Choose $\zeta_0(t) \in C^1[t_0, T]$, such that $\zeta_0(t_0) = 0$, $i = 0$. Choose $\lambda \in (0, 1)$.

Step 2. Compute $\mathbf{w}^-$, $\mathbf{w}^+$.

Step 3. Compute $\nabla I(\zeta_i)(t)$ by (5.4.5), (5.2.12).

Step 4. Check whether the inequality

$$I\left(\zeta_i - \lambda\int_{t_0}^{T}\nabla I(\zeta_i)(t)\,dt\right) - I(\zeta_i) \le -\frac{\lambda}{2}\|\nabla I\|^2_{L^2[t_0, T]} \qquad (5.4.6)$$

is satisfied. If λ does not satisfy the inequality (5.4.6), then reduce λ by the formula $\lambda = \theta\lambda$ where we take, in accordance with the recommendations of [5.17], $\theta \in (0.5; 0.8)$.

Step 5. Set

$$\zeta_{i+1}(t) = \zeta_i(t) - \lambda\int_{t_0}^{T}\nabla I(\zeta_i)(\tau)\,d\tau. \qquad (5.4.7)$$

Step 6. Check the satisfaction of the inequality

$$(\Delta \zeta_{i+1})_{\max} = \max_t |(\zeta_{i+1}(t) - \zeta_i(t))/h| < \delta_1, \qquad (5.4.8)$$

where h is a step of the uniform computing mesh on the x-axis, δ_1 is a user-specified number, and $\delta_1 \in (0, 1)$. If the inequality (5.4.8) is not satisfied, we set $i = i + 1$ and go over to Step 3.

Algorithm 2. That which is presented below differs from Algorithm 1 only by a strategy of the choice of the parameter λ, while computing the next approximation $\zeta_{i+1}(t)$ by formula (5.4.7) [5.4].
Step 1. Choose $\zeta_0(t) \in C^1[t_0, T]$ such that $\zeta_0(t_0) = 0$, $i = 0$.
Step 2. Compute $\mathbf{w}^+$, $\mathbf{w}^-$.
Step 3. Compute $\nabla I(\zeta_i)(t)$ by (5.4.5), (5.2.12).
Step 4. Set $\lambda_0 = 0$, $k = 0$, and give some increment $\Delta \lambda$ to λ.
Step 5. Check whether the inequality

$$I\left(\zeta_i - (\lambda_0 + \Delta\lambda) \int_{t_0}^{T} \nabla I(\zeta_i)(t) \, dt\right) \le I(\zeta_i) \qquad (5.4.9)$$

is satisfied. If (5.4.9) is not satisfied, we set $\Delta\lambda = -\Delta\lambda$ and go over to Step 6.
Step 6. Compute $\lambda_{k+1} = \lambda_k + \Delta\lambda$. Take, in formula (5.4.7), $\lambda = \lambda_{k+1}$ and denote by $I_{(k+1)}(\zeta_i(t))$ the corresponding value of the functional I.
Step 7. If $I_{(k+1)}(\zeta_i(t)) \le I_{(k)}(\zeta_i(t))$, we double $\Delta\lambda$ and go back to Step 6 with $k = k + 1$. If $I_{(k+1)}(\zeta_i(t)) > I_{(k)}(\zeta_i(t))$, we introduce the notation $\lambda_{(m)} = \lambda_{k+1}$, $\lambda_{(m-1)} = \lambda_k$, reduce $\Delta\lambda$ by a factor of 2, and go back to Step 6 for one more (only one) computation.
Step 8. Exclude from the four obtained values of λ $\{\lambda_{(m+1)}, \lambda_{(m)}, \lambda_{(m-1)}, \lambda_{(m-2)}\}$ the value of $\lambda_{(j)}$ at which the value of the functional

$$I_{(j)}(\zeta_i(t)) \equiv I(\zeta_i - \lambda_{(j)} \int_{t_0}^{t} \nabla I(\zeta_i)(t) \, dt), \qquad j = m + 1, m, m - 1, m - 2,$$

is maximal. The remaining three values of the parameter λ are now arranged in an increasing sequence. Denote its elements by $\lambda_{(1)}, \lambda_{(2)}, \lambda_{(3)}$ in such a way that $\lambda_{(1)} < \lambda_{(2)} < \lambda_{(3)}$.
Step 9. Compute an approximate value of λ at the point of minimum of I, by using parabolic approximation of the functional $I(\lambda)$, by the formula

$$\lambda^* = 0.5\{[\lambda_{(2)}^2 - \lambda_{(3)}^2]I_1 + [\lambda_{(3)}^2 - \lambda_{(1)}^2]I_2 + [\lambda_{(1)}^2 - \lambda_{(2)}^2]I_3\}$$

$$/\{[\lambda_{(2)} - \lambda_{(3)}]I_1 + [\lambda_{(3)} - \lambda_{(1)}]I_2 + [\lambda_{(1)} - \lambda_{(2)}]I_3\},$$

where we have used the notation $I_k = I_k(\zeta_i(t))$, $k = 1, 2, 3$. If λ^* and any of the values $\{\lambda_{(1)}, \lambda_{(2)}, \lambda_{(3)}\}$ differ by less than a given small number δ_2, we stop the search for λ^*. Otherwise, we check the satisfaction of the inequality

$$I_{(3)}(\zeta_i(t)) < I(\zeta_i - \lambda^* \int_{t_0}^{t} \nabla I(\zeta_i)(t) \, dt).$$

194   5. Optimization Techniques of Discontinuities Localization

If this inequality takes place, we reduce the increment $\Delta\lambda$ by a factor of two and go back to Step 4.

Step 10. Set, in (5.4.7), $\lambda = \lambda^*$ and compute $\zeta_{i+1}(t)$.

Step 11. Check the satisfaction of the inequality (5.4.8). If this inequality is not satisfied, we set $i = i + 1$ and go back to Step 3.

Note that the question on the convergence of gradient methods for the functional minimization of the steepest descent type realized in Algorithm 1 was considered, for example, in [5.17] where, in particular, the convergence condition (5.4.6) has been obtained. The presentation of the gradient method version realized in Algorithm 2 may be found, for example, in [5.21], [5.22]. Questions on the convergence of the minimization algorithm of the Algorithm 2 form were considered, in particular, in [5.21] (see also [5.4]).

The convexity of the functional is usually required in the theory of optimization (see, for example, [5.19], [5.21]), to ensure uniqueness of the point of the extremum of the functional. Consider a question on the convexity of the functional (5.2.8) in the case when the function $\tilde{Q}$ in (5.2.2) is such that

$$\tilde{Q}(0, 0, \ldots, 0, x, t, h, \tau) = 0. \tag{5.4.10}$$

Let us make use of the definition of a convex functional. Take the functions $\zeta_j(t) \in C^1[0, T]$, $j = 1, 2$, and the numbers $\lambda_1, \lambda_2 \geq 0$, such that $\lambda_1 + \lambda_2 = 1$. According to the convexity definition, we should check on the satisfaction of the inequality

$$I_k(\lambda_1\zeta_1 + \lambda_2\zeta_2) \leq \lambda_1 I_k(\zeta_1) + \lambda_2 I_k(\zeta_2). \tag{5.4.11}$$

Let us introduce the notations

$$\xi_1(t) = \dot{\zeta}_1 - J_k(\zeta(t), t), \qquad \xi_2(t) = \dot{\zeta}_2 - J_k(\zeta(t), t),$$
$$\xi_3(t) = \dot{\zeta}_1 - J_k(\zeta_1(t), t), \qquad \xi_4(t) = \dot{\zeta}_2 - J_k(\zeta_2(t), t), \tag{5.4.12}$$

where

$$\zeta(t) = \lambda_1\zeta_1(t) + \lambda_2\zeta_2(t), \tag{5.4.13}$$

$$J_k(\zeta_j(t), t) \equiv [\varphi_k(w(\zeta_j(t), t)) - \varphi_k(W_2) - Q_k(\zeta_j(t), t)]$$
$$\times [w_k(\zeta_j(t), t) - W_{2k}]^{-1}. \tag{5.4.14}$$

Let us also introduce, by analogy wth (5.1.5), the notation

$$F_k(\zeta(t), \dot{\zeta}(t), t) = J_k(\zeta(t), t) - \dot{\zeta}(t).$$

Employing the notation (5.4.12), we can write

$$F_k^2 = (\lambda_1\xi_1 + \lambda_2\xi_2)^2.$$

Consider the difference

$$R = \lambda_1\xi_3^2 + \lambda_2\xi_4^2 - (\lambda_1\xi_1 + \lambda_2\xi_2)^2. \tag{5.4.15}$$

If the inequality $R \geq 0$ is satisfied at arbitrary real $\xi_1, \xi_2, \xi_3, \xi_4$ in (5.4.15), we then obtain from (5.4.11), (5.4.15) the condition for the convexity of the functional (5.2.8) in the form

$$\int_0^T (\lambda_1 \xi_1 + \lambda_2 \xi_2)^2 \, dt \le \lambda_1 \int_0^T \xi_3^2 \, dt + \lambda_2 \int_0^T \xi_4^2 \, dt. \qquad (5.4.16)$$

We have, on the right-hand side of equation (5.4.15), a Hermitean quadratic form

$$R(\xi_1, \xi_2, \xi_3, \xi_4) = -\lambda_1^2 \xi_1^2 - 2\lambda_1 \lambda_2 \xi_1 \xi_2 - \lambda_2^2 \xi_2^2 + \lambda_1 \xi_3^2 + \lambda_2 \xi_4^2. \quad (5.4.17)$$

Let us make use of the Sylvester criterion to study the nonnegativeness of the quadratic form (5.4.17). A symmetric matrix A associated with (5.4.17) has the form

$$A = \begin{pmatrix} -\lambda_1^2 & -\lambda_1 \lambda_2 & 0 & 0 \\ -\lambda_1 \lambda_2 & -\lambda_2^2 & 0 & 0 \\ 0 & 0 & \lambda_1 & 0 \\ 0 & 0 & 0 & \lambda_2 \end{pmatrix}.$$

The sequential principal minors Δ_j, $j = 1, \ldots, 4$, of the matrix A may easily be computed

$$\begin{aligned} \Delta_1 &= -\lambda_1^2, & \Delta_2 &= \lambda_1^2 \lambda_2^2 - (\lambda_1 \lambda_2)^2 = 0, \\ \Delta_3 &= \lambda_1 \Delta_2 = 0, & \Delta_4 &= \lambda_2 \Delta_3 = 0. \end{aligned} \qquad (5.4.18)$$

It follows from (5.4.18) that $\Delta_1 \le 0$, thus the Sylvester criterion is not satisfied. Consequently, the functional (5.2.8) is not convex. On the other hand, it is not difficult to find such values of the quantities ξ_j in (5.4.17) at which $R \ge 0$. This means that there exist in the (x, t)-plane the domains Ω_k, $k = 1, 2, \ldots$, such that at $(\zeta_j(t), t) \in \Omega_k$, $j = 1, 2, 0 \le t \le T$, the inequality (5.4.16) is satisfied and, thus, the convexity of the functional (5.2.8) takes place. In particular, it can easily be shown that in the case when the functional (5.2.8) is applied to the localization of a front of a stationary shock wave moving from left to right (that is, $\dot{\xi}(t) > 0$), that there then exists a domain Ω_1 in which the functional (5.2.8) is convex,

$$\Omega_1 = \{(x, t) | x_l \le x \le \xi(t), 0 \le t \le T\},$$

where $\xi(t)$ is a function entering the equation of the exact discontinuity trajectory $x = \xi(t)$, and x_l is the abscissa of the left end of the integration interval on the x-axis, $\xi(0) > x_l$. Indeed, let us take two functions $\zeta_j(t) \in C^1[0, T]$, such that

$$x_l \le \zeta_j(t) \le \xi(t), \qquad j = 1, 2, \quad 0 \le t \le T. \qquad (5.4.19)$$

Since we are considering a stationary shock, the quantities u, p, ρ on both sides of the discontinuity are constant, so that the profiles of these quantities have the form of a step. With (5.4.10) in view, $Q_k = 0$ on both sides of a shock.

It is easy to see, with regard to (5.4.13), (5.4.19), that $\zeta(t) \leq \xi(t)$, therefore, $J_k(\zeta(t), t) = \dot{\xi}(t)$. Thus, we have in (5.4.12) that

$$\xi_1 = \dot{\zeta}_1 - \dot{\xi}, \qquad \xi_2 = \dot{\zeta}_2 - \dot{\xi}, \qquad \xi_3 = \xi_1, \qquad \xi_4 = \xi_2. \qquad (5.4.20)$$

Substituting the values of (5.4.20) into (5.4.17), we obtain

$$R = \lambda_1 \lambda_2 (\xi_1 - \xi_2)^2 \geq 0$$

for any real ξ_1, ξ_2, that was to be proved.

Consider now the case when $\zeta_j(t) > \xi(t)$, $j = 1, 2$. Then there arises an uncertainty of the form 0/0 in (5.4.14) in the case of a stationary shock wave moving from left to right, and the computation of the value of $J_k(\xi_j(t), t)$ proves to be difficult.

From the above consideration of the convexity properties of the functional (5.2.8), we can draw the following strategy on the choice of an initial approximation $\zeta_0(t)$ to the discontinuity trajectory when realizing the above-presented gradient methods for the minimization of the basic functional (5.2.11), (5.2.12): it is preferable to set the function $\zeta_0(t)$ in such a way that the points $(\zeta_0(t), t)$ at $0 \leq t \leq T$ lie in the domain behind the shock wave front. This strategy of the choice of $\zeta_0(t)$ was realized by us in practical computations (see below in this section).

Consider now a question on the numerical realization of the above-presented methods for the basic functional (5.2.11), (5.2.12) minimization. Let us introduce in the (x, t)-plane a uniform mesh with steps h and τ along the x- and t-axes, respectively. Denote by I_h a discrete functional approximating I. Let $t_n = t_0 + n\tau$, $n = 0, 1, \ldots, n$, and let $\zeta^n = \zeta(t_n)$, $j_n = \text{entier}(\zeta^n/h + 0.5)$, that is, (j_n, n) are the indices of a mesh point nearest to (ζ^n, t_n); $\text{entier}(x)$ is the integral part of x. In accordance with (5.2.12) let us set

$$J_{k,j}^n \equiv J_k(x_j, t_n) = [\varphi_k(\mathbf{w}^-(x_j, t_n)) - \varphi_k(\mathbf{w}^+(x_j, t_n))$$
$$- Q_k(x_j, t_n)] / [w_k^-(x_j, t_n) - w_k^+(x_j, t_n)], \qquad (5.4.21)$$

where $x_j = (j + 0.5)h$ is the abscissa of a cell center on the x-axis. Consider, by analogy with [5.1], [5.2], the following simplest approximations for the functional (5.2.11) and the gradient (5.4.5)

$$I_h = \sum_{n=0}^{N_1-1} \sum_{k=1}^{3} \alpha_k [(\zeta^{n+\beta} - \zeta^n)/\Delta t - J_{k,j_{n+\beta}}^{n+\beta}]^2 \Delta t, \qquad (5.4.22)$$

$$\nabla I_h \equiv \nabla I_h(t_n) = 2 \sum_{k=1}^{3} \alpha_k \{ (\zeta^{n+\beta} - \zeta^n)/\Delta t - J_{k,j_{n+\beta}}^{n+\beta}$$

$$- \sum_{r=n}^{N_1-1} \Delta t [(\zeta^{r+\beta} - \zeta^r)/\Delta t - J_{k,j_{r+\beta}}^{r+\beta}] (1/h)(J_{k,j_{r+\beta}+1}^{r+\beta} - J_{k,j_{r+\beta}}^{r+\beta}) \}, \qquad (5.4.23)$$

$$\int_{t_0}^{t_0 + n\Delta t} \nabla I(\zeta(t)) \, dt \cong \sum_{k=0}^{n-1} \nabla I_h(t_k) \Delta t. \qquad (5.4.24)$$

In the formulas (5.4.22), (5.4.23) we have used the notations $\Delta t = \beta \tau$ where β is an integer, $\beta \geq 1$; $N_1 = \text{entier}((T - t_0)/\Delta t)$. The step Δt multiple with the step τ has been introduced for the purpose of economy of computer storage required for the realization of the above-presented methods for minimization of the functional I.

The function $\mathbf{w}^-(x, t)$ entering formula (5.4.21) was computed by a well-known finite-difference shock capturing scheme. Thus, the discontinuities in a finite-difference solution $\mathbf{w}^-(x, t)$ were "smeared". The function $\mathbf{w}^+(x, t)$ was taken to be independent of t. We have taken, with regard to the initial conditions (5.2.13), $\mathbf{w}^+(x, t) = \mathbf{u}(x, t_0)$ at $x > x_0$.

It is clear that if $T - t_0 = O(1)$, the formulas (5.4.22)–(5.4.24) have an approximation error $O(\Delta t)$, and it can be established (similar to [5.1], [5.2]) that the discontinuity path $\zeta(t)$ will be determined with an error $O(\Delta t)$ while using the above formulas.

The initial approximation $\zeta_0(t)$ was set by using the fact that the inequality $\partial u/\partial x < 0$ takes place in the zone of a smeared shock wave. Taking this into account we assumed $\zeta_0(t)$ at fixed t to be equal to the abscissa of that end of the smeared shock wave zone which corresponds to a state behind the front. This abscissa was determined by analogy with [5.23]. In some cases we arbitrarily shifted this abscissa to the left by a distance of $10h$ for large values of t (this shift was accomplished by the formula $\zeta_0(t) = x_0 + 0.7 (\zeta_0(t) - x_0)$, that is, only for $t > t_0$). The convergence of the steepest descent method presented above also took place in this case. Four to twelve iterations were needed to achieve the accuracy of the abscissa determination of the order $10^{-3}h$.

5.5. Computational Examples

Let us denote by $x_f(t)$ the shock wave front abscissa computed in accordance with the exact solution, and by $\zeta(t)$ the approximate abscissa of the shock wave front computed with the aid of some optimization algorithm or the differential analyzer presented in Chapter 2. Let us introduce the localization error

$$[x_f(t) - \zeta(t)]/h. \tag{5.5.1}$$

In Figure 5.1(a) (see [5.6]), the error (5.5.1) is depicted as a function of dimensionless time t/τ which is measured along the abscissa axis; the coefficient $a = 3$ in (2.2.18). As was previously noted in [5.11], [5.24]–[5.26], in some cases the shock wave is spread over more than 10 cells. To simulate this situation, and to examine the shock localization accuracy in this case, we have also used in (2.2.18) the value $a = 15$. In this case we find from (2.2.26) that the width of the smeared shock wave zone $X = 9.95h$. In Figure 5.1(b) we present the computational results for the value $a = 15$ in the formula for pseudoviscosity (2.2.18). It follows from Figure 5.1 that the trajectory $x = \zeta(t)$,

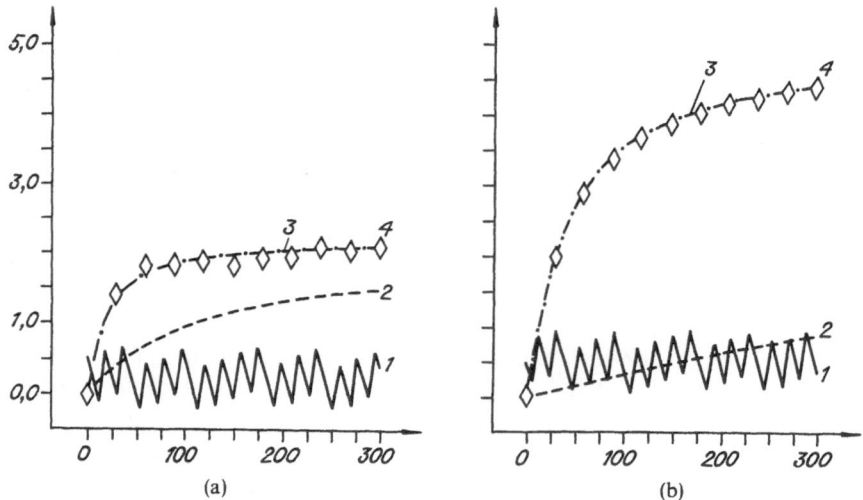

Figure 5.1. The localization error (5.5.1) as a function of time: (a) the coefficient $a = 3$ in (2.2.18); (b) $a = 15$. 1—the localization by max q where q is the pseudoviscosity (2.2.18); 2—the use of the functional (5.2.8) with $k = 2$, $Q_2 = -q$; 3—the function $-v(t)/h$ where $v(t)$ is the function (5.1.40); 4—the use of the original Miranker–Pironneau algorithm.

obtained by numerical minimization of the Miranker–Pironneau functional (5.1.23), (5.1.5), (5.1.6), agrees very well with the analytical solution (5.1.40) at different values of the coefficient "a" in (2.2.18). An especially good agreement between the numerical solution and the exact one (5.1.40) takes place at $a = 15$. We explain this by the fact that with the increase in the coefficient a in (2.2.18) the profiles in the shock wave zone become smoother, and as a consequence of this the size of the truncation errors (caused by the use of one-sided differences for $\partial J(\zeta(t), t)/\partial x$ in (5.4.23) and by the use of the formula of rectangles (5.4.24)) diminishes. Similar considerations explain the increase in discontinuity localization accuracy by means of the new functional (5.2.8), $k = 2$, as the coefficient a in (2.2.18) increases. It should be noted that the use of the functional (5.2.8) provides a higher localization accuracy than in the case of the functional (5.1.23), (5.1.5), (5.1.6).

For the same input data, as in Figure 5.1, we have used the localization method based on minimization of the function (5.3.2). The minimization of the univariate function (5.3.2) was performed by using a standard subroutine MNGGR entering the mathematical software of the computer employed. The application of the Fibonacci method for finding a local minimum of the function is an essential element of the subroutine MNGGR [5.27]. Corresponding results for different values of a in (2.2.18) are presented in Figure 5.2. It is easy to see that the localization results, on the basis of the minimization of the function (5.3.2) and on the basis of the functional (5.2.8) with $k = 2$, practically coincide.

Figure 5.2. The use of the functional (5.2.8) with
$k = 2$, $Q_2 = -q$. 1—$a = 3$ in (2.2.18); 2—$a = 15$
in (2.2.18). The minimization of the function (5.3.2),
(5.3.3) by the Fibonacci method: 3—$a = 3$, 4—
$a = 15$ in formula (2.2.18).

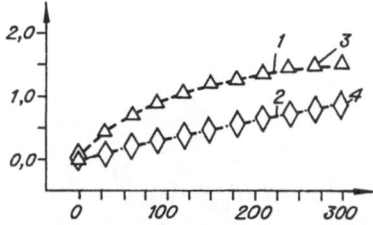

In Figure 5.3 the results of the stationary shock wave localization, obtained
from the difference solution by the "breakdown-of-discontinuity" scheme
[5.13], are shown. It is seen that the error (5.5.1) becomes greater as t increases
in the case of the application of the Miranker–Pironneau functional (5.1.6).
The difference approximation of the function Q_1 (5.2.14) employing one-sided
differences for u_x, p_x, ρ_x, represents a piecewise-constant function; more pre-
cicely, it is constant within each interval $(j - 1) h \leq x \leq jh$, $j = 1, 2, \dots$. In
this connection the shock front obtained by $\max|Q_1|$ will propagate stepwise,
which is seen in Figures 5.1 and 5.3. However, if the front abscissa sought for
enters explicitly into the localization algorithm employed, as takes place in
the case of the functionals (5.1.6), (5.2.8), and the function (5.3.2), then the above
abrupt changes in the located discontinuity abscissa vanish, despite the discrete character of the difference solution on which basis the shock front was
localized. This feature of the localization techniques (5.1.6), (5.2.8), and (5.3.2)
is easily seen in Figures 5.1–5.4. A qualitative explanation of this behavior
can be given using the example of the function (5.3.2). Let us take two such
values ζ_1, ζ_2 of the continuous variable $\zeta(t^{n+1})$, that $\zeta_1 \neq \zeta_2$, and, in addition,

$$\mathbf{w}^+(\zeta_1, t^{n+1}) = \mathbf{w}^+(\zeta_2, t^{n+1}) = \mathbf{b}^+,$$

$$\mathbf{b}^\pm = \{b_1^\pm, b_2^\pm, b_3^\pm\}^{\mathrm{T}},$$

$$\mathbf{w}^-(\zeta_1, t^{n+1}) = \mathbf{w}^-(\zeta_2, t^{n+1}) = \mathbf{b}^-, \tag{5.5.2}$$

$$Q_k|_{\zeta=\zeta_1} = Q_k|_{\zeta=\zeta_2}, \qquad k = 1, 2, 3.$$

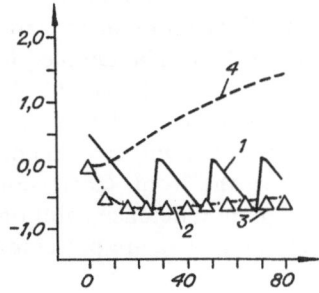

Figure 5.3. The error (5.5.1): 1—localization by $\max|Q_1|$
where Q_1 is the function (5.2.14); 2—the use of the functional (5.2.8) with $k = 1$, Q_1 is the function (5.2.14); 3—the
minimization of the function (5.3.2), (5.3.3) by the golden
section method; 4—the original algorithm of Miranker
and Pironneau.

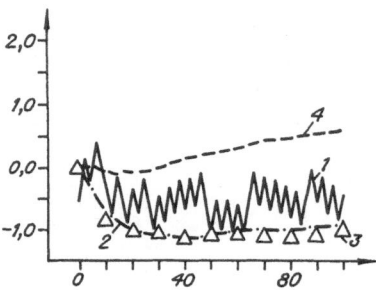

Figure 5.4. The problem on the propagation of a nonstationary shock wave in an inhomogeneous atmosphere (the notation is the same as in Figure 5.3).

Then it is easy to be convinced with the aid of (5.3.2), (5.3.3) that $F(\zeta_1) \neq F(\zeta_2)$. Indeed, introduce the notation

$$a_k = [\varphi_k(\mathbf{b}^-) - \varphi_k(\mathbf{b}^+) - Q_k|_{\zeta=\zeta_1,\zeta_2}]/(b_k^- - b_k^+),$$

$$\zeta_0 \equiv \zeta(t^{n+1} - \tau_1) \tag{5.5.3}$$

In (5.3.2) let $\alpha_k = 1$ at some k, and let $\alpha_j = 0$ for $j \neq k$. Then with notation (5.5.3) and relationships (5.5.2) in view we obtain

$$F(\zeta_1) - F(\zeta_2) = 2\tau_1^{-1}(\zeta_1 - \zeta_2)[((\zeta_1 + \zeta_2)/2 - \zeta_0)/\tau_1 - a_k]. \tag{5.5.4}$$

Let us fix the value ζ_2 and find an interval $\Delta(\zeta_2) = [\zeta_2 - \delta_1, \zeta_2 + \delta_2]$ such that at $\zeta_1 \in \Delta(\zeta_2)$ the equations (5.5.2) are satisfied. Introduce the notation $z \equiv \zeta_1$ and rewrite (5.5.4) in the form

$$F(z) = F(\zeta_2) + 2\tau_1^{-1}(z - \zeta_2)\{[(z + \zeta_2)/2 - \zeta_0]/\tau_1 - a_k\}. \tag{5.5.5}$$

It follows from (5.5.5) that the function (5.3.2) is approximated by a parabola in each of the above-defined intervals $\Delta(\zeta_2)$. Minimization results of the function (5.3.2) shown in Figures 5.3 and 5.4 are obtained by the golden section method [5.20], [5.28]. Only 17 evaluations of the function $F(\zeta^{n+1})$ were needed to compute the abscissa with an error of $0.01h$.

Although the investigation of the accuracy of localization techniques (described in the foregoing sections) was carried out only for the case of a stationary shock wave, it was interesting to see whether the new localization techniques described above are applicable to problems with nonstationary shock waves. We have considered a problem on a self-similar shock wave caused by an impulse plane impact and propagating in an inhomogeneous atmosphere. The detailed formulation and the solution of this problem are contained in Chapter 12 of [5.29]. Here we only note that the profiles of the exact solution in the region behind the front are such that at $t \geq 0$, $\partial u/\partial x > 0$, $\partial p/\partial x > 0$, $\partial \rho/\partial x > 0$; in addition, $\partial \rho/\partial x > 0$ also in the undisturbed medium before the front. Corresponding profiles of the exact solution, as well as the finite-difference solution obtained by the "breakdown-of-discontinuity" scheme [5.13], are presented in Figure 5.5 taken from [5.4]. In this problem

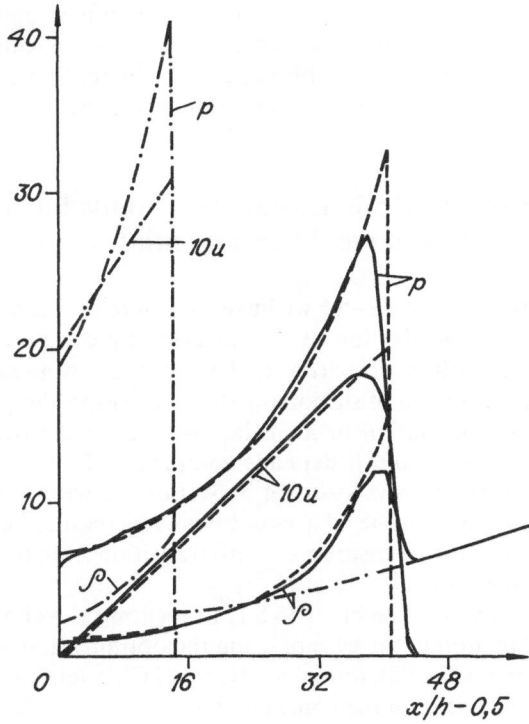

Figure 5.5. Solution profiles in the problem on a self-similar shock wave.

the exact abscissa x_f of the shock front at $t \geq 0$ is computed by the formula $x_f(t) = 6 \ln(t + t_0)$, where t_0 is determined from the condition $x_0 = 6 \ln t_0$, and x_0 is the given abscissa of the shock front at $t = 0$. In computations by the scheme from [5.13] we have used the values $h = 0.1$, $\tau = 0.007$ (cf. [5.4]), which corresponded to the Courant number $K \cong 0.4$. It follows from Figure 5.4 that the optimization algorithms proposed in Sections 5.2 and 5.3 are also applicable in the case of a nonstationary shock wave. It should be noted that in the case of both stationary (Figures 5.1 and 5.3) and nonstationary (Figure 5.4) shock waves, the inclusion in an optimization localization procedure of the information on the approximation and artificial viscosity moves the trajectory $\zeta(t)$ to the right in comparison with the original Miranker–Pironneau algorithm [5.1], [5.2].

In [5.4] the basic functional (5.2.11) was employed at different combinations of the weight coefficients α_1, α_2, α_3. As a result of this the possibility of using any of the three Hugoniot conditions or a combination of them in the form (5.2.11) in the functional (5.2.11) has been shown.

The above-presented gradient methods of the basic functional minimization were realized in the form of a subroutine. This subroutine, or "module", may be applied in combination with various difference schemes of gas dynamics, as was demonstrated in [5.4]. Thus, it is one of the means of solving the problem on the interpretation of the shock-capturing computation results of gas-dynamical problems, and of increasing the accuracy of the numerical solutions in the neighborhood of discontinuities.

5.6. On Optimization Algorithms for the Localization of Contact Discontinuities

In Sections 5.2–5.5 we have considered some optimization methods for shock wave localization by shock-capturing computational results, and the following positive properties of these methods were noted. First, the accuracy of the localization of stationary shock waves in the presence of information on the approximation or artificial viscosity in the basic functional (5.2.11), (5.2.12) increases, and it depends comparatively weakly on the specific form of viscosity Q (as was shown in Section 5.2 with the use of progressive wave-type solutions of the f.d.a. equations). It is required only for the scheme viscosity Q structure to ensure a smooth transition from one state to another in a smeared shock wave.

Second, the error (5.5.1) in locating the shock front abscissa is a smooth function in the case of using the optimization procedures of Sections 5.1–5.4, unlike the differential analyzers of Chapter 2. This corresponds to the physical properties of the considered gas-dynamical flows with shock waves.

Third, a well-developed theory of numerical optimization methods may be used for the convergence analysis of numerical methods for the minimization of the basic functionals, as was demonstrated in Section 5.4. Fourth, the technique for investigating the accuracy of differential analyzers of contact discontinuities (which was presented in Chapter 4) is, strictly speaking, applicable only to the flows with contact discontinuities in which the pressure and velocity are constant (or almost constant) in the neighborhood of contact discontinuities.

In connection with the foregoing it is of present interest to develop the methods for contact discontinuity localization, which are based on principles different from those underlying the method of Chapter 4 for investigating the differential analyzers of contact discontinuities. Below we consider some techniques for constructing optimization algorithms for the localization of contact discontinuities in one-dimensional flows, which are based on the minimization of some integral functionals or on the minimization of certain functions.

It is known that in the one-dimensional flow a contact discontinuity moves at a speed

$$dx/dt = u(x, t), \tag{5.6.1}$$

where u is the gas speed in the neighborhood of a discontinuity. Consider first (as in Chapter 4), the Riemann problem (1.11), (1.9), (1.5), (4.1.1). Let this problem be solved in the interval $0 \leq t \leq T$ where $0 < T < \infty$. Let $\zeta(t) \in C^1[0, T]$. Taking into account formula (5.6.1) let us consider the functional

$$I(\zeta) = \int_0^T (d\zeta/dt - u(\zeta(t), t))^2 \, dt. \tag{5.6.2}$$

We pose for the functional (5.6.2) at the left end $t = 0$ the condition $\zeta(0) = x_0$, and at the right end $t = T$ the transversality condition (5.1.7), where in our case

$$F(\zeta(t), \dot{\zeta}(t), t) = [d\zeta/dt - u(\zeta(t), t)]^2. \tag{5.6.3}$$

Taking into account condition $\zeta(0) = x_0$, as well as the transversality condition (5.1.7) where the function F is determined by (5.6.3), we easily find that the solution of an optimization problem of the form (5.1.8) coincides with the exact trajectory of a contact discontinuity.

Now suppose that there arise at $t > 0$, in the flow under study, shock waves whose trajectories begin at the point $x = x_0$ along a contact discontinuity trajectory emanating from the same point x_0 (see Section 1.1 on the general Riemann problem). We modify the functional (5.6.2) for this situation by the formula

$$I_1(\zeta) = \int_0^T [(\dot{\zeta} - u)^2 + \alpha_1 R_1((\partial p/\partial x)^2)$$
$$+ \alpha_2 R_2((\partial u/\partial x)^2) + \alpha_3 R_3((\partial \rho/\partial x)^2)] \, dt, \tag{5.6.4}$$

where R_1, R_2, R_3 are continuously differentiable functions, such that

$$R_j(z) > 0, \quad z > 0, \quad j = 1, 2, 3; \quad R_1'(z) > 0,$$
$$R_2'(z) > 0, \quad R_3'(z) < 0. \tag{5.6.5}$$

In formula (5.6.4) $\alpha_1, \alpha_2, \alpha_3$ are constant weight coefficients, and $\alpha_1 \geq 0, \alpha_2 \geq 0$, $\alpha_1 + \alpha_2 > 0, \alpha_3 > 0$. The item $\alpha_3 R_3$ has been introduced into formula (5.6.4) to exclude from consideration the subdomains of constant flow in the process of a search for the contact discontinuity trajectory, because there are surely no contact discontinuities in these subdomains. In the zone of a smeared shock wave of finite intensity the gradients $|\partial u/\partial x|, |\partial p/\partial x|$ are usually large at a sufficiently small step h of a uniform mesh along the x-axis. At the same time, the gradients $\partial u/\partial x, \partial p/\partial x$ in the contact discontinuity region are finite by virtue of the fact that the pressure and velocity in a one-dimensional flow are continuous across a contact discontinuity. If the trajectory $\zeta(t) = x$ is in the zone of contact discontinuity smearing, then taking the foregoing considerations and formula (5.6.5) into account, a smaller value of the functional (5.6.4) will correspond to it than in the case of a trajectory $x = \zeta(t)$ located in the zone of shock wave smearing. Thus, the solution of the problem on the functional (5.6.4) minimization will yield (at sufficiently small stepsizes h) an approximation to the contact discontinuity trajectory when there are (in a one-dimensional flow) closely located shock waves and a contact discontinuity. Well-known techniques can be used for the numerical minimization of the functional (5.6.4) (see, for example [5.17], [5.20] as well as Section 5.4).

Following Section 5.3 let us consider a number of optimization methods, for contact discontinuity localization in one-dimensional flows, which are based on the minimization of certain functions. Introduce the functions

$$J(\zeta^{n+1}) = (\dot{\zeta}_{\tau_1}(t^{n+1}) - u(\zeta^{n+1}, t^{n+1}))^2, \tag{5.6.6}$$

$$J_1(\zeta^{n+1}) = [\dot{\zeta}_{\tau_1}(t^{n+1}) - u(\zeta^{n+1}, t^{n+1})]^2$$
$$+ \alpha_1 R_1(p_x^2) + \alpha_2 R_2(u_x^2) + \alpha_3 R_3(\rho_x^2) \tag{5.6.7}$$

taking into account formulas (5.6.2), (5.6.4). In (5.6.6), (5.6.7) $\dot{\zeta}_{\tau_1}(t^{n+1})$ is a difference approximation of the derivative $d\zeta/dt$; it can be taken, for example, in the form (5.3.3).

Note that all functionals (5.6.2), (5.6.4) introduced above, as well as the functions (5.6.6), (5.6.7), contain explicitly the variable ζ sought for. Therefore, based on a consideration similar to that of Section 5.5, we conclude that the contact discontinuity trajectory obtained by numerical minimization of the above functionals and functions will be a monotone function of time, unlike the case of using Algorithm II of the contact discontinuity differential analyzer described in Section 4.1 (see Figure 4.18).

In Figure 4.13(b) we have marked by black points the contact discontinuity trajectory obtained by minimization of the function (5.6.7) in which

$$R_1(z) = R_2(z) = z, \qquad R_3(z) = \exp[30/(1 + z)],$$
$$\alpha_1 = \alpha_2 = \alpha_3 = 1. \tag{5.6.8}$$

The minimization of a function $J_1(\zeta^{n+1})$ was carried out in the interval

$$\zeta(t^{n+1} - \tau_1) - \beta h \leq \zeta^{n+1} \leq \zeta(t^{n+1} - \tau_1) + \beta h.$$

Here it was taken into account that during the time $\tau_1 = \beta\tau$ the contact discontinuity cannot move by a distance larger than βh, since the step τ was computed for the Lax–Wendroff scheme [5.30] with regard to the Courant–Friedrichs–Lewy criterion $(\tau/h) \cdot \max(|u| + c) \leq 1$. In the computation whose result has been presented in Figure 4.13(b) the accuracy of determining the abscissa of the minimum of the function $J_1(\zeta^{n+1})$ was taken to be equal to $0.05h$; $\beta = 3$ in the formula $\tau_1 = \beta\tau$.

One should take into account two limitations of the above optimization methods before using them for the localization of contact discontinuities. First, it is necessary for the implementation of these methods to know the location x of a contact discontinuity at the initial moment of time $t = 0$. Second, the contact discontinuity whose temporal evolution is to be observed should persist during all of the time interval $0 \leq t \leq T$, because this was implicitly assumed in the process of derivation of the functional (5.6.4) and of the function (5.6.7). Thus, the applicability of the optimization algorithms considered in this section depends substantially on the availability of the above *a priori* information on the gas flow under study.

5.7. An Optimization Method for the Localization of Weak Discontinuities

Weak discontinuity surfaces may be present in the flows of an inviscid, non-heat-conducting gas as well as strong discontinuities. The density, pressure, and velocity remain continuous across the weak discontinuity, while at least one of the first derivatives of these quantities, with respect to a spatial variable or to the time t, undergoes a jump [5.31]. Consider a question on the construction of an optimization algorithm for the extraction of weak discontinuities in the numerical solutions obtained with the aid of finite-difference shock-capturing schemes. Let $\psi(x_1, \ldots, x_N, t) = 0$ be the equation of a weak discontinuity surface in the solution of the Euler equation system governing the inviscid gas flow, where $x_1, \ldots, x_N$, $N \geq 1$, are spatial coordinates. Following [5.32], [5.33], let us write this system in the form

$$\partial\rho/\partial t + u_i\rho_{,i} + \rho u_{i,i} = 0;$$

$$\rho(\partial u_i/\partial t + u_k u_{i,k}) + p_{,i} = 0; \qquad (5.7.1)$$

$$\partial p/\partial t + u_i p_{,i} + \rho c^2 u_{i,i} = 0,$$

where c^2 is the square of the isentropic sound speed. The following notations are used in (5.7.1): $\rho_{,i} = \partial\rho/\partial x_i$; the summation from 1 to N is carried out over repeated indices; and $u_1, \ldots, u_N$ are the components of the gas velocity vector along the axes $x_1, \ldots, x_N$, respectively.

Let the superscript $(+)$, by any of the quantities $\rho_{,i}$, $p_{,i}$, and $u_{i,i}$, denote that these quantities have been calculated on that side of the surface $\psi = 0$ which corresponds to the gas state before the front of a weak discontinuity; similarly, the superscript $(-)$ corresponds to the state behind the weak discontinuity front. Let us introduce the jump for the first-order derivatives

$$[f_{,i}] = f_{,i}^- - f_{,i}^+. \qquad (5.7.2)$$

Let n_i and G be the components of the unit vector of a normal to the surface $\psi(x_i, t) = 0$, that is,

$$n_i = \psi_{,i}/|\text{grad }\psi|, \quad i = 1, \ldots, N; \qquad G = (-\partial\psi/\partial t)/|\text{grad }\psi|. \quad (5.7.3)$$

Then the kinematic compatibility conditions on a weak discontinuity in an ideal gas may be written in the form [5.32]–[5.34]

$$[u_{i,j}] = Hn_in_j; \qquad [\partial u_i/\partial t] = -GHn_i; \qquad (5.7.4)$$

$$[p_{,i}] = \pi n_i; \qquad [\partial p/\partial t] = -G\pi; \qquad (5.7.5)$$

$$[\rho_{,i}] = R n_i; \qquad [\partial \rho/\partial t] = -GR. \qquad (5.7.6)$$

In (5.7.4)–(5.7.6) H, π, and R are scalar functions of the variables $x_1, \ldots, x_N$, t which characterize quantitatively the jumps of the derivatives $u_{i,j}$, $p_{,i}$, $\rho_{,i}$ across a weak discontinuity surface.

Besides the kinematic compatibility conditions (5.7.4)–(5.7.6), dynamic compatibility conditions also take place on the weak discontinuity surfaces. The general procedure for the derivation of these conditions is presented in [5.34]: first we write equations of the form (5.7.1) for the functions ρ^-, u_i^-, p^-, then we write equations for ρ^+, u_i^+, p^+. After that the difference of these equations is taken on the surface $\psi(x_i, t) = 0$:

$$[\partial \rho/\partial t] + u_i[\rho_{,i}] + \rho[u_{i,i}] = 0; \qquad (5.7.7)$$

$$\rho([\partial u_i/\partial t] + u_k[u_{i,k}]) + [p_{,i}] = 0; \qquad (5.7.8)$$

$$[\partial p/\partial t] + u_i[p_{,i}] + \rho c^2[u_{i,i}] = 0. \qquad (5.7.9)$$

Equations (5.7.7)–(5.7.9) are called dynamic compatibility conditions on a weak discontinuity in an ideal gas flow. Denote by U the relative velocity of the propagation of the surface $\psi = 0$, that is, U is the vector,

$$U = \{G - u_1 n_1, \ldots, G - u_N n_N\}. \qquad (5.7.10)$$

Note one consequence which can easily be obtained from the kinematic compatibility conditions (5.7.4)–(5.7.6) and from the dynamic compatibility conditions (5.7.7)–(5.7.9) [5.32], [5.33]:

$$U^2 = c^2, \qquad (5.7.11)$$

that is, the magnitude of the vector U is always equal to the sound speed.

Now consider the kinematic and dynamic compatibility conditions on a weak discontinuity in the particular case of one space variable, that is, at $N = 1$. For brevity, the subscript "1" by the space variable x_1 will not be written in the following. Let the exact weak discontinuity trajectory in the (x, t)-plane be given by the equation

$$x = \xi(t). \qquad (5.7.12)$$

Then the function $\psi(x, t)$ entering (5.7.3) has the form

$$\psi(x, t) = x - \xi(t),$$

consequently,

$$n_1 = (\partial \psi/\partial x)/|\text{grad } \psi| = 1/|\text{grad } \psi|,$$

$$G = \dot{\xi}(t)/|\text{grad } \psi|, \qquad |\text{grad } \psi| = (1 + \dot{\xi}(t)^2)^{0.5}, \qquad (5.7.13)$$

$$\dot{\xi}(t) \equiv d\xi(t)/dt.$$

Taking into account (5.7.13), we can write the kinematic compatibilty conditions on a weak discontinuity in one-dimensional flow as

$$[\partial u/\partial x] = Hn_1^2, \qquad [\partial u/\partial t] = -\dot{\xi}(t)H/|\text{grad } \psi|^2, \qquad (5.7.14)$$

$$[\partial p/\partial x] = \pi/|\text{grad } \psi|, \qquad [\partial p/\partial t] = -\dot{\xi}(t)\pi/|\text{grad } \psi|, \qquad (5.7.15)$$

$$[\partial\rho/\partial x] = R/|\text{grad } \psi|, \qquad [\partial\rho/\partial t] = -\dot{\xi}(t)R/|\text{grad } \psi|. \qquad (5.7.16)$$

From (5.7.14)–(5.7.16) we can derive the formulas which do not contain the quantities H, π, R:

$$\dot{\xi}(t) = -[\partial u/\partial t]/[\partial u/\partial x], \qquad (5.7.17)$$

$$\dot{\xi}(t) = -[\partial p/\partial t]/[\partial p/\partial x], \qquad (5.7.18)$$

$$\dot{\xi}(t) = -[\partial\rho/\partial t]/[\partial\rho/\partial x]. \qquad (5.7.19)$$

Let us now turn to the case of a one-dimensional Riemann problem, when there is a centered rarefaction wave in a gas flow arising at $t > 0$. Assume that this wave emanates from the point $x = 0$ at $t = 0$. As is known, the solution of this problem is self-similar and it depends on the variable $\eta = x/t$. Let $\rho = \tilde{\rho}(\eta)$ in (5.7.19). Then we obtain from (5.7.19) by substituting $\rho = \tilde{\rho}(\eta)$:

$$\dot{\xi}(t) = \frac{-[d\tilde{\rho}/d\eta]\cdot(-\xi(t)/t^2)}{[d\tilde{\rho}/d\eta]\cdot(1/t)} = \frac{\xi(t)}{t}. \qquad (5.7.20)$$

Thus, we get from (5.7.20) the differential equation

$$\dot{\xi}(t) = \xi(t)/t$$

for the determination of the weak discontinuity trajectory. This equation may be easily integrated; its solution is $\xi(t) = C\cdot t$, where C is an integration constant. Thus we have obtained, by using the kinematic compatibility condition (5.7.19) and the fact of the self-similarity of the Riemann problem solution, that the weak discontinuity trajectories in the solution of the one-dimensional Riemann problem are straight lines in the (x, t)-plane. This fact was indicated earlier, for example, in [5.9] using a different technique.

Let us write the dynamic compatibility conditions on a weak discontinuity at $N = 1$, employing formulas (5.7.7)–(5.7.9):

$$[\partial\rho/\partial t] + u[\partial\rho/\partial x] + \rho[\partial u/\partial x] = 0; \qquad (5.7.21)$$

$$\rho([\partial u/\partial t] + u[\partial u/\partial x]) + [\partial p/\partial x] = 0; \qquad (5.7.22)$$

$$[\partial p/\partial t] + u[\partial p/\partial x] + \rho c^2[\partial u/\partial x] = 0. \qquad (5.7.23)$$

The derivatives with respect to t entering (5.7.17)–(5.7.19) can be expressed by the derivatives with respect to x with the aid of (5.7.21)–(5.7.23). As a result, we obtain the formulas

$$\dot{\xi}(t) = u + \rho[\partial u/\partial x]/[\partial\rho/\partial x]; \qquad (5.7.24)$$

$$\dot{\xi}(t) = u + (1/\rho)[\partial p/\partial x]/[\partial u/\partial x]; \qquad (5.7.25)$$

$$\dot{\xi}(t) = u + \rho c^2[\partial u/\partial x]/[\partial p/\partial x]. \qquad (5.7.26)$$

Let us now return to the one-dimensional Riemann problem. Suppose that there is a centered rarefaction wave in the flow configuration arising at $t > 0$. Let $x = \zeta(t)$ be some approximation to the trajectory of one of the two weak discontinuities existing in the head and in the tail of the rarefaction wave. Let us construct, with regard to (5.7.24)–(5.7.26), a nonnegative functional

$$I(\zeta) = \int_0^T (\dot{\zeta}(t) - J(\zeta(t), t))^2 \, dt, \qquad (5.7.27)$$

where $T > 0$ is a given value

$$J(\zeta(t), t) = u + \alpha_1 \rho [\partial u/\partial x]/[\partial \rho/\partial x]$$
$$+ (\alpha_2/\rho)[\partial p/\partial x]/[\partial u/\partial x] + \alpha_3 \rho c^2 [\partial u/\partial x]/[\partial p/\partial x]. \qquad (5.7.28)$$

In (5.7.28) α_1, α_2, α_3 are nonnegative weight constant coefficients, and $\alpha_1 + \alpha_2 + \alpha_3 = 1$. All three relationships (5.7.24)–(5.7.26) are valid on the exact weak discontinuity trajectory $x = \xi(t)$, therefore, it is obvious that $I(\xi) = 0$.

Now consider the question of the numerical approximation of the jumps of the derivatives $[\partial u/\partial x]$, $[\partial p/\partial x]$, $[\partial \rho/\partial x]$ entering formula (5.7.28). Suppose that the numerical solution profiles are monotonous in the neighborhood of weak discontinuities at $t > 0$. The corner point in the u, p, ρ profiles in the neighborhood of weak discontinuities is absent, as a consequence of the effects of "smearing" which takes place when using the shock-capturing difference schemes (see Figure 5.6). Therefore, it is necessary to use some special difference approximations for the derivatives $\partial u^-/\partial x$, $\partial u^+/\partial x$, $\partial p^-/\partial x$, $\partial p^+/\partial x$, $\partial \rho^-/\partial x$, $\partial \rho^+/\partial x$ in the neighborhood of weak discontinuities. Since the values of the quantities u, p, ρ on the weak discontinuity were calculated in the difference solution with some error (see Figure 5.7), the quantities, for example, $\partial u^-/\partial x$ and $\partial u^+/\partial x$, were approximated by using one-sided differences of the

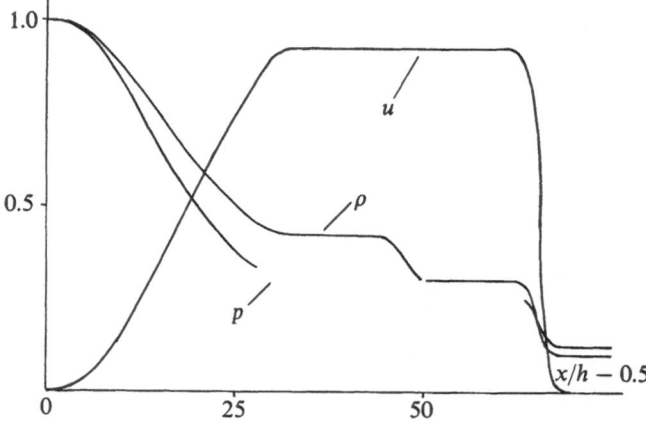

Figure 5.6. Difference solution profiles in the Riemann problem.

Figure 5.7. ($\bullet\!\!-\!\!\bullet\!\!-\!\!\bullet$)—difference solution.

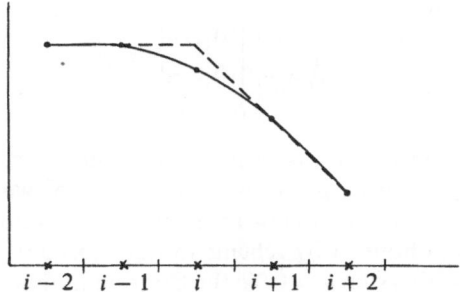

form

$$\partial u^- / \partial x = \begin{cases} (u_{i-1} - u_{i-2})/h, & \zeta(t) \geq 0, \\ (u_{i+2} - u_{i+1})/h, & \zeta(t) < 0; \end{cases}$$

$$\partial u^+ / \partial x = \begin{cases} (u_{i+2} - u_{i+1})/h, & \zeta(t) \geq 0, \\ (u_{i-1} - u_{i-2})/h, & \zeta(t) < 0. \end{cases}$$

(5.7.29)

The implementation of formulas of the form (5.7.29) enables us to reproduce more correctly the specific slopes of the numerical solution profiles in the neighborhood of a weak discontinuity (see the broken line in Figure 5.7).

Since formulas of finite differences are used for the computation of the jumps $[\partial u/\partial x]$, $[\partial p/\partial x]$, $[\partial\rho/\partial x]$, the relations of these jumps (for example, $[\partial u/\partial x]/[\partial\rho/\partial x]$) are very sensitive to the oscillations of the numerical solution, which can occur in the neighborhood of weak discontinuities (see, for example, Figure 4.2(b)). In this connection, we have aimed at the use of a difference scheme which ensured the monotonicity of difference solution profiles in the neighborhood of weak discontinuities. For this purpose, we took the so-called hybrid difference scheme approximating the Euler equation system (1.11), (1.9)

$$\bar{w}_{i+1/2}^{n+1} = (1/2)(w_i^n + w_{i+1}^n) - (\tau_{n+1}/h)(\varphi_{i+1}^n - \varphi_i^n);$$

$$w_i^{n+1} = w_i^n - (\tau_{n+1}/h)(\bar{\varphi}_{i+1/2}^{n+1} - \bar{\varphi}_{i-1/2}^{n+1})$$

$$+ (1/8)(\theta_{i+1/2}^n(w_{i+1}^n - w_i^n) - \theta_{i-1/2}^n(w_i^n - w_{i-1}^n)).$$

(5.7.30)

In (5.7.30) τ_{n+1} is the time step of a scheme, as was computed by the formula

$$\tau_{n+1} = K \cdot h \Big/ \max_i (|u_i^n| + c_i^n), \qquad n = 0, 1, \ldots,$$

(5.7.31)

where K is a given constant (Courant number), $0 < K \leq 1$, and c_i^n is the local speed of sound. The switch function θ in (5.7.30) is usually chosen in such a way that it is small in the domains of small solution gradients, and it is close to 1 in the domains of strong discontinuities of the solution. Following [5.35] we have computed $\theta_{i+1/2}^n$ by the formula

$$\theta_{i+1/2}^n = \max(\hat{\theta}_i, \hat{\theta}_{i+1}),$$

where

$$\hat{\theta}_i = \begin{cases} \dfrac{||\Delta_{i+1/2}| - |\Delta_{i-1/2}||}{||\Delta_{i+1/2}| + |\Delta_{i-1/2}||}, & |\Delta_{i+1/2}| + |\Delta_{i-1/2}| > \varepsilon, \\ \\ 0, & |\Delta_{i+1/2}| + |\Delta_{i-1/2}| \le \varepsilon. \end{cases} \quad (5.7.32)$$

In (5.7.32) $\Delta_{i+1/2} = \rho_{i+1}^n - \rho_i^n$, and ε is the measure of a negligible variation of the density ρ; by analogy with [5.35] we have used $\varepsilon = 0.01 \cdot \max_i |\Delta_{i+1/2}|$. In Figure 5.6 we show the profiles of the velocity u, the pressure p, and the density ρ, obtained by scheme (5.7.30)–(5.7.32) on a uniform mesh along the x-axis with a step $h = 0.02$; the Courant number $K = 0.9$ in (5.7.31); and the moment of time $t = T = 0.4$ in Figure 5.6.

Suppose that u, p, ρ in (5.7.24)–(5.7.26) represent the exact solution of a one-dimensional Riemann problem. There are no jumps of the derivatives $\partial u/\partial x$, $\partial p/\partial x$, $\partial \rho/\partial x$ inside the rarefaction wave, therefore, we obtain uncertainties of the form 0/0. Let us show that there are in reality no such uncertainties when using the difference approximations of the form (5.7.29). Take, for definiteness, the case $\dot{\zeta}(t) < 0$. Then we have from (5.7.29)

$$[\partial u/\partial x] = \partial u^-/\partial x - \partial u^+/\partial x$$

$$= (1/h)(u_{i+2} - u_{i+1} - u_{i-1} + u_{i-2}) = 3hu_{xx} + O(h). \quad (5.7.33)$$

Since the profiles of $u(x, t)$, $p(x, t)$, and $\rho(x, t)$ at fixed t have a nonzero curvature in a centered rarefaction wave (see Figure 1.1), we obtain from (5.7.33) that $[\partial u/\partial x] \ne 0$, as the result of a computation by formulas (5.7.29).

The investigation of the convexity of the functional (5.7.27), (5.7.28) may be carried out in the same way as was done in Section 5.4 in the case of a shock wave. We again obtain a Hermitean quadratic form (5.4.17). It is easy to be convinced with the aid of the Sylvester criterion that this form is not nonnegative, so that the functional (5.7.27), (5.7.28) is generally not convex. From this it follows that the initial approximation $\zeta_0(t)$, for the minimization of the functional (5.7.27), (5.7.28) by the steepest descent method, should be set in such a way that the trajectory $x = \zeta_0(t)$ lies in the neighborhood of the trajectory $x = \xi(t)$ sought for, and which does not contain other weak discontinuities in the solution of a one-dimensional gas-dynamical problem. Let x_f^0 be the given position of a weak discontinuity on the x-axis at $t = 0$. To study the behavior of the integrand in (5.7.27), at fixed $t > 0$, we have computed the values of this integrand for different $x = \zeta$ by the formula

$$F(\zeta) = ((\zeta - x_f^0)/t - J(\zeta(t), t))^2, \qquad 0 \le \zeta \le x_f^0 + 2h. \quad (5.7.34)$$

In Figure 5.8 we present the graph of the function $\log_{10} F(\zeta)$ at $t = 0.3$ and at $\alpha_1 = 1$, $\alpha_2 = \alpha_3 = 0$ in (5.7.28); the numerical solution of the Euler equation system was determined by scheme (5.7.30)–(5.7.32), $K = 0.9$ in (5.7.31). We can see in Figure 5.8 that the function (5.7.34) has two minima; their positions on the x-axis correspond to the positions of the two weak discontinuities in the solution of the Riemann problem under study.

Figure 5.8. The graph of the function $\log_{10} F(\zeta)$.

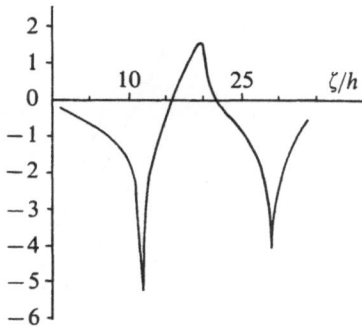

Taking into account the above-established nonconvexity of the basic functional (5.7.27), (5.7.28), we have set the initial approximation $\zeta_0(t)$ to the weak discontinuity trajectory $x = \zeta(t)$ in the form

$$\zeta_0(t) = t \cdot k \cdot 0.5 \cdot (u_1 + u_2 - c_1 - c_2), \qquad (5.7.35)$$

where u_1, u_2, c_1, c_2 are the values of the gas velocity and sound velocity on different sides of a discontinuity at $t = 0$; the constant coefficient k in (5.7.35) was set in such a way that the initial trajectory $x = \zeta_0(t)$ proved to be located in the "domain of attraction" of one of the two weak discontinuity trajectories in the problem under study. As a convergence criterion for the iterations by the steepest descent method (Algorithm 1 from Section 5.4) we have used the inequality

$$\max_t |(\zeta_{k+1}(t) - \zeta_k(t))/h| \leq 0.01,$$

where $k = 0, 1, \dots$. In Figure 5.9 we present the results of the application of the above optimization algorithm for the localization of weak discontinuities; $\alpha_1 = 1$, $\alpha_2 = \alpha_3 = 0$ in (5.7.28). The solid lines I–IV, which show the exact positions of the weak and strong discontinuities, are the left boundary of the rarefaction fan (line I), the right boundary of the rarefaction fan (line II), the contact discontinuity trajectory (line III), and the shock wave front trajectory (line IV). The broken lines I' and II' are the initial approximations $\zeta_0(t) = x$ obtained at $k = 0.8$ and at $k = 0.2$ in (5.7.35), respectively. Six iterations were needed for convergence of the steepest descent algorithm in the case of the left boundary I. Let $I_k = I(\zeta_k(t))$, $k = 0, 1, \dots$. In the case under consideration, we have $I_0 = 0.50983$ and $I_5 = 4.85347 \cdot 10^{-3}$. In the case of the right boundary II, four iterations were needed for the convergence of the process for the functional (5.7.27), (5.7.28) minimization; in this case, $I_0 = 0.28975$ and $I_3 = 0.01086$. It should be noted that the error in determining the locations of weak discontinuities by the above optimization method increases with time t, similar to the case of shock wave localization; however, even at $t = T$, this error has a small magnitude less than $h/2$. For example, at $t = T = 0.4$ and $h = 0.02$, the exact position of the left boundary I is $\xi(T) = 6.836h$; and for boundary

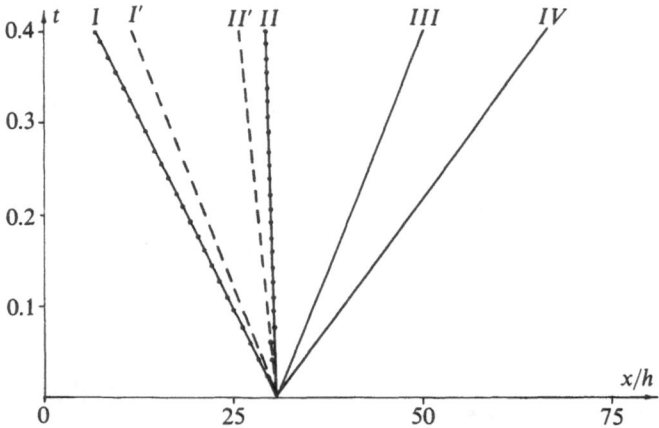

Figure 5.9. ($\cdots$)—weak discontinuity trajectories obtained by optimization method.

II, $\xi(T) = 29.095h$. We have obtained, by the above steepest descent method, that $\zeta_5(T) = 6.481h$ in the case of boundary I, and $\zeta_3(T) = 29.131h$ in the case of boundary II (see also Figure 5.9).

Thus, the above-presented optimization approach to the localization of weak discontinuities, enables us to solve successfully the problem of the extraction of this type of discontinuity on the basis of shock-capturing numerical solutions of one-dimensional problems in inviscid gas dynamics. Here we use *a priori* information on the presence of weak discontinuities in the computational domain during the period of time $0 \leq t \leq T$.

5.8. A Generalization of the Miranker–Pironneau Method for the Case of Polar Coordinates in a Filtration Problem

The processes of filtration of a two-phase fluid (without account being taken of the capillary and gravitation forces in a plane domain) are governed by the equations [5.5], [5.36]

$$\text{div}[f_0(s) \, \text{grad} \, p] = 0, \qquad (5.8.1)$$

$$m \, \partial s/\partial t = \text{div}[f_0(s)\varphi(s) \, \text{grad} \, p], \qquad (5.8.2)$$

where

$$f_0(s) = k[f_1(s)/\mu_1 + f_2(s)/\mu_2],$$

where $k = \text{const}$ is the absolute permeability of a porous medium, $f_i(s)$ and μ_i are the functions of relative permeability and of dynamic viscosity of the ith phase, respectively, $i = 1, 2$, p is the pressure of the two phase-fluid, $s \equiv s_2$ is the saturation (fraction of water in total fluid), m is the porosity of a medium, $\varphi(s) = \mu_0 f_2(s)/[f_1(s) + \mu_0 f_2(s)]$ is the distribution function of the displacement phase, and $\mu_0 = \mu_1/\mu_2$. A systematic consideration of the formulations of two-phase filtration problems in the case of incompressible fluids, with no regard for capillary and gravitation forces in a planar domain, has been carried out in [5.37]. It was shown that the equation (5.8.1) for p is elliptic and the boundary conditions for the pressure should be given as for an elliptic equation, and that the equation (5.8.2) for s is hyperbolic and its solution consists of the motions starting on the lines of initial saturation distribution and on the contours of displacement phase force-pumping. However, by virtue of the fact that the velocities of motion of the lines of saturation isovalues are, as a rule, nonmonotone functions of s for the actual functions $f_1(s)$, $f_2(s)$, there can occur a contact between the different lines of saturation isovalues and an intersection of these lines. To eliminate the arising non-single-valuedness of the solution the formation of certain lines is admitted along which the saturation changes spasmodically [5.37]. However, the pressure p remains a continuous function. Thus, in the solution of system (5.8.1), (5.8.2) only one function, the saturation function s, can undergo strong discontinuities. In this sense system (5.8.1), (5.8.2) is simpler than the Euler equations (1.30) governing the two-dimensional flow of an inviscid compressible non-heat-conducting gas, since four functions ρ, u, v, ε undergo a discontinuity at the front of a gas-dynamical shock wave in two-dimensional flow. In this connection it was decided to carry out first an extension of the above-presented Miranker–Pironneau method (see Sections 5.5 and 5.6) for the two-dimensional case in the example of a simpler system (5.8.1), (5.8.2). In [5.38], [5.39] the method of Miranker and Pironneau was applied to the localization of a jump in saturation functions in the numerical solutions of one-dimensional filtration problems.

5.8.1. Variational Formulation of a Problem on the Localization of the Saturation Function Discontinuity

The discontinuity line is characterized at each moment of time t by certain limiting values $s^+(x, y, t)$ and $s^-(x, y, t)$ of the saturation on both sides of a discontinuity. These values are interconnected by compatibility conditions. One of the compatibility conditions is the condition for the continuity of an individual flux of a displacement phase which may be written in the form

$$D_n = (w_{n2}^+ - w_{n2}^-)/[m(s^+ - s^-)], \qquad (5.8.3)$$

where D_n is a normal velocity of the discontinuity line propagation, and $w_{n2} = -(k/\mu_2) f_2(s) \, \partial p/\partial n$ is the flux of a displacement phase along the normal to the discontinuity line. In the particular case when only the displacement phase is pumped through the contour of force-pumping, and the porous medium initially contains an aggregated displacement phase, the saturation jump arises at the time of force-pumping. The discontinuity line is formed thereat as a line with constant saturation value s_* determined from the relationship

$$\varphi'(s_*) = \varphi(s_*)/(s_* - \underline{s}),\qquad(5.8.4)$$

where $\underline{s}$ is a lower limiting value of s, and $\varphi'(s) = d\varphi/ds$ [5.37].

The filtration problem considered in the following is such that if the equation of the saturation jump front is sought in the form $y = Y(x, t)$ or $x = X(y, t)$, then at fixed t both of these functions can be non-single-valued in some intervals of a spatial coordinate axis. In this connection it is reasonable to use polar coordinates (r, θ), where r is the radius vector and θ is a polar angle. Let the equation

$$r = R(\theta, t)\qquad(5.8.5)$$

describe a line of saturation jump in the (r, θ)-plane at each fixed t. This line in the problem of the surface force-pumping of water (which is considered in Section 5.8.3) has, in the case of a five-point system of the wells disposition, the form of a "tongue" stretched in the direction of the xOy angle bisector (see the dash–dot line in Figure 5.11). Thus, in the case under study, the function $R(\theta, t)$ in (5.8.5) is single-valued. Let us construct the basic functional for determining a discontinuity line by using the condition (5.8.3). Let us introduce the notation $g(r, \theta, t) = r - R(\theta, t)$. Then the discontinuity velocity D_n is computed by a formula [5.40]

$$D_n = (\partial R/\partial t)/|\text{grad } g|.$$

It is clear that the equality

$$(\partial R/\partial t)/|\text{grad } g| = \Phi(R(\theta, t), \theta, t),\qquad(5.8.6)$$

where

$$\Phi(R(\theta, t), \theta, t) = (w_{n2}^+ - w_{n2}^-)/[m(s^+ - s^-)]$$

in accordance with (5.8.3), takes place only when the equation $r = R(\theta, t)$ describes the discontinuity line.

Let the line with the equation $r = Q(\theta, t)$ be an approximation to a discontinuity line. Consider the following nonnegative functional:

$$F(Q) = \int_{t*}^{t_* + \Delta t} U(I(Q, t))\, dt,\qquad(5.8.7)$$

where

$$I(Q, t) = \int_{\theta_1}^{\theta_2} a(Q)[(\partial Q/\partial t)|\text{grad } g| - \Phi(Q, \theta, t)]^2\, d\theta,\qquad(5.8.8)$$

t_*, Δt are constant quantities, and $t_0 \le t_* \le t_* + \Delta t \le T$; t_0 and T are the initial and final moments of time, respectively; $[\theta_1, \theta_2]$ is the domain of the polar angle θ variation, $\theta_1 < \theta_2$. The positive function $a(Q)$ has been introduced for the following purposes: first, to ensure the proper dimensionality of the gradient of the functional $F(Q)$ (see below); second, a higher convergence speed of the computational process of the functional (5.8.7) minimization can be achieved on the appropriate choice of the function $a(Q)$ [5.17]. The function $U(z)$ in (5.8.7) is chosen from the requirements $U(z) > 0$, $U'(z) > 0$ for $z > 0$, and $U(0) = 0$. The simplest function $U(z)$ which meets the above requirements is the function $U(z) = z$.

The solution of a nonstationary system (5.8.1) obtained with the aid of finite-difference schemes (see below in Section 5.8.3) is computed sequentially at the times $t^n = n\tau$, $n = 1, 2, \ldots$. According to this the function $Q(\theta, t)$ in the equation $r = Q(\theta, t)$ determining an approximation to the discontinuity line can also be determined for increasing values of $t = n\tau$. Let the values of the function $Q(\theta, t^n)$ be grouped into a table $\{r_i, \theta_i\}$ for some $t = t^n$ where $r_i = Q(\theta_i, t^n)$, $\theta_1 < \theta_2 < \cdots < \theta_N$. Connecting the neighboring points (r_i, θ_i) and (r_{i+1}, θ_{i+1}), $i = 1, \ldots, N - 1$, by straight-line segments we obtain a certain line $r = R^{(n)}(\theta)$. Since during the step $\Delta t = O(\tau)$ each point (r_i, θ_i) moves by a short distance of the order $O(h)$, if $\tau = O(h)$ where h is a characteristic dimension of a cell of the Eulerian computing mesh, then it is convenient for the purpose of the reduction of a needed number of iterations to seek the position of a discontinuity line at the following moment of time $t = t^{n+1}$ in some ε-neighborhood $\Omega(t^n)$ of the line $r = R^{(n)}(\theta)$ where $\varepsilon = O(h)$. Assume in (5.8.7) $t_* = t^n$, $\Delta t = \beta\tau$ where $\beta = O(1)$, say, $\beta = 2$. Then the discontinuity line $r = R(\theta, t)$ at $t_* \leq t \leq t_* + \Delta t$ may be sought as a solution of the following optimization problem

$$\min_Q \{F(Q)|Q(\theta, t^n) = R^{(n)}(\theta), Q(\theta, t) \in \Omega(t^n)\}. \qquad (5.8.9)$$

Let us construct the basic functional using a simpler jump condition (5.8.4). Rewrite condition (5.8.4) in (r, θ) coordinates

$$J(R) \equiv \varphi'(s(R - 0)) - \varphi(s(R - 0))/[s(R - 0) - s(R + 0)] \equiv 0, \quad (5.8.10)$$

where the arguments $R - 0$ and $R + 0$ mean that the value of s is computed behind and before the discontinuity front, respectively. Let us construct the following nonnegative functional:

$$G(Q) = \int_{t^*}^{t_* + \Delta t} \left\{ \int_{\theta_1}^{\theta_2} b(Q)[J(Q)]^2 \, d\theta \right\} dt. \qquad (5.8.11)$$

A positive function $b(Q)$ has been introduced into the functional (5.8.11) for the same purposes as the function $a(Q)$ in the functional (5.8.8). Then the discontinuity surface $r = R(\theta, t)$ may be considered as a solution of an extremum problem for the functional (5.8.11).

When minimizing the functional (5.8.11) a function of two variables θ, t is sought. The numerical solution of two-dimensional minimization problems requires much more expense in computer time than the solution of one-dimensional problems. On the other hand, it is sufficient to know in practical applications the form of the function $r = R(\theta, t)$ only for the moments of time which are interest. Taking this into account we can simplify substantially the above-proposed two-dimensional optimization problem, by reducing the problem of a search for the discontinuity line at a fixed moment of time, to a problem of minimizing the following one-dimensional functionals (we retain the corresponding former notations)

$$F(Q) = \int_{\theta_1}^{\theta_2} a(Q(\theta, t)) [(\partial Q/\partial t)/|\text{grad } g| - \Phi(Q(\theta, t))]^2 \, d\theta,$$

$$G(Q) = \int_{\theta_1}^{\theta_2} b(Q(\theta, t)) [J(Q(\theta, t))]^2 \, d\theta. \qquad (5.8.12)$$

Since these functionals contain only integration with respect to θ, we shall omit for brevity the second argument in the notation of the function $Q(\theta, t)$. The problem of determining the discontinuity line on the basis of the functional (5.8.12) is then formulated as the problem

$$\min_{Q} \{G(Q)|Q(\theta) \in C^1[\theta_1, \theta_2]\}.$$

5.8.2. A Method of Numerical Minimization of the Basic Functional

Following [5.5] we restrict ourselves to the description of a computational method for the minimization of the functional (5.8.12) designed for the determination of a discontinuity line for a fixed moment in time. The second argument of the function $Q(\theta, t)$ is omitted for brevity.

Take a function $Q(\theta)$ and an increment $\Delta Q(\theta)$ to this function from the space $C^1[\theta_1, \theta_2]$. The corresponding increment of the functional may be represented in the form

$$G(Q + \Delta Q) - G(Q) = \int_{\theta_1}^{\theta_2} \nabla G(Q)\Delta Q \, d\theta + O((\Delta Q)^3), \qquad (5.8.13)$$

where

$$\nabla G(Q) = \{2b(Q)J(Q) \, \partial J(Q)/\partial Q + db(Q)/dQ$$
$$\times [J(Q)]^2\} + \{b(Q)J(Q) \, \partial^2 J(Q)/\partial Q^2$$
$$+ b(Q)[\partial J(Q)/\partial Q]^2 + 2(db(Q)/dQ)J(Q)$$
$$\times \partial J(Q)/\partial Q + 0.5d^2 b(Q)/dQ^2 [J(Q)]^2\}\Delta Q. \qquad (5.8.14)$$

Let us choose the increment ΔQ as follows:

$$\Delta Q = -\lambda \nabla G(Q), \qquad (5.8.15)$$

where λ is some nondimensional positive quantity. Substituting the right-hand side of (5.8.15) into equation (5.8.13) we obtain

$$G(Q + \Delta Q) - G(Q) = -\lambda \int_{\theta_1}^{\theta_2} [\nabla G(Q)]^2 \, d\theta + O((\lambda \nabla G)^3). \quad (5.8.16)$$

It follows from (5.8.16) that at a sufficiently small value $|\lambda \nabla G|$ the right-hand side is negative and, consequently, the functional (5.8.12) decreases. Then, in accordance with (5.8.15), we assume

$$Q_{n+1}(\theta) = Q_n(\theta) - \lambda \nabla G(Q_n). \quad (5.8.17)$$

From (5.8.14), (5.8.17) it follows that the function $b(Q)$ in the functional (5.8.12) should be chosen to have a dimension L^2 where L is the dimension of the length.

Let us present an algorithm for the functional (5.8.12) minimization which is based on formulas (5.8.12)–(5.8.17).

Step 1. Choose $Q_0(\theta) \in C^1[\theta_1, \theta_2]$, $\sigma \in (0.5, 0.8)$, and set $n = 0$.
Step 2. Compute the value $H(Q_n) = -\nabla G(Q_n)$ by formulas (5.8.14) and (5.8.10).
Step 3. If $H(Q_n) = 0$, then stop; otherwise, go over to Step 4.
Step 4. Choose $\lambda > 0$.
Step 5. Compute

$$\Delta = G(Q_n + \lambda H(Q_n)) - G(Q_n) + 0.5\lambda \|H(Q_n)\|^2_{L^2[\theta_1, \theta_2]}. \quad (5.8.18)$$

Step 6. If $\Delta \leq 0$, then assume $\lambda_n = \lambda$ and go over to Step 7; otherwise, set $\lambda = \sigma\lambda$ and go over to Step 5.
Step 7. Compute $Q_{n+1}(\theta)$ by formula (5.8.17).
Step 8. Check the satisfaction of a condition

$$\max_{\theta} \left[|Q_{n+1}(\theta) - Q_n(\theta)|/h \right] < \delta, \quad (5.8.19)$$

where δ is a user-specified constant, $0 \leq \delta \leq 1$. If the condition (5.8.19) is not satisfied, set $n = n + 1$ and go back to Step 2.

Under the above conditions on input data the presented steepest descent algorithm is convergent [5.17].

Consider now a question on the numerical realization of an algorithm for the basic functional (5.8.12) minimization. Suppose that an initial- and boundary-value problem formulated for the equation system (5.8.1), (5.8.2) has been computed on a main Eulerian difference grid in the (x, y)-plane. Let the number "i" of a difference cell be calculated along the x-axis, and the number j along the y-axis. Subdivide the interval $[\theta_1, \theta_2]$ into N_θ equal portions and let $\bar{\theta}_\alpha = \alpha \Delta \theta + \theta_1$, $\Delta \theta = (\theta_2 - \theta_1)/N_\theta$, $\alpha = 0, 1, \ldots, N_\theta$. Take at arbitrary α, $0 \leq \alpha \leq N_\theta$, the point $(Q_n(\bar{\theta}_\alpha), \bar{\theta}_\alpha)$. Let the nearest node of a rectangular grid have the indices (i_α, j_α). The derivative $\partial J/\partial Q$ entering (5.8.14) is computed by the formula

$$\partial J(Q_n(\bar\theta_\alpha), \bar\theta_\alpha)/\partial Q = (\partial J/\partial x)_{i_\alpha, j_\alpha} \cos \bar\theta_\alpha + (\partial J/\partial y)_{i_\alpha, j_\alpha} \sin \bar\theta_\alpha,$$

where the x- and y-derivatives are approximated by using two-point differences. Note also that in the numerical experiments carried out the simplest approximations of the functional (5.8.12) and the gradient (5.8.14), being analogous to formulas (5.4.12), (5.4.13), have been used. The above-mentioned formulas for the functional (5.8.12) and the gradient (5.8.14) have an error $O(\Delta\theta)$ and $O(h_x) + O(h_y)$, respectively, where h_x and h_y are the steps of a uniform difference grid in the (x, y)-plane.

5.8.3. Computational Examples

Let the filtration domain be covered by a five-point system of wells [5.36]. Consider the square 1232′ (see Figures 5.10 and 5.11) in corner 1 of which a force-pumping well Γ_p is located which has a contour 4′54. In corner 2 an operation well Γ_0 is located which has a contour 676′. The diagram of the flow in such a square gives an idea of the flow picture in the overall filtration domain.

Assume further that the force-pumping well supplies only the displacement phase with a volume rate $W_0(t)$, and a constant pressure $p_{op.}$ is provided by the operation well; the symmetry condition is given for the remaining part of the filtration domain boundary. We shall assume further that the porous medium contains, at the initial moment of time $t = t_0 \geq 0$, an aggregated

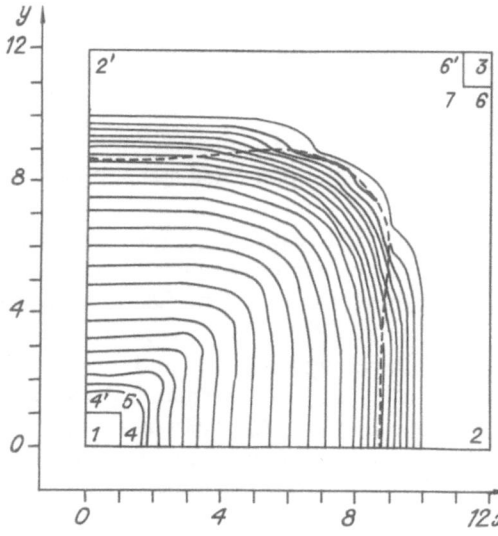

Figure 5.10. The lines of constant saturation in the saturation field.

Figure 5.11. Localization of the front of saturation discontinuity.

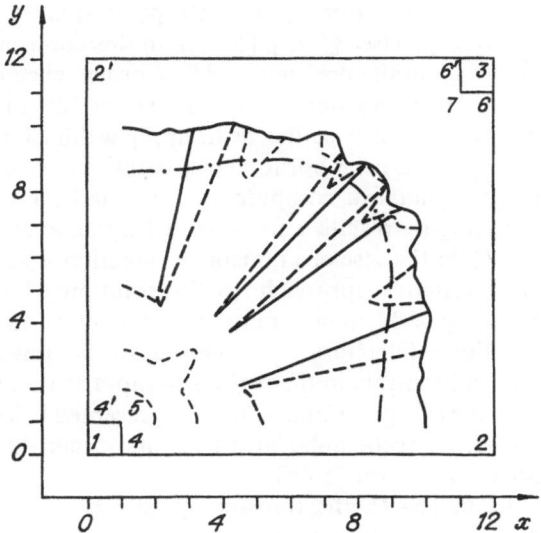

displacement phase. Then the following initial and boundary conditions are formulated for the functions sought:

$$\text{grad}_n \, p|_{\Gamma_p} = -W_0(t)/(k/\mu_2);$$

$$p|_{\Gamma_0} = p_{op.}, \qquad \text{grad}_n p|_{\Gamma_{426}} + \Gamma_{6'2'4'} = 0; \qquad (5.8.20)$$

$$s|_{D_{1232'}} - D_{1454', t=t_0} = \underline{s}, \qquad s|_{\Gamma_p, t \geq t_0} = \bar{s};$$

where we have denoted by $D_{1232'}$, $D_{1454'}$ and Γ_{426}, $\Gamma_{6'2'4'}$ the corresponding domains and contours, $\bar{s}$ is the upper limiting value of saturation.

The numerical computation of the problem (5.8.1), (5.8.2), (5.8.20) was carried out by an algorithm proposed in [5.41]–[5.43]. Equation (5.8.1) was solved by an iteration scheme of alternating directions [5.44]. A constant initial pressure was set throughout the computation domain before starting the iteration process. The iteration process for the determination of the pressure field was continued until the condition $\max_{i,j} |p_{ij}^{\omega+1} - p_{ij}^{\omega}| \leq \varepsilon_p$ was satisfied where ω is the iteration number.

While choosing a difference scheme for the integration of equation (5.8.2) it is necessary to take into account the fact that the function $\varphi(s)$ is not convex (see, for example, [5.42], [5.43]. It was shown in [5.45] that the nonmonotone schemes, such as the Lax–Wendroff scheme, may converge to the solutions which do not satisfy the condition of the entropy increase across a discontinuity in the case of a nonconvex function $\varphi(s)$ in the equation $s_t + (\varphi(s))_x = 0$. On the other hand, only the first-order schemes can be monotone schemes, as was shown, for example, in [5.45]. A shortcoming of these schemes is large numerical diffusion. But this same diffusion plays a positive role: its presence ensures satisfaction of the entropy inequality in the case of a nonconvex function $\varphi(s)$ (see [5.45], [5.46]). In this connection, a conclusion was drawn in [5.46] on the preference of first-order schemes which prevent the appearance of spurious weak solutions. The results of a recent investigation [5.47] indicate that it is sufficient to apply a diffusion of the first order $O(h)$ only locally; in the neighborhood of discontinuities to satisfy the entropy inequality in gas-dynamical computations. It has been communicated in [5.46] that encouraging results were obtained by an extension of the HARTEN method [5.47] for the case of a nonconvex function $\varphi(s)$. As was pointed out in [5.46], an alternative approach, to the solution of a problem on increasing the accuracy of difference solutions of the filtration problems, consists of the use of a finite-difference grid, adapting to the flow in such a way that the mesh lines concentrate in the regions of rapid spatial variation of the solution.

In the computational examples presented below equation (5.8.2) was solved on a fixed rectangular grid by a first-order difference scheme with reduced orientation error [5.48].

As in [5.49], the functions $f_1(s) = (1 - s)^2$, $f_2(s) = s^2$ were used in the numerical computations. For these functions $\underline{s} = 0$, $\bar{s} = 1$. The saturation value behind the discontinuity front at $\mu_0 = 10$ (determined from the relationship (5.8.4)) is equal to $s_* = 0.3015$. In Figure 5.10, solid lines denote the lines of saturation isovalues in the saturation field at some time $t > t_0$ [5.5]. The dashed line in Figures 5.10 and 5.11 shows the line of saturation discontinuity which was obtained by a numerical analytic method, the zonal linearization method of [5.49]. If the line of saturation discontinuity in the numerical solution is determined by concentration of the lines of saturation isovalues, then this line will be determined in Figure 5.10 with an accuracy of about $2h$ (in Figure 5.11 the numbers of the nodes of a difference grid with steps $h_x = h_y = h = 1$ cm are plotted along the coordinate axes). In Figure 5.11 we present the results of locating a line of the saturation function discontinuity by minimization on a numerical solution of the basic functional (5.8.12) with $b(Q) = 1$ cm^2. In accordance with the classical steepest descent method [5.17], [5.50] we have restricted ourselves in the computations by the first item in formula (5.8.14). It should be noted that since the iteration process of the discontinuity line determination is carried out at some finite value of the parameter λ and, accordingly, by the increment ΔQ, the sign of $\nabla G(Q)$ is not, generally speaking, determined only by the sign of the first item in (5.8.14). In some computations this indeed took place, and may be explained by large saturation gradients in the neighborhood of the sought for discontinuity line. As an initial approximation to the sought discontinuity line we have set a circular arc of radius 1 cm. For the considered element of symmetry of the filtration domain the angle θ changes in the interval $[0, \pi/2]$, and the number of subdivisions of this interval has been chosen to be $N_\theta = 40$, $\sigma = 0.5$. The iteration process started with $\lambda = 20$. For the estimation of the accuracy of the discontinuity line, a parameter

$$\delta R_n = \max_\theta |(R_a(\theta, t) - R_{G,n}(\theta, t))/h|$$

has been introduced in [5.5] where $R_a(\theta, t) = r$ is the discontinuity line found by the zonal linearization method of [5.49], and $R_{G,n}(\theta, t) = r$ is a discontinuity line computed by the optimization method where n is the number of iteration. The result of a final iteration is shown in Figure 5.11 by a solid line, and the results of the foregoing iterations are presented by dashed lines. In the computational example of Figure 5.11 the error δR_n was equal to 1.336.

The algorithm of a search for the line of the saturation function discontinuity has been realized in the form of a separate subroutine which works in parallel with the basic program of solving the original problem. This is especially convenient when the discontinuity line is to be determined sequentially for several moments of time, or when the information on the location of the discontinuity line for a given moment of time is used for the refinement of subsequent computations. The processing of the numerical information, with the purpose of determining a discontinuity line by the above optimization method, took an insignificant amount of computer time needed for the calculation of the basic problem; namely, 3–5% of the overall computer time.

The above-presented method for the localization of a discontinuity in multidimensional problems has a number of advantages over the traditional technique of visual determination of a discontinuity line by concentration of isolines. First, it may be subject to a theoretical investigation of convergence (see [5.17], as well as Sections 5.4 and 5.8.2); second, the proposed method enables us to obtain automatically, without subsequent user intervention, the discontinuity surface sought for and which creates the necessary prerequisites for a more accurate modeling of a number of important technological indices.

CHAPTER 6

Difference Solution Refinement in the Neighborhood of Strong Discontinuities

As was already pointed out at the beginning of Chapter 2, the "smearing" of strong discontinuities in the numerical solutions obtained by finite-difference shock-capturing schemes causes the following two problems: a problem on the interpretation of the results of a numerical computation, and a problem on increasing the accuracy of difference solutions in the vicinity of discontinuities. In Chapters 2–5 we were concerned with the problem of the strong discontinuties localization which we consider as one of the basic aspects of a problem on the interpretation of numerical results.

The problem of increasing the accuracy of numerical solutions in the neighborhood of strong discontinuities has been the subject of investigation for a long time. One of the ways of solving this problem consists of an explicit shock tracking, similar to different versions of the "shock fitting" method [6.1]–[6.4]. One more way of the explicit consideration of discontinuities consists of the special construction of a computing mesh, when some nodes of the mesh are placed into the points of the discontinuities [6.5]–[6.7].

Another direction of the investigation of the problem on increasing the numerical solution accuracy in the neighborhood of strong discontinuities is characterized by the fact that the discontinuities are not tracked explicitly in a difference scheme, but that computing meshes are used which automatically concentrate in the domains of large solution gradients (see Sections 2.5 and 3.3, and also [6.8]–[6.11]).

The third direction is characterized by the development of difference schemes which provide a minimum width of the zone of shock "smearing" on the simplest uniform grid. This direction may be represented by the flux-corrected transport method of BORIS and BOOK [6.12]–[6.21], the artificial compression method of HARTEN [6.14], [6.22]–[6.24], the MUSCL scheme of VAN LEER [6.25], [6.26], the PPM method of WOODWARD and COLELLA [6.26]–[6.29], and the TVD method of HARTEN [6.30]–[6.36]. In [6.23] the conclusion was drawn that the method of Boris and Book does not prevent the "smearing" of contact discontinuities. It follows from the data of gas-dynamical test computations presented in [6.14] that the Harten method [6.22]–[6.24] gives results comparable with those obtained by other known difference schemes in Eulerian variables (see Section 6.4). The MUSCL scheme of VAN LEER [6.25], [6.26] and the PPM scheme [6.26]–[6.29] are the

Godunov-type schemes, since nonlinear Riemann problem solvers are used in these schemes for computation of the fluxes of conserved quantities. Compared to the original Godunov scheme these schemes possess a higher (second) order of accuracy. They yield narrow zones of shock smearing without nonphysical oscillations. An explicit–implicit variant of the PPM method [6.28] eliminates a restriction on a time step. The results of one-dimensional tests presented in [6.28] show that the shock waves are smeared over one interval of computing mesh while using the PPM method.

In the TVD method of Harten the difference approximations are constructed from the requirement that the total variation $TV(u^n)$ of a numerical solution should decrease as the number n of a time level increases. The quantity $TV(u^n)$ is determined in the one-dimensional case as

$$TV(u^n) = \sum_{j=-\infty}^{\infty} |u_{j+1}^n - u_j^n|.$$

It should be noted that the solutions of some gas-dynamical problems do not satisfy the requirement $TV(u^{n+1}) \leq TV(u^n)$. For example, the quantity $TV(u^n)$ should, in accordance with the true solution behavior, increase at the collision of two shock waves, at a reflection of a shock wave off the rigid wall, and in the computations of problems with converging spherical shock waves. Thus, the TVD method of Harten is, generally speaking, applicable only to those problems of gas dynamics for which it is known *a priori* that their solutions meet the requirement $TV(u^{n+1}) \leq TV(u^n)$.

The method for the refinement of difference solutions in the neighborhood of discontinuities [6.37] presented in this chapter, makes use of a uniform computing mesh on the x-axis where x is an Eulerian spatial coordinate. Note that the fixed-in-time uniform spatial grids are widely used in computations of fluid flows in the domains with rectangular boundaries (see, for example, [6.38], [6.39]). The specific feature of the method presented below is the use of information on the location of a discontinuity. For its localization a differential analyzer has been used, the construction and investigation of which have been carried out in Chapters 2–4. The position of a discontinuity may not coincide with the nodes of a computing mesh, thus, the points of a discontinuity float freely along the x-axis. In the refinement procedures presented in this chapter the discontinuities are considered explicitly, so that the width of shock "smearing" is finally equal to 1–2 intervals of computing mesh. The advantages of the explicit treatment of shocks have been enumerated in [6.40] (a high solution accuracy in the neighborhood of shocks, a high accuracy in regular zones, and the absence of dissipation across shock waves), and which justify the need for the development of corresponding algorithms. Note that the refinement procedures proposed below are performed locally, only in the neighborhood of discontinuities, and serve only the purpose of correcting the numerical solution obtained by a shock-capturing scheme in the neighborhood of discontinuities. These local procedures make use of an auxiliary local moving fine mesh in the neighborhood of each discontinuity. The transfer of information from this mesh to a fixed uniform mesh, on which a shock-capturing computation of the flow field is performed, provides the conservation of the total fluid mass in the computational domain. This conservation of the mass is enforced by solving an optimization problem, formulated on the basis of the requirement of the minimization of a difference between the fluid masses in the moving mesh and in the fixed mesh. This conservation property of the refinement procedures proposed below establishes their substantial difference from the above-mentioned versions [6.1]–[6.4] of the shock fitting method.

A presentation of the refinement techniques is carried out conformally to the difference solution refinement in the neighborhood of contact discontinuities and shock waves in one-dimensional flows. In Section 7.3.2 we discuss some possible ways of extending the refinement procedures proposed in this chapter for the two-dimensional case.

6.1. Construction of the Basic Functional

As was shown in Section 1.1, one of the kinds of strong discontinuities in one-dimensional flows of an inviscid compressible gas is a contact discontinuity. The pressure and the normal to the discontinuity surface velocity component change continuously across a contact discontinuity, whereas the density, the temperature, and the entropy undergo a discontinuity. In some particular cases the pressure and velocity gradients in the neighborhood of a one-dimensional contact discontinuity are equal to zero; this takes place, for example, in the Riemann problem. In the process of the development of refinement algorithms it appeared reasonable to consider at first the most simple situations, and to learn to construct the refinement algorithms for such situations. In this connection we have chosen the case of a one-dimensional contact discontinuity when the derivatives $\partial u/\partial x$ and $\partial p/\partial x$ on both sides of a contact discontinuity are equal to zero, where u is the velocity and p is the pressure of the gas. As was shown in Section 4.1, in this case the problem of investigating the properties of difference solutions in the vicinity of a contact discontinuity reduces to an investigation of difference schemes for a linear equation

$$\partial \rho/\partial t + u_0\, \partial \rho/\partial x = 0, \tag{4.1.8}$$

where $u_0 = \text{const}$ is the gas speed in the neighborhood of a contact discontinuity. We assume in the following that $u_0 > 0$ (the case $u_0 < 0$ is considered analogously). In what follows we shall use the linear advection equation (4.1.8) for purposes of convenience of the notation in the form

$$\partial u/\partial t + a\, \partial u/\partial x = 0, \tag{1.50}$$

thus equation (1.50) coincides with (4.1.8) if we denote $u \equiv \rho$, $a \equiv u_0$.

Usually the gas dynamics problems are solved in the domains of finite sizes. In this connection we shall seek the solution of (1.50) in a finite domain D determined by the inequalities $0 \leq x \leq l, 0 \leq t \leq T$, where $0 < l < \infty$. Let us pose for the function $u(x, t)$ the following initial and boundary conditions:

$$u(x, 0) = u_0(x), \qquad 0 \leq x \leq l; \tag{6.1.1}$$

$$u(0, t) = \psi(t), \qquad 0 \leq t \leq T; \tag{6.1.2}$$

where the functions ψ and u_0 are such that $\psi(0) = u_0(0)$. Then the solution of an initial- and boundary-value problem (1.50), (6.1.1), (6.1.2) may easily be found in the overall domain and has the form

$$u(x, t) = \begin{cases} \psi(t - x/a), & 0 \leq x < at, \\ u_0(x - at), & at \leq x \leq l. \end{cases} \tag{6.1.3}$$

Let us now go over to obtaining the difference solution of the problem (1.50), (6.1.1), (6.1.2). Introduce in the domain D a uniform mesh G_h with steps h along the x-axis and τ along the t-axis. Let us approximate equation (1.50) by the explicit finite-difference scheme with one-sided differences (the "corner" scheme)

$$u_i^n = u_i^{n-1} - \kappa(u_i^{n-1} - u_{i-1}^{n-1}), \tag{6.1.4}$$

where κ is the Courant number,

$$\kappa = a\tau/h, \tag{6.1.5}$$

n is the number of time steps, $n = 1, 2, \ldots, u_i^n = u(x_i, t_n)$ where $x_i = (i + 0.5)h$, $i = 0, 1, \ldots, i_x, (i_x + 1)h = l, t_n = n\tau$. It was shown in Section 4.1 that the difference scheme (6.1.4) describes, in the vicinity of a one-dimensional contact discontinuity in the solution of a model problem (1.11), (1.9), (4.1.1), the properties of some well-known difference methods of the first order of accuracy, such as the "breakdown-of-discontinuity" scheme [6.41], the FLIC method [6.42], and the "coarse particles" method [6.43].

Since we are interested in a computation of strong discontinuities, we shall assume the presence of at least one strong discontinuity in the initial profile $u_0(x)$ (see formula (6.1.1)). Let N_d be the number of strong discontinuities in a computational domain at $t = 0$; $N_d \geq 1$. Denote by Ω a neighborhood of a strong discontinuity which does not contain other discontinuities. Alongside the mesh G_h we introduce in the neighborhood Ω an auxiliary mesh G_Δ as follows. Let x_f^m be the value of the abscissa of a discontinuity under consideration at $t = t_m$. Denote by x_f^n an approximate value of the discontinuity abscissa at $t = t_n$. On the lines $t = t_m$ and $t = t_n$ the abscissas ξ_k^m, ξ_k^n of the nodes of the grid G_Δ will be computed via formulas

$$\xi_k^\nu = x_f^\nu - (M_1 - k)(\Delta x)_1, \qquad k = 0, \ldots, M_1;$$
$$\xi_k^\nu = x_f^\nu + (k - M_1 - 1)(\Delta x)_2, \qquad k = M_1 + 1, \ldots, M + 1; \tag{6.1.6}$$

where $\nu = m$ or n. In (6.1.6) M_1 and M are integers, $M_1 \geq 1, M \geq M_1 + 1$.

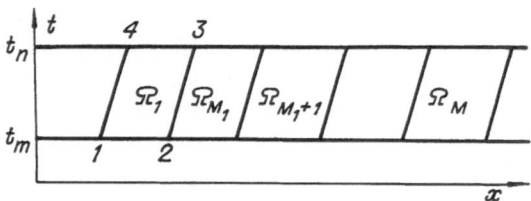

Figure 6.1. The grid in the
neighborhood of a discontinuity.

Thus, the mesh G_Δ contains $(M + 2)$ nodes at $t = t_m$ and the same number of nodes at $t = t_n$. A question on the choice of the numbers M_1 and M and the steps $(\Delta x)_1$, $(\Delta x)_2$ is closely related to the properties of discontinuity "smearing" by a difference scheme and will be discussed in detail below. Now we proceed to the construction of the basic functional. Let us, at $k = 0, \ldots, M + 1$, connect the points (ξ_k^m, t_m) and (ξ_k^n, t_n) by straight-line segments as shown in Figure 6.1. As a result we obtain M subdomains Ω_k. Since such subdomains Ω_k are constructed in the neighborhood of a strong discontinuity at each time step, it can be said that they accompany a strong discontinuity propagating in the (x, t)-plane. The boundary Γ_k of each of the subdomains Ω_k has the shape of a parallelogram in the (x, t)-plane. Now find among the nodes of the grid G_h two such nodes which are nearest to the boundary nodes $x = \xi_0^n$ and $x = \xi_{M+1}^n$ of the mesh G_Δ. Let the numbers of these nodes of the G_h mesh along the x-axis be i_l and i_r, respectively. We require additionally that the coordinates

$$x_{i_l} = (i_l + 0.5)h, \qquad x_{i_r} = (i_r + 0.5)h,$$

satisfy the inequalities

$$\xi_0^n \leq x_{i_l}, \qquad \xi_{M+1}^n \geq x_{i_r}. \qquad (6.1.7)$$

Denote by U_k^n the value of an approximate solution of the problem (1.50), (6.1.1), (6.1.2) on the mesh G_Δ; $U_k^n = U(\xi_k^n, t_n)$. The integral conservation law

$$I_k = \oint_{\Gamma_k} u\, dx - au\, dt = 0, \qquad (6.1.8)$$

being valid for each subdomain Ω_k, corresponds to equation (1.50). Approximate the integral (6.1.8) as follows. Let us enumerate the corner points of the domain Ω_k by figures 1, 2, 3, 4 (see Figure 6.1). Then

$$I_k = \int_1^2 + \int_2^3 + \int_3^4 + \int_4^1.$$

Using the approximate formulas

$$\int_1^2 = [\alpha_1 U_{k-1}^m + (1 - \alpha_1)U_k^m](\Delta x)_1;$$

$$\int_2^3 = [\alpha_2 U_k^m + (1 - \alpha_2)U_k^n]k_f\tau - a\tau[\alpha_3 U_k^m + (1 - \alpha_3)U_k^n];$$

$$\int_{3}^{4} = -[\alpha_4 U_{k-1}^n - (1-\alpha_4)U_k^n](\Delta x)_1;$$

$$\int_{4}^{1} = -k_f\tau[\alpha_5 U_{k-1}^m + (1-\alpha_5)U_{k-1}^n] + a\tau[\alpha_6 U_{k-1}^m + (1-\alpha_6)U_{k-1}^n],$$

$$\text{(6.1.9)}$$

where $k = 1, \ldots, M_1$, α_j are constants, $0 \le \alpha_j \le 1$, $j = 1, \ldots, 6$,

$$k_f = (x_f^n - x_f^m)/(t_n - t_m), \tag{6.1.10}$$

we finally obtain the following difference equations:

$$a^{(1)}U_{k-1}^n + b^{(1)}U_k^n + c^{(1)}U_{k-1}^m + d^{(1)}U_k^m = 0, \tag{6.1.11}$$

where

$$k = 1, \ldots, M_1, \qquad \kappa_1 = a\tau/(\Delta x)_1, \qquad \kappa_2 = k_f\tau/(\Delta x)_1, \tag{6.1.12}$$

$$a^{(1)} = -\alpha_4 - \kappa_2(1-\alpha_5) + \kappa_1(1-\alpha_6);$$

$$b^{(1)} = (1-\alpha_2)\kappa_2 - \kappa_1(1-\alpha_3) - (1-\alpha_4);$$

$$c^{(1)} = \alpha_1 - \kappa_2\alpha_5 + \kappa_1\alpha_6;$$

$$d^{(1)} = 1 - \alpha_1 + \alpha_2\kappa_2 - \alpha_3\kappa_1.$$

$$\text{(6.1.13)}$$

Setting the specific values of the weight coefficients α_j we shall obtain one of the schemes of the multiparametric family (6.1.11)–(6.1.13). Let us now find in the family (6.1.11)–(6.1.13) a scheme having an order of approximation $O(\tau) + O((\Delta x)_1)$, and possessing such a domain of stability which makes possible a stable computation by scheme (6.1.11)–(6.1.13), even at a step $(\Delta x)_1 < h$ with the same value of the time step τ which ensures a stable computation by scheme (6.1.4). Expand the quantities U_{k-1}^n, U_k^n, U_{k-1}^m, U_k^m in (6.1.11) in Taylor series with respect to the point with coordinates (ξ_k^m, t_m), $m = n - 1$, where ξ_k^m is computed by a corresponding formula in the relationships (6.1.6). For example, U_k^n is expanded as follows:

$$U_k^n = U_k^m + (\partial U/\partial x)_k^m \tau k_f + (\partial U/\partial t)_k^m \tau + (1/2)$$

$$\times (\partial^2 U/\partial x^2)_k^m \tau^2 k_f^2 + (\partial^2 U/\partial x\, \partial t)_k^m \tau k_f \tau$$

$$+ (1/2)(\partial^2 U/\partial t^2)_k^m \tau^2 + O(\tau^3). \tag{6.1.14}$$

Substituting the expansions of type (6.1.14) into equation (6.1.11) we obtain after some computations the H-form of the first differential approximation (f.d.a.) of a difference scheme (6.1.11) in the form

$$B_1 U_t + B_2 U_x + B_3 U_{xx} + B_4 U_{xt} + B_5 U_{tt} = 0, \tag{6.1.15}$$

where

$$B_1 = \tau(a^{(1)} + b^{(1)}),$$

$$B_2 = a^{(1)}\tau k_f + b^{(1)}(\tau k_f + (\Delta x)_1) + d^{(1)}(\Delta x)_1,$$

$$B_3 = (1/2)(a^{(1)}\tau^2 k_f^2 + b^{(1)}(\tau k_f + (\Delta x)_1)^2) + d^{(1)}(1/2)((\Delta x)_1)^2,$$

$$B_4 = a^{(1)}\tau^2 k_f + b^{(1)}\tau \cdot (\tau k_f + (\Delta x)_1);$$

$$B_5 = (1/2)\tau^2(a^{(1)} + b^{(1)}). \qquad (6.1.16)$$

In order to ensure the approximation of the original equation (1.50) we require that

$$B_2 = aB_1. \qquad (6.1.17)$$

Assuming that α_j are chosen in such a way that equation (6.1.17) takes place, we find the f.d.a. P-form with the aid of (6.1.15):

$$U_t + aU_x = (1/B_1)(aB_4 - B_3 - B_5 a^2)U_{xx}. \qquad (6.1.18)$$

Let us now study the stability of scheme (6.1.11) by the Fourier method. Substitute into (6.1.11) a particular solution of the form $U_k^n = \lambda^n e^{ik\alpha}$, where $i = \sqrt{-1}$, $\alpha = s(\Delta x)_1$, and s is a real constant. As a result we obtain for the square of a complex number λ the expression

$$|\lambda|^2 = \{[(c^{(1)} + d^{(1)}\cos\alpha)(a^{(1)} + b^{(1)}\cos\alpha) + b^{(1)}d^{(1)}\sin^2\alpha]^2$$

$$+ (a^{(1)}d^{(1)} - b^{(1)}c^{(1)})^2 \sin^2\alpha\}/[(a^{(1)} + b^{(1)}\cos\alpha)^2 + (b^{(1)})^2 \sin^2\alpha]^2.$$

$$(6.1.19)$$

To ensure stability we must require the satisfaction of the inequality

$$|\lambda|^2 \le 1. \qquad (6.1.20)$$

We shall not dwell here on a detailed study of the family of schemes (6.1.11)–(6.1.13), but choose instead the following way: in this family take some particular scheme and show that it meets the requirements of approximation and stability. Namely, let us set in the relationships (6.1.13)

$$\alpha_1 = \alpha_4 = 1/2, \qquad \alpha_2 = \alpha_3 = \alpha_5 = \alpha_6 = 0. \qquad (6.1.21)$$

Such a choice of α_j is based on the general consideration that implicit schemes for the hyperbolic equation (1.50) possess a higher computational stability than the explicit schemes (see also [6.44]). Substituting the values (6.1.21) into (6.1.13) we obtain

$$a^{(1)} = -(1/2) - \kappa_2 + \kappa_1, \qquad b^{(1)} = \kappa_2 - \kappa_1 - (1/2),$$

$$c^{(1)} = d^{(1)} = 1/2. \qquad (6.1.22)$$

It is easy to see with regard to (6.1.16), (6.1.22) that the condition of approximation (6.1.17) is satisfied. Employing the formulas (6.1.16), (6.1.18) we obtain the f.d.a. P-form of a difference scheme (6.1.11), (6.1.22) as the equation

$$U_t + aU_x = (\tau/2)(k_f - a)^2 U_{xx}. \qquad (6.1.23)$$

It is important to note that the difference $k_f - a$ in (6.1.23) is squared; therefore, the coefficient affecting U_{xx} in formula (6.1.23) is always nonnegative, so that we can draw a conclusion on the stability of scheme (6.1.11), (6.1.22) [6.45]. Furthermore, we see from equation (6.1.23) that at $k_f = a$ the difference scheme (6.1.11), (6.1.22) has an order of approximation not less than the second one. Since, in accordance with (6.1.10) k_f depends on x_f^n, it follows from the foregoing that it is desirable to determine the discontinuity abscissa x_f^n with as high an accuracy as possible. Substituting the values (6.1.22) into formula (6.1.19) we obtain

$$|\lambda|^2 = 1/(1 + 4(\kappa_2 - \kappa_1)^2 \text{tg}^2 \, \alpha/2),$$

from where a conclusion follows on the absolute stability of scheme (6.1.11), (6.1.22).

The difference equations for $k = M_1 + 2, \ldots, M + 1$ coincide with equations (6.1.11)–(6.1.13) if one replaces the superscript (1) in formulas (6.1.11), (6.1.13) by the superscript (2), and the quantity $(\Delta x)_1$ in (6.1.12) is replaced by $(\Delta x)_2$. Using the known difference solution u^m on the grid G_h, we obtain with the aid of scheme (6.1.4) the difference solution u_i^n, $i = 0, \ldots, i_x$. On the other hand, we can obtain the solution U^n on the mesh G_Δ by using the difference scheme (6.1.11)–(6.1.13). Introduce the notation

$$\Omega = \bigcup_{k=1}^{M} \Omega_k. \tag{6.1.24}$$

Let us require for the solution U^n to adhere continuously the solution u^n on the boundaries of the domain that was obtained on the mesh G_h. In order to meet this requirement we shall compute U_0^n, U_{M+1}^n by the formulas

$$U_0^n = u^n(\xi_0^n), \qquad U_{M+1}^n = u^n(\xi_{M+1}^n), \tag{6.1.25}$$

where $u^n(\xi_0^n)$, $u^n(\xi_{M+1}^n)$ are determined by interpolation on the basis of the values of the solution u^n in the neighboring nodes of the mesh G_h.

If, for each $t = t^n = n\tau$, $n = 1, 2, \ldots$, the equation $k_f = a$ is satisfied, then the values U_k^n computed by scheme (6.1.11), (6.1.22) are exact, because in this case we simply have a transfer of the values along the characteristics, and scheme (6.1.11), (6.1.22) has an infinite order of approximation (see also [6.44]). However, in practice, the equation $k_f = a$ is not satisfied, because x_f^n in (6.1.10) is determined approximately, and the value U_0^n differs from the exact one by some error whose size depends on a local error of the difference solution u^n. Thus, only the first order of approximation of scheme (6.1.11), (6.1.22) is realized in a practical computation.

It may easily be seen from the construction of the difference equations (6.1.11), (6.1.13) that the differencing across a discontinuity is excluded from these equations. Note that the requirement of the absence of a differencing across a discontinuity is one of the basic requirements in the "shock fitting" method [6.1], [6.4]. Since two, generally speaking, different values $U_{M_1}^n$ and $U_{M_1+1}^n$ are assigned to the node $\xi_{M_1}^n = \xi_{M_1+1}^n = x_f^n$, the solution U^n on the mesh G_Δ will contain a discontinuity, so that the "smearing" of a discontinuity is absent in the profile of $U^n(x)$. On the other hand, in the subdomains of a continuous flow to the left and right of the point $x = x_f^n$ the difference scheme with one-sided differences (6.1.11), (6.1.22) approximates the original equation (1.50) with an order $O(\tau) + O((\Delta x)_1)$ or $O(\tau) + O((\Delta x)_2)$ as was shown above. In connection with the above-listed properties of the grid solution U^n, it can be used instead of a difference solution u^n in the neighborhood Ω of a strong discontinuity. For this purpose, the values u^n are computed at the points of the mesh G_h located in the domain Ω by means of an interpolation by the values of U_k^n on the mesh G_Δ.

As is known [6.44], the requirements of approximation and stability of a difference scheme are generally insufficient for a satisfactory computation of discontinuous solutions. The divergence property or conservative property is also important for the computation of discontinuous solutions [6.44], [6.46]. The difference scheme (6.1.4) is conservative. Denote by $\mathcal{M}_n$ the "mass of gas" in a computational domain at $t = t_n$,

$$\mathcal{M}_n = \sum_{i=0}^{i_x} u_i^n h.$$

Let us present $\mathcal{M}_n$ in the form of a sum of three items:

$$\mathcal{M}_n = \sum_{i=0}^{i_l-1} u_i^n h + S_h + \sum_{i=i_r+1}^{i_x} u_i^n h, \qquad (6.1.26)$$

where we have introduced the notation

$$S_h = \sum_{i=i_l}^{i_r} u_i^n h. \qquad (6.1.27)$$

If the solution u^n obtained by the difference scheme (6.1.4) is used in all the nodes along the x-axis for the computation of the sum (6.1.26), then $\mathcal{M}_n$ has a correct value by virtue of the divergence property of the difference scheme (6.1.4). Suppose now that we use the values of u_i^n determined by a linear interpolation between the values U_k^n while computing the sum S_k defined by (6.1.27). Then we obtain instead of S_h some quantity S_h' which can be presented in the form

$$S_h' = \sum_{k=0}^{M+1} c_k U_k^n h, \qquad (6.1.28)$$

where the coefficients c_k are nonnegative by virtue of the fact that u_i^n is determined by U_k^n with the aid of interpolation. $S_h' \neq S_h$ because of the errors in determining the discontinuity abscissa x_f^n and because of the approximation error of a difference scheme for determining U^n. Therefore, substitution of the quantity S_h' into the sum (6.1.26) instead of S_h leads to the violation of the law of conservation of the quantity $\mathcal{M}_n$. Let us introduce the quadratic functional

$$I(x_f^n) = (S_h - S_h')^2 \qquad (6.1.29)$$

with the purpose of minimizing the difference $|S_h - S'_h|$. The value S_h in (6.1.29) is supposed to be given and fixed, and the value S'_h is a function of the abscissa x^n_f. It is clear that in the process of a search for the minimum of the functional (6.1.29) we can vary x^n_f only within the limits of the domain Ω, because there can occur other (neighboring) discontinuities of the solution outside this domain. Therefore, it is necessary to impose constraints on x^n_f in the form of inequalities

$$x^n_l \le x^n_f \le x^n_r, \qquad (6.1.30)$$

where the quantities x^n_l, x^n_r should satisfy the inequalities $\xi^n_0 \ge x^n_l$, $x^n_r \le \xi^n_{M+1}$. Let us formulate for the functional (6.1.29) the following constrained optimization problem

$$I(x^n_f) \to \min,$$
$$x^n_f \ge x^n_l, \qquad x^n_f \le x^n_r. \qquad (6.1.31)$$

Concluding this section we consider a question on the numerical realization of the difference scheme (6.1.11), (6.1.22). Consider at first the difference equation (6.1.11) at $1 \le k \le M_1$. Show that at sufficiently large values of the quantity M_1 it is impossible to use a recurrence formula

$$U^n_k = (-a^{(1)}U^n_{k-1} - c^{(1)}U^m_{k-1} - d^{(1)}U^m_k)/b^{(1)}, \qquad k = 1, \ldots, M_1, \quad (6.1.32)$$

because of the rounding-off errors accumulation. It is necessary that the inequality

$$|a^{(1)}/b^{(1)}| < 1 \qquad (6.1.33)$$

be satisfied in order to avoid the above accumulation of errors [6.47]. Taking into account (6.1.22) it is easy to show that the inequality (6.1.33) takes place if $k_f < a$ where the value k_f is determined by formula (6.1.10). The inequality $k_f < a$ can be violated because of the errors in determining x^n_f (and this actually took place, in practice, in computations). The instability of a recurrence computation by the formula

$$U^n_{k-1} = (-b^{(1)}U^n_k - c^{(1)}U^m_{k-1} - d^{(1)}U^m_k)/a^{(1)}, \qquad k = M + 1, M, \ldots, M_1 + 2,$$

at sufficiently large values of the quantity $M - M_1$ can be shown analogously. In this connection it is necessary to use at large M_1, $M - M_1$, some other algorithm for solving the difference equations (6.1.11), (6.1.22). Let

$$\mathbf{U} = \{U^n_1, \ldots, U^n_{M_1}\}^T, \qquad \mathbf{F} = \{\gamma^{(1)}_1, \ldots, \gamma^{(1)}_{M_1}\}^T.$$

Then the difference equations (6.1.11) at $k = 1, \ldots, M_1$ may easily be written in the form of an algebraic system

$$H\mathbf{U} = \mathbf{F}, \qquad (6.1.34)$$

where H is a square $M_1 \times M_1$ matrix, the expressions for its elements will not be presented here for brevity. It is easy to see with regard to (6.1.11) that the matrix H is a band matrix, it has nonzero elements only on the principal diagonal and one element immediately to the left of the principal diagonal in each row. Efficient algorithms of the band matrices inversion presented in [6.48] require a positive definiteness of the matrix H. Let us show that the matrix H is not positive definite. For this purpose it is sufficient to show [6.49] that the quadratic form

$$(Hb, b) = \sum_{j=1}^{M_1} \sum_{k=1}^{M_1} \chi_{jk} b_j b_k$$

is not positive definite where χ_{jk} is the element of the matrix H and b_j, b_k are components of an arbitrary column vector $b = \{b_1, \ldots, b_{M_1}\}^{\mathrm{T}}$, such that

$$(b, b) = \sum_{i=1}^{M_1} b_i^2 > 0.$$

After some computations we find

$$(Hb, b) = b^{(1)} b_1^2/2 + \sum_{j=2}^{M_1} (a^{(1)} + b^{(1)}) b_j^2 + \tfrac{1}{2} b^{(1)} b_{M_1}^2 - \tfrac{1}{2} b^{(1)} \sum_{j=1}^{M_1-1} (b_j - b_{j+1})^2.$$

$$(6.1.35)$$

In formula (6.1.35) let $b_j = 1$, $j = 1, \ldots, M_1$. Since we have from (6.1.22) that $a^{(1)} + b^{(1)} = -1 < 0$, we obtain from formula (6.1.35) that $(Hb, b) < 0$.

6.2. Solution Refinement on the Basis of the Least-Squares Method

6.2.1. Formulation of Constrained Optimization Problems

It is well known in the process of constructing implicit and semi-implicit difference schemes for hyperbolic partial differential equations of the first order that there arises the problem of satisfaction of the requirement that the matrices of algebraic systems arising after problem discretization be well conditioned (see, for example, [6.8], [6.50]). This conclusion is in agreement with the above results of studying the implicit difference scheme (6.1.12), (6.1.22) for equation (1.50). It was shown in [6.50] in the case when a quadratic functional

$$I = \int_D (\partial \rho u/\partial x + \partial \rho v/\partial y)^2 \, dx \, dy \qquad (6.2.1)$$

is used as a starting point in the construction of a difference scheme for the stationary equation of mass conservation $\partial \rho u/\partial x + \partial \rho v/\partial y = 0$, that the matrix of an algebraic system obtained as a result of difference approximation of the conditions for the minimum of the functional (6.2.1) is well conditioned (see also [6.51]).

Let Ω_k be one of the subdomains in the neighborhood of a strong discontinuity which was introduced in the foregoing section (see Figure 6.1). Consider a functional

$$I_k = \int_{\Omega_k} (\partial u/\partial t + a\, \partial u/\partial x)^2 \, dx \, dt. \tag{6.2.2}$$

Let us use for the minimization of the quadratic functional (6.2.2) a generalization of the Ritz method described in [6.52]. It consists of an approximation of a function $u(x, t)$ in the functional (6.2.2) by a function $u_{ak}(x, t)$ determined by the formula

$$u_{ak}(x, t) = \sum_{j=0}^{N_x} a_{kj}(t)(x - x_*^{(k)})^j, \qquad (x, t) \in \Omega_k, \tag{6.2.3}$$

where $(x_*^{(k)}, t) \in \Omega_k$, $N_x \geq 1$, $a_{kj}(t)$ are functions to be determined from the requirements of the minimum of the functional (6.2.2). By analogy with [6.53] we also approximate the function $a_{kj}(t)$ in equation (6.2.3) by a formula

$$a_{kj}(t) = \sum_{v=0}^{N_t} a_{kjv}(t - t_*)^v, \qquad j = 0, \ldots, N_x, \tag{6.2.4}$$

where a_{kjv} are constants, $(x_*^{(k)}, t_*) \in \Omega_k$, $N_t \geq 1$.

Let N be the overall number of constant coefficients a_{kjv} in the representation (6.2.3), (6.2.4) of the function $u_{ak}(x, t)$. In [6.54], examples have been presented which indicate an instability of the Ritz process to small errors of intermediate computations at sufficiently large N and when a polynomial approximation of the form (6.2.3) is used. In our case there is no need to take large values of N, because the approximation (6.2.3) is taken locally in a subdomain Ω_k whose size along the x-axis can be made small at the expense of increasing the number M of the subdomains Ω_k in the neighborhood of a discontinuity. Therefore, we can achieve an acceptable accuracy of the solution by taking at an appropriate number M of the subdomains Ω_k the values N_x, $N_t \leq 2$ in formulas (6.2.3), (6.2.4).

Substituting the expansion (6.2.4) into formula (6.2.3) and then the approximation (6.2.3) into formula (6.2.2), and performing the differentiation and integration operations entering formula (6.2.2) we obtain that

$$I_k = I_k(a_{k00}, a_{k01}, \ldots, a_{kN_xN_t}, x_j^n). \tag{6.2.5}$$

The constants $a_{k00}, \ldots, x_j^n$ entering the expression (6.2.5) should be found as a solution of some optimization problem. Let us proceed to the formulation of this problem. Consider a particular case $N_x = N_t = 1$ in formulas (6.2.3), (6.2.4). Then

$$u_{ak}(x, t) = a_{k00} + a_{k01}(t - t_*) + [a_{k10} + a_{k11}(t - t_*)](x - x_*^{(k)}). \tag{6.2.6}$$

Let us assume that at $t = t_m$, $m = n - 1$, the difference solution $u_h(x, t_m)$ has already been refined. Let us approximate $u_h(x, t_m)$ in the domain Ω_k by a straight-line segment:

$$u_h(x, t_m) = b_{0k} + b_{1k}(x - x_*^{(k)}), \tag{6.2.7}$$

where b_{0k}, b_{1k} are known constants. In accordance with the general idea of the Ritz method [6.52] we require the function (6.2.6) to satisfy at $t = t_m$ the initial condition (6.2.7). Equating the coefficients at equal powers of the quantity $(x - x_*^{(k)})$ in (6.2.6), (6.2.7) we obtain the equations

$$a_{k00} + a_{k01}(t_m - t_*) = b_{0k};$$
$$a_{k10} + a_{k11}(t_m - t_*) = b_{1k}. \qquad (6.2.8)$$

Rewriting formula (6.2.6) with regard to the expressions (6.2.8) we obtain

$$u_{ak}(x, t) = b_{0k} + a_{k01}(t - t_m) + [b_{1k} + a_{k11}(t - t_m)](x - x_*^{(k)}). \quad (6.2.9)$$

Denote by $x_{lk}^{(m)}$ and $x_{rk}^{(m)}$ the abscissas of the left and right corner points, respectively, of the Ω_k domain on the line $t = t_m$ (see Figure 6.1). Let us specify the choice of the value $x_*^{(k)}$ in formula (6.2.9) by setting $x_*^{(k)} = x_{lk}^{(m)}$. We shall write in the following a_{0k}, a_{1k} instead of a_{k01}, a_{k11}. Thus, we obtain for $u_{ak}(x, t)$ in the domain Ω_k a representation

$$u_{ak}(x, t) = b_{0k} + a_{0k}(t - t_m) + [b_{1k} + a_{1k}(t - t_m)](x - x_{lk}^{(m)}). \quad (6.2.10)$$

Let us introduce the functional

$$I^{(l)} = \sum_{k=1}^{M_1} \int_{\Omega_k} (\partial u_{ak}/\partial t + a\,\partial u_{ak}/\partial x)^2\, dx\, dt. \qquad (6.2.11)$$

We consider the functional (6.2.11) at a fixed value of the discontinuity abscissa x_f^n. Writing an analogue of the Euler–Lagrange equation for the functional (6.2.11) we obtain

$$\partial I^{(l)}/\partial a_{0k} = \partial I^{(l)}/\partial a_{1k} = 0, \qquad k = 1, \ldots, M_1. \qquad (6.2.12)$$

Since a_{0k}, a_{1k} enter only the integral

$$I_k = \int_{\Omega_k} (\partial u_{ak}/\partial t + a\,\partial u_{ak}/\partial x)^2\, dx\, dt, \qquad (6.2.13)$$

we can rewrite equations (6.2.12) in the form

$$\partial I_k/\partial a_{0k} = \partial I_k/\partial a_{1k} = 0. \qquad (6.2.14)$$

Equations (6.2.14) yield a system of two linear algebraic equations for determining a_{0k}, a_{1k}. Solving this system we can find the value of the function $u_{ak}(x, t)$ at the point $(x, t) \in \Omega_k$ by formula (6.2.10). In particular, let us set $t = t_n$, $x = x_{rk}^{(n)}$. Then

$$u_{ak}(x_{rk}^{(n)}, t_n) = b_{0k} + a_{0k}\tau + (b_{1k} + a_{1k}\tau)(x_{rk}^{(n)} - x_{lk}^{(m)}). \qquad (6.2.15)$$

On the other hand, the point $(x_{rk}^{(n)}, t_n) \in \Omega_{k+1}$, therefore, we can also write

$$u_{a,k+1}(x_{rk}^{(n)}, t_n) = b_{0,k+1} + a_{0,k+1}\tau + [b_{1,k+1} + a_{1,k+1}\tau](x_{rk}^{(n)} - x_{l,k+1}^{(m)}). \quad (6.2.16)$$

It follows from formulas (6.2.15), (6.2.16) that, generally speaking,

$$u_{ak}(x_{rk}^{(n)}, t_n) \neq u_{a,k+1}(x_{rk}^{(n)}, t_n), \qquad k = 1, \ldots, M_1 - 1.$$

This means that the function $u_a(x, t)$ minimizing the functional (6.2.11) has discontinuities at the boundaries between neighboring domains Ω_k and Ω_{k+1}, $k = 1, \ldots, M_1 - 1$. Although the absolute values of the differences of quantities (6.2.15) and (6.2.16) may prove to be small, it is reasonable to eliminate these discontinuities since they should be absent in the subdomains of a continuous flow. For this purpose, let us formulate the following problem of mathematical programming:

$$I^{(l)}(a_{01}, a_{11}, \ldots, a_{0M_1}, a_{1M_1}, x_f^n) \to \min;$$

$$u_{a1}(x_{l1}^{(n)}, t_n) = u_h(x_{l1}^{(n)}, t_n); \tag{6.2.17}$$

$$u_{ak}(x_{rk}^{(n)}, t_n) = u_{a,k+1}(x_{rk}^{(n)}, t_n), \qquad k = 1, \ldots, M_1 - 1.$$

The first of the equality constraints in formulas (6.2.17) ensures a continuous transition from the function $u_{a1}(x, t_n)$ to the difference solution u_h^n, obtained by the scheme (6.1.4) at the left boundary of the domain Ω (see also formulas (6.1.25)).

By analogy with (6.2.11) consider now the functional

$$I^{(r)} = \sum_{k=M_1+1}^{M} (\partial u_{ak}/\partial t + a \, \partial u_{ak}/\partial x)^2 \, dx \, dt. \tag{6.2.18}$$

Formulate for the functional (6.2.18) the mathematical programming problem similar to (6.2.17)

$$I^{(r)}(a_{0,M_1+1}, a_{1,M_1+1}, \ldots, a_{0M}, a_{1M}, x_f^n) \to \min;$$

$$u_{ak}(x_{rk}^{(n)}, t_n) = u_{a,k+1}(x_{rk}^{(n)}, t_n), \qquad k = M_1 + 1, \ldots, M - 1; \tag{6.2.19}$$

$$u_{aM}(x_{rM}^{(n)}, t_n) = u_h(x_{rM}^{(n)}, t_n).$$

In many applications the solution $u(x, t)$ should be nonnegative by its physical meaning: $u(x, t)$ may be the density (see Section 4.1), and the concentration [6.55]. By virtue of the linearity of the function (6.2.10) at fixed t it is easy to write the conditions for nonnegativity of $u_a(x, t)$ in Ω. Taking these conditions into account one can consider a more complicated problem than (6.2.17) of mathematical programming

$$I^{(l)}(a_{01}, a_{11}, \ldots, a_{0M_1}, a_{1M_1}, x_f^n) \to \min;$$

$$u_{a1}(x_{l1}^{(n)}, t_n) = u_h(x_{l1}^{(n)}, t_n);$$

$$u_{ak}(x_{rk}^{(n)}, t_n) = u_{a,k+1}(x_{rk}^{(n)}, t_n), \qquad k = 1, \ldots, M_1 - 1; \tag{6.2.20}$$

$$u_{ak}(x_{rk}^{(n)}, t_n) \geq 0, \qquad k = 1, \ldots, M_1.$$

Problem (6.2.19) may be complicated in a similar way.

Suppose that we have found the solution of problems (6.2.17), (6.2.19) at some fixed value of x_f^n. Introduce the notation

$$U_0^n = u_{a1}(x_{11}^{(n)}, t_n), \qquad U_k^n = u_{ak}(x_{rk}^{(n)}, t_n), \qquad k = 1, \ldots, M_1;$$

$$U_k^n = u_{ak}(x_{lk}^{(n)}, t_n), \qquad k = M_1 + 1, \ldots, M; \qquad (6.2.21)$$

$$U_{M+1}^n = u_h(x_{rM}^{(n)}, t_n).$$

Having determined U_k^n by (6.2.21), (6.2.10), we can find the value of the sum (6.1.28) and then the value of function (6.1.29). Since we require as before the satisfaction of a conservation law, we must solve a problem of constrained optimization (6.1.31). When using nongradient optimization techniques, it is sufficient to find the values of the function (6.1.29) in some number of the points $x = x_f^n$ determined in a certain way (see the next section on the golden section method). Therefore, there arises the question of an efficient method of solving the mathematical programming problems (6.2.17), (6.2.19) or two problems of the type (6.2.10). Problem (6.2.20) is a general nonlinear mathematical programming problem involving the constraints of equality and inequality type. Problem (6.2.17) contains only the equality-type constraints.

6.2.2. Construction of the Discrete Functional and its Minimization

Let us now show that problem (6.2.17) or (6.2.19) may be efficiently solved by a scalar sweep method if one takes into account the specific feature of the problem (6.2.17)—the linearity of constraints with respect to a_{0k}, a_{1k}. First consider the problem (6.2.17). Suppose that the problem (6.2.17) has already been solved for the moment of time $t = t_m$ and, consequently, the equalities of the form (6.2.21) and the equality constraints in the problem (6.2.19) take place where one should replace n by $m = n - 1$. Then we can easily express b_{0k}, b_{1k} in (6.2.10) in terms of U^m

$$b_{0k} = U_{k-1}^m, \qquad b_{1k} = (U_k^m - U_{k-1}^m)/(\Delta x)_1, \qquad k = 1, \ldots, M_1. \quad (6.2.22)$$

Assuming after that the satisfaction of equality constraints at $t = t_n$ in (6.2.17), we can easily find the expressions for the unknown constants a_{0k}, a_{1k} in terms of the new unknowns U_j^n

$$a_{0k} = [U_{k-1}^n - b_{0k} - (U_k^n - U_{k-1}^n)(x_f^n - x_f^m)/(\Delta x)_1]/(t_n - t_m);$$

$$a_{1k} = [U_k^n - U_{k-1}^n - b_{1k}(\Delta x)_1]/[(t_n - t_m)(\Delta x)_1], \qquad k = 1, \ldots, M_1. \qquad (6.2.23)$$

Thus we have eliminated, with the aid of equality constraints in (6.2.17), M_1 unknowns and now we have only M_1 unknowns $U_1^n, \ldots, U_{M_1}^n$ instead of $2M_1$ unknowns a_{0k}, a_{1k}. Computing the integral (6.2.13) by using formula (6.2.10) we obtain the following expression:

$$I_k = D_0[(a_{0k} + ab_{1k} + (1/2)D_1 a_{1k})^2 + D_2 a_{1k}^2], \qquad (6.2.24)$$

where

$$D_0 = (\Delta x)_1 (t_n - t_m), \qquad D_1 = a(t_n - t_m) + k_f(t_n - t_m) + (\Delta x)_1,$$

$$D_2 = (1/12)[(a + k_f)^2(t_n - t_m)^2 + ((\Delta x)_1)^2], \tag{6.2.25}$$

and the quantity k_f is determined by formula (6.1.10), a is a coefficient in the original equation (1.50). Substituting the expressions (6.2.23) instead of a_{0k}, a_{1k} into (6.2.24) we find that

$$I_k = I_k(U_k^n, U_{k-1}^n), \qquad k = 1, \ldots, M_1. \tag{6.2.26}$$

Thus we have obtained M_1 discrete functionals I_k, each of which depends on two quantities U_k^n, U_{k-1}^n calculated at the discrete points—the nodes of the grid G_Δ. Substituting the expressions (6.2.26) into (6.2.11) we obtain for the discrete functional $I^{(l)}$ the following unconstrained minimization problem:

$$I^{(l)}(U_1^n, \ldots, U_{M_1}^n) \to \min, \qquad (U_1^n, \ldots, U_{M_1}^n) \in \mathscr{E}_{M_1}, \tag{6.2.27}$$

where $\mathscr{E}_{M_1}$ is an M_1-dimensional Euclidean space. In order to solve problem (6.2.27) let us write down the discrete Euler equations

$$\partial I^{(l)}/\partial U_k^n = 0, \qquad k = 1, \ldots, M_1, \tag{6.2.28}$$

as is usually done in solving such problems (see, for example, [6.53]). Taking (6.2.26) into account we can rewrite equations (6.2.28) in the form

$$\partial I_k/\partial U_k^n + \partial I_{k+1}/\partial U_k^n = 0, \qquad k = 1, \ldots, M_1 - 1; \tag{6.2.29}$$

$$\partial I_{M_1}/\partial U_{M_1}^n = 0. \tag{6.2.30}$$

Formula (6.2.30) arises because the functional I_{M_1+1} does not enter $I^{(l)}$. To compute the derivatives in formulas (6.2.29), (6.2.30) let us make use of an evident formula

$$\partial I_k/\partial U_k = \sum_{j=0}^{1} (\partial I_k/\partial a_{jk})(\partial a_{jk}/\partial U_k). \tag{6.2.31}$$

Employing formulas (6.2.31), (6.2.29), (6.2.23), (6.2.24), we obtain after some computations the following difference equations:

$$A^{(1)}U_{k-1}^n + B^{(1)}U_k^n + C^{(1)}U_{k+1}^n = F_k, \qquad k = 1, \ldots, M_1 - 1; \tag{6.2.32}$$

$$\alpha^{(1)}U_{k-1}^n + \beta^{(1)}U_k^n + \gamma^{(1)} = 0, \qquad k = M_1; \tag{6.2.33}$$

where

$$A^{(1)} = (1/s)(D_1 - 2k_f s)(1 + k_f s/(\Delta x)_1) - (D_1^2/2 + 2D_2 - D_1 k_f s)/D_0;$$

$$B^{(1)} = -(D_1 - 2k_f s)k_f/(\Delta x)_1 + (4D_2 - D_1 k_f s + D_1 as)/D_0$$

$$+ (D_1 - 2as)(1 + k_f s/(\Delta x)_1)/s; \tag{6.2.34}$$

$$C^{(1)} = -(D_1 - 2as)k_f/(\Delta x)_1 + (D_1^2/2 - D_1 as - 2D_2)/D_0;$$

$$F_k = (D_1 - 2k_f s)(b_{0k}/s - ab_{1k}) + (D_1 - 2as)$$
$$\times (b_{0,k+1}/s - ab_{1,k+1}) + (b_{1k}/s)(D_1^2/2 + 2D_2 - D_1 k_f s)$$
$$+ (b_{1,k+1}/s)(D_1^2/2 - 2D_2 - D_1 as);$$
$$\alpha^{(1)} = (D_1 - 2k_f s)(1 + k_f s/(\Delta x)_1)/s - (D_1^2/2 + 2D_2 - D_1 k_f s)/D_0;$$
$$\beta^{(1)} = -(D_1 - 2k_f s)k_f/(\Delta x)_1 + (D_1^2/2 + 2D_2 - D_1 k_f s)/D_0;$$
$$\gamma^{(1)} = -(D_1 - 2k_f s)b_{0k}/s - (b_{1k}/s)(D_1^2/2 + 2D_2 - D_1 k_f s)$$
$$+ (D_1 - 2k_f s)ab_{1k}, \qquad k = M_1.$$

$$(6.2.35)$$

In formulas (6.2.34), (6.2.35) $s = t_n - t_m$; $s = \tau$ at $m = n - 1$.

Let us show that the coefficients $A^{(1)}$, $B^{(1)}$, $C^{(1)}$ satisfy the inequalities

$$|B^{(1)}| > |A^{(1)}| + |C^{(1)}| + \delta, \qquad (6.2.36)$$

where δ is a positive quantity, $\delta = 2(\Delta x)_1/(3s)$. Substituting into formulas (6.2.34) the expressions for D_0, D_1, D_2 from (6.2.25) we find

$$A^{(1)} = C^{(1)} = [2k_f as^2 + ((\Delta x)_1)^2 - 2a^2 s^2 - 2k_f^2 s^2]/(3s(\Delta x)_1);$$
$$B^{(1)} = 2[(\Delta x)_1/s - A^{(1)}].$$

Now considering separately the cases $A^{(1)} > 0$ and $A^{(1)} < 0$ we easily obtain the inequality (6.2.36).

Taking into account the results of [6.56] and the inequalities (6.2.36) we arrive at the conclusion that the matrix of the difference scheme (6.2.32) is well conditioned. Now consider the question of the numerical solution of the algebraic system (6.2.32), (6.2.33). First consider the case $M_1 = 1$. Employing formulas (6.2.33), (6.2.35) we find

$$U_1^n = -(\alpha^{(1)}U_0^n + \gamma^{(1)})/\beta^{(1)}. \qquad (6.2.37)$$

It only remains to elucidate the question as to whether the quantity $\beta^{(1)}$ in (6.2.37) can vanish. Making use of formulas (6.2.35), (6.2.25) we find that

$$\beta^{(1)} = [as + (\Delta x)_1 - k_f s]^2/(2s(\Delta x)_1)$$
$$+ (a + k_f)^2 s/(6(\Delta x)_1) + (\Delta x)_1/(6s) > (\Delta x)_1/(6s) > 0.$$

Now consider the case $M_1 > 1$. Since the system (6.2.32) is an algebraic system with a three-diagonal matrix we can apply, for the solution of the system (6.2.32), a method of scalar sweep taking into account (6.2.36). The boundary condition for the sweep coefficients at the left end is obtained from the fact that the quantity U_0^n is known, it is computed according to (6.1.25). At the right end equation (6.2.33) is used.

Let us now go over to the problem (6.2.19). Introducing the new unknowns U_k^n, $k = M_1, \ldots, M$, by formulas (6.2.21), we again obtain similarly to the foregoing the system

$$A^{(2)}U^n_{k-1} + B^{(2)}U^n_k + C^{(2)}U^n_{k+1} = \tilde{F}_k, \quad k = M_1 + 2, \ldots, M; \quad (6.2.38)$$

$$\partial I_{M_1+1}/\partial U^n_{M_1+1} = \alpha^{(2)}U^n_k + \beta^{(2)}U^n_{k+1} + \gamma^{(2)} = 0, \quad k = M_1 + 1; \quad (6.2.39)$$

where the coefficients $A^{(2)}$, $B^{(2)}$, $C^{(2)}$, $\tilde{F}_k$ differ from $A^{(1)}$, $B^{(1)}$, $C^{(1)}$, F_k (see (6.2.34)), only by the fact that $(\Delta x)_1$ is replaced by $(\Delta x)_2$, this substitution has also been performed in (6.2.25);

$$\alpha^{(2)} = (D_1 - 2as)(1 + k_f s/(\Delta x)_2)/s - (D_1^2/2 - D_1 as - 2D_2)/D_0;$$

$$\beta^{(2)} = -(D_1 - 2as)k_f/(\Delta x)_2 + (D_1^2/2 - D_1 as - 2D_2)/D_0;$$

$$\gamma^{(2)} = -(D_1 - 2as)b_{0,k+1}/s - (b_{1,k+1}/s)(D_1^2/2 - D_1 as - 2D_2)$$

$$+ (D_1 - 2as)ab_{1,k+1}.$$

While solving the system (6.2.38), (6.2.39) we consider the following two cases: $M_1 + 1 = M$ and $M_1 + 1 < M$. Let $M_1 + 1 = M$. From (6.2.39) we find

$$U^n_M = -(\beta^{(2)}U^n_{M+1} + \gamma^{(2)})/\alpha^{(2)}.$$

The quantity $\alpha^{(2)}$ is always positive, because

$$\alpha^{(2)} = (k_f s + (\Delta x)_2 - as)^2/(2s(\Delta x)_2)$$

$$+ [(as + k_f s)^2 + ((\Delta x)_2)^2]/(6s(\Delta x)_2) > (\Delta x)_2/(6s) > 0.$$

At $M > M_1 + 1$ we solve the system (6.2.38), (6.2.39) by three-diagonal sweep. At the left end formula (6.2.39) is employed, and at the right end the value U^n_{M+1}, determined in accordance with formulas (6.1.25), is employed.

Consider the question on the order of approximation of a difference scheme (6.2.32) which, as we will keep in mind, has been obtained on a parallelogram mesh in the (x, t)-plane (see Figure 6.1). Substituting the Taylor series expansions of the type (6.1.14) into the difference equations (6.2.32) we obtain after some computations the f.d.a. H-form of the difference scheme (6.2.32) in the form

$$U_t + aU_x + \tau k_f(U_{xt} + aU_{xx}) + (\tau/2)(U_{tt} - a^2 U_{xx})$$

$$+ a_1 U_{xxx} + (1/2)k_f\tau^2 U_{xtt} + (1/6)\tau^2 U_{ttt} + a_2 U_{xxt} = 0, \quad (6.2.40)$$

where

$$a_1 = (1/3)k_f a\tau^2(k_f - a) + (1/6)[a((\Delta x)_1)^2 - k_f^3\tau^2];$$

$$a_2 = (2/3)k_f a\tau + (1/3)((\Delta x)_1)^2/\tau - (2/3)a^2\tau + \tau k_f^2/3.$$

Let us go over from the f.d.a. H-form (6.2.40) to the f.d.a. P-form by using recurrence formulas presented on page 16 of [6.45].

$$U_t + aU_x = (1/6)(k_f - a)\tau^2(a^2 + k_f^2)U_{xxx}. \quad (6.2.41)$$

It follows from the expression (6.2.41) for the f.d.a. P-form that the difference scheme (6.2.32) has at $k_f = a$ an order of approximation not lower than the third one. Note that the stencil of the difference scheme (6.2.32) uses three points at the upper level n and three points at the lower level $m = n - 1$. Third-order schemes with such a stencil on a rectangular grid in the (x, t)-plane have previously been constructed for equation (1.50) and investigated in a number of works (see, for example, [6.57]).

Concluding this section we remark that the difference scheme (6.2.32) may also be called a variational difference scheme [6.53] since it was obtained as a result of the minimization of certain functionals.

6.3. On the Difference Solution Refinement in the Neighborhood of a Shock Wave Front

In the computations of flows with shock waves using homogeneous difference schemes the shock waves are smeared, though less intensively than the contact discontinuities. Therefore, we can formulate and solve a problem on the refinement of the difference solution in the neighborhood of the shock wave front. We can mention some more reasons for such a refinement. First, shock waves of weak intensity are strongly smeared by first-order schemes (see Section 1.4). As a consequence of this the information on shock waves present in the flow may be completely "lost" in the process of computation of the problems where there are a number of interacting weak shock waves [6.58]. Second, post-shock parasitic oscillations are present in the numerical solutions obtained by second-order schemes, and it is difficult to get rid of these oscillations. On the other hand, an approximation of the difference solution in the neighborhood of a discontinuity in the refinement procedure by the formula (6.2.3) does not contain oscillations, as has been shown by the practice of computations using the technique of Section 6.2. Third, the treatment of a shock wave as a discontinuity in the refinement procedures, being analogous to those presented in Sections 6.1 and 6.2, restores the true property of a shock wave in an inviscid gas—the property of discontinuity of the flow parameters at the shock wave front. The consideration is carried out in the example of the Cauchy problem (5.1.1)–(5.1.3) for the Burgers equation. Let Ω_k be a domain in the neighborhood of a shock front which was introduced in Section 6.1 (see Figure 6.1). Consider the functional

$$I_k = \int_{\Omega_k} (\partial u/\partial t + \partial \varphi(u)/\partial x)^2 \, dx \, dt. \tag{6.3.1}$$

Introduce along with the domain Ω (see (6.1.24)), the domains

$$\Omega^{(l)} = \bigcup_{k=1}^{M_1} \Omega_k, \qquad \Omega^{(n)} = \bigcup_{k=M_1+1}^{M} \Omega_k, \qquad \Omega = \Omega^{(l)} \cup \Omega^{(r)},$$

and the functionals

$$I^{(l)} = \sum_{k=1}^{M_1} I_k, \qquad I^{(r)} = \sum_{k=M_1+1}^{M} I_k.$$

In Section 6.2 we have considered independently two minimization problems for the functional (6.2.11) and (6.2.18) at a fixed value of x_f^n. In the case of a shock wave such a consideration is impossible because the dynamic compatibility conditions, the Rankine–Hugoniot conditions (2.3.25), should be satisfied at the shock wave. In the case of one equation—the Burgers equation (5.1.1)—the conditions (2.3.25) are represented by a single equation

$$[\varphi(u)] = D[u]. \tag{6.3.2}$$

In connection with the foregoing, let us consider at a given approximate value of the discontinuity abscissa x_f^n a quadratic functional $I_0 = I^{(l)} + I^{(r)}$. By analogy with formulas (6.2.3), (6.2.4) consider in Ω_k local approximations

$$u_{ak}(x, t) = \sum_{j=0}^{N_x} \sum_{\nu=0}^{N_t} a_{kj\nu}(x - x_*^{(k)})^j (t - t_*)^\nu,$$
$$(x, t) \in \Omega_k, \qquad k = 1, \ldots, M_1; \tag{6.3.3}$$

$$u_{bk}(x, t) = \sum_{j=0}^{M_x} \sum_{\nu=0}^{M_t} b_{kj\nu}(x - x_*^{(k)})^j (t - t_*)^\nu,$$
$$(x, t) \in \Omega_k, \qquad k = M_1 + 1, \ldots, M. \tag{6.3.4}$$

In (6.3.3), (6.3.4) $N_x, N_t, M_x, M_t \geq 1$. Setting, in formulas (6.3.3), (6.3.4), $x_*^{(k)} = x_{lk}^{(m)}$, $t_* = t_m$ (as in Section 6.2), let us write the conditions for the continuity of the solution u in the domains to the left and right of a discontinuity (see formulas (6.2.17), (6.2.19))

$$u_{a1}(x_{l1}^{(n)}, t_n) = u_h(x_{l1}^{(n)}, t_n);$$
$$u_{ak}(x_{rk}^{(n)}, t_n) = u_{a,k+1}(x_{rk}^{(n)}, t_n), \qquad k = 1, \ldots, M_1 - 1;$$
$$u_{bk}(x_{rk}^{(n)}, t_n) = u_{b,k+1}(x_{rk}^{(n)}, t_n), \qquad k = M_1 + 1, \ldots, M - 1; \tag{6.3.5}$$
$$u_{b,M}(x_{rM}^{(n)}, t_n) = u_h(x_{rM}^{(n)}, t_n).$$

Suppose that at the moment of time $t = t_m = t_n - \tau$ the equality constraints (6.3.5), in which n should be replaced by $m = n - 1$, are satisfied. Introduce the notation U_k^m for the value of the refined solution in the kth node of the grid G_Δ. Let us require that the functions (6.3.3), (6.3.4) satisfy (as in Section 6.2) the condition

$$u_h(x, t_m) = \begin{cases} u_{ak}(x, t_m), (x, t_m) \in \Omega_k, & 1 \leq k \leq M_1, \\ u_{bk}(x, t_m), (x, t_m) \in \Omega_k, & M_1 + 1 \leq k \leq M, \end{cases} \tag{6.3.6}$$

at $t = t_m$. Since in the domain Ω the values of the solution u_h are known only in the nodes of the grid G_Δ, we obtain from (6.3.6), by analogy with (6.2.21), the following equality constraints:

$$u_{a1}(x_{l1}^{(m)}, t_m) = U_0^m, \qquad u_{ak}(x_{rk}^{(m)}, t_m) = U_k^m, \qquad k = 1, \ldots, M_1;$$

$$u_{bk}(x_{lk}^{(m)}, t_m) = U_k^m, \qquad k = M_1 + 1, \ldots, M; \qquad (6.3.7)$$

$$u_{bM}(x_{rM}^{(m)}, t_m) = U_{M+1}^m.$$

At given x, t and the values of the left-hand sides, equations (6.3.3), (6.3.4) are linear with respect to the coefficients a_{kjv}, b_{kjv}; therefore, we can easily eliminate, with the aid of formulas (6.3.7), $M + 2$ unknown coefficients. Note that we can eliminate some more quantities among the coefficients $\{a_{kjv}\}$, $\{b_{kjv}\}$ by using a spline approximation of the function $u_h(x, t_m)$ [6.59]–[6.61]:

$$u_h(x, t_m) = \begin{cases} \sum_{j=0}^{N_x} c_{kj}(x - x_*^{(k)})^j, \ (x, t_m) \in \Omega_k, & 1 \le k \le M_1, \\ \sum_{j=0}^{M_x} d_{kj}(x - x_*^{(k)})^j, \ (x, t_m) \in \Omega_k, & M_1 + 1 \le k \le M. \end{cases} \qquad (6.3.8)$$

Indeed, equating the coefficients at equal powers of the difference $(x - x_*^{(k)})$ in formulas (6.3.6) and (6.3.8), we obtain the relationships

$$\begin{aligned} a_{kj0} &= c_{kj}, & 1 \le k \le M_1; \\ b_{kj0} &= d_{kj}, & M_1 + 1 \le k \le M; \end{aligned} \qquad (6.3.9)$$

where c_{kj}, d_{kj} are known quantities determined by constructing the splines in formula (6.3.8). Introduce the vectors of the coefficients a_{jk}, b_{jk} sought after: $\mathbf{a} = \{a_{kjv}\}$, $\mathbf{b} = \{b_{kjv}\}$, where a part of the coefficients a_{kjv}, b_{kjv} is of course assumed to be already determined with the aid of formulas (6.3.7) or (6.3.9). Let us introduce for the purpose of brevity of notation an auxiliary function $u_a(x, t)$, determined in the domain $\Omega^{(l)}$ and such that $u_a(x, t) = u_{ak}(x, t)$ at $(x, t) \in \Omega_k$, $k = 1, \ldots, M_1$. The function $u_b(x, t)$ is introduced similarly in the domain $\Omega^{(r)}$. Approximate the functional $I_0 = I^{(l)} + I^{(r)}$ by a formula

$$I_0 = \int_{\Omega^{(l)}} [\partial u_a(x, t)/\partial t + (\partial/\partial x)\varphi(u_a(x, t))]^2 \, dx \, dt$$

$$+ \int_{\Omega^{(r)}} [\partial u_b(x, t)/\partial t + (\partial/\partial x)\varphi(u_b(x, t))]^2 \, dx \, dt. \qquad (6.3.10)$$

Let the equation $x = \zeta(t)$ (determined as in Section 5.1) be some approximation to the true discontinuity trajectory $x = \xi(t)$ in the solution of the problem (5.1.1)–(5.1.3). Let us write the Rankine–Hugoniot condition (6.3.2) in the form of an equality-type constraint

$$\Phi_1(\mathbf{a}, \mathbf{b}, x_f^n) = 0, \qquad (6.3.11)$$

where

$$\Phi_1(\mathbf{a}, \mathbf{b}, x_f^n) = \varphi(u_a(x_f^n, t_n)) - \varphi(u_b(x_f^n, t_n))$$

$$- \dot{\zeta}(t_n)[u_a(x_f^n, t_n) - u_b(x_f^n, t_n)]. \qquad (6.3.12)$$

In formula (6.3.12) $\dot{\zeta}(t_n) = d\zeta(t_n)/dt$. The simplest approximation for $\dot{\zeta}(t_n)$ may be taken (as in Section 6.2) in the form

$$\dot{\zeta}(t_n) = (x_f^n - x_f^m)/(t_n - t_m). \qquad (6.3.13)$$

The entropy condition should be satisfied at the shock wave. This condition in the case of the Burgers equation (5.1.1) has the form (5.1.15). Let us write the inequalities (5.1.15) in the form of two constraints

$$\Psi_1(\mathbf{a}, x_f^n) \geq 0, \qquad \Psi_2(\mathbf{b}, x_f^n) \geq 0, \qquad (6.3.14)$$

where we have introduced the notations

$$\begin{aligned}
\Psi_1(\mathbf{a}, x_f^n) &= \varphi'(u_a(x_f^n, t_n)) - \dot{\zeta}(t_n), \\
\Psi_2(\mathbf{b}, x_f^n) &= \dot{\zeta}(t_n) - \varphi'(u_b(x_f^n, t_n)).
\end{aligned} \qquad (6.3.15)$$

It is assumed that the derivative $\dot{\zeta}(t_n)$ in (6.3.15) has been replaced by its difference approximation, for example, by (6.3.13).

In some problems of mechanics and mathematical physics the function $u(x, t)$ should be nonnegative in accordance with its physical or mechanical meaning. In this case, the constraints (6.3.11), (6.3.14) may be augmented by the following inequality constraints:

$$\begin{aligned}
u_a(x, t_n) &\geq 0, \qquad x_l^n \leq x \leq x_f^n; \\
u_b(x, t_n) &\geq 0, \qquad x_f^n \leq x \leq x_r^n.
\end{aligned}$$

Taking into account the Rankine–Hugoniot condition (6.3.11) and the entropy conditions (6.3.14), we obtain the following nonlinear problem of constrained optimization (the value of x_f^n is fixed):

$$I_0(\mathbf{a}, \mathbf{b}, x_f^n) \to \min;$$
$$\Phi_1(\mathbf{a}, \mathbf{b}, x_f^n) = 0, \qquad \Psi_1(\mathbf{a}, x_f^n) \geq 0, \qquad \Psi_2(\mathbf{b}, x_f^n) \geq 0. \qquad (6.3.16)$$

A widespread approach to solving mathematical programming problems of the form (6.3.16) consists of the fact that one tries to reduce the constrained optimization problem (6.3.16) to some problem of unconstrained optimization [6.62]–[6.64]. One of the variants of reducing the problem (6.3.16) to an unconstrained minimization problem consists in considering an auxiliary function [6.65] using the idea of penalty functions

$$\begin{aligned}
I_1(\mathbf{a}, \mathbf{b}, x_f^n) &\equiv I_0(\mathbf{a}, \mathbf{b}, x_f^n) + \mu_1(\Phi_1(\mathbf{a}, \mathbf{b}, x_f^n))^2 \\
&\quad + \mu_2(\Psi_1(\mathbf{a}, x_f^n))^2[1 - \text{sign } \Psi_1(\mathbf{a}, x_f^n)] \\
&\quad + \mu_3(\Psi_2(\mathbf{b}, x_f^n))^2[1 - \text{sign } \Psi_2(\mathbf{b}, x_f^n)] \to \min. \qquad (6.3.17)
\end{aligned}$$

In formula (6.3.17) μ_1, μ_2, μ_3 are positive penalty constants. By controlling the order of magnitude of the quantities μ_1, μ_2, μ_3, we can control the accuracy of satisfaction of the constraints in the original problem (6.3.16) [6.66]. For example, if we want the Hugoniot condition (6.3.11), (6.3.12) to be satisfied to a higher accuracy than the remaining constraints in the problem (6.3.16), we must then set in formula (6.3.17) a value of the constant μ_1 greater in order of magnitude than the constants μ_2, μ_3.

After problem (6.3.16) has been solved at a fixed value of x_f^n we can find the value of the functional (6.1.29) needed in the process of solving the basic constrained optimization problem (6.1.31). The sum S_h' entering the functional (6.1.29) may be computed either by formula (6.1.28) or by the formula

$$S_h' = \sum_{i=i_l}^{i_{fn}-1} u_a(x_i, t_n)h + u_{i_{fn}}^n h + \sum_{i=i_{fn}+1}^{i_r} u_b(x_i, t_n)h, \qquad (6.3.18)$$

where i_{fn} is the number of a cell of the grid G_h in which the point with coordinates (x_f^n, t_n) is located. Let $x_{i_{fn}-1/2}, x_{i_{fn}+1/2}$ be the abscissas, respectively, of the left and right boundary of the cell i_{fn} of the G_h mesh. Then the formula [6.37]

$$u_{i_{fn}}^n = \frac{1}{h}\left[\int_{x_{i_{fn}-1/2}}^{x_f^n} u_a(x, t_n)\, dx + \int_{x_f^n}^{x_{i_{fn}+1/2}} u_b(x, t_n)\, dx\right] \qquad (6.3.19)$$

may be used for the calculation of the quantity $u_{i_{fn}}^n$ in equation (6.3.18).

Note that the above-presented sequence of computations, as the solution of a series of problems of the form (6.3.16) at fixed values of x_f^n with the purpose of obtaining the solution of problem (6.1.31), is only one of the possible ways for the solution of the constrained optimization problem

$$I_0(\mathbf{a}, \mathbf{b}, x_f^n) \to \min;$$

$$\Phi_1(\mathbf{a}, \mathbf{b}, x_f^n) = 0, \qquad \Psi_1(\mathbf{a}, x_f^n) \ge 0; \qquad (6.3.20)$$

$$\Psi_2(\mathbf{b}, x_f^n) \ge 0, \qquad S_h' - S_h = 0, \qquad x_f^n - x_l^n \ge 0, \qquad x_r^n - x_f^n \ge 0.$$

Employing the penalty functions method we can reduce the number of constraints in the problem (6.3.20) by solving the following constrained optimization problem:

$$I_2(\mathbf{a}, \mathbf{b}, x_f^n) \equiv I_1(\mathbf{a}, \mathbf{b}, x_f^n) + \mu_4(S_h - S_h')^2 \to \min,$$

$$x_f^n - x_l^n \ge 0, \qquad x_r^n - x_f^n \ge 0,$$

where the functional I_1 is determined by (6.3.17). It has been pointed out in [6.65] that the penalty functions method is slow and not very reliable. It is applicable only at a small number of variables $N \le 10$. In this connection, it is useful to consider other methods of constrained optimization for the numerical solutions of problems (6.3.16) or (6.3.20). A comparison of 20 numerical methods of solving a general nonlinear mathematical programming problem has been carried out in [6.67]. As a result of this, the conclusion was drawn that the most efficient methods of solving constrained optimization problems of large dimensionality are: the simplex method as realized by

Nelder and Mead and the method proposed by the author of [6.67], which combines the ideas of the direct search method and the method of feasible directions (see, for example, [6.63] on these methods). Note that in [6.68] the penalty function $\mu_1 \Phi_1^2$ was added to the basic variational functional of the problem in computations of a transonic potential flow around an airfoil by the finite element method (FEM). In the process of numerical discretization this led to the appearance of additional items in some coefficients of the FEM algebraic system.

Employing the results of Section 5.2 we can consider in (6.3.17) an item

$$\mu_1 \int_{t_m}^{t_n} \{\varphi(u_a(\zeta(t), t)) - \varphi(u_b(\zeta(t), t)) - \dot{\zeta}(t)[u_a(\zeta(t), t) - u_b(\zeta(t), t)]\}^2 \, dt$$

instead of the item $\mu_1 \Phi_1^2$ where one can use for $\zeta(t)$, $\dot{\zeta}(t)$, for example, the approximations

$$\zeta(t) = x_f^m + \dot{\zeta}(t_n)(t - t_m),$$

$$\dot{\zeta}(t) = \dot{\zeta}(t_n) = (x_f^n - x_f^m)/(t_n - t_m).$$

Note that the polynomial approximation of the functions u_{ak}, u_{bk} by formulas (6.3.3), (6.3.4) is not uniquely possible. For example, a rational approximation is well known [6.65].

The above method for the reduction of the refinement problem to mathematical programming problems was based on local approximations (6.3.3), (6.3.4) being valid in the domain Ω_k. If the size of the domains $\Omega^{(l)}$, $\Omega^{(r)}$ is not large, then the use of two "global" approximations $u_a(x, t)$ and $u_b(x, t)$ in the domains $\Omega^{(l)}$ and $\Omega^{(r)}$, respectively, instead of M local approximations (6.3.3), (6.3.4), may prove to be more efficient. After the form of the approximation functions u_a and u_b has been specified, we can again formulate the constrained optimization problems of the form (6.3.16) or (6.3.20). In this case it is possible to reduce substantially the dimension of optimization problems, since it does not depend (in the case of a global approximation) on the number of domains Ω_k, but depends only on the number of unknown coefficients in the approximation functions u_a and u_b.

Using the above refinement method we can solve on a uniform spatial grid a problem on the degeneration of a shock wave into a continuous flow. Indeed, if the value of $|u_a(x_f^n, t_n) - u_b(x_f^n, t_n)|$ becomes smaller than some given positive number at some $n = n_0$, then we shall assume that the shock wave has disappeared and gone over to a continuous flow. Then at $t > t_{n_0}$ there is no need to use the refinement procedure. Note that in Section 2.5 we have presented a method of solving a problem on shock wave degeneration which makes use of a nonuniform moving grid adapting to the flow.

Thus, the above method of difference solution refinement in the neighborhood of shock waves is characterized by a very high versatility and a variety of specific forms of numerical realization.

6.4. Computational Examples

Below we discuss some methodical questions of organizing the computations by the difference solution refinement techniques described in Sections 6.1 and 6.2, and we also present computational examples obtained by using the above optimization refinement techniques.

It follows from the results of Sections 6.1–6.3 that we must aim at a reduction of the number of iterations while solving numerically the constrained optimization problem (6.1.31), because the computation of the functional $I(x_f^n)$ (see formula (6.1.29)), is associated in its turn with the solution of some optimization problem. In this connection, it is necessary to set as exactly as possible the initial approximation for the strong discontinuity abscissa x_f^n. For this purpose a differential analyzer was used in [6.37]. Taking into account the results of Section 4.1, for the determination of the discontinuity position in the zone of its smearing by the "corner" scheme (6.1.4), a differential analyzer can be used which is based on determining the abscissa of the maximum of the quantity $|\partial u_h/\partial x|$. This abscissa can be found by examining all the cells of the mesh on the x-axis. This technique was used by us, for example, in [6.69]–[6.72]. However, such a simple technique is not the most economical. The golden section method [6.63] is known as one of the most efficient methods of one-dimensional optimal search. In order to have a possibility of using this method, one must go over from a problem on the search for the maximum of the function $|\partial u_h/\partial x|$ to an equivalent problem on the search for a minimum. Let us introduce for this purpose a function

$$g(x, t) = 1/[1 + (\partial u_h(x, t)/\partial x)^2]. \tag{6.4.1}$$

It is easy to see that $g(x, t)$ has local minima at the points where $|\partial u_h/\partial x|$ has local maxima. It is clear that the value of x_f^n, found with the aid of the function (6.4.1), is only an approximate solution of the problem on the minimization of the basic functional (6.1.29), since the absolute error of localization may reach the value $\pm 0.5h$ (see the computational examples in Sections 4.3 and 5.5). Let $(x_f^n)_{\text{d.a.}}$ be the value of x_f^n determined with the aid of a differential analyzer. Let us set the values x_l^n, x_r^n in formulas (6.1.30), (6.1.31) as follows:

$$x_l^n = (x_f^n)_{\text{d.a.}} - 0.5h; \qquad x_r^n = (x_f^n)_{\text{d.a.}} + 0.5h.$$

For the solution of problem (6.1.31) we use the golden section method. Note that this method is only slightly worse than the Fibonacci method [6.63], being optimal with respect to the needed number of computations of the values of the function $I(x_f^n)$. Let $\delta_3 h$ be a user-specified absolute error in computing x_f^n by the golden section method while solving problem (6.1.31). Denote by δI the corresponding error in the value of the function $I(x_f^n)$. Then in accordance with [6.65]

$$\delta I = O((\delta_3 h)^2) \tag{6.4.2}$$

It follows from the foregoing that in the case when $(x_f^n)_{d.a.}$ is used as x_f^n, and problem (6.1.31) is not solved at all, then one should obviously set in (6.4.2) $\delta_3 = 0.5$. Since the exact value of the functional (6.1.29) at the minimum point should be equal to zero we have, with regard to (6.4.2), that

$$|S_h - S_h'| = O(\delta_3 h).$$

That is, the error of violation of the law of the "mass" (6.1.26) conservation may prove to be considerable. This error can easily be reduced by solving problem (6.1.31) with a reasonably chosen value of δ_3. If we take a very small value of δ_3 (for example, $\delta_3 = 10^{-5}$), then we shall need a rather large number of computations of the values of the function (6.1.29), which would lead to an increase in the needed computing time. On the other hand, in many problems of gas dynamics the relative error in $\mathcal{M}_n$ of the order 0.1% proves to be quite acceptable (see, for example, [6.8]).

Consider the question of taking into account the event when some of the strong discontinuities go out of the computational domain. We consider here the case $a > 0$ in equation (1.50). In this case the discontinuities in the solution of (1.50) propagate from left to right at a speed $dx/dt = a$. Let $x = l$ be the abscissa of the right boundary of the computational domain. Assume that the abscissa x_f^m of a discontinuity under study is known at $m = n - 1$. If $x_f^m + 2h > l$, then at $t \geq t_n$ we do not apply the difference solution refinement algorithm in the neighborhood of this discontinuity. Thus, this discontinuity is computed by the difference scheme (6.1.4) in a "shock-capturing" way during several time steps, which are needed to advance the discontinuity by a distance of $2h$ before leaving the computational domain.

With the purpose of checking the correctness of the golden section subroutine we have compared the number of computations v_c of the function $I(x_f^n)$ (6.1.29) with a theoretical estimate of the number of computations v which may easily be obtained with the aid of [6.63]

$$v \geq v_*, \quad v_* = \ln(\delta_3^{-1})/\ln[(1 + \sqrt{5})/2], \tag{6.4.3}$$

where $\delta_3 h$ is a user-specified error in determining x_f^n by the golden section method. For example, at $\delta_3 = 0.1$ we have from (6.4.3) that $v_* = \ln 10/\ln 1.618 = 8.34$; in a computation by our golden section method subroutine we have obtained $v_c = 9$; at $\delta_3 = 10^{-3}$ — $v_* = 25.02$, $v_c = 22$; and at $\delta_3 = 0.25$ — $v_* = 5.02$, $v_c = 6$.

In Figure 6.2 we show the result of a computation of a "classical" test in which the initial function has the form of a "step". In this example 200 time steps were carried out with the Courant number $\kappa = 0.75$ in the difference scheme (6.2.32)–(6.2.35), (6.2.38), (6.2.39), where $M_1 = 1$, $M = 2$, $h = 0.02$; $\delta_3 = 10^{-3}$. At $t = 200\tau$ the value of x_f^n, obtained by the solution of problem (6.1.31), differed from its exact value by a magnitude of the order 10^{-10}, and the relative error in the "mass" $\mathcal{M}_n$ (see formula (6.1.26)), had a magnitude of the order 10^{-9}. The computational result presented in Figure 6.3 has also been

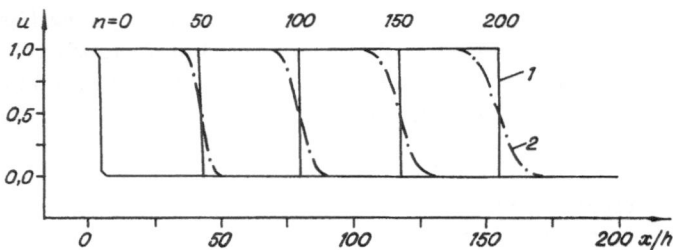

Figure 6.2. The results of the "classical" test calculation. 1—with the refinement; 2—the difference solution by the scheme (6.1.4).

obtained by scheme (6.2.32)–(6.2.35), (6.2.38), (6.2.39) with $\kappa = 0.75$, $M_1 = 1$, $M = 2$. At $t = 150\tau$ the error in the "mass" $\mathcal{M}$ was equal to 0.15%. We can see from Figure 6.3 that in the case of a computation without refinement the information on the specific breaks in the solution profiles is lost. In the computation of Figure 6.3 the difference solution was also refined in the neighborhood of a weak discontinuity by the method of Sections 6.1 and 6.2, that is, this discontinuity was considered in the same way as if it were a strong discontinuity. The abscissa x_f^n of a weak discontinuity was determined in the process of the functional (6.1.29) minimization by the golden section method, and we have set in (6.1.31)

$$x_l^n = x_f^m - 2h, \qquad x_r^n = x_f^m + 2h.$$

The exact solution in Figures 6.2–6.10 coincides within the limits of the accuracy of the graphs with the difference solution obtained with the use of the refinement procedures of Sections 6.1 and 6.2.

It was of interest to compare the above-proposed technique with the "artificial compression" method proposed by A. Harten [6.14], [6.22]–[6.24], for a more accurate shock-capturing computation of gas flows with contact discontinuities and shock waves. In accordance with [6.23] the artificial compression method may be realized as the following three-stage algorithm.

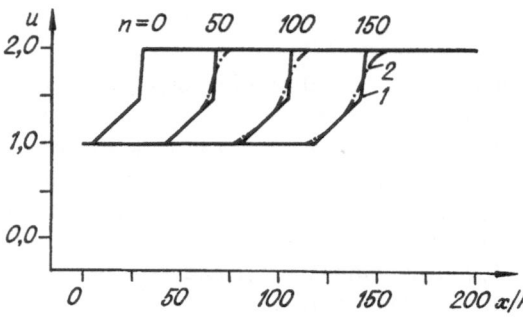

Figure 6.3. The calculation of a profile with jumps. 1—with the refinement in the neighborhood of weak and strong discontinuities; 2—the difference solution by the scheme (6.1.4).

Stage one is the obtaining of a difference solution of equation (1.50) by some conventional difference scheme; we have used the scheme (6.1.4). We shall denote this difference solution by $\tilde{u}^n$.

Stage two consists of the use of the Lapidus artificial viscosity [6.14], or of the "Lapidus smoothing"

$$\tilde{\tilde{u}}_i^n = \tilde{u}_i^n + (v\tau/h)\Delta'[|\Delta'\tilde{u}_{i+1}^n|\Delta'\tilde{u}_{i+1}^n], \tag{6.4.4}$$

where $\Delta'\tilde{u}_i^n = \tilde{u}_i^n - \tilde{u}_{i-1}^n$, and v is the coefficient of the artificial viscosity; we have set $v = 2$ as in [6.14]. This value of the coefficient v (as was pointed out in [6.14]) is needed to ensure that the difference solution to which the artificial compression method is applied has no oscillations (these oscillations in the case of using the scheme (6.1.4) may be generated by the Harten artificial compression scheme (6.4.5)–(6.4.9) (see below), by virtue of a nonlinear character of this scheme).

Stage three is the HARTEN "artificial compression" [6.23], [6.24], [6.14]

$$u_i^n = \tilde{\tilde{u}}_i^n - [\tau/(2h)](\theta_{i+1/2}^n G_{i+1/2}^n - \theta_{i-1/2}^n G_{i-1/2}^n), \tag{6.4.5}$$

where

$$\hat{\theta}_i = \begin{cases} \left|\dfrac{|\Delta_{i+1/2}| - |\Delta_{i-1/2}|}{|\Delta_{i+1/2}| + |\Delta_{i-1/2}|}\right|^p, & |\Delta_{i+1/2}| + |\Delta_{i-1/2}| > \varepsilon_n, \\ 0, & \text{otherwise.} \end{cases} \tag{6.4.6}$$

We have set $p = 1$ in (6.4.6) (as SOD in [6.14]), $\Delta_{i+1/2} = \tilde{\tilde{u}}_{i+1}^n - \tilde{\tilde{u}}_i^n$, and $\varepsilon_n > 0$ is a measure of negligible variation of the solution $\tilde{\tilde{u}}^n$. By analogy with [6.24] we have taken $\varepsilon_n = 0.01 \max_i |\Delta_{i+1/2}|$.

$$\theta_{i+1/2}^n = \max(\hat{\theta}_i, \hat{\theta}_{i+1}). \tag{6.4.7}$$

$$G_{i+1/2}^n = g_i^n + g_{i+1}^n - |g_{i+1}^n - g_i^n|S_{i+1/2}, \tag{6.4.8}$$

$$g_i^n = \alpha_i\Delta_i, \qquad S_{i+1/2} = \text{sgn}(\Delta_{i+1/2}),$$

$$\Delta_i = \tilde{\tilde{u}}_{i+1}^n - \tilde{\tilde{u}}_{i-1}^n, \tag{6.4.9}$$

$$\alpha_i = \max\{0, \min(|\Delta_{i+1/2}|, \Delta_{i-1/2}S_{i+1/2})/(|\Delta_{i+1/2}| + |\Delta_{i-1/2}|))\}.$$

It has been shown in [6.37] in the example of computations that the use of the Lapidus smoothing (6.4.4) with $v = 2$ leads to an excessive smearing of discontinuities. In this connection also, the computations without the Lapidus smoothing (6.4.4) in the artificial compression method have been carried out in [6.37]. As a difference solution $\tilde{u}^n$ we have used the solution obtained by scheme (6.1.4). Comparing Figure 6.4(b) and (a) we see that the artificial compression after Harten enables us to reduce the width of the zone of strong discontinuity smearing. For example, we see in Figure 6.4(b) that at $t = 75\tau$ the width of this zone makes approximately $12h$, whereas in Figure 6.4(a) this width is equal to $17h$. We note for comparison that the use of the refinement algorithm presented in previous sections yielded the width of smearing equal

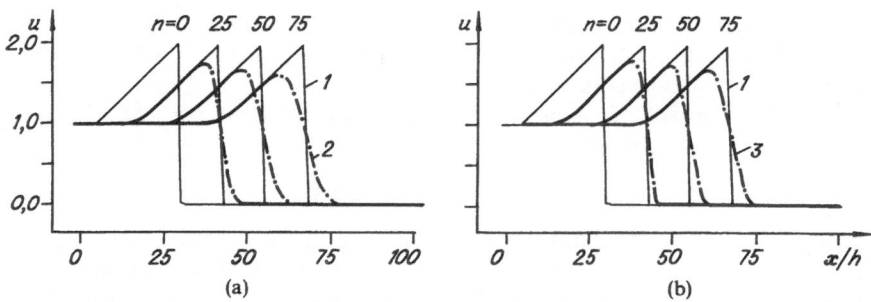

Figure 6.4. Calculation of the shocked profile propagation at nonzero gradients: (a) by the scheme (6.1.4): 1—with the refinement, the scheme (6.1.11), (6.1.22); 2—without the refinement; (b) with the use of the algorithm (6.1.4), (6.4.5)–(6.4.9) (the curve 3).

to $h \div 2h$. Returning to Figure 6.4(b) we clearly see that the width of strong discontinuity smearing in the case of the application of the Harten artificial compression method (6.4.5)–(6.4.9) increases as time t increases. In addition, the height of the difference solution peak diminishes with time when using formulas (6.4.5)–(6.4.9). At the same time, the difference solution peak oscillates with a small amplitude about some value which is very close to the exact solution value immediately behind the front of a strong discontinuity, in the case of the application of the refinement algorithm described in the foregoing sections.

In Figures 6.5 and 6.6 we present an example of a computation of the propagation of a configuration containing seven discontinuities at $t = 0$. The inclination angle to the x-axis of a linear profile behind the front of each discontinuity was taken to be 30°, the distance at $t = 0$ between neighboring discontinuities was equal to $6h$. The scheme (6.1.11), (6.1.22) was used to obtain the solution U^n. The refinement of the solution in the neighborhood of each discontinuity enables us to follow its evolution in time, up to the time when the discontinuity is to leave the computational domain through the right boundary $x = l$. For example, it may be seen from Figure 6.5 that at $t = 50\tau$

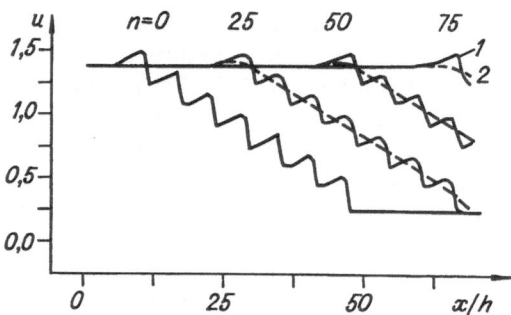

Figure 6.5. The computation of the propagation of a profile with seven discontinuities: 1—with the refinement; 2—without the refinement, by scheme (6.1.4).

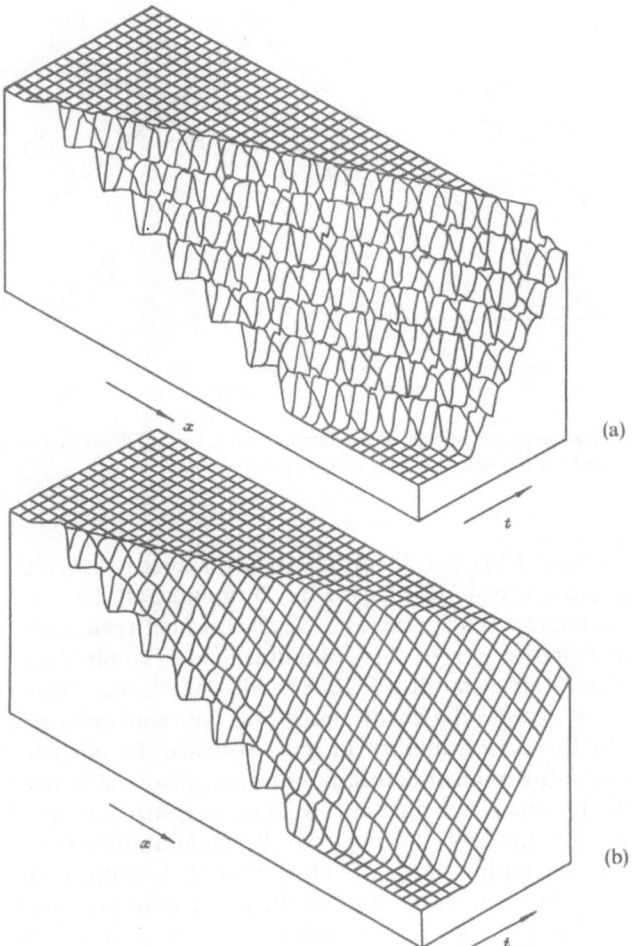

Figure 6.6. The computation of the surface $u = u(x, t)$: (a) with the refinement; (b) without the refinement, by scheme (6.1.4).

four discontinuities are located in the computational domain, whereas at $t = 75\tau$ there is only one discontinuity left—the remaining discontinuities have already left the computational domain. It is seen from Figure 6.6(b) that in the case of a computation by scheme (6.1.4) without refinement the smoothing of steps proceeds very rapidly, during the first 10–15 time steps. After that information on the number of discontinuities and the form of the profiles in the neighborhood of each discontinuity is completely lost. In Figure 6.7 we present an example of a computation on the propagation of a configuration containing 40 discontinuities at $t = 0$. The computation was carried out by

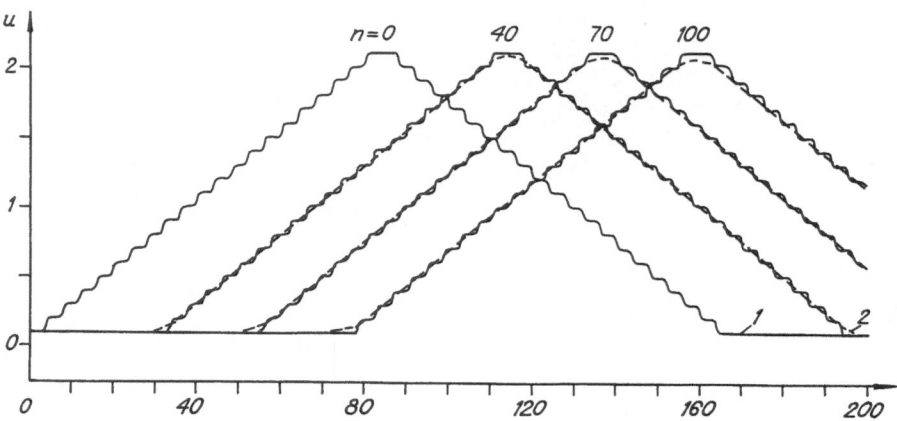

Figure 6.7. The computation of the propagation of a profile with 40 discontinuities: 1—with the refinement; 2—without the refinement, by scheme (6.1.4).

using the scheme (6.1.11), (6.1.22). At $t = 0$ the distance between the neighboring discontinuities was $4h$ (except for the 20th and 21st discontinuity which are located at the top of a "pyramid", the distance between these two discontinuities was equal to $8h$). In this computational example the errors in the discontinuities localization at $t = 100\tau$ made $0.005h$. By virtue of the construction of the above refinement algorithm, the refinement procedure can be applied to the discontinuities in any sequence. In particular, in serial computer calculations the discontinuities were considered in the order "from left to right" by means of an appropriately organized loop. When using multi-processor computers allowing a parallel computation, we can efficiently economize the computing time [6.73] by giving each of the available processors a "task" for the processing of a specific group of discontinuities.

In the case when the derivative $\partial^2 u / \partial x^2$ in the neighborhood of a strong discontinuity is large, the approximation error of the difference scheme (6.1.11), (6.1.22) may prove to be large (see the f.d.a. (6.1.23)). Therefore, in such cases the use of a second-order scheme (6.2.32)–(6.2.35), (6.2.38), (6.2.39) may be more reasonable. To verify this hypothesis we also carried out the computations for the case when the initial profile $u_0(x)$ was composed of three parabolic arcs, in such a way that there were discontinuities at the boundaries between the first and second arc, between the second and third arc, and between the third arc and the constant state $u_0(x) = $ const (the numbering of arcs is from left to right, see Figure 6.8). Note that the sign of the solution profile curvature in the case of using a scheme without refinement (dashed line in Figure 6.8), changes to the sign being opposite with respect to the true curvature in the domain of the first parabolic arc as the time t increases. This is explained by the effect of the rounding off of the difference solution profile obtained by the

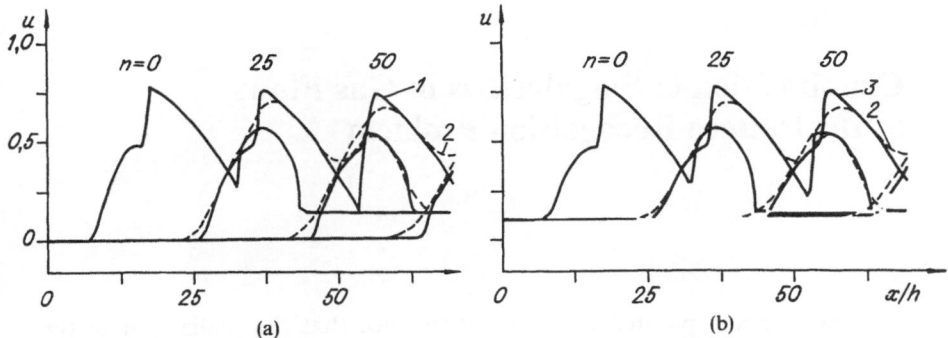

Figure 6.8. A profile of three parabolic arcs: (a) the refinement based on the discrete functional minimization; (b) the refinement on the basis of the integral conservation law (6.1.8) approximation. 1—a computation with refinement, $M_1 = 2$, $M = 4$, in the scheme (6.2.32)–(6.2.35), (6.2.38), (6.2.39); 2—without the refinement, by scheme (6.1.4); 3—with the refinement, $M_1 = 2$, $M = 2$, in the scheme (6.1.11), (6.1.22)

scheme (6.1.4) in the vicinity of a weak discontinuity (the solid line in Figure 6.8 has also been obtained with the refinement of the solution in the neighborhood of a weak discontinuity).

The algebraic system (6.1.34) arising in the process of realization of the scheme (6.1.11), (6.1.22) at $M_1 > 1$, $M - M_1 > 1$, was solved by the Gauss elimination method with column pivoting.

In concluding this section let us present some data on the computing time required for the realization of the above-presented refinement algorithms. First of all, it is clear that this time depends on the number of discontinuities to be considered, as well as on the numbers M_1 and M characterizing the number of nodes of the grid G_Δ. In addition, the computing time depends on the number of time steps to be computed, on the number of nodes of the grid G_h, and on the complexity of the difference scheme used on G_h. For example, a problem with 40 discontinuities (see Figure 6.7), required a computer time which was four times as large as the computer time required for a problem with seven discontinuities (see Figures 6.5 and 6.6(a)).

CHAPTER 7

Classification of Singularities in Gas Flows as the Pattern Recognition Problem

We have already pointed out more than once that the realization of the methods presented in Chapters 2–5 for the localization of strong discontinuity surfaces in the numerical solutions of gas dynamics problems is related closely to the use of *a priori* information on the orientation of these surfaces with respect to spatial coordinate axes. In practice, this leads to the fact that a specific algorithm for the singularities localization which belongs to this rather vast group is only applicable to some restricted class of fluid mechanics problems, and in the course of consideration of new problems one has to search for other algorithms of the singularities localization. This lack of universality in the above-presented localization methods can be observed especially well in the process of numerical simulations of such phenomena or in the processes of gas dynamics, or, speaking more generally, in the macro- and microworld, the investigation of which is not accessible to other techniques (for example, experimental techniques), and where mathematical modeling therefore becomes the only means of studying these phenomena or processes [7.1], [7.2]. Thus, there exists a need to develop methods for the localization of singularities, which do not require for their realization *a priori* information on the absence or on the presence of various types of singularities, and on their approximate orientation with respect to spatial coordinate axes. We present in this chapter a general approach to the solution of the above problem, which makes it possible to carry out on a computer a detailed analysis of the flow picture, in addition to the localization itself.

This approach represents a development of the basic idea (published for the first time in [7.3]) that the higher levels of processing must include recognition algorithms and automatic classification of new objects, i.e., specific capacities of "artificial intelligence" (AI). The works on artificial intelligence make up the content of one of the parts of modern technical cybernetics. In its turn, object recognition is one of the research directions in artificial intelligence. In [7.4] the pattern recognition and mathematical modeling methods are considered as the two basic methods which were introduced by the informatics into the solution of the problem of prognostication (diagnosis) of the phenomena being investigated by descriptive sciences.

Using the terminology of [7.5] we shall call, in the following, the systems of processing the results of gas- and hydrodynamical calculations (which enable one not only to localize the singularities in the multidimensional flows but also to classify them automatically into several types), the intellectual systems of information extraction. As will be seen from the following presentation, the principles of organization of such higher level systems which have been realized by us are similar to those underlying the various intellectual systems of processing and recognition which are presently exploited in various branches of science and technology: in systems of the industrial vision of robots [7.6], [7.7], in biology [7.8], in medicine [7.8]–[7.10], in remote sensing [7.11]–[7.13], etc. The major part of these systems makes use of digital methods of image processing, although analog methods also exist; for example, optical [7.14] and holographic [7.15] methods. It was pointed out in [7.16] that the digital image processing techniques have advantages over the analog techniques—such as flexibility and accuracy; a wide dissemination of digital methods of processing is also favored by the appearance of powerful general-purpose computers as well as highly efficient computers realizing the principles of cellular logic [7.17], [7.18].

It is to be noted that the digital image processing techniques are beginning to penetrate gradually the domain of gas-dynamical experiment [7.19]–[7.24], thus facilitating substantially the processing of experimental data and obtaining the needed quantitative information (for example, on the magnitude and direction of fluid particle velocities, as in [7.19]). In these cases the digital encoding stage of photographic or TV images of the flow under study precedes the stage of digital processing. For this purpose serially produced equipment for image scanning is used. The digital encoding of the image assumes the partitioning of the image brightness into a finite number of grey levels (usually this number is represented by some power of the number 2, for example, 16, 64, 256), and each grey level is encoded by a certain number, say, by a number from the integer number interval $0 \div 255$. At present the number of works dealing with the applications of digital image processing to the analysis of data of gas-dynamical experiments is very small in comparison, for example, with similar works in the domain of computer processing of medical images. It is nevertheless forecast in [7.22] that image processing will become a powerful instrument in the future for investigations into gas- and hydrodynamics.

In the case of the application of digital image processing techniques to the analysis and interpretation of numerical results in the solution of aerohydrodynamics problems the stage of digital encoding is obviously unnecessary, since as the result of computer solution of a problem we obtain digital images which can thus immediately be subject to a goal-oriented processing, which may be based on an appropriate technique or on a combination of techniques from the rich arsenal of modern digital image processing theory. In accordance with the terminology accepted in the theory of digital image processing—by image a two-dimensional array of numbers is meant—the element of this array is usually called a pixel.

Let us enumerate a few aspects of the application of an intellectual system of information extraction from the results of shock-capturing aerohydromechanical computations. First of all, the implementation of computer systems of an automatic analysis of the flow structure enables one to achieve a better understanding of the physics of the flow under study [7.1], [7.2], [7.4], [7.25]. In addition, such systems can aid the computational aerodynamicist to reduce the time devoted to extracting, displaying, and analyzing various features of the solution, thus improving the overall efficiency of the work, especially in cases when three-dimensional fluid flows are studied [7.26], [7.27]. In connection with the foregoing, the authors of [7.27] and [7.28] reached the conclusion that an active or "smart" software display package is needed to search the database for interesting flow phenomena and then display them.

The second aspect of the application of intellectual systems of information extraction in computational fluid dynamics is related to the direct use of the information obtained at the output of such systems, for the control of a computational process of numerical solution of the basic aerohydrodynamic problem with the purpose of increasing the accuracy of computation. In particular, we have presented in sufficient detail in Chapters 2 and 6 some algorithms for difference solution refinement in the neighborhood of singularities which use the information on the location of these singularities obtained with the aid of specialized localization algorithms: local adaptive mesh concentration in the neighborhood of singularities, the refinement of solutions in the neighborhood of discontinuities by the least squares method, and by mathematical programming methods. A number of approaches for the control of different stages of a computational process have been enumerated in [7.29], [7.30]. First of all, we would like to mention a few works [7.31]–[7.33] devoted to the application of AI concepts to the grid adaptation. In particular, DANNENHOFFER and BARON [7.31], [7.32] report on an expert system embedded in a two-dimensional grid adaptation scheme. This system in particular contains a knowledge on when and how to refine or coarsen the computational grid (see a description of this system also in [7.33]). The expert system EZGrid presented in [7.33] is an Expert Zonal Grid Generation system that partitions a two-dimensional flow field into four-sided, well-shaped zones that are then individually discretized. In [7.34] it was proposed to partition the two-dimensional or three-dimensional computational domain with regard to the location of strong and weak discontinuities. In the subdomains of a smooth flow the fields of gas-dynamical quantities were represented in [7.34] as expansions into the Fourier–Chebyshev series. Such a piecewise polynomial representation of the flow fields greatly facilitates the mesh rezoning, or passage to another difference method in the same spatial domain in the process of numerical solution.

As was indicated in [7.29], the AI concepts and methods can also be used for the formulation and choice of a way of solving a problem in aerodynamics;

and the choice and use of a method for flow computation. The importance of the problem on the choice and construction of a corresponding "optimal" (for a given problem) solution technique was also indicated in [7.2]. In this connection a concept of "rational mathematical modeling" was developed in [7.2], and there are enumerated there a number of techniques for the construction of "rational" numerical models as applied to nonlinear and multidimensional problems of continuum mechanics.

One of the first realizations of such a "rational" approach, as applied to the numerical study of jet flows, has been presented in [7.35], [7.36]; the corresponding method of mathematical modeling has been termed (in [7.35]) "identification modeling". Let us briefly present the essence of this approach. Its computer implementation was preceded by a stage of so-called "physical modeling"—being the most laborious. The purpose of this study is to determine a qualitative form of analytic dependencies (for example, of polynomial, exponential form) by which the flow parameters are described in different subdomains of jet flows in processing a large amount of experimental, numerical, and theoretical data on the properties of different subdomains of jet flows. These dependencies are then stored in the computer memory thus forming a database. In the process of a numerical investigation of a specific jet problem, by the method of [7.35], first the domains with monotone distribution of the parameters are extracted. For this purpose the locations of shock waves and contact discontinuities in a flow are determined by using data from the database. Then the identification of the determined subdomains of continuous flow is carried out, with the purpose of determining the form of dependencies being stored in the database which adequately describe qualitatively the behavior of flow parameters at the chosen points. The coefficients of these dependencies are then refined with the help of three–four point interpolation or by the least squares method. Along certain lines in the flow (for example, along the shock waves) the differential equations of the flow may be numerically integrated. The specific feature of the approach of [7.35] is the substantial use in computations of the *a priori* information on the structure of the flows under study. The refusal of the use of a finite-difference method for integrating over all of the flow domain enabled the group of three research workers (including the head of the group I.L. Dobroserdov) to develop a very rapid solver of the applied problems which ensures the accuracy of computation that meets the demands of engineering practice.

Once a numerical technique has been chosen or developed for the solution of a specific flow problem, there arises a problem on the stability analysis of the corresponding numerical algorithm. Here it is also possible to use the AI methods, such as symbolic manipulations on a computer [7.37]–[7.42], and pattern recognition [7.40]–[7.42]. In these works, the symbolic manipulations on a computer, which are performed with the aid of the REFAL programming language, are used to generate a FORTRAN subroutine which enables us to compute the left-hand sides of the inequalities as being equivalent to the satisfaction of the von Neumann stability criterion. The stability analysis method proposed in [7.40]–[7.42] also enables us to analyze the stability of difference schemes in the situations where there is no *a priori* information on the shape of the stability domain of the scheme; in particular, this domain can be multiply connected. The method of [7.40]–[7.42], which is very robust and reliable, employs a procedure (for detecting the stability domain boundaries) which essentially coincides with the procedure for the detection of discontinuity lines presented in Section 7.3.1.

Finally, there exists a third aspect of the application of intellectual systems of information extraction in combination with computational fluid dynamics the development of which has begun quite recently [7.33], [7.43]–[7.46]: the application of the concepts of artificial intelligence and expert systems in combination with the numerical solution of aerohydrodynamic problems for the purposes of the design of new aerodynamic shapes. The expert system is understood as a system which attempts to solve, by means of artificial intelligence, a class of nonanalytical complex problems (of aerohydromechanics) which require specialized knowledge that most people do not possess [7.5], [7.45]. The approach is to imitate those experts who have accumulated the required amount of knowledge. Since the knowledge of expert systems comes from humans, it is almost impossible to build an expert system which performs better than the person, or groups of persons, from whom the knowledge has been extracted. However, if a substantial amount of knowledge is captured from the experts, such a system allows many practical problems of aerodynamic design to be efficiently solved by unskilled persons. In particular, the works [7.33], [7.43], [7.44] present the expert system EXFAN (Expert Cooling Fan Design System). This system performs the aerodynamic design of turbomachinery components by starting with an initial design and then iterating through analysis and redesign until the design goals are met within the specified constraints [7.33]. An example of the efficiency of using the EXFAN expert system, in combinaton with numerical modeling of a flow in the example of the optimal design of an axial cooling fan, has been presented in [7.43]. As a result of this a cooling fan with four blades was obtained. A machine satisfying the same design conditions was developed in parallel by a group of skilled specialists—the experts in the field. They designed an axial machine with five blades. It was recognized by the experts that the cooling fan designed by the automatic system EXFAN is slightly better than the machine designed by them, because it needs one less blade.

Another example of the efficiency of automated design systems is presented in [7.45], where the system MAVR is briefly described (which has been developed by the Computing Center of the U.S.S.R. Academy of Sciences) and which is presently adapted to the design of heating engineering systems. Eight hours of terminal time were needed for the application of this system to the draft design of a two-loop refrigerating unit composed of 12 aggregates. The terminal work-station was operated directly by a heating engineer. According to the assertion of specialists, the design of a system of such complexity takes, under nonsystematic use of computers, 1–2 months by a group of 10–12 design engineers (1,600–3,800 man-hours). A number of the requirements for the structure of future expert systems coupling AI and CFD have been formulated in [7.46]. It is stressed therein that the main goal of such expert systems is to aid in the organization of aerodynamic design.

In the present chapter we emphasize the development of a methodological basis of an intellectual system for the information extraction from the results of shock-capturing numerical computations of multidimensional problems of the dynamics of an inviscid compressible non-heat-conducting gas. Since the information obtained at the output of such a system is rich in content, because it contains both physical and geometrical characteristics of recognized objects, it appears to be quite understandable that this information can be used both for the control of the process of the numerical solution of the basic problem and for decision-making in the expert systems of aerodynamic automated design.

Below we propose to use the methods of pattern recognition theory that belong to the class of discriminant methods for the solution of a problem on the classification of discontinuities into several types (shock waves, contact discontinuities, etc.) in the two-dimensional gas flows computed by shock-capturing finite-difference schemes. Three pattern recognition methods are considered. The first uses the principles of minimum distance and hierarchical classification. The second method realizes the idea of sequential classification, which enables one to construct a decision tree which is optimal in terms of computer expenses for computation of the features. In the third method the automatic classification of patterns is carried out by means of a set of feature functions, in the construction of which the simplest potential functons are employed.

Since curvilinear grids are now widely used for solving gas dynamics problems in the domains of complicated geometric form (see Chapter 3), it is of interest to generalize the methods of the singularities localization, which are developed in the present chapter on the basis of the ideas of digital image processing, for the case when the digital image pixels are placed at the points of a rather arbitrary spatial computing curvilinear mesh. Such a generalization is carried out in the present chapter in the example of the computation and automatic analysis of the transonic potential flow around an airfoil. In this case the classification of singularities has been carried out by a discriminant method of sequential classification.

The algorithms of the singularities localization presented below possess a substantially better flexibility and universality than the algorithms presented in Chapters 2–5. For example, all the examples of shock contour extraction in computations of the flows on a rectangular grid (that are presented below) have been obtained with the same computer code for image segmentation (a presentation of relevant algorithms may be found in Section 7.3).

The presentation of the material of this chapter basically follows our publications [7.41], [7.42], [7.47]–[7.52], which appear to represent the first attempt at a systematic applicaton of the concepts and methods of digital image processing and the modern theory of pattern recognition to problems of the refinement and automatic analysis of the results of shock-capturing multi-dimensional gas-dynamical computations.

7.1. Methodologies of Pattern Recognition

Modern pattern recognition systems realize in various forms two recognition principles: the principle of similarity of the patterns belonging to the same class of patterns, and the principle of dissimilarity between the patterns belonging to different classes. According to [7.53] the pattern recognition techniques may be subdivided into three methods: heuristic, discriminant, and syntactic (linguistic).

The basis of the heuristic approach to pattern recognition is constituted by the experience and intuition of the research worker. The systems built by such methods incorporate specific procedures developed for concrete recognition problems [7.54]. The shortcomings of heuristic methods have been enumerated in [7.53], in particular, the absence of a strict validation, although the efficiency of such methods can be sufficiently high in practice.

The basis of discriminant methods (sometimes they are called mathematical methods [7.55], [7.56] or vector or geometric methods [7.57]) is constituted by the use of certain mathematical formalisms for the description of original data on the pattern, and by the application of well-known mathematical methods for obtaining a solution. This approach may be subdivided into two classes: deterministic and statistical. The methods belonging to the first of these classes do not utilize in an explicit form the statistical properties of the pattern classes under study. The statistical approach is based on the apparatus of mathematical statistics, and the statistical pattern recognition algorithms are based on the application of the Bayesian classification rule or on its varieties [7.53], [7.55].

The syntactic (linguistic, structural) methods [7.5], [7.9], [7.57], [7.58] are applied in the cases when not only the class to which a pattern belongs, but also a description of each pattern with the aid of the elements (subpatterns) and their relationships, are of interest. The pattern can be described with the aid of a hierarchical structure of subpatterns which are similar to some extent to the syntactic structure of a language. This enables us to employ, in the solution of pattern recognition problems, the theory of formal languages developed specifically for communication between man and computer. As was pointed out in [7.53], this approach is especially useful in the work with patterns which cannot be described by numerical data, or are so complicated that their local features are difficult to identify, and then one has to apply to the global properties of the patterns. The work [7.59] appears to represent the first attempt at using the ideas of syntactic pattern recognition to flow analysis. The authors of [7.59] use features such as critical points and dividing streamlines as a basis, which enables them to obtain a representation of the global topology of the flow by a graph (with the various structures represented by the nodes and their relationships in the flow by the connecting lines of the graph). Once the flow field has been placed in this form, it can be studied and compared with other data sets using techniques of syntactic pattern recognition.

In the present chapter we consider the question of the applicability of discriminant methods of pattern recognition for the purposes of recognition and classification of discontinuity surfaces in the two-dimensional flows of an inviscid gas on the basis of numerical solutions obtained by finite-difference shock-capturing schemes. Here a question arises: which of the discriminant methods—deterministic or statistical—should be preferred in the development of such a specialized intellectual system of extraction and interpretation of the informaton? In this connection it is natural to consider digital images obtained by finite-difference methods (for brevity in the following they will be called finite-difference images) from the point of view of the presence of some statistical or probabilistic properties in such images. These images differ substantially from the digital images obtained (for example, in aerial photography, in medicine, and in biology), in that the properties of the noise and blur in finite-difference images have (under deterministic initial and boundary conditions) a deterministic character; they are determined by truncation errors of a difference scheme and by the length of the machine word of the computer employed in the computations [7.60]. This determinism of finite-difference images leads, in practice, to the fact that the results of two repeated computations of the same gas-dynamical problem (obtained on the same computer by the same finite-difference method) coincide in all digits of the mantissas of the numerical results. Note that the same deterministic character is inherent in the results of the application of computer generators of pseudo-random numbers; the sequences of "random" values obtained by these generators are reproducible (see Section 7.6).

Although, as was just pointed out, the finite-difference images obtained in the numerical gas-dynamical simulations have a deterministic character, it is of interest to elucidate a question on the fundamental applicability of statistical pattern recognition methods in the case of finite-difference image recognition, taking into account a comparatively wide acceptance of statistical methods in the pattern recognition systems. Investigations of the properties of finite-difference schemes (carried out by YU.N. VATOLIN [7.60], [7.61] by the methods of mathematical statistics, by the Shannon theory of information, and by probability theory) enable us to answer this question positively. In particular, the well-known Neumann stability condition $a\tau/h^2 \leq 1/2$ was obtained in

[7.60] from the requirement of the growth of the correlation coefficients and the criterion for the uncertainty (entropy) reduction in the solution u^{n+1}, obtained by the numerical solution of the heat equation $u_t = au_{xx}$ with an explicit difference scheme. Applying the method of Yu.N. Vatolin to the difference scheme (6.1.4) approximating the hyperbolic-type equation (1.50) it is easy to obtain the well-known Courant–Friedrichs–Lewy condition $a\tau/h \leq 1$.

7.2. Image Formation

As was already mentioned above, the formation of objects for their presentation to the system of automatic classification is carried out in a lower-level image processing system. In Figure 7.1 we show a block scheme of the process of a machine classification of discontinuities in the two-dimensional flows which we have realized on a general-purpose computer. This scheme reflects the stages which are characteristic of the lower-level processing and the stages which are characteristic of the machine artificial intelligence systems. The block scheme of Figure 7.1 is conceptually close to the block scheme of the process of machine vision of a manipulation robot presented in [7.6].

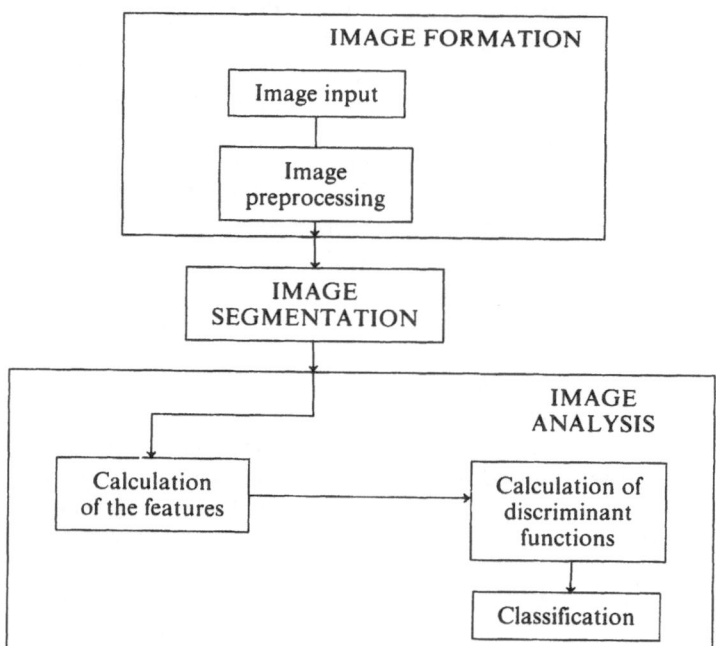

Figure 7.1. Flow chart of the process of the analysis of gas-dynamical computational results.

7.2.1. Image Input

The input of an image into the system of information processing presented below is usually performed from the basic program of the numerical calculation of a gas dynamic flow. Depending on the mode of using the information obtained at the output of an intellectual system of digital image processing various input regimes can be implemented. For example, in the cases when this information is to be used in the process of solving a nonstationary problem of computational control, the processing of the digital images of the flow field may be performed at each time step. In the cases when a stationary problem is solved, and the post-processing of the numerical solution obtained is needed only to obtain information on the flow structure, then the input of an image into the system of information processing may be performed only once.

7.2.2. Image Preprocessing

Image preprocessing includes one or several procedures from the sets: image smoothing, noise filtration, image restoration, etc. (see, for example, [7.62], [7.63]).

The simplest algorithms of image preprocessing are reduced to smoothing. Note that smoothing is applied for a long time in gas-dynamical computations. It is usually performed for two purposes: either the stabilization of computations by a difference scheme, or the reduction of the amplitude of parasitic post-shock oscillations. For example, a four-point smoothing was used by R.D. Richtmyer in his well-known two-step scheme approximating the Euler equation system (1.30) and having the form [7.64]

$$\mathbf{w}_{ij}^{n+1} = \bar{\mathbf{w}}_{ij}^{n} - [\tau/(2h_x)](\mathbf{F}_{i+1,j}^{n} - \mathbf{F}_{i-1,j}^{n}) - [\tau/(2h_y)](\mathbf{G}_{i,j+1}^{n} - \mathbf{G}_{i,j-1}^{n});$$

$$\mathbf{w}_{ij}^{n+2} = \mathbf{w}_{ij}^{n} - (\tau/h_x)(\mathbf{F}_{i+1,j}^{n+1} - \mathbf{F}_{i-1,j}^{n+1}) - (\tau/h_y)(\mathbf{G}_{i,j+1}^{n+1} - \mathbf{G}_{i,j-1}^{n+1});$$

where τ is the time step, h_x, h_y are the steps of a uniform mesh along the x- and y-axes, respectively; $\mathbf{F}_{ij}^{n} = \mathbf{F}(\mathbf{w}_{ij}^{n})$, etc., $\bar{\mathbf{w}}_{ij}^{n}$ is a smoothed value of the numerical solution at the node (i, j) which was computed in [7.64] by the formula

$$\bar{\mathbf{w}}_{ij}^{n} = (1/4)(\mathbf{w}_{i+1,j}^{n} + \mathbf{w}_{i-1,j}^{n} + \mathbf{w}_{i,j+1}^{n} + \mathbf{w}_{i,j-1}^{n}). \tag{7.2.1}$$

A two-step smoothing procedure for the solution $\mathbf{w}_{ij}^{n+1}$ was applied in [7.65]: at first a one-dimensional smoothing along the x-coordinate was performed by the formula

$$\bar{\mathbf{w}}_{ij}^{n+1} = \mathbf{w}_{ij}^{n+1} + \delta(\mathbf{w}_{i-1,j}^{n+1} - 2\mathbf{w}_{ij}^{n+1} + \mathbf{w}_{i+1,j}^{n+1}), \tag{7.2.2}$$

and then the smoothing along the y-coordinate

$$\bar{\bar{\mathbf{w}}}_{ij}^{n+1} = \bar{\mathbf{w}}_{ij}^{n+1} + \delta(\bar{\mathbf{w}}_{i,j-1}^{n+1} - 2\bar{\mathbf{w}}_{ij}^{n+1} + \bar{\mathbf{w}}_{i,j+1}^{n+1}). \tag{7.2.3}$$

In (7.2.2), (7.2.3) δ is the user-specified parameter of smoothing; $0 \leq \delta \leq$

0.25. To control the amplitude of the numerical solution oscillations in the MacCormack scheme (4.2.11), (4.2.12) a third stage was introduced in [7.66] which was a smoothing of the solution $\mathbf{w}^{n+1}$ by the formula

$$\bar{\mathbf{w}}_{ij}^{n+1} = 1/(\varphi + 4) \cdot [\varphi \mathbf{w}_{ij}^{n+1} + \mathbf{w}_{i,j+1}^{n+1} + \mathbf{w}_{i+1,j}^{n+1} + \mathbf{w}_{i,j-1}^{n+1} + \mathbf{w}_{i-1,j}^{n+1}], \quad (7.2.4)$$

where $\bar{\mathbf{w}}_{ij}^{n+1}$ is the vector of the smoothed solution at the node (i, j) at $t = t_{n+1}$, and φ is a parameter to be chosen empirically; the range $200 \le \varphi \le 1{,}000$ was recommended in [7.66]. It is easy to see that (7.2.4) coincides with (7.2.1) at $\varphi = 0$.

K.G. SHKADINSKY [7.67] has shown in 1966 that the smoothing is equivalent to an accuracy of up to $O(\tau)$ for the artificial viscosity. In [7.68] a smoothing procedure was realized in which the constant coefficients of the smoothing operator were replaced by the complexes applied previously by V.B. BALAKIN and V.V. BULANOV [7.69] for the artificial viscosity

$$\bar{\mathbf{w}}_{ij}^{n+1} = \mathbf{w}_{ij}^{n+1} + \Omega \cdot \{(\delta_i \mathbf{w}_{i+1/2,j}^{n+1}) \cdot |\delta_i \mu_{i+1/2,j}| - (\delta_i \mathbf{w}_{i-1/2,j}^{n+1}) \cdot |\delta_i \mu_{i-1/2,j}|$$
$$+ (\delta_j \mathbf{w}_{i,j+1/2}^{n+1}) \cdot |\delta_j v_{i,j+1/2}| - (\delta_j \mathbf{w}_{i,j-1/2}^{n+1}) \cdot |\delta_j v_{i,j-1/2}|$$
$$+ (\mathbf{w}_{i+1,j+1}^{n+1} - \mathbf{w}_{ij}^{n+1})\beta_{i+1/2,j+1/2} - (\mathbf{w}_{ij}^{n+1} - \mathbf{w}_{i-1,j-1}^{n+1})\beta_{i-1/2,j-1/2}$$
$$+ (\mathbf{w}_{i,j+1}^{n+1} - \mathbf{w}_{i-1,j}^{n+1})\beta_{i-1/2,j+1/2} - (\mathbf{w}_{i+1,j}^{n+1} - \mathbf{w}_{i,j-1}^{n+1})\beta_{i+1/2,j-1/2}\}, \quad (7.2.5)$$

where

$$\delta_i \mathbf{w}_{i+1/2,j} = \mathbf{w}_{i+1,j} - \mathbf{w}_{ij}, \qquad \delta_j \mathbf{w}_{i,j+1/2} = \mathbf{w}_{i,j+1} - \mathbf{w}_{ij},$$

$$\mu_{ij} = (\tau/h_x)(|\mu| + c)_{ij}^{n+1}, \qquad v_{ij} = (\tau/h_y)(|v| + c)_{ij}^{n+1},$$

$$\beta_{i+1/2,j+1/2} = 0.5|\mu_{i+1,j+1} - \mu_{ij} + v_{i+1,j+1} - v_{ij}|,$$

$$\beta_{i-1/2,j+1/2} = 0.5|\mu_{i-1,j+1} - \mu_{ij} + v_{i-1,j+1} - v_{ij}|,$$

c is the sound speed, $\Omega \approx 1$. As was found in [7.68] in the computations of two-dimensional shocked flows, the presence of diagonal differences in (7.2.5) leads to the diminution of the influence on the numerical solution of shock wave orientation with respect to the coordinate axes.

The notions of the image window and of the mask are widely used in the theory of digital image processing for different purposes; in particular, for the representation of various smoothing operators. Let f be any of the functions ρ, u, v, p. Then the image window $F_3(f)$ of the 3×3 dimension centred in the pixel f_{ij} is a square matrix of the form

$$F_3(f) = \begin{bmatrix} f_{i-1,j+1} & f_{i,j+1} & f_{i+1,j+1} \\ f_{i-1,j} & f_{ij} & f_{i+1,j} \\ f_{i-1,j-1} & f_{i,j-1} & f_{i+1,j-1} \end{bmatrix}. \quad (7.2.6)$$

Following [7.63], [7.70], let us introduce the notion of the mask H as the matrix, the elements of which are the coefficients by which, in the smoothing operator or in the edge detector, the elements of the image window F_3 are multiplied and then added or subtracted in accordance with the signs of coefficients in the mask; the result of this operation will be denoted $H * F_3$ as in [7.63]. The quantity $H * F_3$ is called the discrete convolution of the image window F_3 with the mask H. One of the simplest smoothing algorithms makes use of smoothing (defocusing) masks, for example, the mask of the form [7.63]

$$H_1 = (1/10) \begin{bmatrix} 1 & 1 & 1 \\ 1 & 2 & 1 \\ 1 & 1 & 1 \end{bmatrix}. \tag{7.2.7}$$

Let us elucidate a question on the influence of the mask (7.2.7) on the smoothing of discontinuities in the image intensity function f. We shall carry out the computations for the case when the edge (or the jump) in the image $f(x, y)$ is oriented parallel with the y-axis and, besides, coincides with one of the lines $x = x_v$, on which, in the nodes (x_v, y_k), the intensities of the discretized image are given. Thus, let

$$f(x, y) = \begin{cases} b + d, & x \le x_v, \\ b, & x > x_v, \end{cases} \tag{7.2.8}$$

where $d \ne 0$. Consider along with (7.2.7) a smoothing mask of the general form

$$H_2 = (1/N_s) \begin{bmatrix} s_1 & s_2 & s_3 \\ s_4 & s_5 & s_6 \\ s_7 & s_8 & s_9 \end{bmatrix}, \tag{7.2.9}$$

where $N_s = \sum_{j=1}^{9} s_j$, s_j, $j = 1, \ldots, 9$, are positive constants. The result of a smoothing of the brightness in the cell (i, j) will be denoted by $\bar{f}_{ij}$. Then, convolving the mask (7.2.9) with the image window (7.2.6), we obtain by definition

$$\bar{f}_{ij} = H_2 * F_3(f) = (1/N_s) \cdot (s_1 f_{i-1,j+1} + s_2 f_{i,j+1} + s_3 f_{i+1,j+1}$$
$$+ s_4 f_{i-1,j} + s_5 f_{ij} + s_6 f_{i+1,j} + s_7 f_{i-1,j-1}$$
$$+ s_8 f_{i,j-1} + s_9 f_{i+1,j-1}). \tag{7.2.10}$$

Taking (7.2.8) into account, we have

$$f_{ij} = \begin{cases} b + d, & i \le v, \\ b, & i > v, \end{cases} \quad \forall j. \tag{7.2.11}$$

Employing (7.2.11), we obtain from (7.2.10)

$$\bar{f}_{ij} = \begin{cases} b + d, & i \le v - 1, \\ ((s_1 + s_2 + s_4 + s_5 + s_7 + s_8)(b + d) + (s_3 + s_6 + s_9)b)/N_s, & i = v, \\ ((s_1 + s_4 + s_7)(b + d) + (s_2 + s_3 + s_5 + s_6 + s_8 + s_9)b)/N_s, & i = v + 1, \\ b, & i \ge v + 2. \end{cases}$$
$$\tag{7.2.12}$$

Let us introduce the following quantities: let v_l be a maximal number which does not exceed v and $\bar{f}_{ij} = b + d, i \le v_l; \bar{f}_{ij} \ne b + d, i > v_l$. Similarly, let v_r be a minimal number, $v_r \ge v$, such that $\bar{f}_{ij} = b, i \ge v_r; \bar{f}_{ij} \ne b, i < v_r$. Let us define the width of the zone of smearing of a ramp edge (7.2.11) by the formula

$$X = (v_r - v_l)h_1. \qquad (7.2.13)$$

We have from (7.2.12) that in the case of using the mask (7.2.9) $v_l = v - 1$, $v_r = v + 2$, so that $X = 3h_1$. In a discrete image determined by formula (7.2.11) the edge is smeared over the width h_1. In a result of the application of the smoothing mask (7.2.9) the width X increases by a factor of three.

In some cases a single application of the mask (7.2.9) is insufficient for an efficient noise smoothing, and then we can perform one more convolution of the mask H_2 with the smoothed image window. Let us investigate the effects of smoothing of the edges of the form (7.2.11) under double application of the defocusing 3×3 masks. Consider two defocusing masks of the form

$$S_1 = \frac{1}{N_1} \begin{bmatrix} a_1 & a_2 & a_3 \\ a_4 & a_5 & a_6 \\ a_7 & a_8 & a_9 \end{bmatrix}, \qquad S_2 = \frac{1}{N_2} \begin{bmatrix} b_1 & b_2 & b_3 \\ b_4 & b_5 & b_6 \\ b_7 & b_8 & b_9 \end{bmatrix}, \qquad (7.2.14)$$

where a_j, b_j are positive constants, $j = 1, \ldots, 9$,

$$N_1 = \sum_{j=1}^{9} a_j, \qquad N_2 = \sum_{j=1}^{9} b_j. \qquad (7.2.15)$$

Let us introduce for further convenience the multi-indices

$$\mathbf{i} = (i, j), \qquad \mathbf{e}_1 = (-1, 1), \qquad \mathbf{e}_2 = (0, 1), \qquad \mathbf{e}_3 = (1, 1),$$

$$\mathbf{e}_4 = (-1, 0), \qquad \mathbf{e}_5 = (0, 0), \qquad \mathbf{e}_6 = (1, 0), \qquad \mathbf{e}_7 = (-1, -1), \qquad (7.2.16)$$

$$\mathbf{e}_8 = (0, -1), \qquad \mathbf{e}_9 = (1, -1),$$

Then we can write

$$g_{\mathbf{i}} \equiv S_1 * F_3(f_{\mathbf{i}}) = (1/N_1) \sum_{k=1}^{9} a_k \cdot f_{\mathbf{i}+\mathbf{e}_k},$$

$$h_{\mathbf{i}} \equiv S_2 * F_3(g_{\mathbf{i}}) = (1/N_2) \sum_{m=1}^{9} b_m \cdot g_{\mathbf{i}+\mathbf{e}_m} \qquad (7.2.17)$$

$$= [1/(N_1 N_2)] \sum_{m=1}^{9} \sum_{k=1}^{9} b_m a_k f_{\mathbf{i}+\mathbf{e}_k+\mathbf{e}_m}.$$

Let

$$\mathbf{e}_k + \mathbf{e}_m = (e_1, e_2), \qquad k, m = 1, \ldots, 9. \qquad (7.2.18)$$

Note the following properties of the multi-index (7.2.18) resulting from (7.2.16): $-2 \le e_j \le 2, j = 1, 2$. This means that the operator $S_2 * F_3(S_1 * F_3)$ may be represented as a discrete convolution of some 5×5 mask S_3 with the image window F_5 centered in the pixel f_{ij}:

$$F_5 = \begin{bmatrix} f_{i-2,j+2} & f_{i-1,j+2} & f_{i,j+2} & f_{i+1,j+2} & f_{i+2,j+2} \\ f_{i-2,j+1} & f_{i-1,j+1} & f_{i,j+1} & f_{i+1,j+1} & f_{i+2,j+1} \\ f_{i-2,j} & f_{i-1,j} & f_{ij} & f_{i+1,j} & f_{i+2,j} \\ f_{i-2,j-1} & f_{i-1,j-1} & f_{i,j-1} & f_{i+1,j-1} & f_{i+2,j-1} \\ f_{i-2,j-2} & f_{i-1,j-2} & f_{i,j-2} & f_{i+1,j-2} & f_{i+2,j-2} \end{bmatrix}. \tag{7.2.19}$$

Let

$$S_3 = \frac{1}{N_3} \begin{bmatrix} c_1 & c_2 & c_3 & c_4 & c_5 \\ c_6 & c_7 & c_8 & c_9 & c_{10} \\ c_{11} & c_{12} & c_{13} & c_{14} & c_{15} \\ c_{16} & c_{17} & c_{18} & c_{19} & c_{20} \\ c_{21} & c_{22} & c_{23} & c_{24} & c_{25} \end{bmatrix},$$

$$N_3 = \sum_{j=1}^{25} c_j, \quad c_j > 0, \quad j = 1, \dots, 25. \tag{7.2.20}$$

Employing formula (7.2.17), we can find in the result of some calculations the following explicit expressions for c_j:

$c_1 = b_1 a_1;$ $\qquad\qquad\qquad$ $c_6 = b_1 a_4 + b_4 a_1;$

$c_2 = b_1 a_2 + b_2 a_1;$ $\qquad\quad$ $c_7 = b_1 a_5 + b_5 a_1 + b_2 a_4 + b_4 a_2;$

$c_3 = b_1 a_3 + b_3 a_1 + b_2 a_2;$ $\quad$ $c_8 = b_1 a_6 + b_6 a_1 + b_2 a_5 + b_5 a_2 + b_3 a_4 + b_4 a_3;$

$c_4 = b_2 a_3 + b_3 a_2;$ $\qquad\qquad$ $c_9 = b_2 a_6 + b_6 a_2 + b_3 a_5 + b_5 a_3;$

$c_5 = b_3 a_3;$

$c_{10} = b_3 a_6 + b_6 a_3;$

$c_{11} = b_1 a_7 + b_7 a_1 + b_4 a_4;$

$c_{12} = b_1 a_8 + b_8 a_1 + b_2 a_7 + b_7 a_2 + b_4 a_5 + b_5 a_4;$

$c_{13} = b_1 a_9 + b_9 a_1 + b_2 a_8 + b_8 a_2 + b_3 a_7 + b_7 a_3 + b_4 a_6 + b_6 a_4 + b_5 a_5;$

$c_{14} = b_2 a_9 + b_9 a_2 + b_3 a_8 + b_8 a_3 + b_5 a_6 + b_6 a_5;$

$c_{15} = b_6 a_6 + b_3 a_9 + b_9 a_3;$

$c_{16} = b_4 a_7 + b_7 a_4;$

$c_{17} = b_4 a_8 + b_8 a_4 + b_5 a_7 + b_7 a_5;$

$c_{18} = b_4 a_9 + b_9 a_4 + b_8 a_5 + b_6 a_7 + b_7 a_6;$

$c_{19} = b_6 a_8 + b_8 a_6 + b_5 a_9 + b_9 a_5;$ $\qquad$ $c_{20} = b_6 a_9 + b_9 a_6;$

$c_{21} = b_7 a_7;$ $\qquad$ $c_{22} = b_7 a_8 + b_8 a_7;$

$c_{23} = b_7 a_9 + b_9 a_7 + b_8 a_8;$ $\qquad$ $c_{24} = b_8 a_9 + b_9 a_8;$ $\qquad$ $c_{25} = b_9 a_9.$ $\quad$ (7.2.21)

In the particular case when $S_1 = S_2 = H_1$, where the mask H_1 is determined by (7.2.7), we obtain with the aid of (7.2.21) the following smoothing mask S_3:

$$S_3 = (1/100) \begin{bmatrix} 1 & 2 & 3 & 2 & 1 \\ 2 & 6 & 8 & 6 & 2 \\ 3 & 8 & 12 & 8 & 3 \\ 2 & 6 & 8 & 6 & 2 \\ 1 & 2 & 3 & 2 & 1 \end{bmatrix}. \qquad (7.2.22)$$

Thus, the convolution $S_3 * F_5$ can be performed by a sequence of two shorter convolutions $S_1 * F_3, S_2 * F_3(S_1 * F_3)$. Applying the convolution $S_3 * F_5(f)$ to all the nodes (i, j) of the discrete digital image determined by formula (7.2.11), we can easily find by analogy with (7.2.12) that in this case in (7.2.13) $v_l = v - 2$, $v_r = v + 3$, so that in the image smoothed with the aid of the mask (7.2.20) the edge is smeared over the width $X = 5h$. This shortcoming of smoothing masks of the type (7.2.9), (7.2.20) leads to the fact that in certain cases there can occur a loss of some fine detail of the image [7.6]: for example, the zones of smearing of two closely located discontinuities can merge into one wide zone of smearing, so that the information on both of the discontinuity surfaces will be lost. Therefore, it is desirable to use, in the analysis of numerical solutions of complicated gas-dynamical problems involving a large number of discontinuities in small subdomains of the flow, some processing procedures which refine the solution. Here the modern practice of the application of finite-difference shock-capturing schemes and of computing meshes suggests, in particular, a way of using dynamic grids adapting to the flow (see Chapter 3); the theory of digital image processing suggests a way of using the image restoration techniques.

It is easy to see that the above smoothing operators (7.2.1)–(7.2.5) may be presented in the form of convolutions of some smoothing masks H with the image window of the form (7.2.6). For example, the smoothing (7.2.4) may be represented as

$$\overline{w}_{ij}^{n+1} = H_3 * F_3(w_{ij}^{n+1}),$$

where the mask H_3 has the form

$$H_3 = [1/(\varphi + 4)] \begin{bmatrix} 0 & 1 & 0 \\ 1 & \varphi & 1 \\ 0 & 1 & 0 \end{bmatrix},$$

thus under the action of the operator (7.2.4) on the image of the form (7.2.11) the edge of this image will be smeared over three cells. Similarly, the smoothing (7.2.2), (7.2.3) can be represented as a result of the sequential application of two defocusing masks H_4, H_5 of the form

$$H_4 = \begin{bmatrix} 0 & 0 & 0 \\ \delta & 1 - 2\delta & \delta \\ 0 & 0 & 0 \end{bmatrix}, \qquad H_5 = \begin{bmatrix} 0 & \delta & 0 \\ 0 & 1 - 2\delta & 0 \\ 0 & \delta & 0 \end{bmatrix},$$

so that

$$\bar{\bar{\mathbf{w}}}_{ij}^{n+1} = H_5 * F_3(H_4 * \mathbf{w}_{ij}^{n+1}) = H_6 * F_3(\mathbf{w}_{ij}^{n+1}),$$

where H_6 is the following 3×3 mask:

$$H_6 = \begin{bmatrix} \delta^2 & \delta - 2\delta^2 & \delta^2 \\ \delta - 2\delta^2 & (1 - 2\delta)^2 & \delta - 2\delta^2 \\ \delta^2 & \delta - 2\delta^2 & \delta^2 \end{bmatrix}.$$

Consider now some questions related to the use of the restoration of finite-difference gas-dynamical images. The problem of restoration of shocked profiles is in itself a highly important problem for the theory and practice of difference schemes, because the effects of discontinuities smearing and of spurious oscillations in the numerical solutions cause substantial difficulties when the difference shock-capturing methods are used for numerical studies of complicated multidimensional problems of aerohydromechanics.

Within the framework of a system for the analysis of numerical gas-dynamical computations presented in Figure 7.1 there can be realized two different strategies of using the methods of the restoration of digital images of a flow field under study. The application of the first of these strategies is preferable in the cases when a difference shock-capturing scheme applied for the solution of a specific problem generates numerical solutions characterized by an excessive smearing of discontinuities or by spurious large-amplitude oscillations. Then it is reasonable to apply the restoration either at each time step, or after every Δn steps, say after every five steps in order to control the level of smearing of discontinuities and the amplitude of parasitic oscillations. The application of restoration, especially at each time step, may lead to a substantial increase in computer time of the computation of a gas-dynamical problem. Therefore, in the cases when the levels of smearing and noise generated by a difference scheme are sufficiently small, one can turn to the second, less expensive strategy of using the restoration methods which assumes that the restoration is carried out only when using the program of automatic analysis of computation results. For example, in the case of computation of a stationary flow it is sufficient to use the restoration only once. Note that the restoration methods used in both of the above strategies may have a quite different structure. The desire to increase the efficiency of the digital processing algorithms has led to the fact that at present the passive strategy is also often used in the processing, which means that one tries to extract the needed information from the real, that is, from the nonrestored, "raw" image [7.63], [7.70].

Let us consider in more detail one of the possible realizations of the image restoration algorithms in finite-difference shock-capturing schemes within the framework of the first strategy. Let $\mathbf{w}^n$ be the solution of a specific initial–boundary-value problem for the Euler equation system (1.30) obtained by a finite-difference scheme for the moment of time $t = t_n$. Let the explicit or implicit difference scheme operated on a fixed uniform spatial grid have the form

$$\mathbf{w}^{n+1} = \Lambda(\mathbf{w}^n, T_{\pm 1}, T_{\pm 2}, T_{\pm 0}, h_1, h_2, \tau), \qquad (7.2.23)$$

where Λ is a vector whose components are difference operators of the form which ensures the approximation of all four equations of the original system (1.30) by scalar equations of the system (7.2.23); $T_{\pm 1}, T_{\pm 2}$ are the shift operators along the x- and y-axes, respectively; $T_{\pm 0}$ is the shift operator along the t-axis, h_1, h_2 are the computing mesh steps in the directions of the x- and y-axes, respectively; and τ is the time step. Let $\mathbf{R}$ be an image restoration operator. One of the ways of its application to the restoration of the solution vector $\mathbf{w}$ components described in [7.47], [7.50] has the form

$$\tilde{\mathbf{w}}^n = \mathbf{R}\mathbf{w}^n;$$
$$\mathbf{w}^{n+1} = \Lambda(\tilde{\mathbf{w}}^n, T_{\pm 1}, T_{\pm 2}, T_{\pm 0}, h_1, h_2, \tau), \qquad n = 1, 2, \dots. \qquad (7.2.24)$$

That is, instead of the solution $\mathbf{w}^n$ we use for the computation of $\mathbf{w}^{n+1}$ by the scheme (7.2.23) the restored image $\mathbf{R}\mathbf{w}^n$.

As is known, in problems of inviscid gas dynamics the solution components should satisfy the conservation laws. In addition, entropy inequalities should be satisfied at the shock waves. Quantities such as density, temperature, and pressure should be nonnegative. Therefore, it is natural that the restoration techniques should be adapted to the gas-dynamical modeling with regard to the above constraints. Let $\mathbf{R} = (R_1, R_2, R_3, R_4)^\mathsf{T}$ where the superscript T denotes transposition. As is known, the entropy inequalities may be satisfied in a difference solution by an appropriate choice of the pseudoviscosity or of a form of conservative difference approximation [7.64], [7.71]. Taking the foregoing into account, we can formulate the constraints for the restoration operator $\mathbf{R}$ in the form of the following equalities and inequalities:

$$\sum_{i,j} R_1 \rho_{ij}^n h_1 h_2 = \sum_{i,j} \rho_{ij}^n h_1 h_2; \qquad (7.2.25)$$

$$\sum_{i,j} R_2 (\rho u)_{ij}^n h_1 h_2 = \sum_{i,j} (\rho u)_{ij}^n h_1 h_2; \qquad (7.2.26)$$

$$\sum_{i,j} R_3 (\rho v)_{ij}^n h_1 h_2 = \sum_{i,j} (\rho v)_{ij}^n h_1 h_2; \qquad (7.2.27)$$

$$\sum_{i,j} R_4 (\rho E)_{ij}^n h_1 h_2 = \sum_{i,j} (\rho E)_{ij}^n h_1 h_2; \qquad (7.2.28)$$

$$R_1 \rho_{ij}^n \geq 0, \qquad \forall i, j; \qquad (7.2.29)$$

$$R_4 (\rho E)_{ij}^n - 0.5\{[R_2 (\rho u)_{ij}^n]^2 + [R_3 (\rho v)_{ij}^n]^2\}/R_1 \rho_{ij}^n \geq 0, \qquad \forall i, j. \quad (7.2.30)$$

The constraint (7.2.30) realizes the requirement of nonnegativeness of the specific internal energy. The existing constrained image restoration techniques are rather laborious, therefore, their incorporation into the image processing systems leads to a substantial increase in computer-time expense [7.72]–[7.74].

Consider some effects of restoration in the example of restoring piecewise constant functions. The problem of restoring such functions is simpler than the general problem of the restoration of arbitrary discontinuous functions. It should be noted that there are in the literature a number of methods for the restoration of piecewise constant functions, for example, Frieden's maximum entropy method [7.72], [7.74], [7.75]. Edge-preserving noise-smoothing techniques [7.76], being much simpler than the above-mentioned methods, are sometimes considered as restoration techniques since their application leads to a steepening of profiles at the edges under simultaneous noise reduction. In the following section we shall discuss in more detail the notion of the edge; now we restrict ourselves to a brief remark that, roughly speaking, the edge in the image brightness is practically the same as a strong discontinuity in the solution of a gas-dynamical problem. In [7.76] seven edge-preserving noise-smoothing techniques have been analyzed. For this analysis specific tests have been designed and a number of quantitative criteria have been proposed. In [7.76] a conclusion was drawn that the KAVE method [7.77], [7.78] is one of the best methods for the restoration of piecewise constant image intensity functions. The abbreviation KAVE means "the K-nearest neighbor averaging". In this method the center point P of an $N \times N$ neighborhood is replaced by the average grey level of the neighbors of P, the grey levels of which are closest to that of P. In other words, this method performs smoothing at the point P using neighboring points drawn from the same "population" as P to reduce the amplitude of the noise fluctuations and to preserve edges. In [7.50] the KAVE method was realized for the case when a 3×3 image window is used. In this case, the number of neighbors of a central pixel ρ_{ij} does not exceed eight. Let us enumerate the pixels ρ_{ml} being neighbors of ρ_{ij} by $\rho_k, k = 1, 2, \dots, 8$, arranging ρ_k in a sequence with increasing $|\rho_{ij} - \rho_k|$. After that we replace the value ρ_{ij} by

$$\bar{\rho}_{ij} = \left(\rho_{ij} + \sum_{k=1}^{K} \rho_k \right) \Big/ (K + 1), \tag{7.2.31}$$

where K is an integer, $1 \le K \le 8$, which is assigned by the program user. The operation (7.2.31) is performed over all the nodes (i, j) of the image. A single application of (7.2.31) may be considered as one iteration for image restoration, after that the smoothing process (7.2.31) can be repeated; this time it is applied to the image obtained in the preceding iteration. Let us denote by N_{it} the number of iterations in the KAVE method. Typically, it is set by the program user. From the results of numerous computations with the KAVE method presented in [7.76] it follows that the optimum number of iterations in the KAVE method is $3 \div 4$, since at $N_{it} > 4$ the results of averaging weakly depend on the iteration number beginning at four. In [7.78], on the basis of examples using the KAVE method for the processing of multispectral images, a conclusion was drawn that iterations beyond five make no great difference to the results.

The results of the KAVE iterations have been directly included in the numerical computation of a problem in accordance with formulas (7.2.24). As can easily be seen from (7.2.31), the use of the KAVE method leads to a violation of the mass conservation law (7.2.25). To reduce this undesirable effect it was proposed in [7.50] to apply the smoothing (7.2.31), (7.2.24) after every Δn time steps, where Δn is a parameter specified by the program user. Thus, when using the KAVE method there are three adjustable parameters $K, N_{it}, \Delta n$ at our disposal. As our computations have shown, these parameters substantially affect the quality of the restored image [7.47], [7.50]. Our further computational experiments with KAVE showed that this method, in some cases, may lead to an excessively large error ($> 30\%$) in the total fluid mass in a computational domain. To reduce this undesirable effect we have slightly modified formula (7.2.31) with regard to a suggestion by the authors of [7.78]: instead of a constant value K we have used a variable value K_{ij} which is determined from the requirements

$$|\rho_{ij} - \rho_{K_{ij}}| < \varepsilon,$$

$$|\rho_{ij} - \rho_{K_{ij}+1}| \geq \varepsilon; \qquad 0 \leq K_{ij} \leq K. \tag{7.2.32}$$

At $|\rho_{ij}| < 0.003 \cdot \{$maximal value of $|\rho_{ij}|$ over all the pixels of the image$\}$ we have assumed $\bar{\rho}_{ij} = \rho_{ij}$, that is, in this case we have made no correction to the value of ρ_{ij}; otherwise, the value ε in (7.2.32) was calculated by the formula

$$\varepsilon = \begin{cases} \varepsilon_1, & \rho_{ij} \leq 0.1, \\ 0.1 \cdot |\rho_{ij}|, & \rho_{ij} > 0.1, \end{cases} \tag{7.2.33}$$

where ε_1 is a positive constant. First, we have used a simpler formula for ε: $\varepsilon = 0.1 \cdot |\rho_{ij}|$. However, in this case the negative undershoots in the ρ'' image have not been eliminated. In the case of formula (7.2.33) the KAVE method (7.2.31)–(7.2.33) smooths out the image regions more efficiently where the value ρ_{ij} is negative or small.

As was shown previously in Section 4.2, equation (4.2.7) can be used in the studies of the properties of difference schemes in the neighborhood of purely contact discontinuities in two-dimensional flows. In this connection, the KAVE method (7.2.31)–(7.2.33) was tested in [7.50] in the example of the numerical integration of (4.2.7) by the MacCormack scheme (4.2.11), (4.2.12) and by the FLIC-type scheme (cf. (3.1.11), (3.1.12))

$$h_1 h_2(\rho_{ij}^{n+1} - \rho_{ij}^n) = \Delta M_{i-1/2,j}^n - \Delta M_{i+1/2,j}^n + \Delta M_{i,j-1/2}^n - \Delta M_{i,j+1/2}^n, \tag{7.2.34}$$

$$\Delta M_{i+1/2,j}^n = \begin{cases} h_2 \rho_{ij}^n u_{10}\tau, & u_{10} > 0, \\ h_2 \rho_{i+1,j}^n u_{10}\tau, & u_{10} \leq 0, \end{cases}$$

$$\Delta M_{i,j+1/2}^n = \begin{cases} h_1 \rho_{ij}^n u_{20}\tau, & u_{20} > 0, \\ h_1 \rho_{i,j+1}^n u_{20}\tau, & u_{20} \leq 0. \end{cases} \tag{7.2.35}$$

In Figure 7.2(b) we present the computational results obtained by the meth-

Figure 7.2. Isometrics of the surfaces $\rho = \rho(x_1, x_2, t)$. (a) $t = 0$; (b) $t = 24\tau = 1.663$.

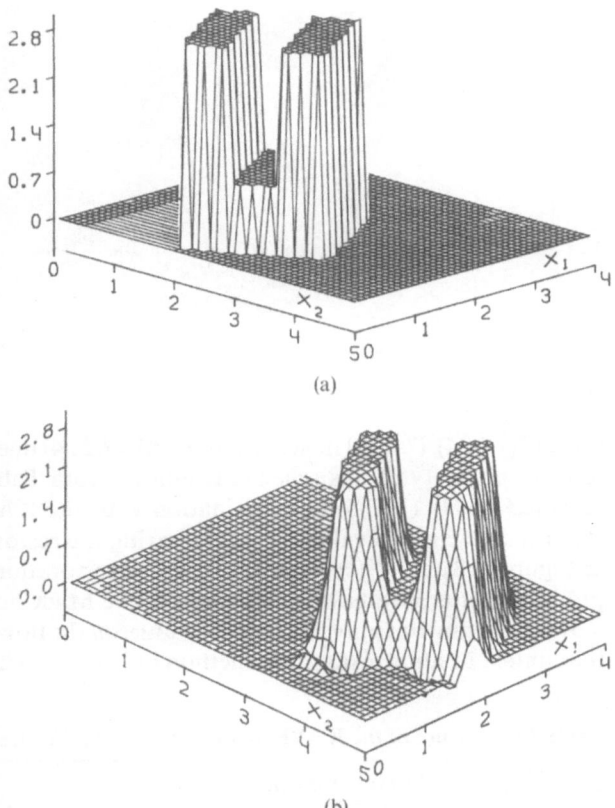

(a)

(b)

od (4.2.11), (4.2.12), (7.2.31)–(7.2.33) for the case when the initial function $\rho(x_1, x_2, 0)$ has the piecewise constant form shown in Figure 7.2(a); here $\varepsilon_1 = 0.1$, $K = 6$, $N_{it} = 8$, $\Delta n = 4$. For comparison we present in Figure 7.3 the form of the surface $\rho = \rho(x_1, x_2, t)$ obtained in a computation without restoration. Some quantitative data obtained while using the KAVE method are presented in Tables 7.1 and 7.2 where $\rho_{\min} = \min_{i,j} \rho_{ij}^n$, $\rho_{\max} = \max_{i,j} \rho_{ij}^n$, δM is the relative error in the total fluid mass. It follows from Table 7.1 that the use of the KAVE method enables us to reduce substantially the amplitude of the oscillations of the numerical solution obtained by the scheme (4.2.11), (4.2.12), and the plateau of constant value $\rho = 3$ (the exact solution) is reproduced in the result of restoration with an error less than 1% (see Figure 7.2(b)).

In Figure 7.4 (similar to Figures 7.2 and 7.3) we present computational results obtained by the first-order scheme (7.2.34), (7.2.35) for the case when the initial function $\rho(x_1, x_2, 0)$ has the form shown in Figure 7.4(a). In this computation we have taken the values $K = 4$, $N_{it} = 4$, $\Delta n = 2$, $\varepsilon = 0.1 \cdot |\rho_{ij}|$ in (7.2.32). Note that the incorporation of KAVE into computations by the

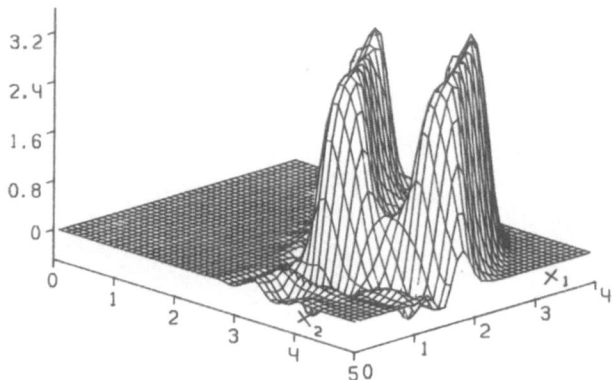

Figure 7.3. Isometrics of
the surface $\rho = \rho(x_1,$
$x_2, t)$ at $t = 1.663$.
Calculation by the
MacCormack scheme
without restoration.

scheme (7.2.34), (7.2.35) in accordance with (7.2.24) does not violate the monotonicity property of this scheme. It follows from Table 7.2 that the use of a
scheme (7.2.34), (7.2.35) in combination with the KAVE method leads to a
slight increase in the value ρ_{max}. Comparing the restoration results presented
in Figures 7.2 and 7.4(c) we can see that the restoration by the KAVE method
produces better results when it is used in the MacCormack scheme.

It is tempting to apply the restored values of the flowfield parameters within
the context of some restoration method only for the computation of the fluxes

Table 7.1. The use of the KAVE method in the MacCormack scheme.

Is the KAVE used?	Formula for ε in (7.2.32)	t/τ	ρ_{min}	ρ_{max}	δM	CPU time
Yes	$0.1|\rho_{ij}|$		-0.28	3.049	0.035	
Yes	(7.2.33), $\varepsilon_1 = 0.1$	24	-0.16	3.048	0.044	
Yes	(7.2.33), $\varepsilon_1 = 0.2$		-0.008	3.049	0.075	
No			-0.25	3.618	$5 \cdot 10^{-11}$	
Yes	$0.1|\rho_{ij}|$		-0.37	3.031	0.069	113.6%
Yes	(7.2.33), $\varepsilon_1 = 0.1$		-0.18	3.018	0.092	110%
Yes	(7.2.33), $\varepsilon_1 = 0.2$	37	-0.021	3.021	0.135	107.3%
No			-0.31	3.741	$7 \cdot 10^{-10}$	100%

Table 7.2. The use of the KAVE method in the scheme (7.2.34), (7.2.35).

Is the KAVE used?	t/τ	ρ_{min}	ρ_{max}	δM	CPU time
Yes	24	0	2.277	0.026	
No		0	2.278	$3 \cdot 10^{-10}$	
Yes		0	2.087	0.054	107.6%
No	37	0	2.069	$3 \cdot 10^{-10}$	100%

7.2. Image Formation

Figure 7.4. Isometrics of the surfaces $\rho = \rho(x_1, x_2, t)$: (a) $t = 0$; (b) $t = 24\tau = 1.663$, calculation without restoration; (c) $t = 24\tau = 1.663$, calculation with restoration. 40×50 mesh, $\beta = \arctan(u_{20}/u_{10}) = \pi/3$, $\theta_F = 0.95$ in the formula (7.2.39).

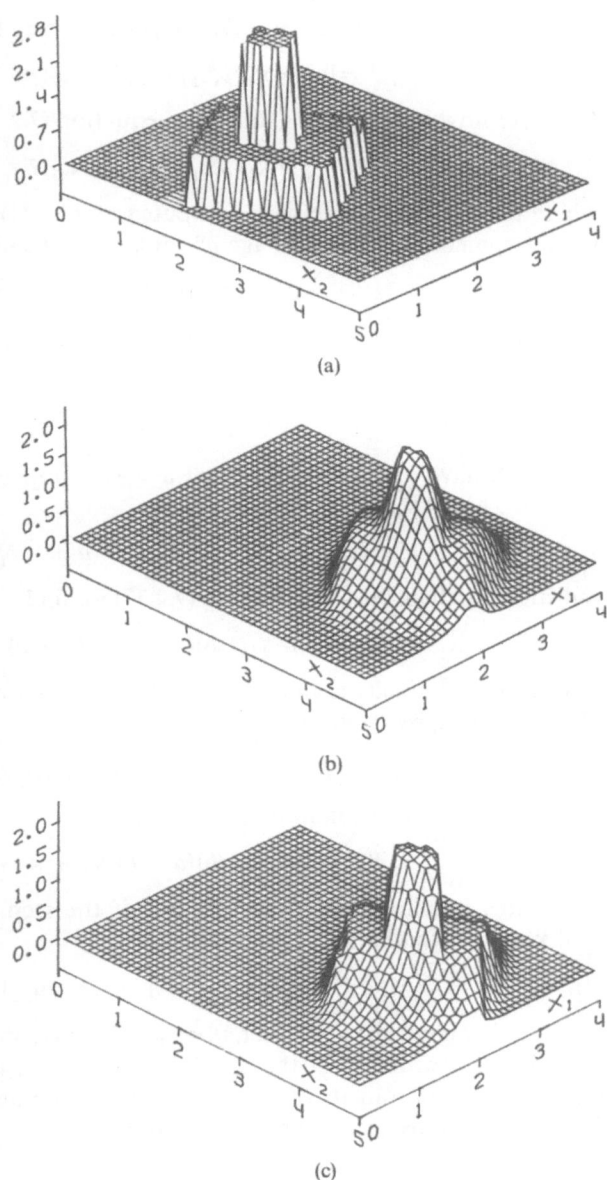

(a)

(b)

(c)

across the boundaries of the cell (i, j). Let us show that the monotonicity of the numerical solution ρ^{n+1} can be violated in the case of using the KAVE method (7.2.31) in the neighborhood of contact discontinuities, and also in the cases when the original difference scheme (7.2.23) was monotone. Suppose that in equation (4.2.7) $u_{10} \geq 0$, $u_{20} \geq 0$. Introduce the notation $\kappa_1 = \tau u_{10}/h_1$,

$\kappa_2 = \tau u_{20}/h_2$. Then the scheme (7.2.34), (7.2.35) may be written as

$$\rho_{ij}^{n+1} = \rho_{ij}^n + \kappa_1(\rho_{i-1j}^n - \rho_{ij}^n) + \kappa_2(\rho_{ij-1}^n - \rho_{ij}^n). \tag{7.2.36}$$

Consider now instead of the difference equation (7.2.34) the equation

$$\rho_{ij}^{n+1} = \rho_{ij}^n + \kappa_1(\bar{\rho}_{i-1j}^n - \bar{\rho}_{ij}^n) + \kappa_2(\bar{\rho}_{ij-1}^n - \bar{\rho}_{ij}^n), \tag{7.2.37}$$

where the values $\bar{\rho}^n$ have been computed by (7.2.31). Suppose that at $t = t_n$ some fragment of a digital image ρ^n containing the pixel ρ_{ij}^n has the form

$$
\begin{array}{cccc}
a & b & b & b \\
a & a & \underline{a} & b \\
a & a & a & a \\
a & a & a & a
\end{array}
$$

where we have underlined the pixel $a = \rho_{ij}^n$. Suppose that $a > b$, $K = 5$, in (7.2.31). Then it is easy to compute that

$$\bar{\rho}_{ij}^n = (5a + b)/6, \qquad \bar{\rho}_{i-1,j}^n = a, \qquad \bar{\rho}_{ij-1}^n = a. \tag{7.2.38}$$

Substituting the values (7.2.38) into (7.2.37) we find

$$\rho_{ij}^{n+1} = a + (1/6)(\kappa_1 + \kappa_2)(a - b) > a,$$

thus the monotonicity is violated. If ρ_{ij}^{n+1} is now computed in accordance with (7.2.24), that is, by the formula

$$\rho_{ij}^{n+1} = \bar{\rho}_{ij}^n + \kappa_1(\bar{\rho}_{i-1j}^n - \bar{\rho}_{ij}^n) + \kappa_2(\bar{\rho}_{ij-1}^n - \bar{\rho}_{ij}^n),$$

then under the same condition (7.2.38) we find

$$\rho_{ij}^{n+1} = a + (1/6)(a - b)(\kappa_1 + \kappa_2 - 1).$$

From this it follows that $a > \rho_{ij}^{n+1} > b$ if the step τ is calculated by the well-known formula

$$\tau = \theta_F \cdot h_1/(|u_{10}| + (h_1/h_2)|u_{20}|), \tag{7.2.39}$$

where θ_F is a constant multiplier, $0 < \theta_F < 1$. This simple example shows that there exists a problem of the development of a restoration method which ensures the monotonicity of the numerical solution under simultaneous satisfaction of the difference conservation laws (7.2.25)–(7.2.28).

7.3. Image Segmentation

According to [7.79], by an image segmentation is meant, as a rule, the extraction of some connected regions. After [7.6], by segmentation is meant a process of a subdivision of an image of a scene into constituent parts: objects, their fragments, or specific features.

In the theory of digital image processing there is a notion of "edge in the image brightness" which is analogous to the notion of "strong discontinuity". Localization of edges in brightness is an important constituent part of many digital image processing algorithms, especially of the segmentation algorithms. However, the notion of "edge in the brightness" in the theory of digital image processing is in itself not defined in a strict sense and is more intuitive [7.70]. The problem of a formal definition of an edge in the brightness (in the following we shall write "edge" for brevity) is complicated by its discrete nature, because a sequence of pixels representing the edge is often not connected (by virtue of the noise action), and contains irregularities which are the result of digital encoding, variation of brightness, texture, etc.

A review of image segmentation techniques may be found, for example, in [7.5], [7.7], [7.63], [7.70], [7.80]–[7.82]. As was pointed out in [7.82], in recent years there have been proposed hundreds of segmentation algorithms. There exist two general approaches to the solution of the segmentation problem. The first approach makes use of the discontinuous behavior of image pixels at a boundary between different regions of the image. In this approach the segmentation problem is reduced to a problem of extracting the region's boundaries. The second approach is based on the extraction of image points which are homogeneous with respect to their local properties; these points are assembled into a region to which a name or a label will then be assigned. Within the framework of the first approach we can distinguish three methods for the extraction (or detection) of the boundaries (or of the edges): (1) the spatial differencing; (2) the functional approximation; and (3) high-frequency filtration. The most complicated of these methods is the method of high-frequency filtration. The method of functional approximation (for example, the Hueckel operator [7.63]) is also rather complicated (see Section 7.3.2).

The methods implementing the second approach can also be subdivided into three groups: (1) thresholding and cluster analysis; (2) region growing; and (3) relaxation labeling. In their turn, the methods belonging to the first of these groups are realized with the aid of parametric algorithms and non-parametric algorithms; the methods of region growing may be subdivided into two groups: the methods of centroid linking and the split-and-merge algorithms. The methods realizing the second approach have an advantage over the methods of edge detection in that they enable us to assign different labels to different regions. However, in our case, for the localization and classification of singularities in inviscid gas flows, this labeling of different regions is generally unnecessary; the only thing which is needed in our case is the set of points (x, y) belonging to the lines of strong discontinuities and (optionally) to other flow singularities. The computer implementation of the segmentation methods realizing the second approach (region's labeling) is generally more complicated than the implementation of the edge extraction techniques, in the respect that the corresponding computer codes have a more complicated logic, they often use the elements of graph theory, in particular, the quadtrees.

In the literature on image processing the edges are subdivided into two

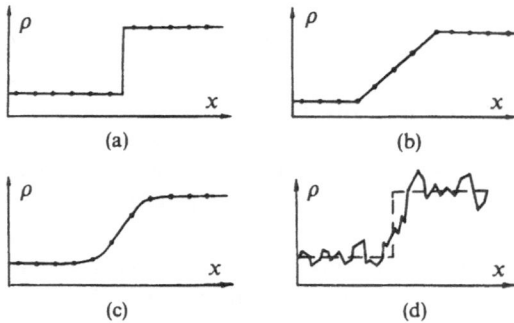

Figure 7.5. (a) Step edge [7.79]; (b) ramp edge [7.63], [7.83]; (c) blurred edge; (d) step edge (dashed line) degraded by both blur and noise (solid line) [7.63], [7.79], [7.85].

classes: ideal and real edges. In Figure 7.5 three forms of ideal edges (Figure 7.5(a)–(c)) and one form of the real edge (Figure 7.5(d)) are shown in a section which is perpendicular to the edge orientation. Thus, from the point of view of this classification the profiles shown in Figures 2.3, 2.4, 2.11, 4.3, 4.6, and 4.7 are ideal edges, but the profiles of Figures 2.7, 2.8, 4.2, 4.10, and 4.11 are real edges.

The algorithms designed for edge detection in a given image are usually called edge detectors [7.63], [7.79]. In [7.70], [7.79], [7.81] there are formulated a number of requirements which should be met by edge detectors. Here we enumerate those requirements which we consider to be most important when these detectors are to be used for the automatic analysis of gas-dynamical computation results:

(1) the computer time needed for the computations in an edge detector should be small;
(2) the required storage should not be large, which is especially important in work with large-size images which may cause a shortage of memory;
(3) isotropy which is a property in possession of which the algorithm detects the edge regardless of its orientation;
(4) noise immunity which in practice is difficult to attain in the general case;
(5) universality which means the independence of the edge detector algorithm from the specific problem; and
(6) the level and quality of the output information which determine the degree of its usefulness for higher level processing systems (in particular, for the systems of interpretation and classification of detected objects).

If in the computed gas dynamics problem there are many different singularities of the shock wave, contact discontinuity, type, etc., then from the viewpoint of digital image processing it is more convenient to rank the corresponding digital images to the class of texture images [7.8], [7.79], [7.86]. Textures are typically characterized by the presence of a large number of structures of the same type, the shape and number of which are usually determined in the process of digital processing of images of this type. A review of various criteria for textures may be found in [7.86]. For example, ROSENFELD [7.79] proposed as a measure of texture the amount of edges per unit image area. Specific methods are developed for digital processing of textures. Many of these methods are also successfully employed for processing of images which do not have textural properties.

Local methods based on calculation of the gradient of the image intensity function are the oldest and, at the same time, they now have the widest acceptance among the various segmentation algorithms listed above. Let $f(x_1, x_2)$ be the continuous image intensity function. Then $\nabla f(x_1, x_2)$ is the vector of the gradient with coordinates $\partial f/\partial x_1, \partial f/\partial x_2$, its magnitude is

$$g(x_1, x_2) = |\nabla f(x_1, x_2)| = [(\partial f/\partial x_1)^2 + (\partial f/\partial x_2)^2]^{0.5}, \qquad (7.3.1)$$

and its orientation angle α may be determined by the formula

$$\alpha = \arctan[(\partial f/\partial x_2)/(\partial f/\partial x_1)]. \qquad (7.3.2)$$

7.3.1. Image Segmentation on a Rectangular Grid

Formula (7.3.1) is usually avoided in the numerical realizations of gradient edge detectors because it is computationally expensive. For example, in the well-known Roberts edge detector [7.87], [7.63] the magnitude $g(x_1, x_2)$ of the gradient at a point (x_1, x_2) is computed by the formula

$$g(x_1, x_2) = |H_1 * F_2| + |H_2 * F_2|, \qquad (7.3.3a)$$

or by the formula

$$g(x_1, x_2) = \max(|H_1 * F_2|, |H_2 * F_2|), \qquad (7.3.3b)$$

where

$$F_2 = \begin{bmatrix} f_{i,j+1} & f_{i+1,j+1} \\ f_{i,j} & f_{i+1,j} \end{bmatrix}, \quad H_1 = \begin{bmatrix} 0 & -1 \\ 1 & 0 \end{bmatrix}, \quad H_2 = \begin{bmatrix} -1 & 0 \\ 0 & 1 \end{bmatrix}. \quad (7.3.4)$$

Another example of the edge detector employing the 2×2 image window from (7.3.4) is the Mérö–Vassy operator in which the masks H_1, H_2 are [7.88]

$$H_1 = \begin{bmatrix} 1 & 1 \\ -1 & -1 \end{bmatrix}, \quad H_2 = \begin{bmatrix} -1 & 1 \\ -1 & 1 \end{bmatrix}.$$

The above two operators are simple, but at the same time they are very noise-sensitive because of the small size of the image window. This fact was noted in a number of works (see, for example, [7.80], [7.89]). In this connection a large number of edge detectors have been developed which use the image windows of the dimension $N \times N$ where $N \geq 3$. The Sobel edge detector [7.84], [7.63] became very popular among these algorithms. This edge detector makes use of a 3×3 image window of the form (7.2.6) and of the masks

H_1, H_2 of the form

$$H_1 = \begin{bmatrix} 1 & 2 & 1 \\ 0 & 0 & 0 \\ -1 & -2 & -1 \end{bmatrix}, \qquad H_2 = \begin{bmatrix} -1 & 0 & 1 \\ -2 & 0 & 2 \\ -1 & 0 & 1 \end{bmatrix}. \tag{7.3.5}$$

Let us introduce the notations

$$S_1 = H_1 * F_3 = (f_{i-1,j+1} + 2f_{i,j+1} + f_{i+1,j+1})$$
$$- (f_{i-1,j-1} + 2f_{i,j-1} + f_{i+1,j-1}), \tag{7.3.6}$$

$$S_2 = H_2 * F_3 = (f_{i+1,j+1} + 2f_{i+1,j} + f_{i+1,j-1})$$
$$- (f_{i-1,j+1} + 2f_{i-1,j} + f_{i-1,j-1}). \tag{7.3.7}$$

Then the magnitude g_{ij} of the gradient at a point (i, j) is computed in the Sobel edge detector by the formula

$$g_{ij} = |S_1| + |S_2|, \tag{7.3.8}$$

and the orientation of the gradient vector is computed as follows:

$$\alpha_{ij} = \arctan(S_1/S_2). \tag{7.3.9}$$

Some general rules for the construction of isotropic edge detectors have been formulated in [7.79] in the class of continuous images, that is, images described by a continuous function of continuous arguments $f(x_1, x_2)$. With regard to these rules, the edge detector employing formula (7.3.1) is isotropic; however, the edge detector based on the formula

$$g(x_1, x_2) = |\partial f/\partial x_1| + |\partial f/\partial x_2| \tag{7.3.10}$$

is anisotropic.

Following [7.79] let us consider an ideal step edge of the form (7.2.8). It may easily be seen that $S_1 = 0$ at the edge (7.2.8), whereas $S_2 \neq 0$. Thus, in the case of an ideal edge (7.2.8) being parallel to the x_2-axis this edge may be detected with the aid of the mask H_2, but it cannot be detected with the aid of the mask H_1. In this connection the mask H_1 is also called a horizontal mask, and H_2 a vertical mask [7.90]. Note that the mask H_2 in (7.3.5) can be obtained from the mask H_1 by a 90° clockwise rotation of H_1.

Relatively many papers [7.63], [7.90]–[7.93] have been devoted to the study of the Sobel edge detector accuracy in cases of real edges degraded by both blur and noise. It was shown in [7.91], [7.92] that the amplitude of the Sobel gradient (7.3.6)–(7.3.8) increased from the normalized value 1 up to 1.5, with the variation of the edge orientation angle β in radians from 0 to 1 (the normalization of the value g was carried out with respect to the gradient amplitude for the vertical direction of the edge). In [7.92] also, a version of the Sobel edge detector was considered, in which the formula

$$g_{ij} = (S_1^2 + S_2^2)^{1/2} \tag{7.3.11}$$

was used instead of (7.3.8). The amplitude of the Sobel gradient (7.3.6), (7.3.7), (7.3.11) increased from the normalized value 1 up to 1.1, with the variation of the edge orientation angle β in radians from 0 to 1 according to [7.92]. The best accuracy in determining the angle β was achieved with the aid of the Sobel operator [7.92]. In [7.92] tests were also carried out for vertically and dia- gonally oriented edges to which noise was added, with the signal-to-noise ratios being equal to 1 and 10. It was shown on a statistical basis that the Sobel and Prewitt operators are the best ones. Comparisons of a number of edge detectors in the case of a straight edge degraded by blur and by random noise have also been made in [7.90] where the edge orientation angle was varied within $0°-(5°)-90°$. The conclusion was drawn that in total, if the data obtained on the basis of different quantitative accuracy characteristics of the edge detectors are summaried, the Sobel edge detector is uniformly superior to the rest of the detectors considered in [7.90]. It was pointed out in [7.94] that the gradient methods are the most suitable for the determination of the orientation of edges (see above for a presentation of some of these local methods).

Let us consider the question on the approximation error of the formula (7.3.6), (7.3.7). Suppose that $f(x_1, x_2)$ is a four-times continuously differentia- ble function of both arguments in the domain of its definition. Consider a general 3×3 mask of the form

$$H = \begin{bmatrix} a & b & c \\ d & e & g \\ k & l & m \end{bmatrix}.$$

Expand the quantities $f_{\alpha\beta}$ entering the image window (7.2.6) with respect to the point (x_{1i}, x_{2j}) into Taylor series, where (x_{1i}, x_{2j}) are coordinates of the point at which the value f_{ij} is given. As a result we obtain

$$H * F_3 = R_0 f(x_{1i}, x_{2j}) + \sum_{\alpha=1}^{2} R_\alpha h_\alpha \, \partial f(x_{1i}, x_{2j})/\partial x_\alpha$$

$$+ R_3 (h_1^2/2) \, \partial^2 f(x_{1i}, x_{2j})/\partial x_1^2 + R_4 h_1 h_2 \, \partial^2 f(x_{1i}, x_{2j})/\partial x_1 \, \partial x_2$$

$$+ R_5 (h_2^2/2) \, \partial^2 f(x_{1i}, x_{2j})/\partial x_2^2 + R_6 (h_1^3/6) \, \partial^3 f(x_{1i}, x_{2j})/\partial x_1^3$$

$$+ R_7 (h_1^2 h_2/2) \, \partial^3 f(x_{1i}, x_{2j})/\partial x_1^2 \, \partial x_2$$

$$+ R_8 (h_1 h_2^2/2) \, \partial^3 f(x_{1i}, x_{2j})/\partial x_1 \, \partial x_2^2$$

$$+ R_9 (h_2^3/6) \, \partial^3 f(x_{1i}, x_{2j})/\partial x_2^3 + O(\cdot), \qquad (7.3.12)$$

where

$$R_0 = a + b + c + d + e + g + k + l + m,$$

$$R_1 = c - a - d + g - k + m, \qquad R_2 = a + b + c - k - l - m,$$

$$R_3 = a + c + d + g + k + m,$$

$$R_4 = c - a + k - m, \qquad R_5 = a + b + c + k + l + m,$$

$$R_6 = R_1, \qquad R_7 = a + c - k - m, \qquad R_8 = c - a - k + m, \qquad R_9 = R_2,$$

$$O(\cdot) = O(h_1^4) + O(h_1^3 h_2) + O(h_1^2 h_2^2) + O(h_1 h_2^3) + O(h_2^4). \qquad (7.3.13)$$

For the quantities (7.3.6), (7.3.7) we obtain with regard to (7.3.5), (7.3.11)–(7.3.13) the following expansions:

$$S_1 = 8h_2 \, \partial f/\partial x_2 + 2h_1^2 h_2 \, \partial^3 f/\partial x_1^2 \, \partial x_2 + (8/6)h_2^3 \, \partial^3 f/\partial x_2^3 + O(\cdot);$$
$$S_2 = 8h_1 \, \partial f/\partial x_1 + 2h_1 h_2^2 \, \partial^3 f/\partial x_1 \, \partial x_2^2 + (8/6)h_1^3 \, \partial^3 f/\partial x_1^3 + O(\cdot). \qquad (7.3.14)$$

In gas dynamic computations uniform grids are often used with the steps $h_1 \neq h_2$. For these cases it is necessary to modify some of the formulas (7.3.6), (7.3.7) in order that the edge orientation be correctly calculated by formula (7.3.9) in the case of nonsquare cells. In this connection one can use instead of (7.3.7) the formula

$$S_2 = \lambda H_2 * F = \lambda \cdot [(f_{i+1,j+1} + 2f_{i+1,j} + f_{i+1,j-1})$$
$$- (f_{i-1,j+1} + 2f_{i-1,j} + f_{i-1,j-1})], \qquad (7.3.15)$$

where $\lambda = h_2/h_1$. It follows from (7.3.14) that owing to the choice 1,2,1 of the coefficients in the Sobel edge detector (7.3.5) there are in (7.3.14) no second-order derivatives. This circumstance was indicated previously in [7.90] as a motivation for the choice of the coefficients 1,2,1 in the Sobel edge detector.

Let us now describe a multistep procedure for edge detection on the basis of the information obtained with the aid of the Sobel edge detector. This information is stored in two two-dimensional arrays of dimension equal to the dimension of the initial image. One array is filled by the magnitudes of the gradient g_{ij}, the second array represents the values of the angle α_{ij} computed by formula (7.3.9). An edge detection procedure presented below is carried out in several stages and combines the ideas of [7.63], [7.79], [7.89], [7.95]. First, let us enumerate these stages and then we shall describe the purpose and method of realization of each stage.

1. Computation of the gradient magnitude g_{ij} and gradient orientation α_{ij} at all points (i, j).
2. Thresholding.
3. Edge thinning.
4. Tracing contour segments.
5. Approximating contour segments by straight line or parabolic segments.
6. Compression of data obtained at the previous stage.

In fact, we have already presented above the first stage of this procedure. Let us make a few remarks on this stage. First, the choice of 3×3 masks is not the most optimal and universal. It is known, for example, that the 4×4 masks produce more reliable results in the case of small jumps of the edge brightness, that is, when the difference of the image intensity function values on both sides of an edge is small in its absolute value [7.92]. On the other hand, the application of masks of increased dimension $N > 3$ leads to an increase in computer expense. As was noted in [7.89], the optimal mask size should vary for different images, and within the limits of the same image.

The second stage enables us to find among the pixels (i, j) those which may be potential edge regions. Here the correct choice of the threshold T of the gradient g is of crucial importance [7.63]. Several techniques are known for determining the threshold value T among which we mention only two: global and locally adaptive techniques. The simplest global technique for computing T (which was presented in [7.95]) consists of computing T as an arithmetic mean of the gradient magnitudes over all nodes (i, j). In [7.49] we applied an iteration technique for computing T. Denote by $T^{(v)}$ the value of T at the vth iteration, $v = 1, \ldots, N_g$, N_g is the user-specified number of iterations. Then the initial approximation was calculated with regard to [7.95] as the arithmetic mean

$$T^{(1)} = (1/N_1) \sum_{i,j} g_{ij}, \tag{7.3.16}$$

where N_I is the overall number of pixels in the initial image. Let $\mathscr{S}_1^{(v)}$ be a set of pixels f_{ij} of the image for which the inequality $g_{ij} > T^{(v)}$ is satisfied, and let $N_I^{(v)}$ be the number of pixels entering $\mathscr{S}_1^{(v)}$, $v = 0, 1, \ldots$, $N_I^{(0)} = N_I$. Then the next approximation $T^{(v)}$ was computed by the formula

$$T^{(v)} = [1/N_I^{(v-1)}] \sum_{(i,j) \in \mathscr{S}_1^{(v)}} g_{ij}, \qquad v = 1, \ldots, N_g. \tag{7.3.17}$$

Usually 1–2 iterations were performed. In practice, we always obtained that $T^{(v)} < T^{(v+1)}$, $N_I^{(v)} > N_I^{(v+1)}$. This is understandable, because the inequality $g_{ij} > T^{(v)}$ is satisfied at a larger value of the threshold $T^{(v)}$ by a lesser number of image pixels than in the case of a smaller value of the threshold $T^{(v)}$. The increase in the value of $T^{(v)}$ with v is related to the fact that when v increases, then the value $T^{(v+1)}$ is computed according to (7.3.17) as an arithmetic mean over the population of pixels possessing, on average, higher values of the gradients g_{ij} for higher v. Since $N_I^{(2)} < N_I^{(1)}$, a smaller number of pixels is used for processing at subsequent stages of a segmentation procedure, than in the case of using only formula (7.3.16), which increases the speed of processing and contributes to the diminution of the number of artefacts, that is, the pixels corresponding to a spurious edge detection. On the other hand, in the case of too large a number of iterations N_g it may easily be seen that information on small jumps, which in the case of gas-dynamical numerical modeling correspond to the shock waves and contact discontinuities of weak intensity, can be lost. The user should select a value of N_g with account being taken as to his final goal. For example, if he is not interested in weak shock waves, he can set $N_g = 2$ or 3. Otherwise, he must take the least possible value of N_g, that is, $N_g = 1$. In the case of the application of a locally adaptive technique the value T proves to be different for each node (i, j); this technique typically takes into account the local average brightness of the image in the vicinity of a pixel with indices (i, j) (see, for example, [7.95], [7.96]). Denote by $\mathscr{S}_1$ a set of points (i, j) at which the inequality $g_{ij} > T^{(N_g)}$ is satisfied; let N_1 be the number of points of this set.

Since in the real image (and in particular in the images obtained by numerical computation of discontinuous solutions by shock-capturing schemes) the edges are usually smeared, the condition $g_{ij} > T$ will be satisfied not only by the pixels lying at real edges but also by neighboring pixels. Therefore, it is necessary to eliminate from the set $\mathscr{S}_1$ those points which do not belong to the edges. A corresponding operation is called contour thinning [7.63], [7.79]. A thinning technique proposed by ROSENFELD [7.79], [7.97] has received wide acceptance. This is the "nonmaximum suppression" technique. Let us describe this technique. Let (i, j) be some node belonging to the set $\mathscr{S}_1$. The angle α_{ij} determines the direction of a normal to the orientation of a potential edge passing through the cell (i, j). Let us take two neighboring pixels (i_l, j_l) and (i_r, j_r) closest to (i, j) and lying on the normal to the edge and on different sides of it. If the inequalities

$$g_{ij} - g_{i_l, j_l} > 0, \qquad g_{ij} - g_{i_r, j_r} > 0, \qquad (7.3.18)$$

are simultaneously satisfied, the node (i, j) is included in the set of points $\mathscr{S}_2$. Let N_2 be the number of points of this set. It is clear that $N_2 < N_1$ since not all the points of the set $\mathscr{S}_1$ satisfy the inequalities (7.3.18). In practical computations the value of the angle α_{ij} was approximated by one of eight principal directions in the image window centered in the (i, j) pixel, as in [7.95], [7.98]. The scheme of numbering the eight principal directions is presented in Figure 7.6. The angles with the values $\alpha_{ij} < 0$ were replaced by $\alpha'_{ij} = 2\pi + \alpha_{ij}$. As a consequence of this the quantity α'_{ij} undergoes a jump of magnitude 2π when passing from negative values of α_{ij} to positive values. In order to determine more accurately the number of a principal direction in this situation, we have introduced an auxiliary ninth direction determined by the angle of magnitude 2π. After that the direction determined by the angle α_{ij} was encoded by the number m, $0 \leq m \leq 8$, of the principal direction which is the closest one to α_{ij}. As soon as the number of the direction corresponding to a given α_{ij} is determined, it is replaced by a number "0" if it proved to be equal to eight. When the number of a principal direction is known, it is not difficult to

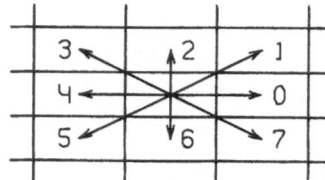

Figure 7.6. Principal directions and their numbering.

determine the indices i_l, j_l, i_r, j_r. For example, in cases when $m = 3$ or $m = 7$ we have set in (7.3.18)

$$i_l = i - 1, \qquad j_l = j + 1, \qquad i_r = i + 1, \qquad j_r = j - 1.$$

The use of the inequalities (7.3.18) is based on the consideration that the gradient achieves its maximum in the direction normal to the edge orientation just at the point of a true edge. This approach eliminates the competition of neighboring pixels located on the edge [7.79], that is, the node (i, j), at which the inequalities (7.3.18) are satisfied, will not be eliminated from the set $\mathscr{S}_2$ in the case when the neighboring pixel lying on the edge possesses a higher value of g than the pixel (i, j).

Generally speaking, in the set $\mathscr{S}_2$ there are also those pixels (i, j) in which isolated peaks of the image intensity f_{ij} take place. Since the edge contour should possess the property of lineal connectivity, it is reasonable to exclude from $\mathscr{S}_2$ the points corresponding to the isolated peaks. For this purpose it was proposed in [7.89] that we also check, at each point (i, j), the satisfaction of the inequalities

$$|\alpha_{i_l, j_l} - \alpha_{ij}| < \Delta\alpha, \qquad |\alpha_{i_r, j_r} - \alpha_{ij}| < \Delta\alpha, \qquad (7.3.19)$$

where the subscripts i_l, j_l, i_r, j_r have the same meaning as in (7.3.18). We have set $\Delta\alpha$, by analogy with [7.89], equal to the maximum difference between the magnitudes of the angles of two neighboring principal directions in Figure 7.6, that is,

$$\Delta\alpha = \max\{\arctan(h_2/h_1), \pi/2 - \arctan(h_2/h_1)\}.$$

Let us denote by $\mathscr{S}_3$ a subset of those points (i, j) of the set $\mathscr{S}_2$ at which the inequalities (7.3.19) are satisfied; let N_3 be the number of points of the set $\mathscr{S}_3$.

In the following, the fourth stage of tracing contour segments or linking of edge points is carried out [7.63], [7.89]. Here the basic operation is the determination of the neighbors of each node $(i, j) \in \mathscr{S}_3$ that lie on the same edge as the node (i, j). Relatively simple linking algorithms have been described in [7.63], [7.89], [7.95]. Isolated artefacts can be removed by the following simple algorithm. We shall consider a point $(x_{ij}, y_{ij}) \in \mathscr{S}_3$ an isolated artefact, if in a circle of radius $R = \vartheta \cdot (h_1^2 + h_2^2)^{0.5}$ there are no other members of the set $\mathscr{S}_3$. The multiplier ϑ is chosen in an appropriate way in order that the points (x_{ij}, y_{ij}) are not eliminated from $\mathscr{S}_3$, which may be identified as belonging to some segment of an edge contour as the result of the operation of one-pixel gaps filling the contours (see below). In our computations we have used the value $\vartheta = 1.5$.

The filling of missing contour points may also be performed at the same stage, as well as the removal of "hairs" which represent short lines branching off the true contour. Here the algorithms using the notions and methods of the theory of graphs [7.89], [7.98], [7.99] have the widest acceptance. Along with graph-theoretical methods of contour tracing we also mention the methods based on cluster analysis [7.100]–[7.102] and the use of fuzzy sets [7.103], [7.104]. We shall denote the set of points (x_{ij}, y_{ij}) obtained as a result of the execution of the fourth stage by $\mathscr{S}_4$; let N_4 be the number of points of this set.

Some irregularities of the form of contour segments which are the result of random noise effects and of image blur are eliminated at the fifth stage. For this purpose an approximation of contour segments by straight-line segments (as in [7.89], [7.101], [7.105]), or parabolas, is often used; for determination of the coefficients of these lines the least squares method may be used [7.63].

If the amount of information obtained at the preceding stages is very large, then we can realize the sixth stage, information compression, before transferring the edge information to a higher-level processing system—in our case the system of singularities classification. At the sixth stage each edge segment is encoded in a certain way, for example, by setting the coordinates of the starting and terminal points in the case of a linear segment. Here the notions and methods of graph theory are usually employed [7.98].

In order to illustrate the above image segmentation procedure we now present some computational results from [7.47]–[7.50]. In these computations we have used the FLIC method [7.106] and the MacCormack scheme (4.2.11), (4.2.12). These schemes, with the incorporation of some modifications, are still used in the numerical solution of various applied problems (see, for example, [7.107]–[7.110]). It should be noted that in recent years there have appeared schemes which enable us to compute more accurately the flows with strong discontinuities as compared to the schemes [7.106] and (4.2.11), (4.2.12) (see a review of such schemes at the beginning of Chapter 6). We have not chosen the best schemes for our computations, with the purpose of elucidating the question as to whether it is possible to apply the above-presented edge detectors for the localization of strong discontinuities in difference solutions characterized either by an intensive smearing of contact discontinuities or by oscillations in the vicinity of shocks.

Figure 7.7 illustrates the influence of the iterations number N_g on the computation of the threshold value $T^{(v)}$ by the formula (7.3.17) in the computational example; one of the results of which has been presented in Figure 7.3. The points of the sets $\mathscr{S}_1$, $\mathscr{S}_2$, and $\mathscr{S}_3$ are marked in Figure 7.7, and in subsequent figures by black points, crosses, and stars, respectively. The dashed lines in Figure 7.7 and in subsequent figures show the exact positions of the discontinuity fronts. Figure 7.9 illustrates the results of the action of the first four stages of the above segmentation procedure applied to numerical processing of the results of the numerical solution of equation (4.2.7) by the FLIC-type scheme (7.2.34), (7.2.35) with the piecewise constant initial function $\rho(x_1, x_2, 0)$ depicted in Figure 7.8; $\theta_F = 0.95$ in formula (7.2.39), $\beta = \arctan(u_{20}/u_{10}) = \pi/4$. Note that the initial function $\rho(x_1, x_2, 0)$, of the form shown in Figure 7.8, was also used in [7.111] while testing a new FCT-type scheme. Only the isolated artefacts were removed at the fourth stage; the

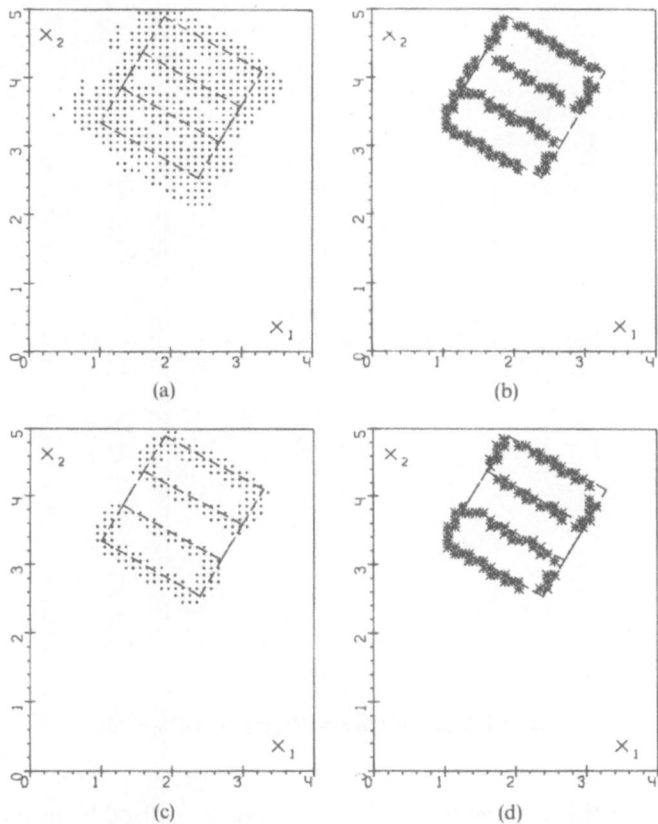

Figure 7.7. (a), (b) The points of the sets $\mathscr{S}_1$ and $\mathscr{S}_4$, $N_g = 1$ in (7.3.17); $N_1 = 484$, $N_2 = 142$, $N_3 = 105$, $N_4 = 98$; (c), (d) the points of the sets $\mathscr{S}_1$ and $\mathscr{S}_4$, $N_g = 2$ in (7.3.17); $N_1 = 225$, $N_2 = 109$, $N_3 = 94$, $N_4 = 92$.

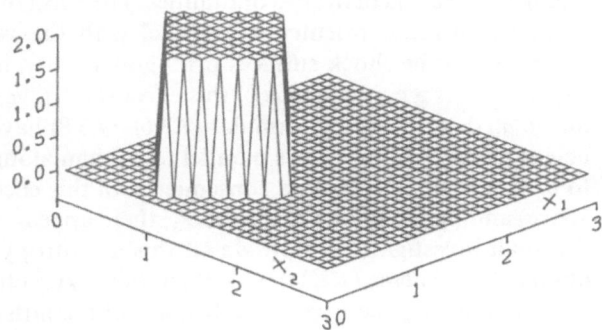

Figure 7.8.   Isometrics of the surface $\rho = \rho(x_1, x_2, 0)$. 30×30 mesh.

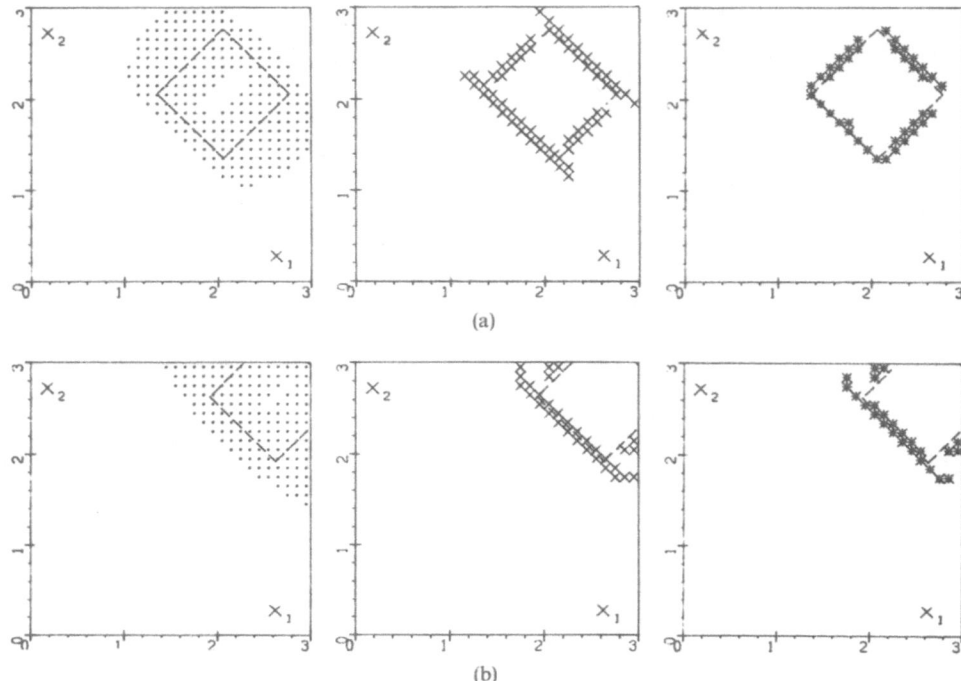

Figure 7.9. (a) $t = 1.612 = 24\tau$; (b) $t = 2.418 = 36\tau$.

algorithm for the removal of "hairs" had not been realized by us until now. It may be seen from Figure 7.9 that the procedure for contour thinning enables us to exclude from the set $\mathscr{S}_1$ a large number of the (i, j) points that do not belong to the shock lines. For example, in the case of Figure 7.9(a), $N_1 = 256$, $N_2 = 60$, $N_3 = 40$.

Since the set $\mathscr{S}_1$ includes all those (i, j) points at which the difference solution gradients are sufficiently large, then the set $\mathscr{S}_1$ approximately describes the region of smearing of the discontinuities. The effect of more intensive smearing of shock surfaces, oriented in parallel with the vector $\mathbf{u}_0 = \{u_{10}, u_{20}\}$ as compared to the shock surfaces orthogonal to $\mathbf{u}_0$, may clearly be seen from Figure 7.9. As a consequence of this effect the "ridges" in the gradient surface image g_{ij} determined by formulas (7.3.6)–(7.3.8) have a substantially smaller height along the shock lines parallel to $\mathbf{u}_0$ than along the lines perpendicular to $\mathbf{u}_0$ (see Figure 7.10). As a consequence of this effect, at $t = 36\tau$ (see Figure 7.9(b)) among the points of the set $\mathscr{S}_3$, there appear artefacts forming "hairs". We shall investigate in more detail this anisotropy of a difference solution obtained by scheme (7.2.34)–(7.2.35) in the next section.

Denote by δr_m the Euclidean distance of the mth point of the set $\mathscr{S}_3$ from

Figure 7.10. Gradient
surfaces: (a) $t = 1.612$;
(b) $t = 2.418$.

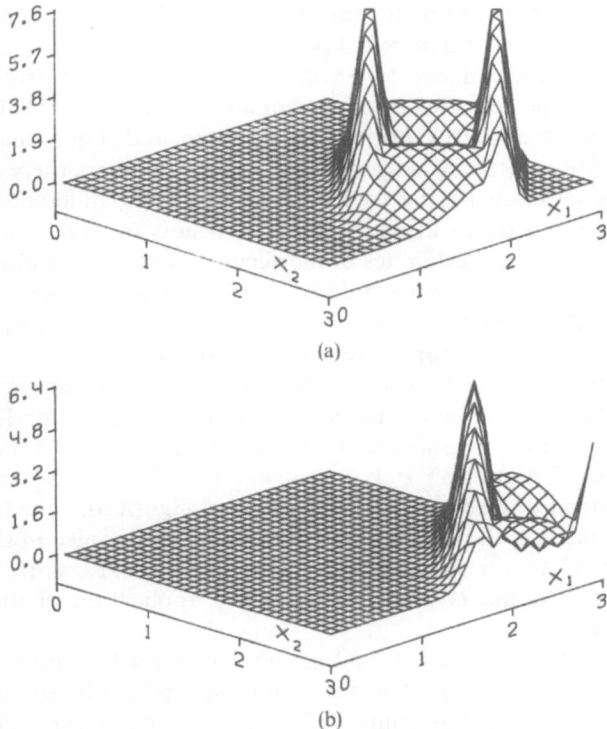

(a)

(b)

the exact discontinuity contour and let us introduce the arithmetic mean

$$\delta r = (1/h_1) \sum_{i=1}^{N_3} \delta r_i.$$

In the case of Figure 7.9, $\delta r = 0.349$ at $t = 24\tau$ and $\delta r = 0.849$ at $t = 36\tau$. Thus, the arithmetic mean absolute error in determining the shock contour location here does not exceed the size of the step h_1 despite the presence of artefacts at $t = 36\tau$.

7.3.2. Accuracy Assessment of the Image Segmentation Techniques

We have already presented above some information on the accuracy of edge detectors while discussing the gradient edge detectors. This information is obtained in the theory of digital image processing on the basis of the calculation of some quantitative characteristics (see, for example, [7.90]–[7.92]). Since the noise and blur are usually introduced into the image when testing the edge detectors, it is clear that quantitative data on the accuracy of edge detectors thus obtained also retain their meaning in the case of the application of these detectors for the localization of strong discontinuities on the basis of smeared gas-dynamical profiles.

The usefulness of the above-mentioned quantitative criteria for the accuracy of edge detectors is unquestionable, since one is able to compare with the aid of these criteria various edge detectors and choose the best ones among them. After the "best" edge detector has been chosen, there arises a question on the conditions on the parameters entering the difference scheme as well as the edge detector at which the best results of shock localization are achieved. Theoretical estimates of the accuracy of shock localization algorithms based on edge detectors would be very useful here. Comparing formula (3.2.21) for a differential analyzer of a two-dimensional shock wave and formula (7.3.10) for the anisotropic edge detector, one can easily see a resemblance between both formulas. This resemblance is not fortuitous, since the same idea of edge localization on the basis of function gradients underlies both the algorithms. The first investigations of the differential analyzers of shock waves were carried out by V.F. Kuropatenko [7.112] as early as the 1960s. The development of essentially similar gradient algorithms of edge localization was initiated in the digital image processing theory also in the 1960s [7.87]. Nevertheless, the present authors were not aware, until 1983, of these parallel developments carried out for the application of the digital processing of images.

It follows from the foregoing that an analytic investigation of the accuracy of gradient edge detectors can be carried out by the same methods which we have applied in Chapters 2–4, while studying the accuracy of the differential analyzers of strong discontinuities. We reproduce in the following (as an example) an analysis of [7.50] for the accuracy of the Sobel edge detector, in the case where this algorithm is applied for the localization of purely contact discontinuities in two-dimensional gas flows calculated by an explicit first-order scheme of the FLIC-type (7.2.34), (7.2.35). In our analysis equation (4.2.7) will be employed as the basic one. As was pointed out in Section 7.3.1, in the Sobel edge detector the position of a strong discontinuity is identified by the position of a point of maximum of the image intensity function gradient in a section which is perpendicular to the edge orientation. In this connection it is convenient to consider the first differential approximation (f.d.a.) (4.2.10) in a Cartesian coordinate system (n_1, n_2), the n_1-axis of which is directed along a normal to the edge and the n_2-coordinate increases in the direction of the shock contour advection. Let us consider, for definiteness, a closed contour of a contact discontinuity which has the form of a rectangle in the (x, y)-plane (see Figure 7.11). Suppose that the two sides of this rectangle are parallel with the On_2-axis which makes an angle β with the positive direction of the Ox_1-axis. Let us choose the positive direction of the On_1-axis in such a way that the coordinate system n_1, n_2 is the right coordinate system. Introduce the notations

$$r_{11} = -\sin \beta, \qquad r_{12} = r_{21} = \cos \beta, \qquad r_{22} = -r_{11}. \qquad (7.3.20)$$

Figure 7.11. Contact discontinuity contour and the axes of coordinates n_1,
n_2.

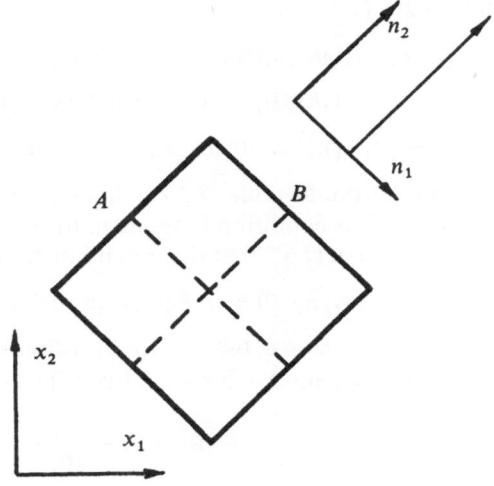

Then the following formulas are valid

$$\partial/\partial x_\alpha = \sum_{\lambda=1}^{2} r_{\alpha\lambda}\, \partial/\partial n_\lambda, \qquad \alpha = 1, 2;$$

$$u_{\alpha 0} = \sum_{\lambda=1}^{2} r_{\alpha\lambda} u_{n_\lambda}^0, \qquad \alpha = 1, 2,$$

(7.3.21)

where $u_{n_1}^0$, $u_{n_2}^0$ are components of the velocity vector $\mathbf{u}_0 = \{u_{10}, u_{20}\}$ in the (n_1, n_2)-coordinate system. Employing formulas (7.3.20), (7.3.21) let us transform equation (4.2.10) to a form depending on the variables n_1, n_2, t. The result may be written in the form

$$\frac{\partial\rho}{\partial t} + u_{n_1}^0 \frac{\partial\rho}{\partial n_1} + u_{n_2}^0 \frac{\partial\rho}{\partial n_2} = q_1 \frac{\partial^2\rho}{\partial n_1^2} + q_2 \frac{\partial^2\rho}{\partial n_2^2} + q_3 \frac{\partial^2\rho}{\partial n_1\, \partial n_2}, \qquad (7.3.22)$$

where

$$q_1 = \sum_{\alpha=1}^{2} (h_\alpha/2)|u_{\alpha 0}|r_{\alpha 1}^2 - (\tau/2)u_{\alpha 0}r_{\alpha 1}u_{n_1}^0;$$

$$q_2 = \sum_{\alpha=1}^{2} (h_\alpha/2)|u_{\alpha 0}|r_{\alpha 2}^2 - (\tau/2)u_{\alpha 0}r_{\alpha 2}u_{n_2}^0;$$

(7.3.23)

$$q_3 = \sum_{\alpha=1}^{2} h_\alpha |u_{\alpha 0}|r_{\alpha 1}r_{\alpha 2} - (\tau/2)u_{\alpha 0}(r_{\alpha 1}u_{n_2}^0 + r_{\alpha 2}u_{n_1}^0).$$

Consider further a particular case when

$$u_{n_1}^0 = 0, \qquad u_{n_2}^0 = |\mathbf{u}_0| = (u_{10}^2 + u_{20}^2)^{0.5}. \qquad (7.3.24)$$

Then $u_{\alpha 0} = r_{\alpha 2}u_{n_2}^0$, $\alpha = 1, 2$, and formulas (7.3.23) may be rewritten with regard to (7.3.20), (7.3.24) as

$$q_1 = |\mathbf{u}_0| \cdot [(h_1/2)|\cos \beta| \sin^2 \beta + (h_2/2)|\sin \beta| \cos^2 \beta]; \qquad (7.3.25)$$

$$q_2 = |\mathbf{u}_0| \cdot [(h_1/2)|\cos \beta| \cos^2 \beta + (h_2/2)|\sin \beta| \sin^2 \beta - \tau |\mathbf{u}_0|/2]; \quad (7.3.26)$$

$$q_3 = |\mathbf{u}_0| \cdot [h_2 |\sin \beta| \cos \beta \sin \beta - h_1 |\cos \beta| \sin \beta \cos \beta]. \qquad (7.3.27)$$

It follows from formula (7.3.27) that $q_3 = 0$ under the condition $h_1 |\cos \beta| = h_2 |\sin \beta|$. This condition is assumed to be satisfied in the following. We shall solve equation (7.3.22) at the initial function $\rho(n_1, n_2, 0)$ of the form

$$\rho(n_1, n_2, 0) = b \cdot \text{Rect}[(n_1 - n_1^0)/a_1] \cdot \text{Rect}[(n_2 - n_2^0)/a_2], \quad (7.3.28)$$

where $b = \text{const} \neq 0$, (n_1^0, n_2^0) are coordinates of the center of a rectangle with sides $2a_1$, $2a_2$, and the function $\text{Rect}(\xi)$ is defined as follows [7.63]:

$$\text{Rect}(\xi) = \begin{cases} 1, & |\xi| \leq 1, \\ 0, & |\xi| > 1. \end{cases}$$

The product form of the initial function (7.3.28) suggests the idea of searching for the solution of the problem (7.3.22), (7.3.28) by analogy with [7.113] with the aid of the separation of variables in the form

$$\rho(n_1, n_2, t) = R_1(n_1, t)R_2(n_2, t), \qquad t \geq 0. \qquad (7.3.29)$$

Substituting the right-hand side of the relationship (7.3.29) for ρ in equation (7.3.22) we obtain the solution of problem (7.3.22), (7.3.28) in the form [7.50]

$$\rho(n_1, n_2, t) = (b/4) \prod_{j=1}^{2} [\text{erf}(\xi_{1j}) - \text{erf}(\xi_{2j})], \qquad (7.3.30)$$

where $\text{erf}(\xi)$ is the error function (see (4.1.23)),

$$\xi_{kj} = (n_j - n_j^{(k)} - U_j t)/(2\sqrt{q_j t}), \qquad k, j = 1, 2, \qquad (7.3.31)$$

$$U_j = \begin{cases} 0, & j = 1, \\ u_{n_2}^0, & j = 2, \end{cases} \quad n_j^{(1)} = n_j^0 - a_j, \quad n_j^{(2)} = n_j^0 + a_j, \quad j = 1, 2.$$

Let us show that in the case when the time step τ satisfies the stability condition (7.2.39) the value q_2 determined by formula (7.3.26) is positive if

$$|\mathbf{u}_0| > 0, \qquad h_1 > 0, \qquad h_2 > 0, \qquad \theta_F < 1, \qquad \sin 2\beta \neq 0. \quad (7.3.32)$$

For this purpose let us rewrite the relationship (7.2.39) as follows:

$$\tau = (\theta_F h_1/|\mathbf{u}_0|)/(|\cos \beta| + (h_1/h_2)|\sin \beta|). \qquad (7.3.33)$$

Let $\lambda = h_2/h_1$. Since

$$h_1 |\cos \beta| \cos^2 \beta + h_2 |\sin \beta| \sin^2 \beta$$

$$= h_1/(|\cos \beta| + (1/\lambda)|\sin \beta|)$$

$$+ h_1 |\sin \beta| |\cos \beta| (|\cos \beta|/\sqrt{\lambda} - \sqrt{\lambda}|\sin \beta|)^2/(|\cos \beta| + |\sin \beta|/\lambda),$$

$$(7.3.34)$$

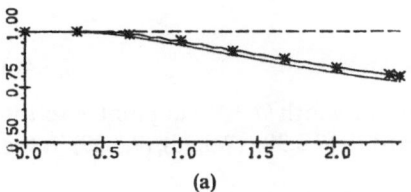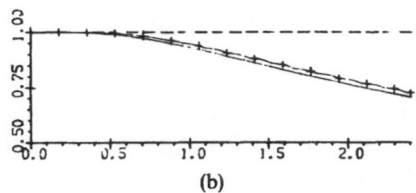

(a)   (b)

Figure 7.12. Comparison of the quantity $\max_{i,j} \rho_{ij}^n$ (solid lines with markers) with the function (7.3.35) (solid lines). (a) $\theta_F = 0.95$; (b) $\theta_F = 0.50$.

with (7.3.33) in view $q_2 > 0$ under the conditions (7.3.32). Making use of (7.3.30) we can find from the equation $\partial\rho/\partial n_1 = \partial\rho/\partial n_2 = 0$ that

$$\rho_{\max}(t) = \max_{n_1, n_2} |\rho(n_1, n_2, t)| = |\rho(n_1^0, n_2^0 + u_{n_2}^0 t, t)|$$
$$= |b| \operatorname{erf}(a_1/(2\sqrt{q_1 t})) \cdot \operatorname{erf}(a_2/(2\sqrt{q_2 t})). \tag{7.3.35}$$

In Figure 7.12 the maximum value of ρ obtained from computations by the scheme (7.2.34), (7.2.35) is compared with the value $\rho_{\max}(t)$ determined by formula (7.3.35). It is seen that formula (7.3.35) describes very well the actual behavior of the quantity $\max_{i,j} \rho_{ij}^n$. By analogy with (4.1.12) one might introduce the notion of a width X_j after Prandtl of the zone of smearing of a discontinuity which is orthogonal to the direction of the n_j-axis, $j = 1, 2$;

$$X_j = \rho_{\max}(t)/\max_{n_j} |\partial\rho/\partial n_j|. \tag{7.3.36}$$

However, the exact determination of the (n_1, n_2)-coordinates of the points where $|\partial\rho/\partial n_j|$ achieves its maximum is impossible in analytic form because of a complicated transcendental character of the equation $\partial^2\rho/\partial n_j^2 = 0$. Therefore, in the following we restrict ourselves to calculation of the widths

$$X_j = \rho_{\max}(t)/|\partial\rho/\partial n_j| \tag{7.3.37}$$

at some chosen points. As one of these points it is natural to take the point A (see Figure 7.11) with coordinates $n_1 = n_1^0 - a_1, n_2 = n_2^0 + u_{n_2}^0 \cdot t$. Then the quantities ξ_{kj} are calculated at point A in accordance with (7.3.31) by the formulas

$$\xi_{11}(A) = 0; \qquad \xi_{21}(A) = -a_1/\sqrt{q_1 t},$$
$$\xi_{12}(A) = -\xi_{22}(A) = a_2/(2\sqrt{q_2 t}). \tag{7.3.38}$$

As another point let us take point B with coordinates $n_1 = n_1^0, n_2 = n_2^0 + a_2 + u_{n_2}^0 \cdot t$, that is, located on a discontinuity line which is perpendicular to the velocity vector $\mathbf{u}_0$. At this point

$$\xi_{11}(B) = -\xi_{21}(B) = -\xi_{21}(A)/2;$$
$$\xi_{12}(B) = 2\xi_{12}(A); \qquad \xi_{22}(B) = 0. \tag{7.3.39}$$

Let us introduce the coefficient

$$k = (a_2/a_1)(q_1/q_2)^{0.5}. \qquad (7.3.40)$$

Then the expression for the relationship of the width (7.3.37) at point A to the width at point B may be written with regard to (7.3.30), (7.3.38), (7.3.39) in the form

$$X_2(B)/X_1(A) = (q_2/q_1)^{0.5} \cdot G(\zeta, k), \qquad (7.3.41)$$

where

$$G(\zeta, k) = \mathrm{erf}(\zeta \cdot k)[1 - \exp(-4\zeta^2)]\{\mathrm{erf}(\zeta) \cdot [1 - \exp(-4k^2\zeta^2)]\}^{-1}, \quad (7.3.42)$$

$$\zeta = a_1/(2\sqrt{q_1 t}). \qquad (7.3.43)$$

It follows from (7.3.41), (7.3.42) that at $k = 1$

$$X_2(B)/X_1(A) = (q_2/q_1)^{0.5} = a_2/a_1.$$

Employing formulas (7.3.25), (7.3.26), (7.3.34) it is easy to find that

$$q_2/q_1 = \lambda^{-1} \cdot [1 - \theta_F + |\sin \beta||\cos \beta|(\sqrt{\lambda}|\sin \beta| - |\cos \beta|/\sqrt{\lambda})^2]$$
$$\times \{[(1/\lambda)|\cos \beta| \sin^2 \beta + |\sin \beta| \cos^2 \beta]$$
$$\times (|\cos \beta| + |\sin \beta|/\lambda)\}^{-1}. \qquad (7.3.44)$$

In particular, at $\beta = \pi/4$, $\lambda = 1$, we obtain from (7.3.44)

$$q_2/q_1 = 1 - \theta_F. \qquad (7.3.45)$$

Let us now consider the case when $k > 0$ and the quantity t is small. Since with regard to (7.3.43)

$$\lim_{t \to 0} \mathrm{erf}(\zeta \cdot k)/\mathrm{erf}(\zeta) = 1,$$

we obtain from (7.3.41) that at small t

$$X_2(B)/X_1(A) \cong (q_2/q_1)^{0.5}.$$

Let us now consider the case of finite $t > 0$ and $k > 0$. Let us fix ζ and consider the behavior of the function

$$G_1(z) = \mathrm{erf}(z) \cdot [1 - \exp(-4z^2)]^{-1}$$

in the domain $z > 0$ where $z = k\zeta$. Investigating the behavior of the derivative $G_1'(z)$ it is not difficult to find that $G_1'(z) < 0$ at $0 < z < z_1$, $G_1'(z) > 0$ for $z > z_1$, where z_1 is the only positive root of the equation $dG_1(z)/dz = 0$, $z_1 \cong 0.65880827$. Thus, in a section $\zeta = \mathrm{const}$ there are intervals of both increasing and decreasing the function $G(\zeta, k)$ with a variation of k. In Figure 7.13 we present the graphs of the function $G(\zeta, k)$ for some values of k with a variation of ζ in the interval $0.01 \leq \zeta \leq 3$. It follows from these graphs that the quantity G weakly depends on k at sufficiently large ζ (at $\zeta \gtrsim 1.5$). Therefore, at $\zeta \gtrsim 1.5$ we obtain from formulas (7.3.41), (7.3.44) that the value q_2/q_1 increases with

Figure 7.13. The function (7.3.42) graphs for some values of k: $(-\cdot-\cdot-\cdot-)$—k = 0.25; $(----)$—k = 0.50; (———)—k = 1.00; $(-\square-\square-)$—k = 2.00; $(-\bullet-\bullet-\bullet-)$—$k$ = 4.00.

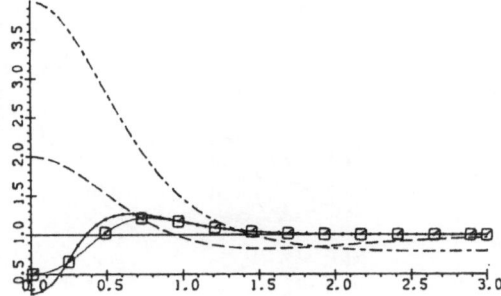

decreasing θ_F. Figure 7.14 supports this conclusion: at $\theta_F = 0.5$ the width of a zone of smearing of the discontinuity lines which is perpendicular to the velocity vector $\mathbf{u}_0$ is substantially larger than at $\theta_F = 0.95$ (compare with Figure 7.9(a)). As a result of this the width of the zones of smearing of differently oriented shock lines becomes almost the same, that is, the scheme (7.2.34), (7.2.35) at $\theta_F = 0.5$ becomes almost isotropic. However, with decreasing θ_F the intensity of smearing of shocks increases, as a result of this $\max_{i,j}|\rho_{ij}^n|$ decreases more rapidly with increasing t for smaller θ_F (see Figure 7.12). Although the edge detector response (7.3.8) at $\theta_F = 0.5$ depends comparatively weakly on the orientation of edges (see Figure 7.15), an effect of rounding the corners of the discontinuity contour takes place (see Figure 7.14(b), (c)), in consequence of a more intensive smearing.

Let us now derive the estimates of contact discontinuity lines localization accuracy when using the Sobel edge detector. The equation $\partial^2\rho/\partial n_2^2 = 0$ leads to a problem of calculating the roots of the transcendental equation

$$f(z) = (z + c)\exp\{-(z + c)^2\} - (z - c)\exp\{-(z - c)^2\} = 0, \quad (7.3.46)$$

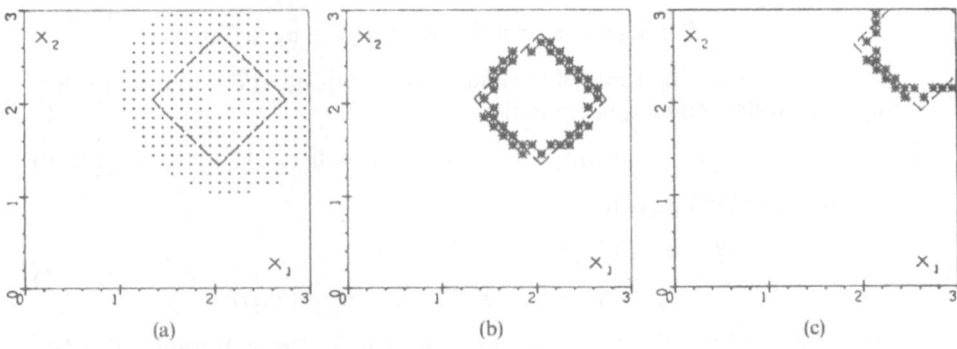

(a)   (b)   (c)

Figure 7.14. (a) The set $\mathscr{S}_1$ at $t = 45\tau = 1.591$, $N_1 = 318$; (b) the set $\mathscr{S}_3$ at $t = 1.591$, $N_3 = 42$. $\delta r = 0.447$; (c) the set $\mathscr{S}_3$ at $t = 68\tau = 2.404$, $N_3 = 19$, $\delta r = 0.520$.

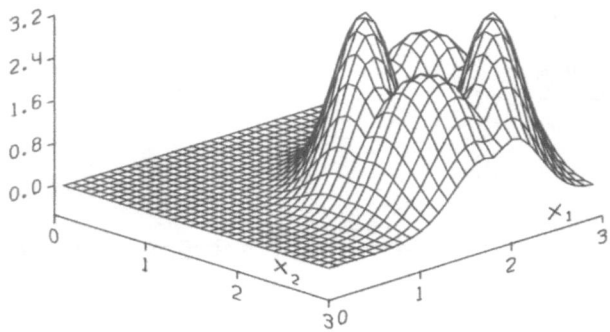

Figure 7.15. Isometrics
of the gradient surface
(7.3.8) at $\theta_F = 0.5$, $t =$
$45\tau = 1.591$.

where

$$z = (n_2 - n_2^0 - U_2 t)/(2\sqrt{q_2 t}), \qquad c = a_2/(2\sqrt{q_2 t}).$$

Considering the behavior of the functions

$$f_1(z) = (z + c)\exp[-(z + c)^2], \qquad f_2(z) = (z - c)\exp[-(z - c)^2], \quad (7.3.47)$$

it can be shown that there exist two, and only two, points of intersection of the graphs of these two functions. In proof of this fact the inequality ([7.114])

$$\exp(x) < 1/(1 - x), \qquad x < 1,$$

has proved to be very useful. Let us denote the abscissa of these points by z_1 and z_2. Omitting for brevity an analysis of the quantitative behavior of the functions (7.3.47) we note that the roots z_1 and z_2 of equation (7.3.46) satisfy the inequalities

$$-c - 2^{-1/2} < z_1 < -c, \qquad c < z_2 < c + 2^{-1/2}. \qquad (7.3.48)$$

Returning to the variable n_2 we obtain from (7.3.48) the inequalities

$$n_2^0 + U_2 t - a_2 - (2q_2 t)^{0.5} < n_2^{(1)} < n_2^0 + U_2 t - a_2;$$
$$n_2^0 + U_2 t + a_2 < n_2^{(2)} < n_2^0 + U_2 t + a_2 + (2q_2 t)^{0.5}; \qquad (7.3.49)$$

where $n_2^{(1)}$, $n_2^{(2)}$ are the roots of the equation $\partial^2 \rho(n_1^0, n_2, t)/\partial n_2^2 = 0$. Considering, in a similar manner, the equation

$$\partial^2 \rho(n_1, n_2^0 + U_2 t, t)/\partial n_1^2 = 0, \qquad (7.3.50)$$

we can derive the inequalities

$$n_1^0 - a_1 - (2q_1 t)^{0.5} < n_1^{(1)} < n_1^0 - a_1;$$
$$n_1^0 + a_1 < n_1^{(2)} < n_1^0 + a_1 + (2q_1 t)^{0.5}; \qquad (7.3.51)$$

where $n_1^{(1)}$, $n_1^{(2)}$ are the roots of equation (7.3.50). From formulas (7.3.49), (7.3.51) it follows that the absolute error in determining the coordinates n_1, n_2 of the contact discontinuity points increases proportionally to $\sqrt{t}$. The error in the localization of a discontinuity line perpendicular to the vector $\mathbf{u}_0$ may be reduced by setting a maximal value of θ_F in (7.2.39), that is, $\theta_F \approx 1$. This conclusion follows from formulas (7.3.49), (7.3.26), (7.3.33), (7.3.34). The quantity q_1 entering (7.3.51) is always positive according to (7.3.25). In addition, $q_1 > q_2$ at a value of θ_F being slightly different from unity. Consequently, the discontinuity lines parallel to the fluid velocity vector $\mathbf{u}_0$ will be localized with less accuracy than the lines perpendicular to the vector $\mathbf{u}_0$. This conclusion is confirmed by the computation results presented in Figure 7.9, as well as by other examples presented in [7.50].

Thus, we have shown that the f.d.a. of a difference scheme enables us to estimate the shock localization accuracy as well as to explain various computational effects. A difficulty of application of this approach is related mainly to the difficulties of the analytic investigation of the properties of solutions of differential approximation equations. In the case when second-order schemes (for example, the MacCormack scheme (4.2.11), (4.2.12)), are used in gasdynamical computations an analysis which is similar to the above analysis was not carried out by us. The interested reader can carry out this analysis by using the solution (4.1.29) which is oscillatory, as was shown in Section 4.1.

It follows from the above analysis that the accuracy of shock localization by isotropic edge detectors can be substantially nonuniform if the difference scheme used in computations is not isotropic. In its turn, as we have shown above in the example of the FLIC-type scheme (7.2.34), (7.2.35) [7.106], [7.107], the difference solution isotropy can depend substantially on the time step size. In particular, it follows from the foregoing analysis that in those cases when on the lines of the contact discontinuities in the two-dimensional flows there is no discontinuity in the tangential velocity component, an excessive smearing of these discontinuities may take place in computations by the "break-down-of-discontinuity" scheme of S.K. Godunov [7.115], [7.116] and by the FLIC method [7.106], if the above discontinuity lines are parallel to the fluid velocity vector. The "coarse particles" method [7.107] has the same shortcoming since this method makes use of the same type of splitting as the FLIC method, in terms of physical processes.

Thus, there exists the problem of the development of shock-capturing difference schemes providing an isotropic localization of strong discontinuities in multidimensional problems. In [7.117] an attempt was undertaken to solve this problem by constructing difference schemes whose f.d.a. is invariant with respect to the same group of transformations which is allowed by the original Euler equation system (1.30). The results of a test for two-dimensional calculations by these schemes presented in [7.117] are encouraging, although the difference schemes themselves appear to be rather complicated. In this connection the use of the experience which has now been accumulated on the theory of digital image processing (in connection with the development of isotropic edge detectors, smoothing algorithms, noise filtration, and image restoration), can prove useful in the solution of a problem on the development of sufficiently efficient isotropic difference schemes. We can make a small "vocabulary" interpreting some notions of the digital image processing theory in the terms employed in modern computational fluid dynamics.

Theory of digital image processing	Computational fluid dynamics
Digital image	Numerical solution of a two-dimensional problem
Image window	Stencil of a difference scheme
Pixel	The cell of a spatial grid
Edge	Strong discontinuity
Edge blur	Smearing of a strong discontinuity
Image noise	Parasitic oscillations of the numerical solution
Edge detection	Localization of discontinuities
Smoothing of the image intensity function	Smoothing of the numerical solution

The variety of methods and techniques used in the modern theory of digital image processing enables us to expect that their adaptation to the construction of isotropic difference schemes for aerohydrodynamics problems will prove to be quite successful, although the numerical algorithms thus obtained can have a rather unusual form in comparison with the currently used difference schemes.

Let us now briefly discuss the following question: Can we indeed reach such an accuracy of shock localization on the basis of two-dimensional computations on a uniform grid that the absolute error of localization is much less than the step sizes h_1, h_2? In the one-dimensional case in Section 2.4, we have demonstrated a positive answer to this question by presenting computational examples. One of the conditions for attaining a high localization accuracy was the existence of a smeared shock wave center. In the two-dimensional case the theoretical study of a question on the existence and uniqueness of a smeared shock wave center is a very complicated problem. If we turn to the theory of digital image processing, a positive answer to the above question may be found there and also in the case of the two-dimensional difference solutions. As a matter of fact, there have appeared, in recent years in the theory of digital image processing, algorithms of edge localization the purpose of which is edge localization to subpixel accuracy. While not pretending to have a complete list of references we enumerate, in this connection, [7.118]–[7.121], [7.123]–[7.126]. The algorithm proposed by HUECKEL [7.118], [7.63] is the earliest. In this algorithm the quantitative data on the local orientation of an edge, as well as on the edge height H, are determined as the result of minimization of a quadratic functional

$$\mathcal{E} = \int_{\mathcal{D}} \int [f(x, y) - S(x, y)]^2 \, dx \, dy,$$

where $f(x, y)$ is the original digital image, $S(x, y)$ is the image intensity function determining an ideal step edge in the domain $\mathscr{D}$ as

$$S(x, y) = \begin{cases} b, & x \cos \theta + y \sin \theta < \rho, \\ b + H, & x \cos \theta + y \sin \theta \geq \rho, \end{cases}$$

where θ and ρ are polar coordinates of the edge point nearest to the center of a considered circular domain $\mathscr{D}$. The diameter of the circle $\mathscr{D}$ is usually much less than the image size. A more efficient algorithm of this kind has been proposed in [7.119]. Note that the Hueckel algorithm [7.118] has some resemblance to the algorithms of difference solution refinement in the neighborhood of the discontinuities presented in Chapter 6. Therefore, some ideas underlying the computer implementation of the algorithms of [7.118], [7.119] may prove useful in the generalization of the algorithms of Chapter 6 for the two-dimensional case. A theoretical proof of the fact that the gradient edge detection techniques, similar to the those presented in Section 7.3.1, cannot be applied in cases when there is a need in subpixel edge detection has been given in [7.120]. In this connection the authors of [7.120] proposed a new algorithm where the edge detection is based not on the derivatives, but makes use of the integrals in order to reduce the influence of the noise. In [7.121] moments have been used for edge detection. Previously, the idea of using moments for shock localization in the numerical profiles of a "step" form had been considered in [7.122].

In [7.123] two edge detectors were proposed which were based on using the sets of edge detectors, each of which is sensitive to a certain group of edge types. In [7.124] an algorithm was constructed on the basis of a recursive edge detector and a minimum-distance classifier.

We have already mentioned, at the beginning of Chapter 2, the works [7.127], [7.128] in which the location of a shock wave was determined by an inflection point in the pressure profile along a beam $x = $ const. Essentially the same idea was considered in the theory [7.129] and realized in the practice of digital image processing [7.129]–[7.131] in the method of zero crossings for edge detection. In this method the position of an edge is determined, in a section perpendicular to the edge orientation, as a point at which the second derivative of the image intensity function (with respect to the coordinate measured along a normal to the edge orientation) vanishes. To reduce the number of artefacts it was recommended in [7.129] to perform an image smoothing before proceeding to the localization of zero crossings.

It follows from the results of Section 4.2 that the lines of purely contact discontinuities in two-dimensional flows can be localized as the zero crossings of a quantity coinciding with the right-hand side of equation (4.2.9). For example, in the case of the FLIC method [7.106] the expression for this quantity with regard to (4.2.10) may be written in the form

$$Q = \sum_{\alpha=1}^{2} \{(h_\alpha/2)(|u_\alpha| \rho_{x_\alpha})_{x_\alpha} - (\tau/2) u_\alpha [(u_1 \rho)_{x_1} + (u_2 \rho)_{x_2}]_{x_\alpha}\}. \quad (7.3.52)$$

It is clear that the algorithm of edge detection based on the localization of zero crossings of a quantity of the type (7.3.52) has a limited domain of applicability in the computer analysis of gas-dynamical computational results, because it is applicable only for the localization of purely contact discontinuities. In addition, in the case of oscillatory difference schemes, it is natural to expect (with regard to the results of Section 4.1) a large number of artefacts while using the zero-crossings algorithm which increases computational expense. Part of the artefacts are removed at the concluding stages of the segmentation procedure, for which graph-theoretical methods are usually employed. A crucial "cleaning" of the artefacts is performed at the classification stage (see the examples in Section 7.6).

It was found in Section 4.1, as the result of an analysis of the f.d.a.s of a number of second-order schemes, that the best accuracy in the localization of contact discontinuities in one-dimensional flows is to be expected in the case when using a differential analyzer based on determining the points of maxima of the quantity $|\partial^2\rho/\partial x^2|$. One of the simplest two-dimensional differential operators which in the one-dimensional case are reduced to the operator $\partial^2/\partial x^2$ is the Laplace operator $\Delta \equiv \partial^2/\partial x^2 + \partial^2/\partial y^2$. In the practice of digital image processing the algorithms of edge detection are applied for a long time in which the edges are determined as the loci of points of local maxima of the quantity $|\Delta f(x, y)|$ where $f(x, y)$ is the image intensity function [7.63], [7.70], [7.79], [7.132]. Similar to the mask representation of the operators $\partial/\partial x$, $\partial/\partial y$, the Laplace operator can also be approximated by convolving the image array with a mask ensuring a difference approximation of this operator. Several forms of the masks for computing the Laplacian are enumerated in [7.63], [7.70], [7.132]:

Mask 1

$$H_1 = \begin{bmatrix} 0 & 1 & 0 \\ 1 & -4 & 1 \\ 0 & 1 & 0 \end{bmatrix}. \tag{7.3.53a}$$

Mask 2

$$H_2 = \begin{bmatrix} 1 & 1 & 1 \\ 1 & -8 & 1 \\ 1 & 1 & 1 \end{bmatrix}. \tag{7.3.53b}$$

Mask 3

$$H_3 = \begin{bmatrix} 1 & -2 & 1 \\ -2 & 4 & -2 \\ 1 & -2 & 1 \end{bmatrix}. \tag{7.3.53c}$$

It has been pointed out in [7.70] that the edge detectors based on a Laplacian are also noise-sensitive. In particular, the masks (7.3.53) can yield a substantial nonzero response in the node (i, j) containing an isolated peak of the image intensity $f(x, y)$.

Employing formulas (7.3.12), (7.3.13) it is easy to find that in the case of using the mask (7.3.53b) the formula

$$H_2 * F_3(f) = 3h_1^2 \, \partial^2 f/\partial x_1^2 + 3h_2^2 \, \partial^2 f/\partial x_2^2 + O(\cdot)$$

is valid, so that the approximation of the Laplace operator takes place, strictly speaking, only for $h_1 = h_2$. In the case when $\lambda = h_2/h_1 \neq 1$ it is easy to obtain, with the aid of (7.3.12), (7.3.13), the following more general expression for the mask H_2:

$$H_2 = \begin{bmatrix} 1 & 1 & 1 \\ 3\lambda^2 - 2 & -2 - 6\lambda^2 & 3\lambda^2 - 2 \\ 1 & 1 & 1 \end{bmatrix}. \qquad (7.3.54)$$

Similarly, the mask (7.3.53a) generalizes to the form

$$H_1 = \begin{bmatrix} 0 & 1 & 0 \\ \lambda^2 & -2(1 + \lambda^2) & \lambda^2 \\ 0 & 1 & 0 \end{bmatrix}. \qquad (7.3.55)$$

In the case of using the masks (7.3.54), (7.3.55) the approximation of the Laplace operator also takes place for $h_2 \neq h_1$.

7.3.3. Image Segmentation on a Curvilinear Grid

We have already discussed, in Section 3.3, some questions related to the use of curvilinear grids in the numerical computations of multidimensional problems. We have made, in Section 3.3, an emphasis on moving grids adapting to the solution with regard to its gradients. At present, the fixed curvilinear grids are widely used in the solution of problems in the domains with complex geometric form along adaptive grids [7.133], [7.134]. Taking this into account it appears that the problem of development of the methods for automatic analysis of difference solutions computed on arbitrary curvilinear grids (both adaptive and nonadaptive) is of present interest. In this section we present an extension of the image segmentation algorithm, presented in Section 7.3.1 for the case of a rectangular uniform grid, for the case of a curvilinear grid the coordinates of the points of which, with the aid of the functions

$$x = x(\xi, \eta), \qquad y = y(\xi, \eta), \qquad (7.3.56)$$

can be uniquely mapped onto the points of the lines of a uniform rectangular grid in the plane of the curvilinear coordinates (ξ, η). Let J be the Jacobian of the transformation (7.3.56). Suppose that the functions $x(\xi, \eta)$ and $y(\xi, \eta)$ in (7.3.56) are differentiable. Then

$$J = 1/(x_\xi y_\eta - x_\eta y_\xi), \qquad (7.3.57)$$

where $x_\xi = \partial x(\xi, \eta)/\partial \xi$, $y_\eta = \partial y(\xi, \eta)/\partial \eta$, etc. Let $f(x, y)$ be also a differentiable function of both arguments. Then

$$f_x = J \cdot (f_\xi y_\eta - f_\eta y_\xi), \qquad f_y = J \cdot (f_\eta x_\xi - f_\xi x_\eta). \qquad (7.3.58)$$

Employing the image window (7.2.6), the masks of the Sobel edge detector (7.3.5), and formulas (7.3.12), (7.3.13), we can obtain the following difference approximations of formulas (7.3.58):

$$f_x = (1/8)J \cdot (y_\eta \cdot S_2 - y_\xi \cdot S_1);$$
$$f_y = (1/8) \cdot J \cdot (x_\xi \cdot S_1 - x_\eta \cdot S_2); \qquad (7.3.59)$$

where

$$S_1 = H_1 * F_3(f) \cong 8 \, \partial f/\partial \eta;$$
$$S_2 = H_2 * F_3(f) \cong 8 \, \partial f/\partial \xi. \qquad (7.3.60)$$

For the computation of the derivatives x_ξ, y_η in (7.3.59) we can use the formulas

$$x_\xi = (1/8)H_2 * F_3(x); \qquad y_\eta = (1/8)H_1 * F_3(y);$$

etc. Taking into account the relative complexity of these formulas we have used in practical computations in the interior nodes (i, j) the more simple formulas [7.133]

$$x_\xi = 0.5(x_{i+1,j} - x_{i-1,j}); \qquad y_\eta = 0.5(y_{i,j+1} - y_{i,j-1}); \qquad (7.3.61)$$

etc. Let the nodes (x_{ij}, y_{ij}) be numbered as follows: $1 \le i \le i_{max}$, $1 \le j \le j_{max}$. We can use one-sided and central differences of the form

$$(x_\eta)_{i,1} = 0.5(4x_{i,2} - 3x_{i,1} - x_{i,3}), \qquad i = 1, \ldots, i_{max};$$
$$(x_\xi)_{i,1} = 0.5(x_{i+1,1} - x_{i-1,1}), \qquad i = 2, \ldots, i_{max} - 1; \qquad (7.3.62)$$
$$(f_\eta)_{i,1} = 0.5(4f_{i,2} - 3f_{i,1} - f_{i,3}), \qquad i = 1, \ldots, i_{max};$$

etc. in the boundary nodes.

The basic idea of the segmentation algorithm for an image given on a curvilinear grid will be illustrated in the example of a problem on the computation of a two-dimensional transonic potential flow around an airfoil NACA 0012. The merits and shortcomings of the mathematical model of potential flow have been discussed in a number of surveys (see, for example, [7.135]–[7.139]). Here we note only the following limitation of this model: it is generally applicable only for the description of those transonic flows in which the maximal value of the Mach number $M = |\mathbf{u}|/c$ ($|\mathbf{u}|$ is the magnitude of the gas velocity, c is the local speed of sound) does not exceed the value 1.2 [7.140], [7.141]. This means that only shock waves of a relatively weak intensity are allowed in a potential transonic flow around an airfoil. At larger values of the Mach number in the supersonic flow subdomains the Rankine–Hugoniot curve deviates substantially from the corresponding curve for an isentropic shock wave [7.142]. In addition, the further increase of the intensity of a shock wave impinging on an airfoil surface leads to the increase in the role of viscous effects, which serve as an additional source for the discrepances between the results of computations by the potential model and experimental data [7.140], [7.143], [7.144]. Interesting considerations have been presented in [7.145] in favor of using the numerical modeling on the basis of the full equation for the velocity potential in the design of efficient shockless transonic profiles. It is pointed out in [7.145] that the solutions of the full-potential equation in the case of strong shock waves on an airfoil reflect inaccurately the flow picture as compared to the computations on the basis of the Euler equations (1.30). In each case of a presence of an intensive shock wave, the intensity of a shock obtained by the solution of the full-potential equation and the position of this shock will, respectively, be stronger and more downstream than in the solution obtained with the aid of the Euler equations. This is a well-known result, at least for the case of using the conservative form of the full-potential equation (see, for example, [7. 145]). For the airfoil designers the very fact of a presence of shock waves near the profile at transonic cruise flight speed is undesirable because of a wave drag caused by shock waves. Therefore, in the process of refinement of shockless profiles we can effectively use numerical modeling on the basis of a full-potential equation. If in the process of such a refinement this numerical modeling begins to yield sufficiently weak shock waves, the solution of the full-potential equation will already be accurate in this case. If this design procedure is carried out on the basis of the Euler equations, it will be much more expensive. Thus, we deal here with a situation when the computer should give the researcher an answer of the form "yes" or "no" to the following question: Are there shock waves in a given transonic flow around an airfoil? Although the answer produced by the computer is very brief, to obtain it the computer performs rather complicated multistage computations: numerical generation of a curvilinear computing mesh around an airfoil, a relaxation solution of the full equation for the velocity potential, and finally, an automatic analysis of the computer-generated flow field by means of the airtificial intelligence algorithms being developed in the present chapter. Note that the possibility of a situation of this kind, when the computer needs to produce only a brief answer of the "yes" or "no" type, was forecast by J. von Neumann as early as 1949 [7.146].

The full equation for the velocity potential in a conservative form may be written as follows:

$$(\rho\varphi_x)_x + (\rho\varphi_y)_y = 0, \tag{7.3.63}$$

where x, y are Cartesian rectangular coordinates, and the components u, v of the velocity vector $\mathbf{u}$ are expressed by the potential φ as

$$u = \partial\varphi/\partial x, \qquad v = \partial\varphi/\partial y. \tag{7.3.64}$$

The density ρ is expressed with the aid of the Bernoulli integral (3.2.17) in terms of φ_x, φ_y by the formula

$$\rho = \left[1 - \frac{\gamma - 1}{\gamma + 1}(\varphi_x^2 + \varphi_y^2)\right]^{1/(\gamma-1)}, \tag{7.3.65}$$

where the density ρ and the velocity components φ_x, φ_y are nondimensionalized with respect to the stagnation density ρ_s and the critical speed of sound c_*, respectively. The transformation (7.3.56) retains the conservative form of equation (7.3.63). The full-potential equation written in ξ, η coordinates has the form [7.145]

$$(\rho U/J)_\xi + (\rho V/J)_\eta = 0, \tag{7.3.66}$$

where

$$\rho = \left[1 - \frac{\gamma - 1}{\gamma + 1}(U\varphi_\xi + V\varphi_\eta) \right]^{1/(\gamma-1)}, \tag{7.3.67}$$

$$U = A_1\varphi_\xi + A_2\varphi_\eta, \qquad V = A_2\varphi_\xi + A_3\varphi_\eta,$$

$$A_1 = \xi_x^2 + \xi_y^2, \qquad A_2 = \xi_x\eta_x + \xi_y\eta_y, \qquad A_3 = \eta_x^2 + \eta_y^2,$$

$$\xi_x = Jy_\eta, \qquad \xi_y = -Jx_\eta, \qquad \eta_x = -Jy_\xi, \qquad \eta_y = Jx_\xi,$$

J is the Jacobian of the transformation (7.3.56) calculated by formula (7.3.57).

For the generation of a curvilinear computing mesh around an airfoil we have used the method of [7.147]. This method enables us to generate a C-type grid with the aid of the transformation

$$x = B + A \cosh(\eta) \cos \xi,$$
$$y = A \sinh(\eta) \sin \xi, \tag{7.3.68}$$

where A, B are constants. The properties of the transformation (7.3.68) had been considered in more detail previously in [7.148] where it was applied in the numeical integration of the heat equation in an ellipse. Employing the algorithm of [7.147], we can create with the aid of a number of user-specified parameters the concentrations of the curvilinear grid lines in the domains near the profile, and near its leading and trailing edges (see Figure 7.16).

We have used the modified approximate factorization scheme AF2 of HOLST [7.149] for the numerical solution of equation (7.3.66). Before proceeding to the difference equations of this scheme, let us introduce some notations. Let us number the indices i, j of the coordinates (x_{ij}, y_{ij}) of the nodes of a grid depicted in Figure 7.16(b) as follows: let the ξ direction correspond to the index i and change from 1 to i_{max}, so that the corresponding point (x_{ij}, y_{ij}) moves clockwise. The η direction corresponds to the index j which changes from 1 to j_{max}. The points with the index $j = 1$ lie either on the airfoil surface or on the cut line behind the profile. As j increases, the corresponding (x_{ij}, y_{ij}) points move away from the airfoil surface. Denote by $\overleftarrow{\delta}_\xi(\ )$, $\overleftarrow{\delta}_\eta(\ )$ the backward differencing operators of the first-order accuracy in the directions ξ and η, respectively notations $\overrightarrow{\delta}_\xi(\ )$, $\overrightarrow{\delta}_\eta(\ )$ correspond to the forward differencing operators. The AF2-type scheme which we have employed is realized in two steps

$$[\alpha + \overleftarrow{\delta}_\eta \bar{\rho}_j^n (A_3/J)_{i,j+1/2}]f_{ij}^n = \alpha\omega L\varphi_{ij}^n, \tag{7.3.69}$$

$$[-\alpha\overrightarrow{\delta}_\eta \mp \alpha\beta\overrightarrow{\delta}_\xi - \overleftarrow{\delta}_\xi \bar{\rho}_i^n (A_1/J)_{i+1/2,j}\overrightarrow{\delta}_\xi]C_{ij}^n = f_{ij}^n. \tag{7.3.70}$$

Figure 7.16. (a) The
partial view of the mesh;
(b) the C-type grid
around the NACA 0012
airfoil. 95 × 15 mesh.

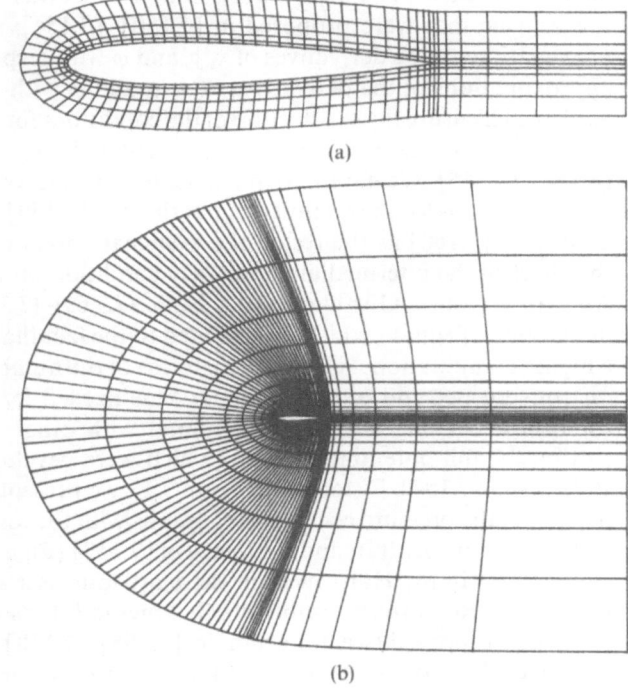

(a)

(b)

In (7.3.69) $L\varphi_{ij}^n$ is the residual operator at the nth iteration,

$$L\varphi_{ij}^n = \overleftarrow{\delta}_\xi[\tilde{\rho}_i^n(U/J)_{i+1/2,j}] + \overleftarrow{\delta}_\eta[\bar{\rho}_j^n(V/J)_{i,j+1/2}], \qquad (7.3.71)$$

$$\tilde{\rho}_i^n = [(1 - v)\rho]_{i+1/2,j} + v_{i+1/2,j}\rho_{i+k+1/2,j},$$

$$\bar{\rho}_j^n = [(1 - v)\rho]_{i,j+1/2} + v_{i,j+1/2}\rho_{i,j+l+1/2},$$

$$k = \begin{cases} -1, & U_{i+1/2,j} > 0, \\ 1, & U_{i+1/2,j} < 0, \end{cases} \qquad l = \begin{cases} -1, & V_{i,j+1/2} > 0, \\ 1, & V_{i,j+1/2} < 0, \end{cases}$$

$$v_{i+1/2,j} = \begin{cases} \max[(M_{i,j}^2 - 1)C, 0], & U_{i+1/2,j} > 0, \\ \max[(M_{i+1,j}^2 - 1)C, 0], & U_{i+1/2,j} < 0, \end{cases}$$

where $M_{i,j}$ is the local Mach number and C is the user-specified constant (we
have taken the value $C = 1.0$). Further, in (7.3.69), α is a parameter used to
speed up the convergence; ω is a relaxation factor which should satisfy the
requirement of computational stability, $0 \le \omega \le 2.0$ [7.150], [7.151] (we have
used the value $\omega = 1.8$). In (7.3.70) $C_{ij}^n = \varphi_{ij}^{n+1} - \varphi_{ij}^n$, β is the user-specified
parameter (we have set $\beta = 0$ in the domains of subsonic flow, and $\beta = 0.5$ in
the domains of supersonic flow). The operator $\overrightarrow{\delta}_\xi$ in (7.3.70) means that the
difference is taken as a backward one near the upper surface of the airfoil and
as a forward one near the lower surface [7.145], [7.152]. We have imple-

mented, in scheme (7.3.69)–(7.3.71), some of the ideas on the improvement of
this scheme that were proposed in [7.150]–[7.154]; namely, consistent approximations of the derivatives of x, y, and φ with respect to ξ, and consistent
approximations of the derivatives of x, y, and φ with respect to η have been
used, which ensure the satisfaction of the conditions for the exact reproduction
of a homogeneous flow when it is calculated by scheme (7.3.69)–(7.3.71)
[7.153], [7.155]. We have also implemented the improved intermediate boundary conditions for the quantity f_{i1}^n obtained in [7.150]. These conditions were
derived in [7.150] as the result of a systematic treatment of the problem on
the effect of the intermediate boundary condition on the convergence of the
AF-type schemes. Our choice of scheme (7.3.69)–(7.3.71) for the numerical
integration of equation (7.3.66) is related to the fact that this scheme possesses
a higher computational efficiency (in terms of the number of iterations needed
for convergence and of the computer time needed for the calculation of the
solution φ^n at one iteration), compared with other numerical methods of
solving the full-potential equation which were developed during the period
1978–1985 [7.154], [7.155]. In Figure 7.17 we present examples of the comparison of the pressure coefficient c_p graphs with the results obtained by LOCK
[7.156] (Figure 7.17(a)) and by JAMESON [7.157] (Figure 7.17(b)).

Note that in the Holst method there are four user-specified parameters α,
β, ω, and C. Recently, a number of other methods for solving equation (7.3.66)
have been proposed (see, for example, [7.155], [7.158]–[7.165]). In [7.155] a
parameter-free conjugate gradient-type method was introduced to avoid the

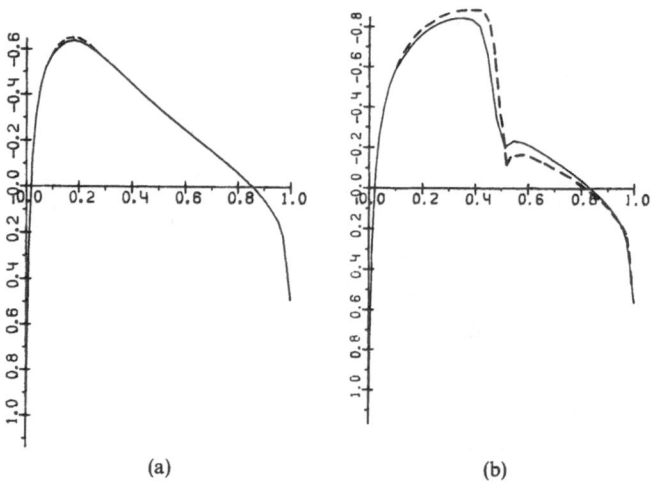

(a)   (b)

Figure 7.17. Pressure coefficient c_p as a function of x. Airfoil NACA 0012. (a) The freestream
Mach number $M_\infty = 0.72$, the angle of attack $\alpha = 0$: (———)—the present method, (– – – –)—
(From Lock, 1970: Test Cases for Numerical Methods in Two-Dimensional Transonic Flow,
AGARD Report R575-70); (b) $M_\infty = 0.8$, $\alpha = 0$. (———)—the present method, 95 × 15 mesh,
(– – –)–– (From Jameson, 1981: *Numerical Methods for the Computation of Inviscid Transonic Flows
with Shock Waves*, eds. A. Rizzi and H. Viviand, Vieweg, Braunscht Weig), 256 × 64 mesh.

user-specified relaxation factor ω in the Holst original work. The parameter C in the Holst Code, controlling the amount of artificial density, can be replaced by a flux-biasing scheme, satisfying an entropy inequality [7.158]. The authors of [7.159]–[7.161] also avoid using the artificial compressibility method in AF-type schemes, because they have shown that this method gives a significant overestimation of the gas flow rate while calculating the nozzle flows. The AF schemes proposed in [7.159] contain only one user-specified parameter $\omega = 4/3$. The stability of these schemes was studied in [7.159], by considering their first differential approximations (or modified equations, in another terminology). The authors of [7.162] solve equation (7.3.63) with the aid of the finite-element method and the multi-grid method of R.P. Fedorenko. In addition, the equations of the shock polar are used in [7.162] to increase the domain of agreement between the potential equation solutions and the solutions of the Euler equations, in cases of transonic flows with shock waves. The AF2 scheme of HOLST was also used in [7.163]–[7.165] for the computations of transonic airfoils and cascade flows on the spatial adaptive curvilinear grids of the H-, C-, and O-types. For example, in the computation of a cascade flow on a 64×32 H-type mesh, the residual dropped by five orders after 90 iterations [7.164].

To reduce the level of classification errors, while detecting the shock waves in transonic potential flows with the aid of a pattern recognition method whose different stages are presented in this section and in Section 7.5.2, it is desirable to choose a numerical method for solving equation (7.3.63) which captures shocks without wiggles. The AF2 scheme meets this requirement, as may be seen in Figure 7.17(b). This requirement can be weakened, if one makes use of an image preprocessing stage (see Section 7.2.2). However, it should be kept in mind that the implementation of an image preprocessing, especially image restoration, involves additional computer time expense (see Section 7.2.2).

At the first stage of a procedure for the segmentation of an image given in the nodes of a curvilinear grid the gradient $g(x, y)$ of the image was computed by the formulas of the Sobel edge detector (7.2.18), (7.3.5), (7.3.58), (7.3.10) where one should set $f \equiv \rho$, ρ is the gas density. In Figure 7.18 we present the gradient surface $g(x(\xi, \eta), y(\xi, \eta))$ mapped onto a rectangular computational domain in the (ξ, η)-plane. The shaded parallelogram is placed in Figure 7.18 under the nodes (x_{i1}, y_{i1}) lying on the airfoil surface. We can see from Figure 7.18 that the use of one-sided and central differences of the form (7.3.62) in the boundary nodes leads to a smooth transition in the image from the boundary points to the interior points which justifies the use of formulas of the form (7.3.62).

Since the shock waves which may be present in a transonic flow around an airfoil are characterized by a relatively weak intensity, it is reasonable to perform only one iteration while calculating the threshold $T^{(\nu)}$ by formula (7.3.17); that is, one must set $N_g = 1$.

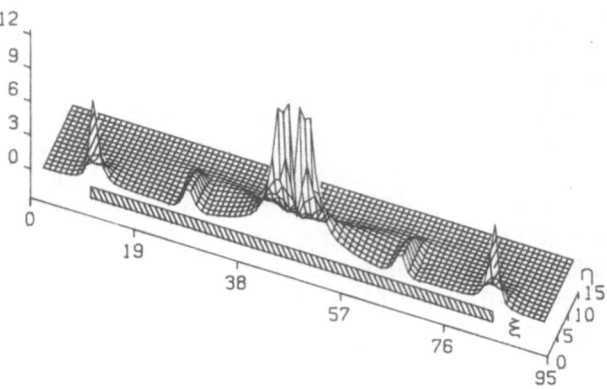

Figure 7.18. Dimetrics of the gradient surface $g(x(\xi, \eta), y(\xi, \eta))$.

For the application of the Rosenfeld nonmaximum suppression method it is at first necessary to number the principal directions. The scheme of numbering the principal directions in the case of a rectangular uniform mesh was presented in Figure 7.6. In the case of a curvilinear grid we associated the kth principal direction with a certain node of a curvilinear grid, independently of the orientation of the grid lines in the neighborhood of a node (x_{ij}, y_{ij}) under consideration with respect to the axes of a Cartesian coordinate system (x, y) (see Figure 7.19). For example, by the number 3 we encoded the direction from the node (i, j) to the node $(i - 1, j + 1)$, etc. Denote the corresponding angles by $\alpha_{ij}^{(k)}$, $k = 0, 1, \ldots, 7$. For example,

$$\alpha_{ij}^{(3)} = \arctan[(y_{i-1,j+1} - y_{ij})/(x_{i-1,j+1} - x_{ij})];$$

if $\alpha_{ij}^{(k)} < 0$, then we replace the value $\alpha_{ij}^{(k)}$ by $\alpha_{ij}^{(k)} + 2\pi$. By analogy with Section

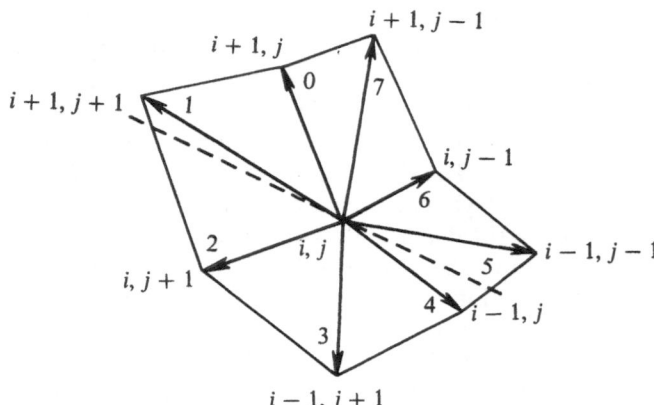

Figure 7.19. Principal directions numbering in the case of a curvilinear grid.

7.3.1 we have introduced an auxiliary ninth direction by the formula

$$\alpha_{ij}^{(8)} = \begin{cases} \alpha_{ij}^{(0)} + 2\pi & \text{if } \alpha_{ij}^{(0)} \le \Delta\alpha/2, \\ 0 & \text{if } 2\pi - \alpha_{ij}^{(0)} \le \Delta\alpha/2, \\ \alpha_{ij}^{(0)} & \text{otherwise.} \end{cases} \qquad (7.3.72)$$

The increment $\Delta\alpha$ in (7.3.72) was set as an arithmetic mean of the differences $\alpha_{ij}^{(k)} - \alpha_{ij}^{(k-1)}$, $k = 1, \ldots, 8$. It is obvious that this value $\Delta\alpha = \pi/4$. Let us turn to Figure 7.19 in order to explain how the introduced principal directions are used in the edge-thinning process. The dashed line in Figure 7.19 shows the orientation of a normal to the shock wave front passing through the point (i, j). The orientation angle α_{ij} of this normal is computed via the formula

$$\alpha_{ij} = \arctan[(\partial\rho/\partial y)/(\partial\rho/\partial x)] \qquad (7.3.73)$$

where the derivatives $\partial\rho/\partial y$, $\partial\rho/\partial x$ are approximated by formulas (7.3.59), (7.3.60), in which one should set $f \equiv \rho$. Similar to the case of the rectangular grid, the angles $\alpha_{ij}^{(k)}$, $k = 0, \ldots, 7$, are used for the determination of two

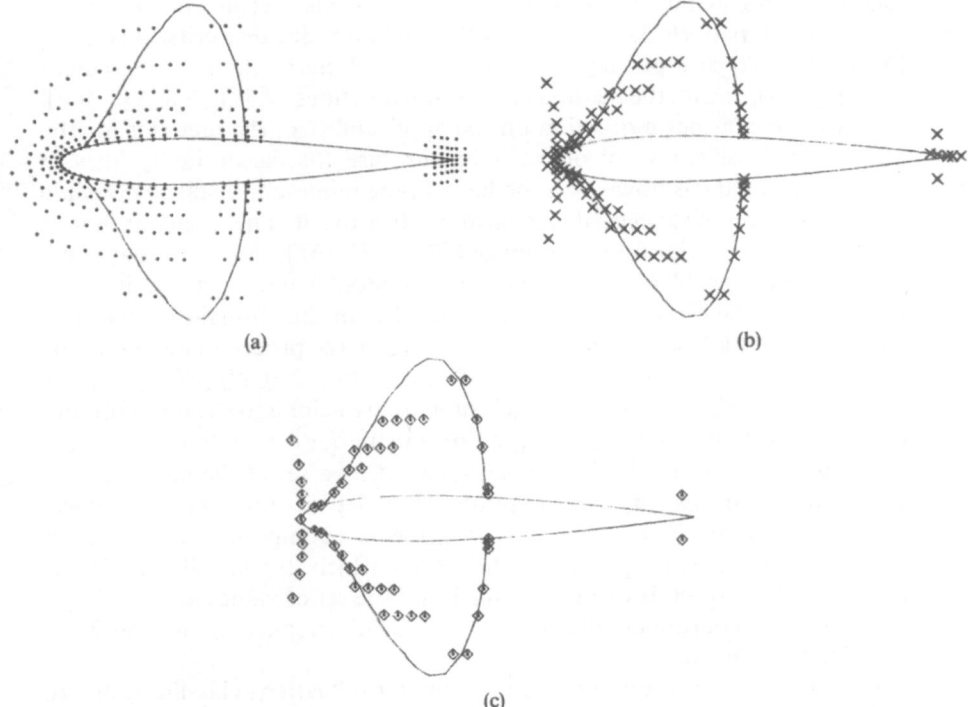

(a)   (b)

(c)

Figure 7.20. The sets $\mathscr{S}_1$, $\mathscr{S}_2$, and $\mathscr{S}_3$ obtained at $M_\infty = 0.8$, $\alpha = 0$: (a) the set $\mathscr{S}_1$, $N_1 = 309$; (b) the set $\mathscr{S}_2$, $N_2 = 87$; (c) the set $\mathscr{S}_3$, $N_3 = 54$. (———)—sonic lines and the airfoil contour.

neighboring nodes with indices (i_l, j_l) and (i_r, j_r) closest to (i, j), and approximately lying on the normal to the edge and on different sides of it. For example, in the case shown in Figure 7.19, we should set in (7.3.18) $i_l = i + 1$, $j_l = j + 1$, $i_r = i - 1$, $j_r = j$. Figure 7.20 shows the points of the sets $\mathscr{S}_1$, $\mathscr{S}_2$, and $\mathscr{S}_3$ obtained as a result of the first three stages of the segmentation procedure applied to a digital image given on a 95×15 curvilinear grid. The fourth stage (see Section 7.3.1) was not realized by us in the case of a curvilinear grid. As we shall see in the following, the classification stage enables us to exclude from the set $\mathscr{S}_3$ a large number of points, and as a result of this a set of the points of shock wave fronts is obtained in which there are no isolated artefacts. This is related to the fact that the scheme AF2 used by us enables us to obtain numerical solutions which do not contain parasitic oscillations in the neighborhood of a shock wave (see Figure 7.17(b)).

7.4. Feature Space

In the monographic literature on pattern recognition there has already been established a classification of deterministic methods. Let us enumerate the basic ones: (1) pattern classification by the minimum-distance criterion [7.55], [7.84], [7.5]; (2) perceptron [7.5], [7.55], [7.84]; (3) the method of the gradient [7.55], [7.56]; and (4) the method of potential functions [7.55], [7.84], [7.166].

Below we consider a question on the applicability of the methods for the automatic classification of objects—in our case the discontinuity lines in two-dimensional gas flows—on the basis of the minimum-distance principle and the method of sequential classification. For this it is necessary at first to carry out the description of an image ([7.53], [7.167]). Suppose that some gas-dynamical problem is computed by a finite-difference scheme in some domain D in the plane of spatial variables. Let in this domain be detected, with the aid of a lower-level system of digital image processing like the one presented above (see also Chapters 2–5 and Section 7.3), N_4 points (x_m, y_m), $m = 1, \ldots, N_4$, $N_4 \geq 1$, some of which supposedly belong to the discontinuity lines in the solution. These (x_m, y_m) points are, in our case, the objects which are to be classified. Each object is represented by a set of the measurement results which are called its description. Thus, let us assign to each object (x_m, y_m) a vector $\mathbf{a}_m = (a_{1m}, \ldots, a_{Mm})$, $M \geq 1$, whose components are the result of measuring some properties of the object which are usually called the features of the object. It can be assumed that the set of values $(a_{1m}, \ldots, a_{Mm})$ determines the coordinates of a point in the Euclidean space E_M in a Cartesian coordinate system $(a_1, \ldots, a_M)$.

Let us now turn to the discussion of the term "pattern classification", In [7.167] a difference has been made between the classification and the recognition of patterns. In [7.167] the pattern classification procedure is understood as a procedure which includes an object in the class ω_j, if and only if its description falls into the domain R_j of the space E_M that corresponds to this class. The pattern recognition procedure is understood in [7.167] as a procedure for determining these domains $\{R_j\}$ by investigating the descriptions of the sets of objects for which it is known to which classes they indeed belong. A similar definition may be found in [7.6] where the pattern recognition is understood as the assignment of a pattern of a certain type (which is known in advance in the case of classification). In [7.168] it is suggested that the process of pattern recognition be considered as a variant of information compression, since its aim is the elimination of information which is not related to the properties of the classes.

As was pointed out in [7.55], the choice and extraction of features play a central role in pattern recognition. Various classifications of features have been proposed in the literature. For example, it is suggested in [7.55] to subdivide the features into three sufficiently arbitrary groups: physical, structural, and mathematical. It was proposed in [7.169] to estimate the information content of the features by the methods of information theory. According to [7.169], the metric form of the information representation is the most appropriate for engineering applications. This form is in turn subdivided into parametric, topological, and abstract information. From other terms relevant to the feature characteristic let us mention logical, verbal, and geometric features [7.6] and binary features [7.84].

Let us enumerate a number of the requirements for the features which (in the literature on pattern recognition) are taken into account while determining a system of features ensuring the possibility of an efficient object classification by the algorithms of automatic recognition. While formulating these requirements we shall immediately try to "attach" them to our specific problem of recognition and automatic classification of singularities on the basis of the results of a shock-capturing calculation of two-dimensional gas dynamics problems.

1. Feature invariance [7.53]. This property means that all the objects belonging to the same class (for example, to the shock wave class) should possess this feature.

2. Rich information content of the features. There are various quantitative and qualitative definitions of the information content of the features. It was proposed in [7.169] to use, for the quantitative characteristic of the information content of the features, the probabilistic measures of the information quantity (the measures of Hartley, Fano, Kotelnikov, Shannon, Kullback, etc.). In [7.53] the feature of an object is considered to have a high information content—if the removal of this feature from the classification algorithm increases substantially the measure of uncertainty in the classification—which is similar to the information criteria underlying the definitions of information content in [7.169]. An estimate of the average risk of the Bayesian classifier has been proposed in [7.170] as a measure of the utility of a set of features.

3. Independence of the form of the functional dependencies, which the features obey, of the specifics of the gas-dynamical problem being solved. This property determines in the end the universality of a system of pattern classification.

4. The requirement of a minimal overall number of features [7.171], [7.172]. As was already mentioned above, the features may be subdivided into features with rich information content and those with little information content. The inclusion into a recognition system of too large a number of features: first, increases computer time expense and the requirement for the computer memory resources; and second, it may lead in some cases to the deterioration of classification [7.84], [7.172].

5. Feature invariance with respect to the location and orientation of the objects with respect to the spatial coordinate axes [7.6].

6. The features should possess a sufficiently high noise immunity in order to provide the correct classification of singularities in the presence of the numerical solution oscillations in the neighborhood of discontinuities.

The above six requirements are directed eventually to the goal of diminution of the level of classification errors under the simultaneous satisfaction of the requirement of universality and computational efficiency of a recognition system.

Let us now proceed to the derivation of the functional dependencies with the aid of which we may be able to compute the features appropriate to the recognition of discontinuities in the two-dimensional gas flows modeled on the basis of finite-difference shock-capturing schemes. Assume that there are, in the flow under study, K discontinuity surfaces $\Sigma_k(t)$, $k = 1, \ldots, K$, $K \geq 1$. Let D be the magnitude of the velocity vector of the propagation of the kth surface Σ_k (for brevity the subscript "k" at D will be omitted in the following) which is oriented along a normal to Σ_k in the considered point of the Σ_k surface. Taking into account the consideration of Section 1.2 of the properties of the solution components at the gas-dynamical discontinuities, it is reasonable to take, as the image intensity function f in the window (7.2.6), the gas density ρ. Suppose that we have obtained, with the aid of a lower-level system of processing the results of gas-dynamical computation, a set of objects—the points (x_m, y_m), $m = 1, \ldots, N_4$. Among these points we can also find points which are indeed located in the subdomains of a continuous flow, for example, in the domain of a rarefaction wave. From this it follows that we should also take into account, in a system of recognition and automatic classification of the objects (x_m, y_m), the object classes corresponding to such subdomains. Figure 7.21 presents both these classes and the classes corresponding to the types of shock waves and contact discontinuities described in Section 1.2. Since the normal and oblique shock waves are types of shock wave, it is generally possible to unite the classes ω_1 and ω_2 into one "big" class Ω_1 of shock waves (see, in Figure 7.21, the upper frame formed by a dashed line).

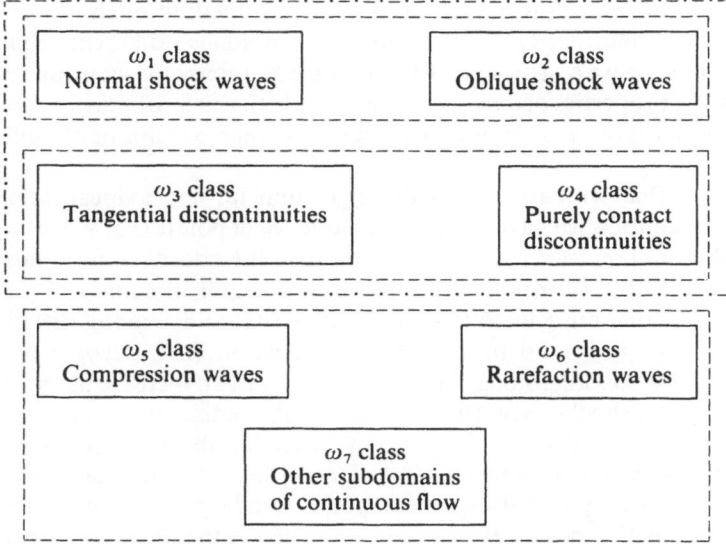

Figure 7.21. The classes of singularities in the two-dimensional inviscid gas flows.

Similarly, the classes ω_3 and ω_4 may be united into one class Ω_2 of contact discontinuities, and it is possible to interpret ω_3 and ω_4 as the subclasses of the Ω_2 class. Finally, the classes ω_5, ω_6, ω_7 can be united into a class Ω_3 of the points belonging to the subdomains of continuous flow. In turn, the classes Ω_1 and Ω_2 have the common property that they include only the points belonging to the discontinuity surfaces Σ_k. Therefore, it is worthwhile considering a class C_1 of all the points of the discontinuity surfaces. Then Ω_1 and Ω_2 prove to be subclasses of the class C_1 and ω_1, ω_2, ω_3, ω_4 are subsubclasses of C_1. Such a hierarchical grouping of classes, as we shall see in the following, may be put into the basis of hierarchical systems of the discontinuities classification.

The classes presented in Figure 7.21 are determined by the mathematical properties of the solutions of a differential system (1.30) which take place in the subdomains of a continuous flow, and of the algebraic system (1.31)–(1.34) coupling the quantities on both sides of the discontinuity surfaces. But since our goal is the classification of discontinuities on the basis of finite-difference solutions rather than the above ideal solutions, it is also necessary to take into account in the construction of features the mathematical properties of the solutions of a system of finite-difference equations approximating the Euler equation system (1.30). The necessity for such a treatment is related to the fact that the solutions obtained by existing difference schemes are far from being ideal: the discontinuities in these solutions are "smeared", and often parasitic oscillations are present in the neighborhood of discontinuities. As was shown in Chapters 1–6, the differential approximation method often proves useful for obtaining information on the properties of nonlinear difference equations. In the following we construct relatively simple features without using information on fine properties of difference schemes which can be obtained, in particular, on the basis of an analysis of the differential approximations of a difference scheme.

Suppose that with the aid of the algorithm for shock localization in a difference solution we have obtained a sequence of points (x_m, y_m), $m = 1, \ldots,$ N_4. Take the point (x_m, y_m) and assume that the coordinates of this point coincide with the coordinates of a geometric center of some cell (i, j) of a rectangular computing mesh in the (x, y)-plane. Consider, along with the (i, j) cell, the cells (i_l, j_l) and (i_r, j_r) defined in Section 7.3.1 devoted to image segmentation. Taking into account the property of an increase in the entropy (1.37) across a shock wave, let us ascribe to the indices (i_l, j_l) and (i_r, j_r) the following meaning: if $S_{i_l, j_l} > S_{i_r, j_r}$, then we leave the above indices unchanged; otherwise, the pairs of indices (i_l, j_l) and (i_r, j_r) should be interchanged. Thus, the pair of indices (i_l, j_l) always corresponds to a larger entropy. In practical computations this procedure, in the case of using the ideal gas equation of state (1.8), was realized as follows. Consider the well-known relationship

$$\exp[(S - S_0)/c_V] = p/(\rho^\gamma), \tag{7.4.1}$$

where S_0 is an arbitrary constant and c_V is the gas specific heat under constant volume. Let S, p, ρ in (7.4.1) be differentiable functions of some variable ξ and let $A = p/(\rho^\gamma)$. Then it follows from (7.4.1) that $\text{sign}(\partial S/\partial \xi) = \text{sign}(\partial A/\partial \xi)$. Thus, from the inequality $A_{i_l, j_l} > A_{i_r, j_r}$ follows the inequality $S_{i_l, j_l} > S_{i_r, j_r}$, and conversely.

Strictly speaking, the inequality $S_{i_l, j_l} > S_{i_r, j_r}$ is ensured on a shock wave, if $S_{i_l, j_l} = S_1$, $S_{i_r, j_r} = S_2$, where the subscripts "1" and "2" refer to the medium states behind and before the front of an original nonsmeared shock wave, respectively. Let us show that the inequality $A_{i_l, j_l} > A_{i_r, j_r}$ can be violated, if the values of the quantities ρ, ε at the points with indices (i_l, j_l) and (i_r, j_r) deviate substantially from the values ρ_1, ε_1 and ρ_2, ε_2, respectively. In the following we shall write for brevity A_l and A_r instead of A_{i_l, j_l} and A_{i_r, j_r}. It is not difficult to show that

$$A_l - A_r = (\gamma - 1)\varepsilon_r \rho_l^{1-\gamma} \cdot [\varepsilon_l/\varepsilon_r - (\rho_l/\rho_r)^{\gamma-1}]. \tag{7.4.2}$$

Rewrite equation (1.28) of the Hugoniot adiabat in the form

$$\varepsilon_1 - \varepsilon_2 = -(1/2)(V_1 - V_2)(p_1 + p_2). \tag{7.4.3}$$

Since in the case of a shock wave $\rho_1 > \rho_2$, we have that $V_1 < V_2$, therefore, we obtain from (7.4.3) that at a shock wave $\varepsilon_1 > \varepsilon_2$. Suppose that the components ρ, ε of a difference solution have monotone behavior in the zone of smearing of a shock wave. Then in this zone also the inequalities

$$\varepsilon_l > \varepsilon_r, \qquad \rho_l > \rho_r, \qquad (7.4.4)$$

take place. The relationships (7.4.4) are insufficient to provide the satisfaction of the inequality $A_l > A_r$. In fact, it follows from (7.4.2) that in the case when

$$1 < \varepsilon_l/\varepsilon_r < (\rho_l/\rho_r)^{\gamma-1},$$

we have that $A_l < A_r$. Such a phenomenon was observed in the computations of some two-dimensional shock flows. In connection with the above we have realized a search procedure for determining such values as p_l, ρ_l, p_r, ρ_r, which provide the satisfaction of the entropy inequality $A_l > A_r$. A search for the values p_l, ρ_l and p_r, ρ_r is subdivided into two stages. At the first stage the quantities p_l and ρ_l are sought; if the values p_l, ρ_l such that $A_l > A_r$ are not found, one proceeds to the second stage, the stage of a search for p_r and ρ_r. Let us introduce the image windows $F_m^-(f, i, j)$ and $F_m^+(f, i, j)$ of the form

$$F_m^-(f, i, j) = \begin{bmatrix} f_{i,j+v_y(m)} & f_{i+v_x(m),j+v_y(m)} \\ f_{i,j} & f_{i+v_x(m),j} \end{bmatrix},$$

$$F_m^+(f, i, j) = \begin{bmatrix} f_{i,j+\mu_y(m)} & f_{i+\mu_x(m),j+\mu_y(m)} \\ f_{ij} & f_{i+\mu_x(m),j} \end{bmatrix},$$

where

$$v_x(m) = m \cdot (i_l - i), \qquad v_y(m) = m \cdot (j_l - j),$$

$$\mu_x(m) = m \cdot (i_r - i), \qquad \mu_y(m) = m \cdot (j_r - j).$$

Let us also make use of a smoothing mask H_5 of the form

$$H_5 = (1/4) \begin{bmatrix} 1 & 1 \\ 1 & 1 \end{bmatrix}. \qquad (7.4.5)$$

Then we compute sequentially the quantities

$$p_l^{(1)} = p_{i_l,j_l}, \qquad \rho_l^{(1)} = \rho_{i_l,j_l};$$

$$p_l^{(2)} = H_5 * F_{-1}^-(p, i, j), \qquad \rho_l^{(2)} = H_5 * F_{-1}^-(\rho, i, j);$$

$$p_l^{(3)} = H_5 * F_1^-(p, i, j), \qquad \rho_l^{(3)} = H_5 * F_1^-(\rho, i, j);$$

$$p_l^{(4)} = H_5 * F_2^-(p, i, j), \qquad \rho_l^{(4)} = H_5 * F_2^-(\rho, i, j); \qquad (7.4.6)$$

$$p_l^{(5)} = 0.5(p_{i_l,j_l} + p_{ij}), \qquad \rho_l^{(5)} = 0.5(\rho_{i_l,j_l} + \rho_{ij});$$

$$p_l^{(6)} = 0.5(p_{i_l,j_l} + p_{i+v_x(1),j+v_y(1)}),$$

$$\rho_l^{(6)} = 0.5(\rho_{i_l,j_l} + \rho_{i+v_x(1),j+v_y(1)}).$$

As soon as the value $p_l^{(k)}$, $\rho_l^{(k)}$ is found, such that $A_l^{(k)} > A_r$, any further search is stopped. If such a value is not found, we set $p_l = p_l^{(k_0)}$, $\rho_l = \rho_l^{(k_0)}$, where k_0 is a number, $1 \le k_0 \le 6$, such that $A_l^{(k_0)} > A_l^{(k)}$ at $k \ne k_0$, $1 \le k \le 6$, then we proceed to the second stage at which we compute sequentially the quantities

$$p_r^{(1)} = p_{i_r, j_r}, \qquad \rho_r^{(1)} = \rho_{i_r, j_r};$$

$$p_r^{(2)} = H_5 * F_{-1}^+(p, i, j), \qquad \rho_r^{(2)} = H_5 * F_{-1}^+(\rho, i, j);$$

$$p_r^{(3)} = H_5 * F_1^+(p, i, j), \qquad \rho_r^{(3)} = H_5 * F_1^+(\rho, i, j);$$

$$p_r^{(4)} = H_5 * F_2^+(p, i, j), \qquad \rho_r^{(4)} = H_5 * F_2^+(\rho, i, j); \qquad (7.4.7)$$

$$p_r^{(5)} = 0.5(p_{i_r, j_r} + p_{ij}), \qquad \rho_r^{(5)} = 0.5(\rho_{i_r, j_r} + \rho_{ij});$$

$$p_r^{(6)} = 0.5(p_{i_r, j_r} + p_{i + \mu_x(1), j + \mu_y(1)}),$$

$$\rho_r^{(6)} = 0.5(\rho_{i_r, j_r} + \rho_{i + \mu_x(1), j + \mu_y(1)}).$$

As soon as the value $p_r^{(k)}$, $\rho_r^{(k)}$ is found, such that $A_l > A_r^{(k)}$, any further search is stopped. If we connect by straight-line segments the cell centers of a spatial grid involved in the computations by the formulas (7.4.6), (7.4.7), we obtain the array of rectangular contours shown in Figure 7.22. In cases when the edge passing through the point (i, j) is parallel to the x- or y-axis, the rectangular contours depicted in Figure 7.22 degenerate into straight-line segments. Of course, the use of the contours shown in Figure 7.22 is justified only in the cases when all the points of a contour involved in the computation of the quantity, for example, $p_l^{(4)}$ or $\rho_r^{(3)}$, lie on one side of a generally curvilinear discontinuity line.

The above search procedure was applied only in the cases when $\varepsilon_l > \varepsilon_r$, $\rho_l > \rho_r$, but $A_l < A_r$. In the remaining cases we have used for the computation of A_l and A_r simpler formulas

$$A_l = p_l/(\rho_l^\gamma), \qquad A_r = p_r/(\rho_r^\gamma). \qquad (7.4.8)$$

If $A_l > A_r$, the pairs of indices (i_l, j_l) and (i_r, j_r) are left unchanged, otherwise they are interchanged.

A relatively wide acceptance was received in gas-dynamical computations

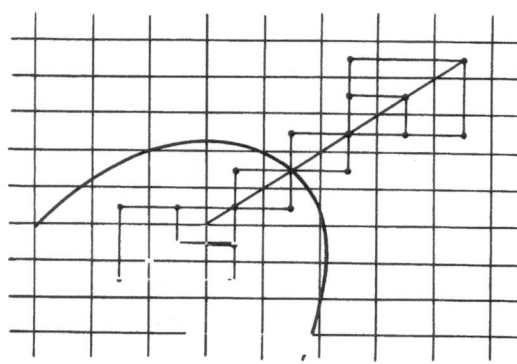

Figure 7.22. Contours for the search interpolation procedure.

for a method of introducing the artificial viscosity on the basis of various criteria [7.64], [7.71], [7.106], [7.173]. Consider, for example, the pseudo-viscosity (2.2.18) introduced additively into the pressure. The implementation of formula (2.2.18) in the computer code may be considered as the simplest variant of the decision-making process. Indeed, it is reasonable to assume with regard to Section 2.2.1 that in the zone of a smeared shock wave in a one-dimensional flow the inequality $\partial u/\partial x < 0$ takes place. The same inequality is also satisfied in the compression waves in a one-dimensional flow along the x-axis. Consequently, in the case of using (2.2.18) the artificial viscosity is introduced in the zones of shock waves smearing and in the zones of compression waves, and is not introduced in other subdomains of the flow. In the case of two-dimensional flows we shall compute the features a_{1m} of the mth object (x_m, y_m) with regard to the equality (1.39) by the formula

$$a_{1m} = 1 - \text{sign}(\partial u_n/\partial n)_{ij}, \qquad (7.4.9)$$

where n is a coordinate measured along a normal to the discontinuity surface at the point (x_m, y_m). The quantity $(\partial u_n/\partial n)_{ij}$ was computed by the formula

$$(\partial u_n/\partial n)_{ij} = (\partial u_n/\partial x)_{ij} \cos \alpha_{ij} + (\partial u_n/\partial y)_{ij} \sin \alpha_{ij}. \qquad (7.4.10)$$

For the calculation of $\partial u_n/\partial x$, $\partial u_n/\partial y$ we have used two versions of the computational formulas. In the first version, which is simple in terms of computational effort, the derivatives $\partial u_n/\partial x$ and $\partial u_n/\partial y$ were approximated by the simplest one-sided differences:

$$\begin{aligned} (\partial u_n/\partial x)_{ij} &= (1/h_1)[(u_n)_{i+1,j} - (u_n)_{ij}], \\ (\partial u_n/\partial y)_{ij} &= (1/h_2)[(u_n)_{i,j+1} - (u_n)_{ij}], \end{aligned} \qquad (7.4.11)$$

where

$$(u_n)_{k,l} = u_{kl} \cos \alpha_{ij} + v_{kl} \sin \alpha_{ij}.$$

In the other version the edge detector of MÉRÖ and VASSY [7.88] was used for the computation of $\partial u_n/\partial x$, $\partial u_n/\partial y$:

$$\begin{aligned} (\partial u_n/\partial x)_{ij} &= (1/(2h_1))H_6 * F_2(u_n, i, j); \\ (\partial u_n/\partial y)_{ij} &= (1/(2h_2))H_7 * F_2(u_n, i, j); \end{aligned} \qquad (7.4.12)$$

where H_6 and H_7 are the masks of Mérö and Vassy

$$H_6 = \begin{bmatrix} -1 & 1 \\ -1 & 1 \end{bmatrix}, \qquad H_7 = \begin{bmatrix} 1 & 1 \\ -1 & -1 \end{bmatrix}.$$

Expanding the quantities $(u_n)_{kl}$ entering (7.4.12) with respect to the point with indices (i, j), it is easy to find that

$$H_6 * F_2(u_n, i, j) = 2h_1 \, \partial u_n/\partial x + h_1^2 \, \partial^2 u_n/\partial x^2 + h_1 h_2 \, \partial^2 u_n/\partial x \, \partial y + O(\cdot);$$

$$H_7 * F_2(u_n, i, j) = 2h_2 \, \partial u_n/\partial y + h_1 h_2 \, \partial^2 u_n/\partial x \, \partial y + h_2^2 \, \partial^2 u_n/\partial y^2 + O(\cdot);$$

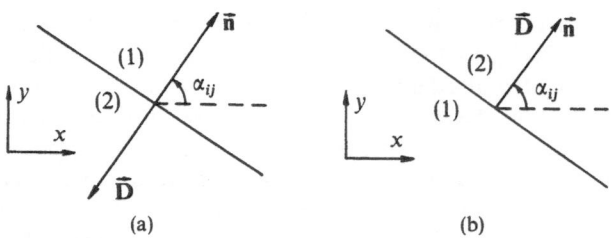

Figure 7.23. (a) $\hat{j} > 0$; (b) $\hat{j} < 0$.

where

$$O(\cdot) = O(h_1^2 h_2) + O(h_1 h_2^2) + O(h_2^3).$$

The practical use of the formulas (7.4.10)–(7.4.12) in the smeared shock waves in the two-dimensional flows showed that the inequality $\partial u_n/\partial n < 0$ was always satisfied independently of the sign of the quantity $\hat{j}$ in (1.39). To explain this phenomenon let us turn to the formula

$$\tan \alpha_{ij} = (\partial \rho/\partial y)/(\partial \rho/\partial x). \tag{7.4.13}$$

Consider the two cases of the propagation of a shock wave (oblique or normal wave) depicted in Figure 7.23. Let the point (x, y) in the case of Figure 7.23(a) lie on the shock front line. Suppose that at this point there is no intersection of the shock wave under study with any other shock wave or with a contact discontinuity. Note that $\rho_1 > \rho_2$ since the density ρ increases behind the shock wave front. Let us set two small positive quantities Δx and Δy and consider the differences

$$\Delta_x \rho = \rho(x + \Delta x, y) - \rho(x - \Delta x, y);$$

$$\Delta_y \rho = \rho(x, y + \Delta y) - \rho(x, y - \Delta y).$$

If in a small neighborhood of the point (x, y) there are no discontinuities of the flow, the function $\rho(x, y)$ is continuous near the different sides of the shock wave under study, therefore,

$$\lim_{\Delta x \to 0} \Delta_x \rho = \lim_{\Delta y \to 0} \Delta_y \rho = \rho_1 - \rho_2 > 0.$$

From this it follows that at sufficiently small values of the steps h_1 and h_2 of the spatial grid in (7.4.13) we will always have $\partial \rho/\partial y > 0$, $\partial \rho/\partial x > 0$, if the difference solution for ρ changes monotonously in the zone of a smeared shock wave. Thus, the unit normal vector $\mathbf{n} = \{\cos \alpha_{ij}, \sin \alpha_{ij}\}$ will have a direction opposite to the direction of the velocity vector $\mathbf{D}$ of the shock wave (see Figure 7.23(a)). In the case under study $\hat{j} > 0$ in (1.39), so that $u_{n2} - u_{n1} > 0$. In the image segmentation procedure described in Section 7.3.1 the coordinate n always increases in the direction indicated by the vector $\mathbf{n}$. Let the value $n = 0$

correspond to the (x, y) point of a shock front. Let us set a small increment $\Delta n > 0$. Then

$$\text{sign}(\partial u_n/\partial n) = \text{sign}\{[u_n(\Delta n) - u_n(-\Delta n)]/(2\Delta n)\}$$

$$= \text{sign}\ (u_{n1} - u_{n2}) = -1,$$

that is, $\partial u_n/\partial n < 0$ in the case shown in Figure 7.23(a). Considering similarly the case $\hat{j} < 0$ presented in Figure 7.23(b), we can write with regard to (1.39) that

$$\text{sign}(\partial u_n/\partial n) = \text{sign}(u_{n2} - u_{n1}) = -1.$$

Thus, we have shown that in the zones of smeared shock waves $\alpha_{1m} = 2$, under the condition of a monotone behavior of the grid functions ρ, u, v in these zones and in the absence of other discontinuities in some neighborhood of the point (x, y) under consideration.

Let us now proceed to the construction of the second feature a_{2m}. Let the function $F(\rho, \varepsilon)$ in the equation of state (1.5) be continuously differentiable at $\rho \geq 0$, $\varepsilon \geq 0$, and let

$$\partial F/\partial \rho > 0, \qquad \partial F/\partial \varepsilon > 0, \qquad \varepsilon > 0, \quad \rho > 0. \qquad (7.4.14)$$

Taking (1.41) into account, at the original "ideal" (that is, nonsmeared contact discontinuity) the condition for the pressure continuity $p_1 = p_2$ is satisfied. Employing the Taylor formula we obtain

$$F(\rho_1, \varepsilon_1) = F(\rho_2, \varepsilon_2) + \frac{\partial F(\bar{\rho}, \bar{\varepsilon})}{\partial \rho}(\rho_1 - \rho_2) + \frac{\partial F(\bar{\rho}, \bar{\varepsilon})}{\partial \varepsilon}(\varepsilon_1 - \varepsilon_2), \quad (7.4.15)$$

where

$$\bar{\rho} = \rho_1 + \theta(\rho_2 - \rho_1), \qquad \bar{\varepsilon} = \varepsilon_1 + \theta(\varepsilon_2 - \varepsilon_1), \qquad 0 < \theta < 1.$$

Since at the contact discontinuity $F(\rho_1, \varepsilon_1) = F(\rho_2, \varepsilon_2)$ we obtain from (7.4.15) the relationship

$$\frac{\partial F(\bar{\rho}, \bar{\varepsilon})}{\partial \rho} \cdot (\rho_1 - \rho_2) + \frac{\partial F(\bar{\rho}, \bar{\varepsilon})}{\partial \varepsilon} \cdot (\varepsilon_1 - \varepsilon_2) = 0. \qquad (7.4.16)$$

In the case of the satisfaction of the inequalities (7.4.14) it follows from (7.4.16) that at the contact discontinuity

$$\text{sign}(\rho_1 - \rho_2) = -\text{sign}(\varepsilon_1 - \varepsilon_2). \qquad (7.4.17)$$

Let us elucidate the question as to whether the relationship (7.4.17) will be satisfied in the case of a shock wave. Since in the case of a shock wave $\rho_1 > \rho_2$, $\varepsilon_1 > \varepsilon_2$, the equality

$$\text{sign}(\rho_1 - \rho_2) + \text{sign}(\varepsilon_1 - \varepsilon_2) = 2 \qquad (7.4.18)$$

is satisfied at the shock wave. At the same time, at a contact discontinuity, we have according to (7.4.17) that

$$\text{sign}(\rho_1 - \rho_2) + \text{sign}(\varepsilon_1 - \varepsilon_2) = 0. \qquad (7.4.19)$$

This consideration leads to a conclusion that it is reasonable to calculate the second feature by the formula

$$a_{2m} = \text{sign}(\bar{\rho}_{i_l,j_l} - \bar{\rho}_{i_r,j_r}) + \text{sign}(\bar{\varepsilon}_{i_l,j_l} - \bar{\varepsilon}_{i_r,j_r}), \qquad (7.4.20)$$

where

$$\bar{\varepsilon}_{i_l,j_l} = \bar{p}_{i_l,j_l}/(\bar{\rho}_{i_l,j_l}(\gamma - 1)), \qquad \bar{\varepsilon}_{i_r,j_r} = \bar{p}_{i_r,j_r}/(\bar{\rho}_{i_r,j_r}(\gamma - 1)),$$

in the case when the equation of state (1.8) is used, and the quantities $\bar{p}$, $\bar{\rho}$ at the points with the numbers (i_l, j_l) and (i_r, j_r) are calculated in the general case by the formulas (7.4.6), (7.4.7). Note that the situation $\bar{\rho}_{i_l,j_l} = \bar{\rho}_{i_r,j_r}, \bar{\varepsilon}_{i_l,j_l} = \bar{\varepsilon}_{i_r,j_r}$, is excluded automatically, because at the points (x_m, y_m) a threshold limitation $g_{ij} > T^{(v)}$ is satisfied where $T^{(v)} > 0$, which means that there are among the points (x_m, y_m) no points located in a subdomain of constant flow. In this sense we can assert that the classification of pixels of the original digital image already starts at a stage of image segmentation, by the aid of which the domains with nonzero gradients of the gas density are separated from the domains of constant flow.

Let us proceed to the construction of the features a_{3m}, a_{4m}. We have presented in Chapter 5 optimization algorithms for the contact discontinuity localization, which used the fact of the pressure continuity across a contact discontinuity and the discontinuity of the normal velocity component u_n at a shock wave. By virtue of the continuity of p across a contact discontinuity the quantity $(1/p)|\partial p/\partial n|$ should remain finite. The quantity $|\partial u_n/\partial n|$ should also be finite at a contact discontinuity. At a shock wave of finite intensity of the order $O(1)$ the quantity $(h_1/p)|\partial p/\partial n|$, computed on the basis of the difference solution, will also be a quantity of the order $O(1)$. At the same time, at a contact discontinuity, $|\partial p/\partial n| = O(1)$, consequently $(h_1/p)|\partial p/\partial n| = O(h_1)$. Similarly, at a shock wave of finite intensity $\tau|\partial u_n/\partial n| = O(\tau/h_1)$, and at a contact discontinuity $\tau|\partial u_n/\partial n| = O(\tau)$. Therefore, let us introduce the features

$$a_{3m} = (h_1/\bar{p}_{ij}) \cdot |\partial p/\partial n|_{ij}, \qquad (7.4.21)$$

$$a_{4m} = \tau|\partial u_n/\partial n|_{ij}, \qquad (7.4.22)$$

keeping in mind that we are going to use in the following a classification technique based on the minimum distance principle. In (7.4.21)

$$\bar{p}_{ij} = (1/4)(p_{i+1,j+1} + p_{i-1,j+1} + p_{i+1,j-1} + p_{i-1,j-1}). \qquad (7.4.23)$$

The averaging (7.4.23) was chosen rather arbitrarily, that is, we can in principle also use other formulas for $\bar{p}_{ij}$. The quantity $h_1|\partial p/\partial n|_{ij}$ was computed with the aid of the formulas of the Sobel edge detector

$$h_1|\partial p/\partial n|_{ij} = (1/8) \cdot [(H_1 * F_3(p)) \cdot (h_1/h_2) \cdot \sin \alpha_{ij} + (H_2 * F_3(p)) \cdot \cos \alpha_{ij}],$$
$$(7.4.24)$$

where $F_3(f)$ is a 3×3 image window of the form (7.2.6); the masks H_1, H_2 are determined according to (7.3.5). Finally, the feature (7.4.22) was computed with the aid of (7.4.10), (7.4.11) or (7.4.10), (7.4.12). Note that the quantity (7.4.21) may also be interpreted as the approximation of the quantity $h_1 \cdot |\partial \ln p/\partial n|_{ij}$.

In order to distinguish between the points of discontinuity surfaces and the points belonging to the subdomains of continuous flow (for example, to compression waves), let us introduce the binary feature

$$a_{5m} = \begin{cases} 1, & |A_l - A_r|/A_{\max} > \delta_1, \\ 0, & |A_l - A_r|/A_{\max} \le \delta_1. \end{cases} \qquad (7.4.25)$$

The quantities A_l and A_r in (7.4.25) are calculated in accordance with (7.4.8), $A_{\max} = \max(A_l, A_r)$, δ_1 is a user-specified positive constant. This constant is set with regard to the following considerations. First, $|A_l - A_r| = O(1)$ at the shock waves and contact discontinuities of finite intensity; in the domains of continuous flow $|A_l - A_r| = O(\Delta n)$, where Δn is the distance in the (x, y)-plane between the points with indices (i_l, j_l) and (i_r, j_r); in the case of a square mesh $\Delta n \le 2\sqrt{2}h_1$.

In order to distinguish between the points of purely contact discontinuities and the points of tangential discontinuities, we make use, with (1.40), (1.41) in view, of the binary feature

$$a_{6m} = \begin{cases} 1, & (\tau/h_1)|u_{\tau l} - u_{\tau r}| > \delta_2, \\ 0, & (\tau/h_1)|u_{\tau l} - u_{\tau r}| \le \delta_2. \end{cases} \qquad (7.4.26)$$

In (7.4.26) δ_2 is a user-specified positive constant,

$$u_{\tau l} = v_{i_l,j_l} \cos \alpha_{ij} - u_{i_l,j_l} \sin \alpha_{ij},$$
$$u_{\tau r} = v_{i_r,j_r} \cos \alpha_{ij} - u_{i_r,j_r} \sin \alpha_{ij}. \qquad (7.4.27)$$

In order to distinguish between the points of normal shock waves and the points of oblique shock waves, let us introduce the binary feature

$$a_{7m} = \begin{cases} 1, & (\tau/h_1)(|u_{\tau l}| + |u_{\tau r}|) > \delta_3, \\ 0, & (\tau/h_1)(|u_{\tau l}| + |u_{\tau r}|) \le \delta_3, \end{cases} \qquad (7.4.28)$$

where δ_3 is a user-specified positive constant. To substantiate the use of this feature in the case of a normal shock wave it is necessary to prove that the magnitude of the tangential component of the gas velocity will be small in the zone of smearing of such a discontinuity. In the practical computation of the quantities $u_{\tau l}$ and $u_{\tau r}$ by formulas (7.4.27) we have used the grid values of the velocity components u, v in the nodes lying on a straight line corresponding to one of the eight principal directions shown in Figure 7.6. Let us number these grid values of the quantities u and v as $u_1, \ldots, u_N$; $v_1, \ldots, v_N$ where the subscript "1" refers to one end of the zone of a smeared normal shock wave, and the subscript "N" refers to the other end of the same zone. Assume that the sequences $u_1, \ldots, u_N$ and $v_1, \ldots, v_N$ are monotone

$$\text{sign}(u_j - u_{j+1}) = s_1; \qquad \text{sign}(v_j - v_{j+1}) = s_2; \qquad j = 1, \ldots, N - 1, \quad (7.4.29)$$

where $s_k = 1$ or -1, $k = 1, 2$. The relationships (7.4.29) are valid while using monotone difference schemes. In the case of using many well-known schemes of the second and higher orders of approximation we are usually able to indicate such a neighborhood of a shock wave front in which the relationships (7.4.29) are satisfied. Outside this neighborhood (usually behind the front of a smeared shock wave) the second-order schemes generate parasitic oscillations (see, for example, Figures 2.7 and 2.8). Thus, let us assume that the relationships (7.4.29) are satisfied. Assume also that the equations

$$u_{\tau 1} = 0, \qquad u_{\tau N} = 0, \qquad (7.4.30)$$

take place at the ends of a zone of smearing of a normal shock wave, that is, at $j = 1$ and $j = N$, where

$$u_{\tau j} = v_j \cos \alpha - u_j \sin \alpha, \qquad j = 1, \ldots, N, \qquad (7.4.31)$$

where α is the angle between the normal to the wave front and the positive direction of the x-axis. Let us show that the only condition for $u_{\tau j}$, which is noncontradictory to the requirements (7.4.29), (7.4.30), is the vanishing of the quantity $u_{\tau j}$ throughout the zone of smearing of a normal shock wave. Since u_j, v_j in (7.4.29) may have arbitrary signs, it is convenient in the following to employ the formula

$$u_{\tau i} = \sum_{j=2}^{i} (u_{\tau j} - u_{\tau j-1}), \qquad i = 2, \ldots, N. \qquad (7.4.32)$$

This formula is obtained with regard to the first relationship in (7.4.30). Substituting the right-hand side of formula (7.4.31) instead of $u_{\tau j}$ in (7.4.32), we obtain

$$u_{\tau i} = \left[\sum_{j=2}^{i} (v_j - v_{j-1}) \right] \cos \alpha - \left[\sum_{j=2}^{i} (u_j - u_{j-1}) \right] \sin \alpha, \qquad i = 2, \ldots, N.$$
$$(7.4.33)$$

Consider now the case

$$v_j - v_{j-1} > 0, \qquad u_j - u_{j-1} > 0, \quad \forall j, \qquad \cos \alpha > 0, \qquad \sin \alpha < 0.$$

Then we obtain from (7.4.33): $u_{\tau N} > 0$, which contradicts the second of the relationships (7.4.30). Assume now that

$$v_j - v_{j-1} > 0, \qquad u_j - u_{j-1} > 0, \quad \forall j, \qquad \cos \alpha > 0, \qquad \sin \alpha > 0.$$

The orientation of the front line of the considered shock wave, with respect to the axes x', y' of some other coordinate system $x'Oy'$ whose axes Ox', Oy' are rotated by a certain angle with respect to the axes of the original coordinate system xOy, will change in the numerical computation of the same gas-dynamical problem in the $x'Oy'$ system. Let us rotate the axes Ox and Oy by an angle such that in the new coordinate system $x'Oy'$ the inequalities $\cos \alpha > 0$, $\sin \alpha < 0$ are satisfied. Then we arrive at a case already considered. All other possible combinations of the signs of the quantities s_1, s_2 and $\cos \alpha$, $\sin \alpha$ are considered in a similar way. In all of these cases the only requirement for $u_{\tau j}$, which does not contradict the conditions (7.4.29), (7.4.30), is the condition $u_{\tau j} = 0$, $\forall j$.

Thus, we have introduced, for the characteristic of the objects (x_m, y_m) in the feature space, seven features $a_{1m}, \ldots, a_{7m}$. The nondimensional character of the chosen functional dependencies for a_{jm} gives reason to expect that the results of the singularities classification with the aid of these features will comparatively weakly depend on the specifics of a gas-dynamical problem under study.

We do not usually employ, in the systems of automatic pattern classification, the original features $a_{1m}, \ldots,$ but rather the features $b_{1m}, \ldots$ obtained from the features a_{jm} by means of transformations. A number of such transformations has been presented in [7.167]. The purpose of these transformations is to maximize in the feature space the interclass distances, and to minimize the intraclass distances. A linear transformation of these features

$$\mathbf{b} = W\mathbf{a}, \tag{7.4.34}$$

where W is a matrix of weight coefficients, enables us to minimize the intraclass distances with the aid of a proper choice of the weights. In the case of a general matrix W the solution of this minimization problem is cumbersome, as was pointed out in [7.55]. For a particular case when W is a diagonal matrix,

$$W = \text{diag}(w_{11}, \ldots, w_{MM}), \tag{7.4.35}$$

in [7.55] a solution of the form

$$w_{kk} = \left[\sigma_k^2 \sum_{k=1}^{M} (1/\sigma_k^2) \right]^{-1} \tag{7.4.36}$$

has been obtained where σ_k is the dispersion of the kth feature a_k. The multiplier $\sum_{k=1}^{M} \sigma_k^{-2}$ in (7.4.36) ensures the satisfaction of the condition

$$\sum_{k=1}^{M} w_{kk} = 1. \tag{7.4.37}$$

The application of formula (7.4.36) corresponds to the scaling of the kth feature a_{km} of the objects (x_m, y_m), $m = 1, \ldots, N_4$, as a result of which the intraset distances in the feature space are reduced. Let us analyze the dispersions σ_k of our features a_{km}. If it can be elucidated in the result of this analysis that the quantity σ_k is small for some k, then we can in principle refuse the scaling of the kth coordinate a_k. Otherwise, the scaling will be carried out in accordance with (7.4.34) where we shall assume (following [7.55]) that W is a diagonal matrix (7.4.35). The intraset distances between the patterns of the objects in the feature space will be reduced in this way, though they will not be minimal, since we want to refuse the use of the constraint (7.4.37) with the purpose of a maximal simplification of the computational procedure of the features transformation.

Consider at first the features (7.4.9) and (7.4.20). Some smoothed values of the quantities $\rho, u, v, p, \varepsilon$ are used in these features. Therefore, it can be expected that if the noise level in an image does not exceed certain limits (the reader can obtain an index of these limits by reading Section 7.6 in which the computational examples are presented), then the relationships $a_{1m} = 2, a_{2m} = 2$ will be satisfied at the shock waves, and at the contact discontinuities the relationships $\{a_{1m} = 2 \text{ or } a_{1m} = 0\}$ and $a_{2m} = 0$ will be satisfied. Thus, the dispersions of the values of a_{2m} corresponding to the points of shock waves and contact discontinuities are equal to zero and, consequently, the coordinate a_2 may not be scaled. The dispersion of the features a_{1m} of the objects (x_m, y_m) can be considered at the contact discontinuities to be relatively small. Thus, let us set in (7.4.35) $w_{11} = w_{22} = 1$. Considering, similarly, the features a_{5m} and a_{6m} we set $w_{55} = w_{66} = w_{77} = 1$.

Let us now estimate the dispersion of the feature (7.4.22). In practical gas-dynamical computations the quantity $(\tau/h_1)(|\mathbf{u}| + c)$—the Courant number—is always bounded,

$$(\tau/h_1)(|\mathbf{u}| + c) \leq K, \qquad (7.4.38)$$

where $|\mathbf{u}| = (u^2 + v^2)^{1/2}$; for example, in the case of the two-dimensional flow computation by the two-step Lax–Wendroff scheme with the steps $h_1 = h_2$, $K = 2^{-1/2}$ [7.64]. Employing (7.4.38), it is easy to obtain an upper bound for the quantity a_{4m}. Consider for definiteness the case of using formulas (7.4.10), (7.4.12). Then

$$a_{4m} \leq 2K(|\cos \alpha_{ij}| + |\sin \alpha_{ij}|)(|\cos \alpha_{ij}| + (h_1/h_2)|\sin \alpha_{ij}|)$$

$$\leq 2K\sqrt{2}[1 + (h_1/h_2)^2]^{0.5}. \qquad (7.4.39)$$

In the case when implicit difference schemes are used in the computations the value K can be relatively large, for example, $K = 10$. Therefore, it is reasonable, with (7.4.39) in view, to perform a scaling of the coordinate a_4. Taking (7.4.39) into account this can be simply done by setting in (7.4.35)

$$w_{44} = \{2K\sqrt{2}[1 + (h_1/h_2)^2]^{0.5}\}^{-1}. \qquad (7.4.40)$$

In the computations, the results of which will be presented below, we have used an algorithmically more complicated technique

$$w_{44} = 1 \Big/ \max_m a_{4m}. \qquad (7.4.41)$$

The application of any of the weight multipliers (7.4.40) or (7.4.41) guarantees the satisfaction of the inequality $b_{4m} \leq 1$.

In the case of the feature (7.4.21) it is easy to obtain, with regard to (7.4.24),

(7.2.6), (7.3.5), the inequality

$$a_{3m} < \left\{ 8 \max_{k,l} p_{kl} \cdot [(h_1/h_2)|\sin \alpha_{ij}| + |\cos \alpha_{ij}|] \right\}$$

$$\times [2(p_{i+1,j+1} + p_{i-1,j+1} + p_{i+1,j-1} + p_{i-1,j-1})]^{-1}. \qquad (7.4.42)$$

Since the pressure jumps across the shock waves can be arbitrary, we obtain from (7.4.42) that the dispersion σ_3 can be sufficiently large. In this connection we have applied a scaling of the coordinate a_3 by a formula similar to (7.4.41): $w_{33} = 1/\max_m a_{3m}$. Thus, in the following, we shall use a vector **b** of the features determined by the formula (7.4.34) where

$$W = \text{diag}(1, 1, w_{33}, w_{44}, 1, 1, 1). \qquad (7.4.43)$$

Note that since the shock wave speed D does not enter explicitly into the expressions for the features (7.4.9), (7.4.19), (7.4.21), (7.4.22), (7.4.25), (7.4.26), (7.4.28), then these features may be used for the classification of singularities both in stationary and in nonstationary gas flows.

7.5. Algorithms of Pattern Classification

7.5.1. Minimum-Distance Classifier

Pattern classification on the basis of the minimum-distance criterion is one of the first ideas of automatic pattern recognition. As was pointed out in [7.55], this simple classification method proves to be a very efficient tool in the solution of the problems in which the classes are characterized by a degree of variability which is within reasonable limits. For a realization of this method it is at first necessary to construct a training set of patterns. In our case we shall assume that there do not exist any other classes of objects besides those which have been presented in Figure 7.21. Thus, we assume that we know in advance a set of classes to which the points (x_m, y_m) under study may be assigned. Such classification problems are called supervised classification [7.84]. The set of training patterns is composed of those patterns which are typical or representative of the corresponding class. Following [7.55] we shall call these patterns (in what follows) the samples, and denote the vectors of samples in the feature space $(b_1, \ldots, b_7)$ by $\mathbf{z}_i$, $i = 1, \ldots, N_\omega$, where N_ω is the number of classes. In the case of parallel classification (see Figure 7.24), the assignment of the object (x_m, y_m) to some class is carried out independently of the fact as to which classes the other objects were assigned; therefore, the classification of all the objects (x_m, y_m) can be carried out simultaneously, in this case, with the aid of the minimum-distance principle, which especially can be efficiently implemented on a computer with parallel processors. Consider at first, for definiteness, the case when there is only one sample pattern $\mathbf{z}_i$ in

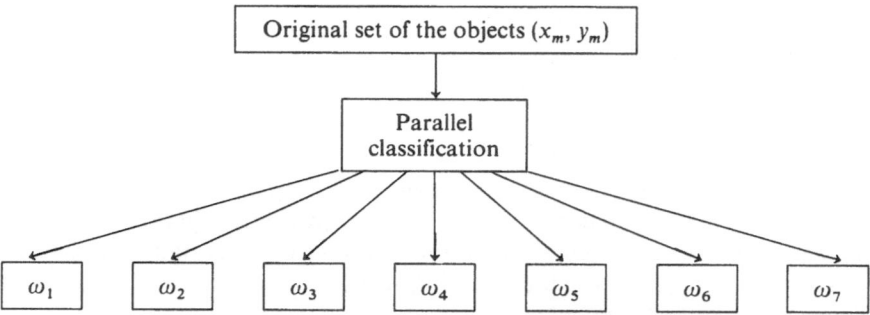

Figure 7.24. The scheme of parallel objects classification.

each class ω_i. It will sometimes be convenient to consider the vectors $\mathbf{b}_m$, $\mathbf{z}_i$ in the following as row matrices. We shall use along with $\mathbf{b}_m$, $\mathbf{z}_i$ also the column matrices $\mathbf{b}_m^T$, $\mathbf{z}_i^T$ obtained by the transposition of $\mathbf{b}_m$, $\mathbf{z}_i$. The quantity $D_{mi} = \|b_m - \mathbf{z}_i\|$ is understood here and in what follows as the Euclidean distance in the feature space between the points having the radius vectors $\mathbf{b}_m$ and $\mathbf{z}_i$. In accordance with the minimum-distance principle the mth object is assigned to the class ω_i if $D_{mi} = \min_k D_{mk}$. As is known (see, for example, [7.55]), the choice of the minimal D_{mi} is equivalent to the choice of the maximum value of the function

$$d_i(\mathbf{b}) = \mathbf{b}^T\mathbf{z}_i - (1/2)\mathbf{z}_i^T\mathbf{z}_i, \qquad i = 1, \ldots, N_\omega. \tag{7.5.1}$$

The functions $d_i(\mathbf{b})$ are called the decision or discriminant functions. The pattern $\mathbf{b}$ is assigned to the class ω_i, if the condition $d_i(\mathbf{b}) > d_j(\mathbf{b})$ is valid for all $j \neq i$.

Consider now an important question on the determination of the co-ordinates $z_1, \ldots, z_7$ of sample patterns for each of the classes $\omega_1, \ldots, \omega_7$. Let us make use of the properties of the objects belonging to $\omega_1, \ldots, \omega_7$ as well as the properties of the features $b_{1m}, \ldots, b_{7m}$ constructed in the foregoing section. Consider at first a question on determining the sample values of the coordinate z_1 for the classes ω_1, ω_2 and ω_3, ω_4. As was shown in Section 7.4, $a_{1m} = 2$ at the shock waves, consequently, we can take the value $z_1 = 2$ as a sample value of the coordinate z_1 for the classes ω_1 and ω_2.It is well known that at the contact discontinuities three possibilities can be realized: $\partial u_n/\partial n < 0$, $\partial u_n/\partial n > 0$, $\partial u_n/\partial n = 0$. The situation $\partial u_n/\partial n = 0$, that is, when the normal velocity component is constant in the neighborhood of a contact discontinuity, occurs very seldom in the actual applied problems of gas dynamics (see the examples in Sections 7.6.2–7.6.5). However, in computations of model problems on the fluid mass advection by the constant velocity field such a situation may take place (see, for example, the problem (4.2.7), (4.2.8)). Since $u_{10} = u_0 \cos \beta$, $u_{20} = u_0 \sin \beta$, in (4.2.7), in the case of a machine computation of the quantities u_{10}, u_{20}, the exact equality $\partial u_n/\partial n = 0$ will not be satisfied because of the truncation errors in the computation of the values of the functions $\cos \beta$, $\sin \beta$. The equality $\partial u_n / \partial n = 0$, where the derivative $\partial u_n / \partial n$ was computed by formulas (7.4.10), (7.4.11) or (7.4.10), (7.4.12), was really never realized in the numerical computations of the model problem (4.2.7), (4.2.8) (see also Figure 7.41 in Section 7.6.1). In connection with the foregoing we have taken into account, in the process of the development of a minimum-distance classifier, only two sample values of the coordinate z_1 for each of the classes ω_3 and ω_4: $z_1 = 2$ and $z_1 = 0$.

Taking into account (7.4.18), (7.4.19), the sample values of the coordinate z_2 for the classes ω_1, ω_2 and ω_3, ω_4 are determined as $z_2 = 2$ and $z_2 = 0$, respectively. Consider now the formulas (7.4.21), (7.4.22). It follows from the properties of the functions $p(x, y, t)$ and $u_n(x, y, t)$ at the contact discontinuities that

$$\lim_{h_1 \to 0} a_{3m} = 0, \qquad \lim_{\tau \to 0} a_{4m} = 0.$$

From this we obtain that at sufficiently small steps h_1, h_2, and τ we can take, as sample values of the coordinates z_3 and z_4 for the contact discontinuities classes ω_3 and ω_4, the values $z_3 = 0$ and $z_4 = 0$. At the shock wave of finite intensity $a_{3m} = O(1)$, $a_{4m} = O(\tau/h_1)$ in accordance with Section 7.4. Taking into account (7.4.34), (7.4.41), (7.4.43), we have that at the shock waves

$$b_{3m} = a_{3m} \bigg/ \max_m a_{3m} = O(1)/O(1) \leq 1,$$

$$b_{4m} = a_{4m} \bigg/ \max_m a_{4m} = O(\tau/h_1)/O(\tau/h_1) = O(1) \leq 1.$$

From this it follows that in cases when the intensities of shock waves in a problem under study vary insignificantly, then the quantities b_{3m} and b_{4m} are close to 1. In the case of shock waves with an intensity which strongly varies in space, the scatter of the values b_{3m}, b_{4m} is more significant. Nevertheless, it is also possible to assume in this case that at sufficiently small τ, h_1, h_2 there will occur in the (b_3, b_4)-plane the grouping of the (b_{3m}, b_{4m}) points around a point with the coordinates $z_3 = 1$, $z_4 = 1$. This property of clustering of the (b_{3m}, b_{4m}) points is confirmed below in the computational examples presented in Section 7.6.1. From the above consideration of the features b_{3m}, b_{4m} a conclusion may be drawn that the minimum-distance classifier presented below in this section will have a better performance at the smaller steps h_1, h_2, τ.

The sample value of the binary feature a_{5m} for the classes ω_1, ω_2, ω_3, ω_4 is obviously the value $z_5 = 1$. In its turn, the decision on the assignment of the value of 1 to the feature a_{5m} depends on the value of a user-specified constant δ_1. In Section 7.5.2 we present a general procedure for determining the constants of the δ_1 constant type, with account being taken of the well-known procedures from the pattern recognition theory (see, for example, [7.55]).

Table 7.3

z_j \ ω_i	ω_1	ω_2	ω_3	ω_3	ω_4	ω_4	ω_4	ω_4	ω_5	ω_5	ω_5	ω_5	ω_6	ω_7
z_1	2	2	0	2	0	2	0	2	2	2	2	2	0	0
z_2	2	2	0	0	0	0	0	0	2	2	2	2	2	0
z_3	1	1	0	0	0	0	0	0	1	1	1	1	0	0
z_4	1	1	0	0	0	0	0	0	1	1	1	1	0	0
z_5	1	1	1	1	1	1	1	1	0	0	0	0	0	0
z_6	0	0	1	1	0	0	0	0	0	1	0	1	0	0
z_7	0	1	1	1	1	1	0	0	0	0	1	1	0	0

We present in Table 7.3 the coordinates $z_1, \ldots, z_7$ of the samples not only for the classes of discontinuities $\omega_1, \ldots, \omega_4$, but also for the classes of domains of continuous flow ω_5, ω_6, ω_7. In accordance with the foregoing analysis we have included, in the ω_3 class, two samples $z_3^{(1)}$, $z_3^{(2)}$ which correspond to different values of the coordinate z_1. We have included in the ω_4 class four samples $z_4^{(1)}, \ldots, z_4^{(4)}$ with regard to possible differences in the values of the coordinates z_1 and z_7. We were able to find only one sample in the ω_6 class. This sample was obtained in the analysis of the properties of the solutions of of gas dynamics equations in the domain of a rarefaction wave in the one-dimensional flow (see Section 1.1.3). It appears that in the case of two-dimensional rarefaction waves there exist also other samples. We have also included in the class ω_7 only one sample; it is not difficult to obtain several additional samples belonging to the class ω_7. However, as we shall see in the following, the samples of the class ω_7 may not be needed at all in a proper organization of the pattern classification system.

From the point of view of the optimality of the classification systems [7.174] the scheme of parallel classification presented in Figure 7.24 is not optimal, since it requires for the classification a calculation of all the coordinates $b_1, \ldots, b_7$ of the pattern vectors, which is not economical in terms of computer expense. In this connection let us turn to the idea of hierarchical classification according to which the data subject to classification are taken from the only class which may be subdivided into subclasses and sub-subclasses [7.167]. In our case such a subdivision was discussed above in the analysis of the inter-relations between the classes presented in Figure 7.21. In Figure 7.25 we present one of the possible schemes of hierarchical classification. Its application enables us, in particular, to observe in the process of a study of non-stationary gas flows the evolution of compression waves which in some cases can, as is known, go over into the shock waves.

In the case of the scheme of Figure 7.25 one feature b_{1m} is not sufficient for the assignment of the object (x_m, y_m) to one of the two classes Ω_4 or Ω_5, because the equality $b_{1m} = 2$ can be satisfied both at the shock waves and at the contact discontinuities. Similarly, the equality $b_{2m} = 2$ can be satisfied both at the

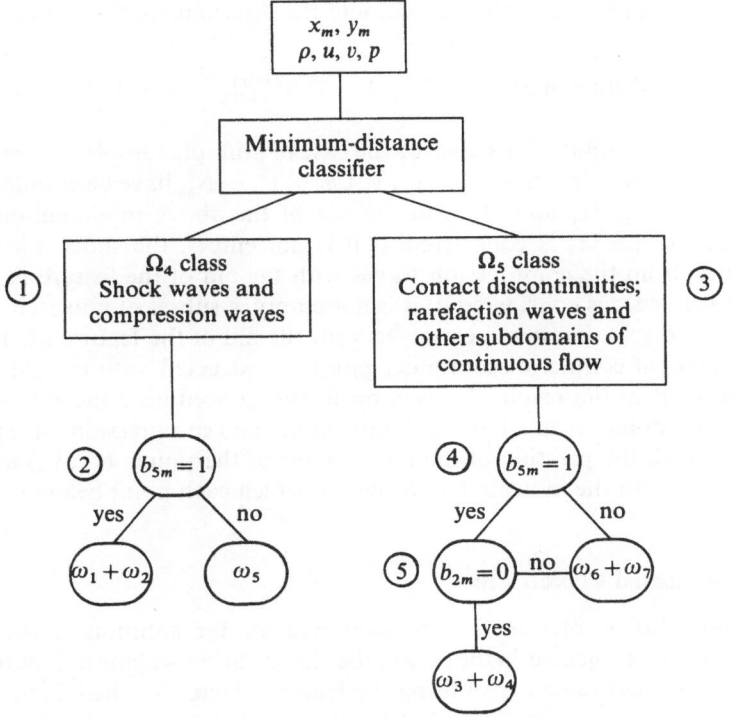

Figure 7.25. The scheme of hierarchical classification of singularities in gas flows.

shock waves and in some subdomains of the continuous flow. In this connection we have used at the initial stage of the scheme of Figure 7.25 a minimum-distance classifier in which four features b_{1m}, b_{2m}, b_{3m}, b_{4m} are involved. Let

$$\boldsymbol{\beta} = (b_1, b_2, b_3, b_4), \qquad \boldsymbol{\beta}_m = (b_{1m}, b_{2m}, b_{3m}, b_{4m}),$$

$$\boldsymbol{\zeta}_1 = (z_{11}, z_{12}, z_{13}, z_{14}), \qquad \boldsymbol{\zeta}_2^{(k)} = (z_{21}^{(k)}, z_{22}^{(k)}, z_{23}^{(k)}, z_{24}^{(k)}), \qquad k = 1, 2. \tag{7.5.2}$$

In accordance with Table 7.3, we place in the class Ω_5 two samples $\boldsymbol{\zeta}_2^{(1)}$, $\boldsymbol{\zeta}_2^{(2)}$ having the coordinates

$$z_{21}^{(1)} = 0, \qquad z_{22}^{(1)} = 0, \qquad z_{23}^{(1)} = 0, \qquad z_{24}^{(1)} = 0;$$

$$z_{21}^{(2)} = 2, \qquad z_{22}^{(2)} = z_{23}^{(2)} = z_{24}^{(2)} = 0. \tag{7.5.3}$$

Further, with Table 7.3 in view,

$$z_{11} = 2, \qquad z_{12} = 2, \qquad z_{13} = z_{14} = 1. \tag{7.5.4}$$

A linear discriminant function $d_1(\boldsymbol{\beta})$ is then written by analogy with (7.5.1) in the form

$$d_1(\boldsymbol{\beta}) = \boldsymbol{\beta}^{\mathrm{T}} \cdot \boldsymbol{\zeta}_1 - (1/2)\boldsymbol{\zeta}_1^{\mathrm{T}}\boldsymbol{\zeta}_1.$$

Since two samples enter Ω_5, we compute the discriminant function $d_2(\beta)$ by the formula

$$d_2(\beta) = \max_{l} \{\beta^T \cdot \zeta_2^{(l)} - (1/2)(\zeta_2^{(l)})^T \zeta_2^{(l)}\}, \qquad l = 1, 2, \qquad (7.5.5)$$

in accordance with the formulas for the case of multiple samples presented in [7.55]. After the original objects $(x_m, y_m), m = 1, \ldots, N_4$, have been subdivided into two classes Ω_4 and Ω_5 with the aid of the above minimum-distance classifier, the set Ω_4 is considered. If it is not empty, the shock waves are separated from the compression waves with the aid of the feature b_5. After that the class Ω_5 is considered. If it is not empty, a subset of potential points of contact discontinuities is extracted with the aid of the feature b_5. Then a subsubclass of contact discontinuity points is extracted with the aid of the feature b_2. If in the result of execution of this procedure nonempty sets of shock wave points, contact discontinuity points, and compression wave points are obtained, the printing of the coordinates of the points (x_m, y_m) may be carried out with the indication of a class to which each point belongs.

7.5.2. Sequential Classification

In Figure 7.26 is presented a decision tree for the solution of the same classification problem as in the case of the classification scheme of Figure 7.25, but at a different sequence of using the features. Here the idea of sequential classification has been realized [7.167]. We have denoted by ω_8 in Figure 7.26 a class of rarefaction shock waves which in principle should not arise in the case of using a difference scheme ensuring the increase in the entropy at the shock waves. Nevertheless, in the case of substantial oscillations in the numerical solution the class ω_8 can prove to be nonempty upon completing the process of classification of the set (x_m, y_m). We shall call for convenience the classification schemes presented in Figure 7.25 and 7.26 "scheme I" and "scheme II", respectively. Following [7.174] denote by c_j the cost of the measurement of the jth feature $b_{jm}, j = 1, \ldots, M$. In our case we mean by c_j the computing time needed for the calculation of the jth feature. The scheme presented in Figure 7.26 is indeed a decision tree for the analysis of gas flows. There are in this tree both interior and terminal nodes (a definition of the tree as a graph possessing certain properties may be found, for example, in [7.167]). The less the number of interior nodes in a tree, the shorter is the way in the graph in recognition of the mth object (x_m, y_m). In this connection in [7.174], the questions of optimization of the decision trees were considered both from the point of view of minimization of the cost of the objects classification and from the point of view of minimization of the number of interior nodes. In Figure 7.25 and 7.26 the interior nodes are numbered by circled figures. We can see that there are in scheme I five interior nodes and in scheme II eight such nodes. However, it is required, in the multistage sequential

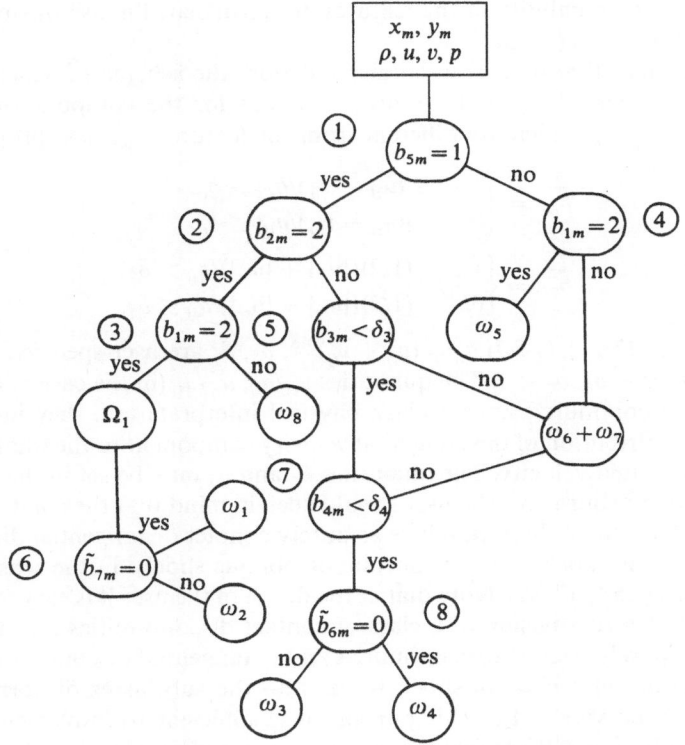

Figure 7.26. Decision tree in the sequential classification scheme.

classification scheme II at each stage, to compute only one feature. There are in [7.175] formal proofs of the optimality of such classification procedures from the point of view of measurement costs (in our case of computer time needed for the computation) of the features.

Thus, from the point of view of the "cost," scheme I is worse than scheme II. However, we should keep in mind, along with the factor of cost, also the reliability of the classification schemes. In the case of scheme I a decision on the assignment of the object (x_m, y_m) to one of the two classes Ω_4 or Ω_5 is made on the basis of ensemble of the four features b_1, b_2, b_3, b_4, so that the random errors in one or two features may not affect the classification result. In the case of scheme II the object (x_m, y_m) will be classified incorrectly in the case of an error in one feature. Thus, the classification reliability makes increased demands on the features used in the systems of sequential classification. In particular, the features should be less sensitive to noise (or they should possess a higher noise immunity) than in the case of a classification system based on the minimum-distance principle. We have discussed above some measures for increasing the reliability of the features, in particular, the use of smoothing masks of the form (7.4.5).

In the algorithm of sequential classification, the scheme of which is presented in Figure 7.26, we have used formulas for the computation of the features $\tilde{b}_{6m}$, $\tilde{b}_{7m}$ which are different from the features a_{6m}, a_{7m} proposed in Section 7.4:

$$\tilde{b}_{6m} = \begin{cases} 1, & |(u_{\tau l} - u_{\tau r})/q_{ij}| > \delta_6, \\ 0, & |(u_{\tau l} - u_{\tau r})/q_{ij}| \le \delta_6, \end{cases} \tag{7.5.6}$$

$$\tilde{b}_{7m} = \begin{cases} 1, & (1/2)(|u_{\tau l}| + |u_{\tau r}|)/q_{ij} > \delta_7, \\ 0, & (1/2)(|u_{\tau l}| + |u_{\tau r}|)/q_{ij} \le \delta_7. \end{cases} \tag{7.5.7}$$

In formulas (7.5.6), (7.5.7) $q_{ij} = (u_{ij}^2 + v_{ij}^2)^{0.5}$, δ_6, δ_7 are user-specified positive constants, $0 < \delta_6, \delta_7 < 1$. The quantities $u_{\tau l}/q_{ij}$, $u_{\tau r}/q_{ij}$ (in the case of a purely contact discontinuity) have a clear physical interpretation: they indicate a relative contribution of the tangential velocity component to the total magnitude of the fluid velocity. The quantities δ_6 and δ_7 may be set in the interval $]0, 1[$ rather arbitrarily. The user should keep in mind that the computer will classify the contact discontinuities as purely contact or tangential discontinuities, and the shock waves as normal or oblique shocks in the sense of the definitions (7.5.6), (7.5.7). Note that at the desire of the user it is easy to realize in scheme I a subdivision of a class of contact discontinuities into the subclasses of purely contact discontinuities and of tangential discontinuities, and a subdivision of a class of shock waves into the subclasses of normal and oblique shock waves. For this purpose it is sufficient to implement in the scheme of Figure 7.25 the features a_{6m} and a_{7m} (see formulas (7.4.26), (7.4.28)) or the features $\tilde{b}_{6m}$, $\tilde{b}_{7m}$ (see formulas (7.5.6), (7.5.7)) in the same way as the features $\tilde{b}_{6m}$ and $\tilde{b}_{7m}$ have been implemented in the scheme of Figure 7.26.

The above algorithm of sequential classification also requires the specification of positive constants δ_3 and δ_4 (see Figure 7.26). The choice of these constants should be carried out in accordance with the general procedure for determining the values of features for the sample patterns (see, for example, [7.55]). In accordance with this procedure a relatively well-studied problem was at first chosen while considering some class of gas-dynamical problems, such that for this problem there is information on the types of singularities which may be present in a given problem. After that such sample values of the constants δ_3 and δ_4 were determined by test computations at which the level of classification errors was minimal. As a rule, for each of the constants δ_3 and δ_4 it is possible to indicate a certain interval within which the classification result depends very weakly on the specific values δ_3 and δ_4 from the corresponding intervals. It appears that owing to this fact it proves to be rather easy to determine the values δ_3 and δ_4 when going over to an analysis of other classes of aerohydromechanics problems. In computations of problems where the intensity of shock waves varies quickly with time it proves useful to set δ_3 and δ_4 as some functions of time (for example, as linear functions).

Let us now describe a scheme of sequential classification for the extraction of shock waves in the computations of stationary potential transonic flows. The conditions at the shock waves in such flows are obtained as a particular case from formula (1.31) at $D = 0$ and from (1.35):

$$\rho_1 u_{n1} = \rho_2 u_{n2}, \tag{7.5.8}$$

$$u_{\tau 1} = u_{\tau 2}. \tag{1.35}$$

In addition, we obtain from the inequalities (1.38) at $D = 0$:

$$|u_{n2}|/c_2 > 1, \qquad |u_{n1}|/c_1 < 1. \tag{7.5.9}$$

The components u, v of the velocity vector in a two-dimensional potential flow were computed by formulas (7.3.64). The orientation angle α_{ij} of the vector of a normal to the edge at the node (x_{ij}, y_{ij}) was computed by the formula (7.3.73). After that the normal velocity component u_n and the tangential velocity component u_τ, at some node (x_{kl}, y_{kl}) being a neighbor of the node (x_{ij}, y_{ij}), were determined by the formulas

$$(u_n)_{k,l} = u_{kl} \cos \alpha_{ij} + v_{kl} \sin \alpha_{ij},$$

$$(u_\tau)_{k,l} = v_{kl} \cos \alpha_{ij} - u_{kl} \sin \alpha_{ij}.$$

Let (i_l, j_l) and (i_r, j_r) be the indices of the grid nodes (x_{ij}, y_{ij}) lying on different sides of the edge and closest to the line of a normal to the edge (see Section 7.3). The subscripts "1" and "2" in (7.5.9) refer to the quantities behind and before the shock wave front, respectively. Therefore, to check the inequalities (7.5.9) it is at first necessary to determine which of the points $(x_{i_l,j_l}, y_{i_l,j_l})$ and $(x_{i_r,j_r}, y_{i_r,j_r})$ is located behind the front of a shock wave in the potential flow. In the case of nonisentropic shock waves we have checked for this purpose the inequality $A_l > A_r$ (see formulas (7.4.5)–(7.4.8)). This criterion is inapplicable in the case of isentropic shock waves. There are known in the literature several analogues of the entropy condition at a compression shock wave in a potential flow. In particular, it was shown in [7.176] that if the pressure function $p(V)$, $V = 1/\rho$, satisfies the Weyl condition, namely, the function $p(V)$ is convex, then at a discontinuity across which the pressure and the density increase (shock wave) the relationship

$$p_1 + \rho_1 u_{n1}^2 - (p_2 + \rho_2 u_{n2}^2) \geq 0 \tag{7.5.10}$$

is satisfied where the subscripts "1" and "2" refer to the gas state behind and before the shock wave front, respectively. For an approximate determination of the quantities p_1, ρ_1, u_{n1}, p_2, ρ_2, u_{n2} entering (7.5.10) we have realized a search interpolation procedure similar to the one presented in Section 7.4 (see formulas (7.4.6), (7.4.7)). This procedure enabled us to find the values p_l, ρ_l, u_{nl} and p_r, ρ_r, u_{nr}. Setting then in (7.5.10) $p_1 = p_l$, $\rho_1 = \rho_l$, $u_{n1} = u_{nl}$, $p_2 = p_r$, $\rho_2 = \rho_r$, $u_{n2} = u_{nr}$, we have checked the satisfaction of the inequality (7.5.10). If it was not satisfied, the pairs of indices (i_l, j_l) and (i_r, j_r) were interchanged.

To increase the reliability of classification we have used along with the condition (7.5.10) also a simpler criterion enabling us to exclude from further consideration the rarefaction shock waves, and having the form [7.177]

$$u_{n1} \cdot [|\mathbf{u}|] < 0, \qquad u_{n2} \cdot [|\mathbf{u}|] < 0, \qquad (7.5.11)$$

where the symbol $[|\mathbf{u}|]$ denotes the jump of the quantity $|\mathbf{u}|$ across a shock wave. The conditions (7.5.11) were implemented in the classification computer code as follows. At first the values p_l, ρ_l, u_{nl}, p_r, ρ_r, u_{nr} were found with the aid of a search interpolation procedure (the corresponding points with the indices (i_l, j_l) and (i_r, j_r) were interchanged, if needed), such that $p_l + \rho_l u_{nl}^2 - p_r - \rho_r u_{nr}^2 \geq 0$. After that the inequalities

$$
\begin{aligned}
u_{nr} \cdot [(u_{nl}^2 + u_{\tau l}^2) - (u_{nr}^2 + u_{\tau r}^2)] &< 0; \\
u_{nl} \cdot [(u_{nl}^2 + u_{\tau l}^2) - (u_{nr}^2 + u_{\tau r}^2)] &< 0;
\end{aligned}
\qquad (7.5.12)
$$

were checked in accordance with (7.5.11). If we consider a flow around an airfoil, the chord of which is oriented along the x-axis and the flow moves in the direction of the x-axis, it can be assumed that $u_{nr} > 0$, $u_{nl} > 0$, near the shock wave front, and then the formulas (7.5.12) are reduced to a trivial inequality $|\mathbf{u}_l|^2 < |\mathbf{u}_r|^2$ which corresponds to the well-known fact that the flow slows down behind the shock wave front.

In Figure 7.27 we present the scheme of sequential classification that we have realized for the extraction of shock wave points in two-dimensional stationary potential transonic flows. Note that by virtue of the fact that the mathematical model of potential flow is simpler than the model of a nonisentropic rotational flow, then the scheme of Figure 7.27 is simpler than the scheme of Figure 7.26. The features a_{1m}, a_{2m} were computed with regard to (7.5.8), (1.35) by the formulas

$$
\begin{aligned}
a_{1m} &= |[\rho_{i_r,j_r} \cdot (u_n)_{i_r,j_r} - \rho_{i_l,j_l} \cdot (u_n)_{i_l,j_l}]/[\rho_{ij} \cdot (u_n)_{ij}]|, \\
a_{2m} &= |[(u_\tau)_{i_r,j_r} - (u_\tau)_{i_l,j_l}]/(u_n)_{ij}|.
\end{aligned}
$$

The quantities δ_1, δ_2 in the scheme of Figure 7.27 are user-specified positive constants. In order to determine the sample values of these constants we have carried out a number of computations of the transonic flows around an airfoil by the AF2 scheme for different values of the free stream Mach number M_∞, for which the presence of shock waves is established in the literature. As a result we have found a range of the variation of the quantities a_{1m}, a_{2m} at the points (x_{ij}, y_{ij}) belonging to the shock waves: $0.0014 \leq a_{1m} \leq 0.013$; $0.0012 \leq a_{2m} \leq 0.0049$. Taking this into account we have set $\delta_1 = 0.02$, $\delta_2 = 0.01$. Note that the choice of the specific values δ_1, δ_2 generally depends on the difference scheme employed, because the shock waves are smeared over a different number of mesh intervals in the numerical solutions obtained by different difference schemes.

After completion of the work on the classification system presented in Figure 7.27, we can use the data obtained on the presence or absence of shock

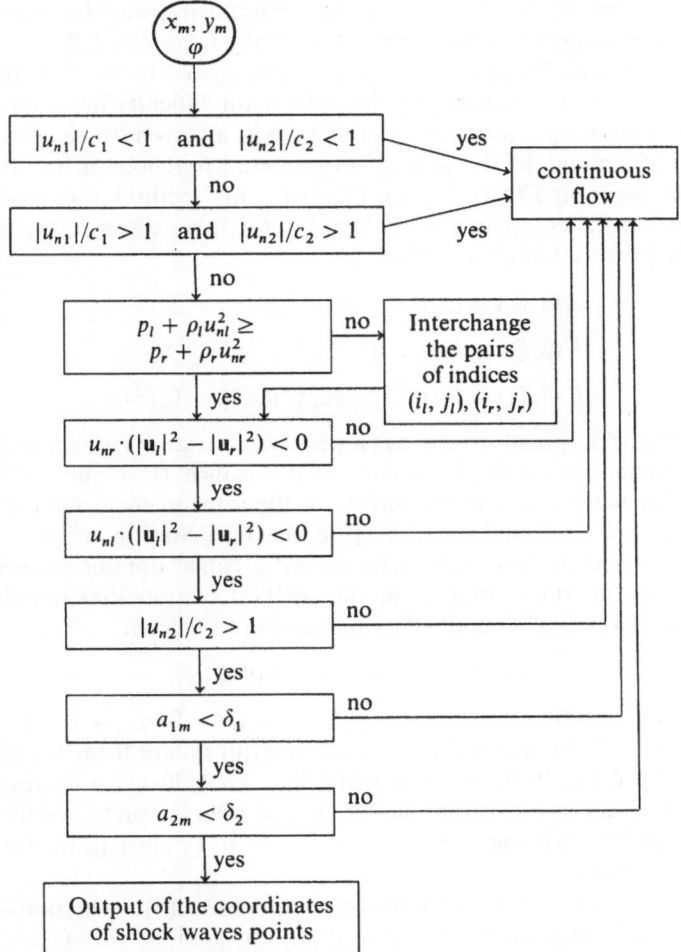

Figure 7.27. The scheme of sequential classification of singularities in the stationary transonic potential flows.

waves in the problem under study in an expert system for the automation of the design of aerodynamic bodies which meet some set of requirements, in particular, the requirement for the absence of shock waves in a transonic flow around an airfoil.

7.5.3. Classification by Feature Functions

Suppose that the feature functions $f_1(\boldsymbol{\beta})$ and $f_2(\boldsymbol{\beta})$ have been determined on the basis of the training information $\zeta_1, \zeta_2^{(1)}, \zeta_2^{(2)}$. Then the simple classification scheme described in [7.55] is realized in the following way: for a pattern $\boldsymbol{\beta}_m$ with an unknown classification the values of the functions $f_1(\boldsymbol{\beta}_m)$ and $f_2(\boldsymbol{\beta}_m)$ are computed, and the pattern in question is assigned to the class of feature function which takes the biggest value. The main difficulty here consists of a good choice of feature functions. Let us turn in a search for the expressions for feature functions to the method of potential functions in the theory of pattern recognition [7.55], [7.84], [7.166]. This method uses explicitly a hypothesis on the compactness of classes in the feature space. According to [7.55], the potential functions of the forms

$$K(\boldsymbol{\beta}, \zeta_k) = \exp\{-\alpha\|\boldsymbol{\beta} - \zeta_k\|^2\}; \tag{7.5.13}$$

$$K(\boldsymbol{\beta}, \zeta_k) = (1 + \alpha\|\boldsymbol{\beta} - \zeta_k\|^2)^{-1}; \tag{7.5.14}$$

$$K(\boldsymbol{\beta}, \zeta_k) = |(\sin \alpha\|\boldsymbol{\beta} - \zeta_k\|^2)/(\alpha\|\boldsymbol{\beta} - \zeta_k\|^2)|; \tag{7.5.15}$$

are the most widespread, where α is a positive constant. These functions are inversely proportional to the square of a distance $D^2 = \|\boldsymbol{\beta} - \zeta_k\|^2$ which serves, in particular, as a characteristic of the force in the potential gravity field. A remark was made in [7.55], in the presentation of the potential functions method, that the decision function $d(\boldsymbol{\beta})$ can be obtained directly from the potential functions. Indeed, in the method of potential functions the decision function $d(\boldsymbol{\beta})$ is obtained by iterations of the form

$$d_{k+1}(\boldsymbol{\beta}) = d_k(\boldsymbol{\beta}) + r_{k+1}K(\boldsymbol{\beta}, \zeta_{k+1}), \tag{7.5.16}$$

where k is the iteration number, $k = 0, 1, \ldots, d_0(\boldsymbol{\beta}) = 0, r_{k+1} \neq 0$. At $k = 0$ we obtain from (7.5.16) that the first iteration $d_1(\boldsymbol{\beta})$ differs from the potential function $K(\boldsymbol{\beta}, \zeta_1)$ only by a scalar multiplier. Thus, if one uses directly the functions $K(\boldsymbol{\beta}, \zeta_k)$, as the feature functions $f_k(\boldsymbol{\beta})$, then the presented method of classification into two classes Ω_4 and Ω_5 proves to be close to the method of potential functions.

We have chosen for the computations formula (7.5.14) as the simplest one. Let us set the feature functions $f_1(\boldsymbol{\beta})$ and $f_2(\boldsymbol{\beta})$ by analogy with (7.5.14) in the form

$$f_1(\boldsymbol{\beta}) = \alpha_1 \left/ \left[\gamma_1 + \sum_{j=1}^{4}(b_j - z_{1j})^2\right]\right.;$$

$$f_2(\boldsymbol{\beta}) = \alpha_2 \left/ \left[\gamma_2 + (b_1 - 0.5(z_{21}^{(1)} + z_{21}^{(2)}))^2 + \sum_{j=2}^{4}(b_j - z_{2j}^{(1)})^2\right]\right.; \tag{7.5.17}$$

where $\alpha_1, \gamma_1, \alpha_2, \gamma_2$ are positive constants. We determine these constants as one of the solutions of the following system of inequalities:

$$f_1(\zeta_1) - f_2(\zeta_1) > 0; \tag{7.5.18}$$

$$f_2(\zeta_2^{(1)}) - f_1(\zeta_2^{(1)}) > 0; \tag{7.5.19}$$

$$f_2(\zeta_2^{(2)}) - f_1(\zeta_2^{(2)}) > 0. \tag{7.5.20}$$

Note that $f_2(\zeta_2^{(1)}) = f_2(\zeta_2^{(2)})$ with regard to (7.5.3). Further, $f_1(\zeta_2^{(2)}) > f_1(\zeta_2^{(1)})$ by virtue of (7.5.4), therefore, it is sufficient to consider two inequalities (7.5.18) and (7.5.20). Let $\alpha_1/\alpha_2 = \lambda$. Rewrite (7.5.18), (7.5.20) in the form

$$\lambda > \gamma_1/(\gamma_2 + 7); \qquad 1/(\gamma_2 + 1) > \lambda/(\gamma_1 + 6). \tag{7.5.21}$$

Rewrite the inequalities (7.5.21) in the form

$$\gamma_1/(\gamma_2 + 7) < \lambda < (\gamma_1 + 6)/(\gamma_2 + 1). \tag{7.5.22}$$

To enhance the difference between the classes Ω_4 and Ω_5, in the process of their characterization with the aid of the feature functions (7.5.17), it is necessary for the quantity

$$G(\gamma_1, \gamma_2) = (\gamma_1 + 6)/(\gamma_2 + 1) - \gamma_1/(\gamma_2 + 7)$$

to be maximal. A direct computation shows that $\partial G/\partial \gamma_1 > 0$, $\partial G/\partial \gamma_2 < 0$ at $\gamma_1 > 0$, $\gamma_2 > 0$, that is, the function G has no extrema at positive γ_1, γ_2. To increase G one must take large values of $\gamma_1 > 0$ and small values of $\gamma_2 > 0$. Let us set with (7.5.22) in view

$$\lambda = (1/2)[\gamma_1/(\gamma_2 + 7) + (\gamma_1 + 6)/(\gamma_2 + 1)], \qquad \alpha_1 = \alpha_2\lambda, \tag{7.5.23}$$

where $\gamma_1, \gamma_2, \alpha_2$ are positive constants specified by the user of the classification program.

7.6. Examples of Pattern Classification in the Numerical Solutions of Two-Dimensional Gas Dynamics Problems

7.6.1. Tests of Pattern Classification Algorithms

Assume that N_4 points, with coordinates (x_m, y_m), are introduced into the singularities classification system along with the ρ, u, v, p images. Assume further that N_s points of this (x_m, y_m) set $(0 \le N_s \le N_4)$ should belong to the shock waves, and N_c points should belong to the contact discontinuities $(0 \le N_c \le N_4)$, so that $N_s + N_c \le N_4$. Now assume that N'_s points $(0 \le N'_s \le N_4)$ are assigned as the result of classification to the shock waves; it can then occur in the general case that $N'_s \ne N_s$. Similarly, let N'_c be the number of points assigned by the classification system to the contact discontinuities, $0 \le N'_c \le N_4$, $N'_s + N'_c \le N_4$. Then the levels of the classification errors of shock waves, δN_s, and of the contact discontinuities, δN_c, may be determined quantitatively as follows:

$$\delta N_s = (|N'_s - N_s|/N_s) \cdot 100\%;$$
$$\delta N_c = (|N'_c - N_c|/N_c) \cdot 100\%. \tag{7.6.1}$$

It is of interest to obtain analytic estimates of the quantities (7.6.1) while using a specific classification scheme. Having such estimates, we should be able to compare the reliability of different classification schemes. However, it was pointed out in [7.84] that an exact calculation of the level of classification errors is very difficult, even in the case when the probabilistic structure of a classification problem is known completely. In this connection it was proposed in [7.84] to apply an empirical approach which enables us to avoid the above difficulties, and which consists of experimental tests of a classifier. It was shown, in [7.84] in the examples, that the reliability of the estimation of a level of the classification error with the aid of formulas (7.6.1) depends substantially on the numbers N_s and N_c of the objects introduced into the classifier: the greater the values N_s and N_c, then the closer the values δN_s and δN_c are to their true values. In our case the quantities N_s and N_c depend substantially on the size of the steps of a spatial computing mesh. Indeed, the finer the mesh, then the more the number of the cells of this mesh will be traversed by a given discontinuity line. For example, in the case when a shock wave front is oriented along the x-axis, then we obviously have $N_s = O(1/h_1)$.

To carry out an estimation of the level of classification errors by formulas (7.6.1), it is necessary to have a control set of the objects (x_m, y_m) for which the true values of N_s and N_c are known. Here we can use two approaches. The first approach consists of obtaining the above set (x_m, y_m) as a result of the segmentation of a digital image $\rho(x, y, t)$ determined by the solution of a specific gas-dynamical problem with the aid of a specific shock-capturing difference scheme. Another approach consists of the generation of some artificial numerical solution of a hypothetical two-dimensional gas-dynamical problem. Note that such a method of testing is widespread in the theory of digital pattern recognition (see, for example, [7.63], [7.72], [7.79], [7.84], [7.167], [7.174]). The first of the above approaches to the testing of classifiers appears to be more natural, since the classification schemes presented in Section 7.5 have been designed only for the classification of singularities on the basis of shock-capturing numerical solutions. However, it should be noted that each specific difference scheme has: (i) its own intensity of discontinuities smearing (or of edge blur); (ii) its own level of parasitic oscillations; in addition, it depends (for one and the same scheme) on the value of the Courant number K, on the magnitude of the solution gradients in the domains adhering to the shock fronts, on the relation h_2/h_1, and on some other factors [7.64], [7.71]. In this connection it is of interest to perform the tests of classifiers within the framework of the second approach. This subsection is devoted to a brief presentation of the results of such testing which was carried out in [7.49]. Subsequent subsections of Section 7.6 may be regarded as a presentation of the results of testing the classifiers of Section 7.5 within the framework of the first approach, that is, on the basis of actual finite-difference shock-capturing solutions of a number of applied problems of gas dynamics.

When constructing the test in [7.49], we aimed at the satisfaction of the following requirements:

(1) both shock waves and contact discontinuities should be present in the "gas flow";

(2) the noise and blur with controlled characteristics should be present in the numerical solution;
(3) the shock wave intensities should be variable, changing from some finite magnitude to small values corresponding to a continuous flow;
(4) the angles of orientation of the discontinuity lines with respect to the x-axis should vary within a wide range, that is, the discontinuities fronts should be curvilinear.

Taking into account these requirements we have placed, in the flow field, a shock wave with a cylindrical front and the contact discontinuity line in the form of an ellipse. Let (x^0, y^0) be the coordinates of some point in the (x, y)-plane. Note that this point does not necessarily lie in the domain of the analyzed digital image. Let us introduce the polar coordinates r, θ by the formulas $x - x^0 = r \cos \theta$, $y - y^0 = r \sin \theta$. Let U be the gas velocity in a one-dimensional shock wave. We then assume in the domain behind the shock wave front that $u = U_1 \cos \theta$, $v = U_1 \sin \theta$. Let us determine the intensity of a shock wave by the relationship $\eta = p_1/p_2$. At $\eta \to 1$ the shock wave goes over into a continuous flow. Let the quantities U_2, p_2, ρ_2, $u_2 = U_2 \cos \theta$, $v_2 = U_2 \sin \theta$ before the shock wave front be given. Then we set $\rho_1, u_1, p_1, \varepsilon_1$ by the formulas known for the ideal gas case [7.112]

$$\eta = \eta_{max} + (1 - \eta_{max}) \cdot (\theta/\theta_0)^2, \qquad |\theta| \le \theta_0;$$

$$p_1 = \eta p_2, \qquad \rho_1 = \rho_2(p_1 + \mu p_2)/(p_2 + \mu p_1), \qquad \mu = (\gamma - 1)/(\gamma + 1); \tag{7.6.2}$$

$$D = U_2 + [(1/\rho_2)(((\gamma + 1)/2)p_1 + ((\gamma - 1)/2)p_2)]^{1/2};$$

$$\varepsilon_1 = p_1/(\rho_1(\gamma - 1)), \qquad U_1 = U_2 + (2/(\gamma + 1))(|U_2 - D| - \gamma p_2/(\rho_2|U_2 - D|)).$$

The quantities η_{max}, θ_0 in (7.6.2) are assumed to be given. Now let (x^1, y^1) be some point lying in the domain before the shock wave front. Consider two ellipses e_1 and e_2 having the common center (x^1, y^1), and all the points of the ellipse e_1 lie inside the ellipse e_2. Along the beams $\tilde{\theta} = $ const emanating from (x^1, y^1) the values ρ, u, v, p were given inside the e_2 ellipse as linear functions of the polar coordinate $\tilde{r}$, where $x - x^1 = \tilde{r} \cos \tilde{\theta}$, $y - y^1 = \tilde{r} \sin \tilde{\theta}$, from the requirement of a continuous transition to the state $\{U_2, p_2, \rho_2\}$. Then the density ρ inside a "small" ellipse e_1 was recalculated by a linear formula $\rho = a_1(\tilde{\theta}) + a_2(\tilde{\theta})\tilde{r}$ in such a way that there is a discontinuity in the profile of ρ within the "big" ellipse e_2 along each beam $\tilde{\theta} = $ const. In this way we have obtained the contact discontinuity line in the form of an ellipse.

Consider the question of introducing the blur into the original image. For convenience, denote by $\tilde{S}$ the blur operator, and by $I(\rho)$, $I(u)$, $I(v)$, $I(p)$ the digital images of the flow field created by the numerical arrays of the values $\rho_{ij}, u_{ij}, v_{ij}, p_{ij}$. The resulting image obtained after the action of the operator $\tilde{S}$ (for example, on $I(\rho)$), will be denoted by $\tilde{S}I(\rho)$. The blur was introduced by convolving the smoothing mask H of the form (7.2.7) with the image window (7.2.6), where f is any of the functions ρ, u, v, p. As a result of this, four digital images $\widetilde{S}I(\rho)$, $\widetilde{S}I(u)$, $\widetilde{S}I(v)$, $\widetilde{S}I(p)$ were obtained. As was shown in the foregoing, the discontinuous profiles are smeared in these images over three cells of a spatial mesh.

Since we want to introduce into our artificial image a noise with a given signal-to-noise ratio (we shall denote it by S/N in the following), we must elucidate at first what is usually meant by the ratio S/N. In [7.63] the S/N ratio was determined as

$$S/N = d^2/\sigma_n^2, \tag{7.6.3}$$

where σ_n is a mean quadratic deviation of an independent Gaussian noise, and d is the height of a step edge determined by formula (7.2.11). The value S/N was taken in [7.100] to be $S/N = 10 \log(d^2/\sigma_n^2)$. In our case the profiles of the quantities ρ, u, v, p differ substantially from the step edge profile, therefore, there arise certain difficulties in determining the "height" of the edges of these quantities. In practical computations the height $d(f)$, where f is any of the functions ρ, u, v, p, was computed by the formula

$$d(f) = \left(\frac{1}{s_1} \int \int_{\mathcal{D}_1} [f(x, y) - f_2]^2 \, dx \, dy\right)^{0.5}, \tag{7.6.4}$$

where $\mathcal{D}_1$ is a domain in the (x, y)-plane in which $f(x, y) \neq f_2$, f_2 is the value of the function $f(x, y)$ in an undisturbed medium before the shock wave front, and s_1 is the area of the domain $\mathcal{D}_1$. It is easy to see that in a particular case of a step edge (7.2.15) formula (7.6.4) yields the exact value of edge height α. In the numerical computations the integral in (7.6.4) was approximated by the formula of rectangles. Let

$$v(f) = (d(f))^2/(\sigma_n(f))^2. \tag{7.6.5}$$

Suppose that the quantity v in (7.6.5) is given, for example, $S/N \equiv v = 50$. Then we have from (7.6.5)

$$\sigma_n(f) = d(f)/\sqrt{v}. \tag{7.6.6}$$

To introduce the noise into the image we used a standard generator GSN1R of an array of pseudorandom numbers normally distributed with a zero mean value and with a unit dispersion. This generator, realizing the algorithm of [7.178], was used as follows. Let ξ_k, ξ_{k+1}, ξ_{k+2}, ξ_{k+3} be the four sequential values of a random quantity obtained with the aid of the subroutine GSN1R. Then we set

$$\bar{\rho}_{ij} = \rho_{ij} + \xi_k \cdot \sigma_n(\rho), \qquad \bar{u}_{ij} = u_{ij} + \xi_{k+1} \cdot \sigma_n(u),$$

$$\bar{v}_{ij} = v_{ij} + \xi_{k+2} \cdot \sigma_n(v), \qquad \bar{p}_{ij} = p_{ij} + \xi_{k+3} \cdot \sigma_n(p). \tag{7.6.7}$$

Denote an operator of the additive noise introduction in accordance with (7.6.7) by N, and denote the result of its action on the function f by Nf. In the computations the image Nf was usually subject to a smoothing with the aid of the above operator $\widetilde{S}$ before its input into the recognition system. In some

Figure 7.28. Isometrics
of the surfaces; (a) $I(\rho)$;
(b) $I(p)$; (c) $I(\sqrt{u^2 + v^2})$

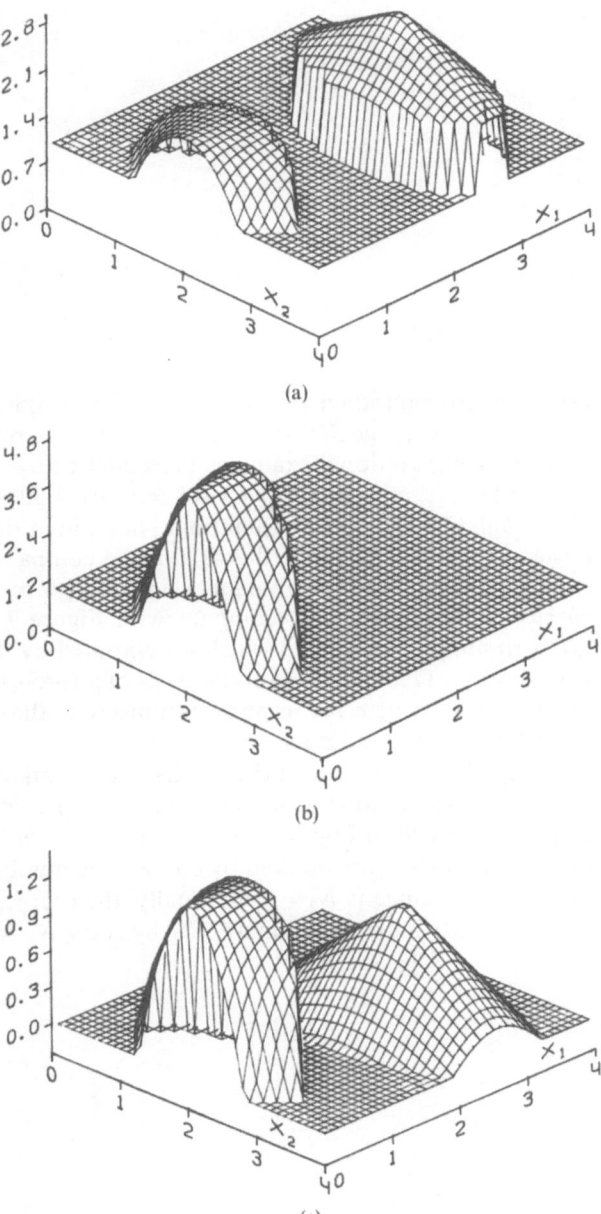

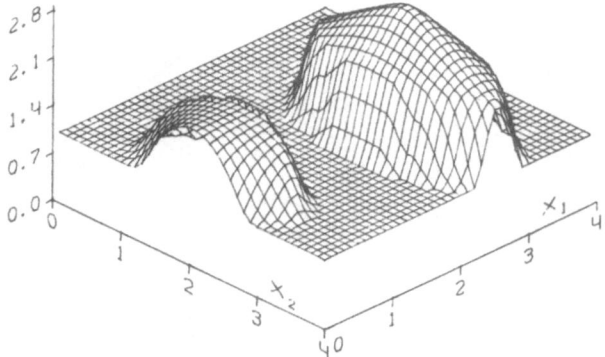

Figure 7.29. Isometrics
of the surface $\tilde{S}I(\rho)$.

cases we also smoothed with the aid of $\tilde{S}$ the original image $I(f)$, so that in
these cases the image $\tilde{S}N\tilde{S}I(f)$ was input into the recognition program.

In the computational examples presented below $\gamma = 2$ in the equation of
state (1.8), $\eta_{\max} = 5$, $p_2 = \rho_2 = 1$, $U_2 = 0$, $\theta_0 = 50°$ in (7.6.2). $h_1 = h_2 = 0.1$,
$K = 0.5$ in (7.4.38). The computational domain in the (x, y)-plane is a square
covered by a 40×40 mesh, that is, this is a comparatively rough mesh.

In Figure 7.28 and in subsequent figures $x_1 \equiv x$, $x_2 \equiv y$. In Figure 7.29 a
surface $\tilde{S}I(\rho)$ is presented (compare with Figure 7.28(a)). In Figure 7.30 a
gradient image is presented which is computed by the formulas of the Sobel
edge detector (7.3.6)–(7.3.8) on the basis of a smoothed image $\tilde{S}I(\rho)$. We can
see that the edge detector response is nonzero in the domains of a shock wave
and a contact discontinuity.

In Figure 7.31 we present the results of the work of the first three stages of
the image segmentation procedure. The (x_m, y_m) points obtained after the first
stage are marked in Figure 7.31 by dots; their number $N_1 = 246$; the results
of the second stage are marked by crosses; the number of the points obtained
at the second stage is $N_2 = 102$. Finally, the (x_m, y_m) points obtained at the
third stage are marked in Figure 7.31 by stars; $N_3 = 97$. This sequence of 97

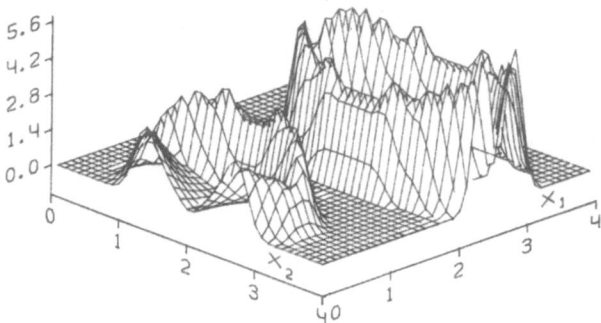

Figure 7.30. Sobel edge
detector response.

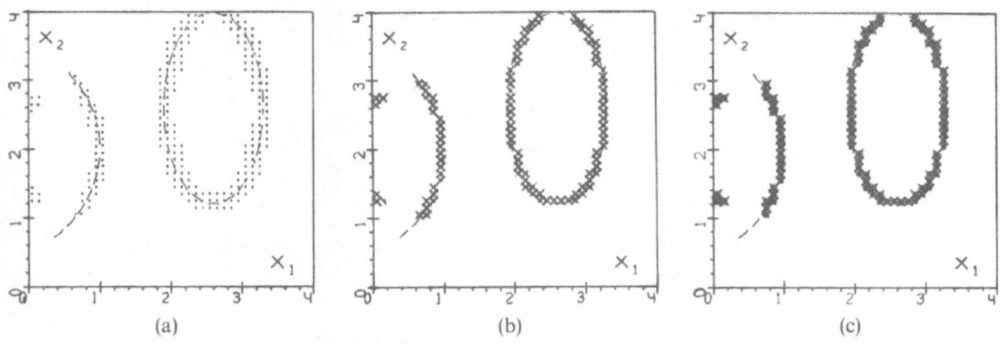

Figure 7.31. The sets (a) $\mathscr{S}_1$, (b) $\mathscr{S}_2$, (c) $\mathscr{S}_3$, obtained at $N_g = 2$ in (7.3.17).

objects (x_m, y_m) was an input information for the system of pattern classification. In Figure 7.32 we present the results of the work of the hierarchical classification system described above (see Figure 7.25). We have set in (7.4.25) $\delta_1 = 0.01$. The stars and rectangles mark (in Figure 7.32) the points of a shock wave and a contact discontinuity, respectively. The feature a_{1m} was computed here by formulas (7.4.5)–(7.4.7). Comparing Figures 7.31 and 7.32, we see that some of the points (x_m, y_m), namely, six points adhering to the x_2-axis, were in the classification assigned to the class ω_7 of the subdomains of a continuous flow and, thus, from the 97 points, 91 points were assigned to the discontinuity surfaces, of which 20 points were assigned to a shock wave and 71 points were assigned to a contact discontinuity. Note that the assignment of the above six points to the class ω_7 was performed after it was found that at these points $(x_m, y_m) \in \Omega_5$ the relationship $b_{2m} = 0$ was not satisfied. Thus, the above constructed feature (7.4.20) enables us to carry out, with sufficient efficiency, the classification of the objects belonging to the class Ω_5.

In Figure 7.33 are shown (by different symbols) the projections of the points $(b_{1m}, \ldots, b_{4m})$ on the b_1- and b_2-axes (Figure 7.33(a)) and on the b_3- and b_4-axes (Figure 7.33(b)). We can see from Figure 7.33 that the patterns of the points

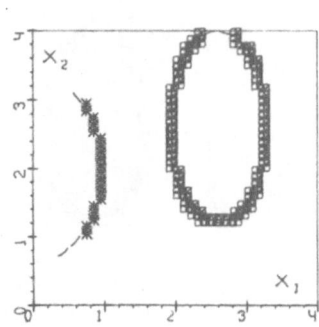

Figure 7.32. Results of classification by the minimum-distance classifier.

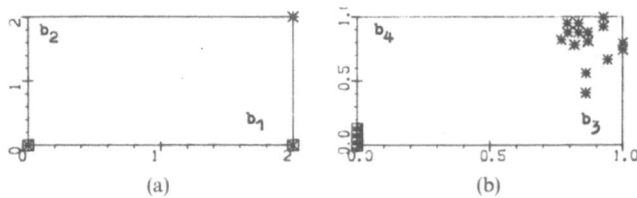

Figure 7.33. (∗∗∗)—The patterns of the points belonging to the shock wave; (⊠⊠⊠)—the patterns of points belonging to the contact discontinuity.

(x_m, y_m) in the four-dimensional feature space have a trend to grouping or clusterization. There are in the (b_1, b_2)-plane three clusters of points $(b_{1m}, \ldots, b_{4m})$: one cluster has as its central point the shock wave sample with the coordinates (7.5.4), and the two other clusters have been formed around two sample patterns of the contact discontinuity points (see formulas (7.5.3)). In the (b_3, b_4)-plane we can indicate the presence of two clusters.

In Figure 7.34 the graphs of the potential functions (7.5.17) are presented as depending on the number m of the point (x_m, y_m), $m = 1, \ldots, N_4$, $\alpha_2 = 1$, $\gamma_1 = 3$, $\gamma_2 = 1$, in (7.5.23). The results of classification of the points (x_m, y_m), shown in Figure 7.31(c), proved to be identical to those presented in Figure 7.32. In Figure 7.35 similar to Figure 7.32, we show the results of the work on

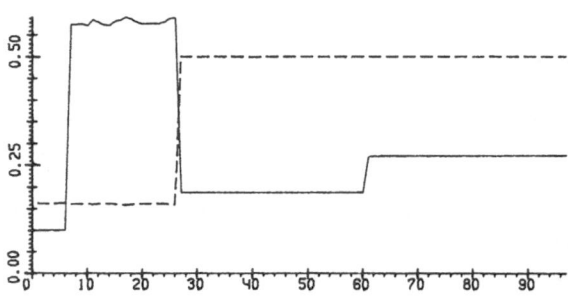

Figure 7.34. (———)—The function $f_1(\beta_m)$; (– – – –)—the function $f_2(\beta_m)$.

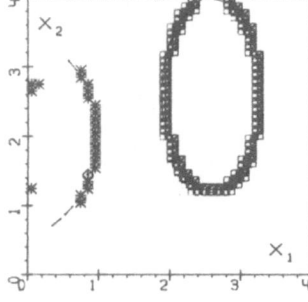

Figure 7.35. (∗∗∗)—The shock wave points; (⊠⊠⊠)—the tangential discontinuity points; (× × ×)—the points of purely contact discontinuities; (◊ ◊ ◊)—the points of compression waves.

the sequential classification technique presented in the foregoing section; the decision tree of this technique has been presented in Figure 7.26; $\delta_2 = 0.001$ in (7.4.26). Comparing Figures 7.35 and 7.32, we can see a difference in the results in the domain behind the shock wave front. Using the classification results presented in Figures 7.32 and 7.35, we can obtain the following rough estimates for the quantities (7.6.1): $\delta N_s = \delta N_c = 0$ for scheme I; $\delta N_s = 20\%$, $\delta N_c = 0$ for scheme II of Section 7.5.

We have considered above the case when the original image is subject to the action of the blur only. In gas-dynamical numerical modeling such images with smoothed discontinuity surfaces are obtained while using monotone difference schemes. As was already pointed out in Chapter 2, nonmonotone difference schemes are also presently used in the practice of gas-dynamical computations. These schemes generate parasitic oscillations, the amplitude of which is usually larger in the neighborhood of strong discontinuities. In this connection we have carried out a simulation of such a situation with the aid of introducing parasitic oscillations into the original digital image by means of superposing an additive noise in accordance with formulas (7.6.7). In Figure 7.36(a) we present the surface $N\tilde{S}I(\rho)$ at $S/N = 100$. The values of the heights

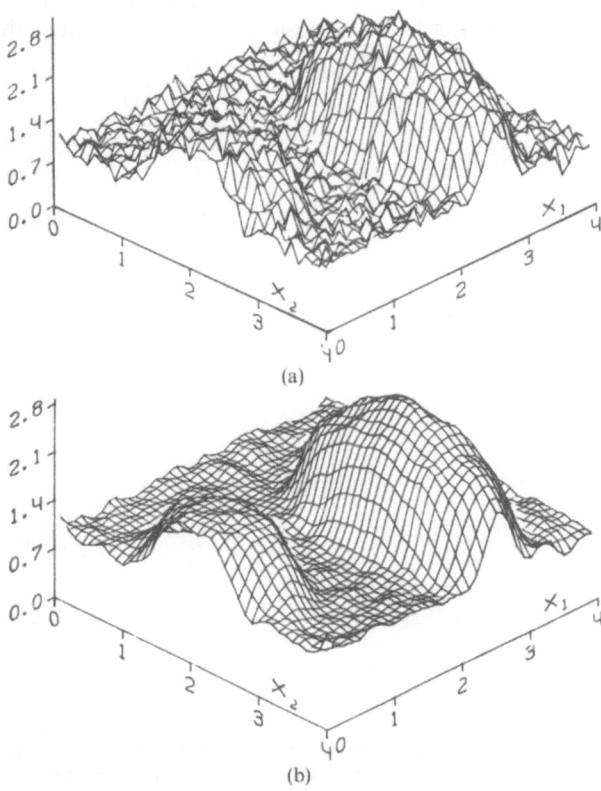

Figure 7.36. (a) The surface $N\tilde{S}I(\rho)$; (b) $\tilde{S}N\tilde{S}I(\rho)$.

of the profiles $d(f)$ calculated with the aid of (7.6.4) are $d(p) = 2.529$, $d(\rho) = 1.015$, $d(u) = 0.507$, $d(v) = 0.273$. Before the segmentations, the images $N\tilde{S}I(\rho)$, $N\tilde{S}I(p)$, $N\tilde{S}I(u)$, $N\tilde{S}I(v)$ were smoothed with the aid of the above-defined operator $\tilde{S}$ to reduce the number of artefacts. In Figure 7.36(b) we show the surface $\tilde{S}N\tilde{S}I(\rho)$ obtained from the surface of Figure 7.36(a) by a smoothing with the aid of $\tilde{S}$. Besides the classification results for $S/N = 100$ we shall present below the data for $10 \le S/N \le 50$. The picture of the clusters disposition in the (b_1, b_2)-plane does not change and remains the same as in Figure 7.33(a). In this connection we present in Figure 7.37 the clusters in the (b_3, b_4)-plane only. The notation of the type $\tilde{S}^2 NI(f)$ in Figures 7.37 and 7.38 means that the classification was carried out, in this case, on the basis of an ensemble of the digital images $\tilde{S}^2 NI(\rho)$, $\tilde{S}^2 NI(p)$, $\tilde{S}^2 NI(u)$, $\tilde{S}^2 NI(v)$, obtained as a result of superposing a noise, and of a subsequent double application of the smoothing mask (7.2.7) which is equivalent (as was shown above) to a single application of the smoothing mask (7.2.22). Comparing the clusters presented in Figure 7.33 and 7.37, we can see that with the diminution of the S/N ratio the clusters become more rarefied, "friable". At $S/N = 100$ (Figure 7.37(a)) there arises, in the (b_3, b_4)-plane along with the two basic clusters located near the points with coordinates $(0, 0)$ and $(1, 1)$, one more cluster corresponding to the (⊠⊠⊠) points detected near the axis $x = 0$ (see Figure

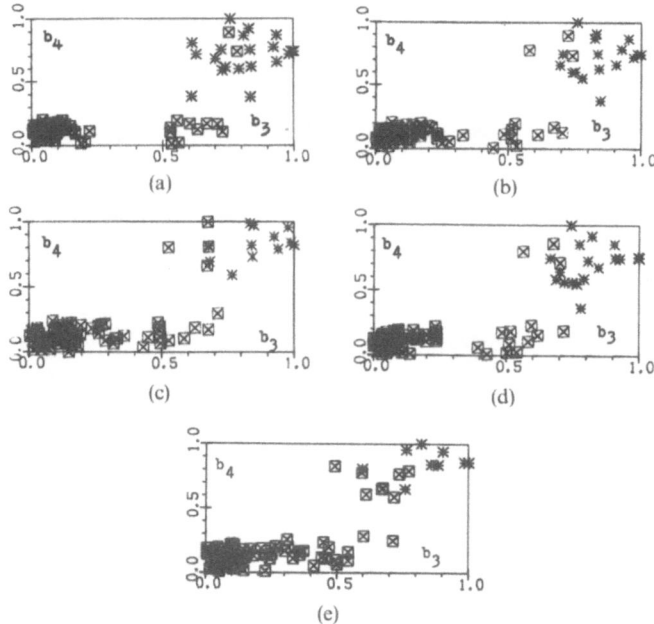

(a)

(b)

(c)

(d)

(e)

Figure 7.37. Clusters in the (b_3, b_4)-plane: (a) $S/N = 100$, $\tilde{S}N\tilde{S}I(f)$; (b) $S/N = 50$, $\tilde{S}N\tilde{S}I(f)$; (c) $S/N = 25$, $\tilde{S}N\tilde{S}I(f)$; (d) $S/N = 25$, $\tilde{S}^2 NI(f)$; (e) $S/N = 10$, $\tilde{S}^2 NI(f)$.

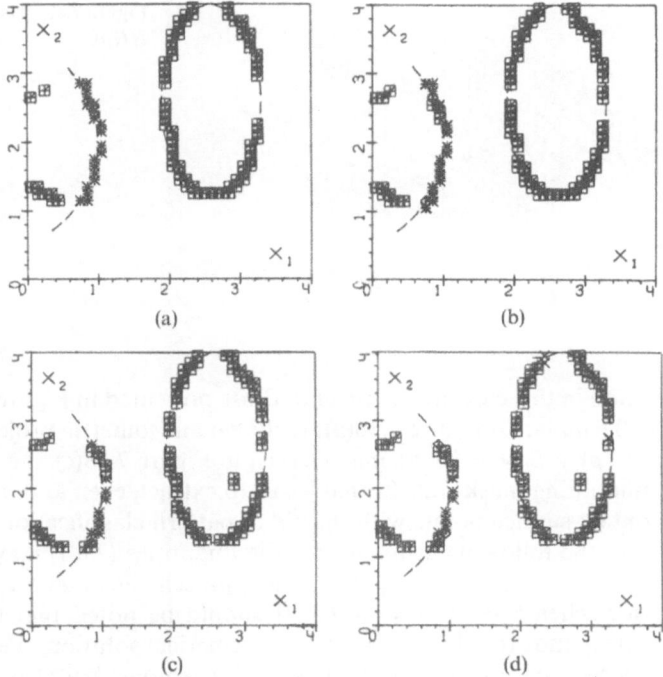

Figure 7.38. Classification results for different S/N ratios: (a) $S/N = 100$, $\tilde{S}N\tilde{S}I(f)$; (b) $S/N = 25$, $\tilde{S}^2NI(f)$; (c) $S/N = 10$, $\tilde{S}^2NI(f)$ (minimum-distance classifier); (d) $S/N = 10$, $\tilde{S}^2NI(f)$ (sequential classification).

7.38(a)). In the process of further reduction of the S/N ratio these two clusters of (⊠⊠⊠) points merge into one cluster; in addition, the boundary between the clusters of points (∗∗∗) and (⊠⊠⊠) becomes fuzzy, blurred.

In Figure 7.39 (similar to Figure 7.34) we present the graphs of the functions (7.5.17) as depending on the number m of the point (x_m, y_m), $m = 1, \ldots, N_4$, at $S/N = 10$, the classification is made on the basis of $\tilde{S}^2NI(f)$. The results of

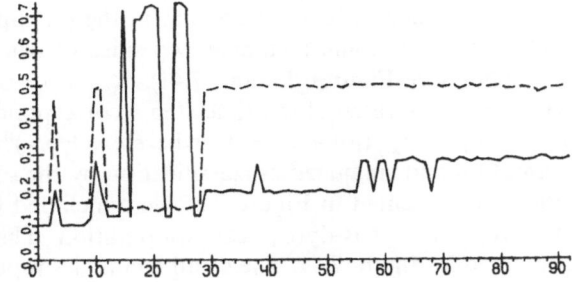

Figure 7.39. The graphs of the functions $f_1(\beta_m)$ (solid line) and $f_2(\beta_m)$ (dashed line) at $S/N = 10$, $\tilde{S}^2NI(f)$.

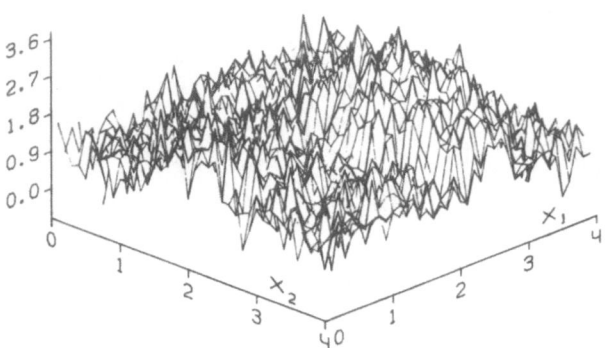

Figure 7.40. The surface $NI(\rho)$.

the classification in this case are identical to those presented in Figure 7.38(c). In Figure 7.40 (similar to Figure 7.36(a)) is shown an isometric projection of the surface $NI(\rho)$ at $S/N = 10$. As may be seen in Figure 7.38(c), the application of the smoothing mask (7.2.22) enables us to extract, even at this level of noise, the contact surface points with the aid of pattern classification.

At $S/N = 10$ the following estimates for the quantities (7.6.1) may be obtained from Figure 7.38: $\delta N_s = 50\%$, $\delta N_c = 20\%$ for scheme I and $\delta N_s = 45\%$, $\delta N_c = 20\%$ for scheme II of Section 7.5. It should be noted that the case $S/N = 10$ corresponds to a highly oscillatory numerical solution; the oscillations were superposed not only in the post-shock regions, but also in other spatial subdomains including both the zones of smearing of the discontinuities and the domains of undisturbed flow (see Figure 7.40).

Let us now present an example of the computation of a two-dimensional model flow by the MacCormack scheme (4.2.11) (4.2.12), approximating the system (1.30), (1.8) with the second order of accuracy. As initial data we have taken the following piecewise constant distribution:

$$u(x, y, 0) = |\mathbf{u}_0|\cos \beta, \qquad v(x, y, 0) = |\mathbf{u}_0|\sin \beta,$$

$$|\mathbf{u}_0| = 1, \qquad \beta = \pi/3, \qquad p(x, y, 0) = 1,$$

$$\rho(x, y, 0) = \rho_2, \qquad (x, y) \in \mathscr{D} - \mathscr{D}_1, \tag{7.6.8}$$

$$\rho(x, y, 0) = \rho_1, \qquad (x, y) \in \mathscr{D}_1, \qquad \rho_1 = 2, \quad \rho_2 = 0.25.$$

Here $\mathscr{D}$ is a computational domain in the (x, y)-plane representing a 3×4 rectangle, $\mathscr{D}_1$ is some rectangle two sides of which are parallel to the line $y = x \text{ tg } \beta$ (see Figures 7.8 and 7.41), $h_1 = h_2 = 0.1$, $\gamma = 2$ in (1.8), and $\tau = 0.013856$. This value of the time step τ corresponds to the Courant number $K = (\tau/h_1)\max_{x, y}(|\mathbf{u}| + c) = 0.5$ where $c = (\gamma p/\rho)^{0.5}$. The subroutines of segmentation and singularities classification by the scheme of hierarchical classification presented in Figure 7.25 were included without any alterations in the program of gas-dynamical computation. The classification results are presented in Figure 7.41. The small rectangles represent (in Figure 7.41 as in

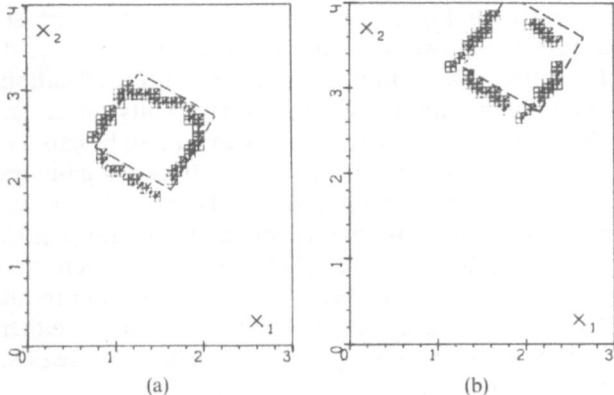

Figure 7.41. Contact discontinuity contours: (a) $t = 76\tau$; (b) $t = 151\tau$.

(a) (b)

Figures 7.35 and 7.38) the (x_m, y_m) points, classified by the pattern recognition program as the points of the contact discontinuities. Dashed lines in Figure 7.41 indicate the positions of a contour of a purely contact discontinuity in accordance with the exact solution. We can see from Figure 7.41 that the recognition program classifies correctly the points of the edges in the digital image brightness that were detected by the segmentation program. Note that we have not applied here a preliminary smoothing of the digital images $I(p)$, $I(\rho)$, $I(u)$, $I(v)$ with the aid of the mask (7.2.7).

The main conclusion which may be drawn from the above results is that, in the case of using oscillatory difference schemes for which the quantity (7.6.3) is a finite number of the order $10 \le S/N \le 100$, both schemes I and II from Section 7.5 yield comparable levels of classification errors. The second conclusion is that the most favorable situation for the classification of singularities with the aid of schemes I and II is a situation when a monotone difference scheme is used in computations of a gas-dynamical problem, and in the numerical solutions obtained with this scheme all the discontinuities are smeared over not more than three to four cells of a spatial grid.

Let us now turn to the computational example, the results of which are presented in Figure 7.40. This example serves as an illustration of a general problem on the interpretability of gas- and hydrodynamic computations, and of a problem on determining the applicability of a certain numerical method NM for the investigation of a specific problem P. We shall say that the method NM is applicable to the numerical investigation of the problem P, if by means of an available system of pattern recognition PRS we can draw, from the numerical solution obtained by the method NM, information on those features of the phenomenon or of the process under study which are of interest to the research worker.

For example, let us assume that in the process of a numerical study of a gas-dynamical problem we obtain the result of computation like the one presented in Figure 7.40. If we are mainly interested in the shock wave behavior, then we can say that the algorithms of the hierarchical (Figure 7.38(b)) and sequential (Figure 7.38(d)) classifications do not permit us to obtain information of a satisfactory quality on the shock wave under study. Here the researcher can go two ways. First, he can try to improve the quality of the output information in PRS by improving the system. If these efforts are in vain, the researcher can go another way: he can take for the solution of a problem P another numerical method NM. Suppose that this other numerical method enables the researcher to obtain a difference solution as depicted in Figure 7.36(a). The corresponding classification result is presented in Figure 7.38(a). If this result does not suit the researcher either, he should then resort to the third method NM. As a result of this the researcher can obtain an "ideal" picture like the one shown in Figure 7.32.

Let us now assume that the researcher is interested in the process of a study of the same gas-dynamical problem (see Figure 7.40), but only in the behavior of a contact boundary. Then he will probably be satisfied with the information on the contact boundary which is obtained by the above classification algorithms (see Figure 7.38(c), (d)).

7.6.2. Recognition of Shock Waves in Transonic Flow Around an Airfoil

In Figure 7.42(a) we present the results of the application of the scheme of sequential classification of Figure 7.27 to the set of the points (x_m, y_m) obtained as the result of the image segmentation shown in Figure 7.20(c). Instead of checking the inequality (7.5.10), we have used at a stage of debugging of the classification program a simpler criterion, taking into account the flow direction over the airfoil, the chord of which is parallel to the x-axis. Since this flow

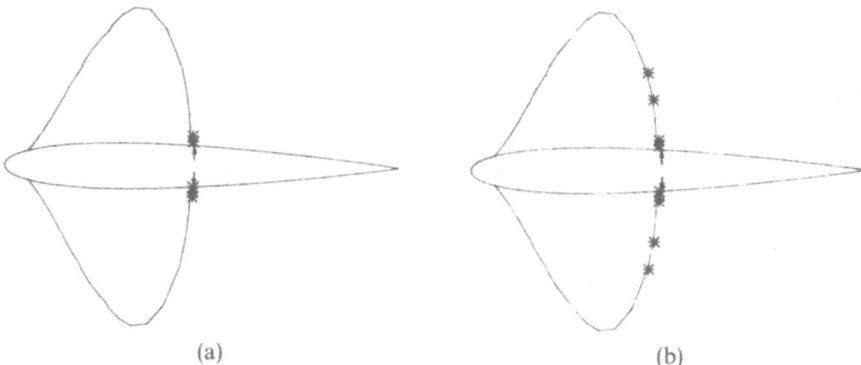

(a)   (b)

Figure 7.42. The application of the sequential classification scheme of Figure 7.27 in the problem on potential flow around the NACA 0012 airfoil. (———)—sonic lines and the airfoil contour; (∗∗∗)—the points of shock waves.

moves in the direction of the x-axis, the inequality $x^{(2)} < x^{(1)}$ should be satisfied for the points lying before and behind the shock wave front, respectively. Taking this into account we assumed that $x^{(2)} = x_{i_l, j_l}$, if the inequality $x_{i_l, j_l} < x_{i_r, j_r}$ was satisfied. Otherwise, the pairs of indices (i_l, j_l) and (i_r, j_n) were interchanged. After that we checked the analogues (7.5.12) of the entropy inequality. The result of the work of such a simplified classification program (it does not need a search interpolation procedure as described in Section 7.5.2) is presented in Figure 7.42(b). Comparing Figure 7.42(b) and (a), we can see that the weakening of the classification criteria leads to an increase in the number of points included by the classification program in the class of the shock waves points. We can get an increase in the number of shock waves points, while using the criterion (7.5.10), by performing a more accurate computation of the potential flow. For example, to obtain a stationary solution of the AF2-type scheme (7.3.69)–(7.3.71) we made, by analogy with [7.151], a number of iterations providing a three-order-of-magnitude drop in the residual, as compared to the residual value for the initial iteration. It appears that the error in the quantities p, ρ, u_n, describing the states on both sides of a weak shock wave in a potential flow, may be slightly reduced by performing a larger number of iterations.

In [7.142] the coordinate x_s of the shock position on the airfoil surface, measured from the leading edge, was defined as the middle point of the jump in the pressure coefficient c_p curve. The arrows in Figure 7.42 indicate the shock positions on the airfoil surface, and were found by this method for the same problem in [7.142] on the basis of the computational results of [7.157] (see Figure 7.17(b)).

If only the fact of the shock waves presence near the airfoil surface is of practical interest, and there is no need for accurate information on the extent of the shock front lines, then it is sufficient (for example, for the expert systems of automated design) to have information on the shock waves of the form presented in Figure 7.42(a).

There may be present, in a transonic flow around an airfoil, two types of shock waves. The first type is characterized by supersonic gas speeds on both sides of a shock (supersonic-to-supersonic shock wave). In the case of a shock of the second type the flow is supersonic on one side of the shock front and subsonic on the side of the shock front (supersonic-to-subsonic shock wave). For example, in a transonic flow around the NACA 0012 airfoil, at the freestream Mach number of 0.95 and at zero degrees angle of attack, there are shock waves of both types [7.145]. In the case of supersonic-to-subsonic shock waves there arises the question as to whether it is possible to use, in the gaps between the points (x_m, y_m) detected as the result of classification, the points of sonic lines to fill the above gaps. A study carried out in [7.179], with the aid of matched asymptotic expansions for the case of viscous transonic flows, shows that the sonic line position does not depend in the first approximation on small dissipative effects but corresponds to the shock front. In Figure 7.43

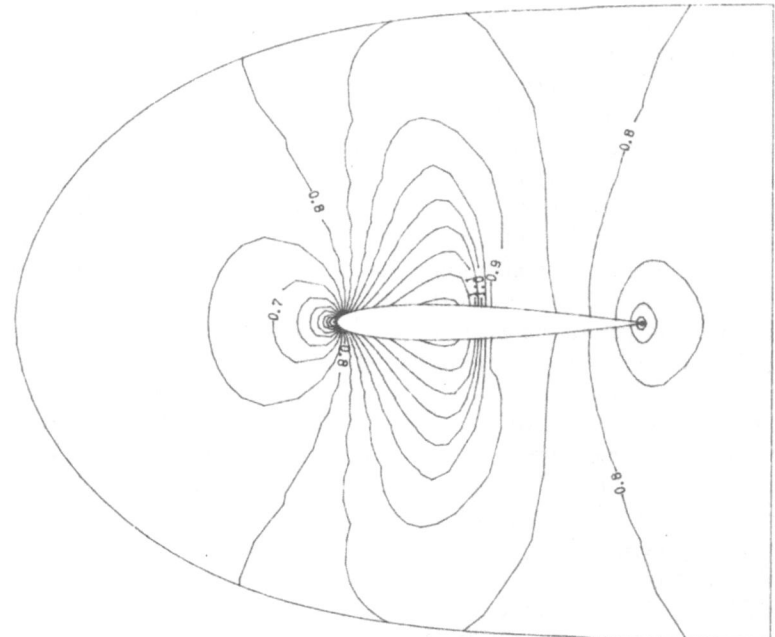

Figure 7.43. The lines of constant Mach number.

we present the lines of constant Mach numbers for the same flow problem as in Figure 7.42. The qualitative behavior of the lines $M =$ const coincides with that known in the literature (see, for example, [7.180]).

7.6.3. Cylindrical Shock Problem

Let us now present three more examples of gas-dynamical flows for the automatic analysis of which we have used the pattern recognition algorithms described in the present chapter. A common feature of these three examples is that the solution gradients in the neighborhood of the discontinuities are nonzero; in addition, there are (in the second and third examples presented below) the lines of the strong discontinuities along which the jumps of gas-dynamical parameters vary within rather wide limits, so that there may be present in the same problem, for example, both weak and intensive shock waves.

We have already considered the cylindrical shock problem in Section 3.1.2, while presenting a method for investigating the properties of the curvilinear shock waves smearing. In order to estimate quantitatively the accuracy of the localization of a shock wave and a contact discontinuity, by the image segmentation procedure presented in Section 7.3, it was necessary to find exact values of the abscissas of the points of strong discontinuities for certain chosen moments of time. For this purpose we have considered the problem within the framework of a mathematical model of a one-dimensional axisymmetric flow, the equations of which may be written in the form [7.64], [7.181]

$$w_t + [F(w)]_r = \Psi(w, r), \tag{7.6.9}$$

where

$$w = \begin{pmatrix} R \\ M \\ E \end{pmatrix}, \qquad F = \begin{pmatrix} M \\ (\gamma - 1)E + \dfrac{3 - \gamma}{2}\left(\dfrac{M^2}{R}\right) \\ \dfrac{M}{R}\left[\gamma E + \dfrac{1 - \gamma}{2}\left(\dfrac{M^2}{R}\right)\right] \end{pmatrix},$$

$$\tag{7.6.10}$$

$$\Psi = \dfrac{(\gamma - 1)}{r}\begin{pmatrix} 0 \\ E - \dfrac{M^2}{2R} \\ 0 \end{pmatrix} = \begin{pmatrix} 0 \\ P \\ 0 \end{pmatrix},$$

$$R = r\rho, \qquad M = r\rho u, \qquad E = r\rho(\varepsilon + u^2/2) \equiv re,$$

$$P = [(\gamma - 1)/r](E - M^2/(2R)) \tag{7.6.11}$$

$$= (\gamma - 1)(e - m^2/(2\rho)), \qquad m = \rho u.$$

In (7.6.9)–(7.6.11) r is a spatial coordinate, the symmetry axis corresponds to the value $r = 0$, and γ is a constant in the equation of state (1.8). We shall determine the "exact" solution of the system (7.6.9)–(7.6.11) (by analogy with [7.181]) with the aid of a numerical integration of this system with very small steps h and τ along the r- and t-axes, respectively. At the points of a spatial grid for which $r \geq h$ let us approximate the system (7.6.9) by the MacCormack scheme (cf. (4.3.44), (4.3.45))

$$\tilde{w}_i^{n+1} = w_i^n - (\tau/h)(F_{i+1}^n - F_i^n) + \tau\Psi_i^n;$$

$$w_i^{n+1} = (1/2)[(w_i^n + \tilde{w}_i^{n+1}) - (\tau/h)(\tilde{F}_i^{n+1} - \tilde{F}_{i-1}^{n+1}) + \tau\tilde{\Psi}_i^{n+1}]. \tag{7.6.12}$$

Let us also use by analogy with (7.2.4) the smoothing

$$\bar{w}_i^{n+1} = (1/(\varphi + 2))(\varphi w_i^{n+1} + w_{i+1}^{n+1} + w_{i-1}^{n+1}), \tag{7.6.13}$$

where $\bar{w}_i^{n+1}$ is a vector of the smoothed solution and φ is an empirically chosen parameter. In all the examples presented below of computations by scheme (7.6.12), (7.6.13) we have used the value $\varphi = 200$. In order to avoid spurious fluxes of the material across the boundaries of a computational domain on the r-axis, while using the smoothing (7.6.13), we have applied formula (7.6.13) in a slightly modified form

$$\bar{w}_i^{n+1} = w_i^{n+1} + (1/(\varphi + 2))(Q_{i+1/2}^{n+1} - Q_{i-1/2}^{n+1}),\qquad(7.6.14)$$

where $Q_{i+1/2}^{n+1} = w_{i+1}^{n+1} - w_i^{n+1}$. We have assumed the quantity $Q_{i+1/2}^{n+1}$ to be equal to zero at the boundaries of the computational domain, that is, $Q_{-1/2}^{n+1} = Q_{i_{max}+1/2}^{n+1} = 0$, where we have taken $r_{max} = i_{max} \cdot h = 1.5$ (the subscript "i" in (7.6.12), (7.6.13) refers to the cell boundary, so that $r_i = ih$).

It follows from (7.6.11) that on the symmetry axis where $r = 0$, $i = 0$, it is necessary (for the computation of ρ and P) to use formulas different from (7.6.12). At the same time, the boundary conditions for the components of the vector w in (7.6.10) have a simple form [7.181]

$$u_0^n = 0, \qquad R_0^n = [r\rho_0^n]_{r=0} = 0,$$

$$M_0^n = R_0^n u_0^n = 0,\qquad(7.6.15)$$

$$E_0^n = (r p_0^n/(\gamma - 1)]_{r=0} + M_0^n u_0^n/2 = 0.$$

The stable extra boundary formulas, in the case of using the MacCormack scheme in the interior nodes, are simple extrapolation formulas [7.117], [7.182]

$$\rho_0^n = 2R_1^n/r_1 - R_2^n/r_2; \qquad e_0^n = 2E_1^n/r_1 - E_2^n/r_2;$$

$$p_0^n = (\gamma - 1)e_0^n.\qquad(7.6.16)$$

The computational domain $0 \le r \le 1.5$ was covered by a uniform mesh of 300 nodes. For the time advancement of the solution from $t = 0$ to $t = 0.3$, 458 time steps were needed at such a mesh. Denote by $r_{sw}(t)$ and $r_c(t)$ the abscissas of a shock wave front and of a contact discontinuity, respectively. The values of r_{sw}, r_c were found, on the basis of a numerical solution by scheme (7.6.12), (7.6.14), (7.6.15), as the abscissas of the nodes at which the gradients $|\partial\rho/\partial r|$ reached their local maxima in the zones of smearing of a shock wave and a contact discontinuity. The initial data for the system (7.6.9)–(7.6.11) were set as in [7.183] in the form

$$\rho_i^0 = \begin{cases} 4, & r \le r_{sw}(0), \\ 1, & r > r_{sw}(0), \end{cases} \qquad p_i^0 = \begin{cases} 5, & r \le r_{sw}(0), \\ 1, & r > r_{sw}(0), \end{cases}\qquad(7.6.17)$$

where $r_{sw}(0) = 0.4$; $\gamma = 2$ in the equation of state (1.8). As a result of integration of the system (7.6.9)–(7.6.11), by the method (7.6.12), (7.6.14), (7.6.15) at a initial distribution (7.6.17), the following values of the quantities r_{sw}, r_c were obtained:

$$r_{sw}(0.2) = 0.770; \qquad r_c(0.2) = 0.510;$$

$$r_{sw}(0.3) = 0.940; \qquad r_c(0.3) = 0.570.\qquad(7.6.18)$$

The two-dimensional computations of the problem on the axisymmetrical breakdown of discontinuity were carried out as in [7.183] by the modified particle-in-cell method of Harlow (for the calculation of the pressure with the aid of the equation of state (1.8) the density ρ was used that was obtained by a recalculation by the FLIC-type scheme approximating the continuity equation). A uniform spatial 26×26 mesh with the steps $h_1 = h_2 = 0.05$ was used in these computations. Sixteen particles were placed in each cell lying to the left of a discontinuity at $t = 0$, and four particles were placed in each cell lying to the right of a discontinuity at $t = 0$ taking into account the initial condition (7.6.17). In Figures 7.44(b), (e) we present the results of using a hierarchical classification technique, the scheme of which has been presented in Figure 7.25. A traditional technique of determining the location of discontinuities by a concentration of isolines would be difficult to apply in the case under study, as may be seen in Figures 7.44(a), (d). It is to be noted that the classification scheme of Figure 7.25 has a shortcoming which manifests itself at $t = 0.3$ (see

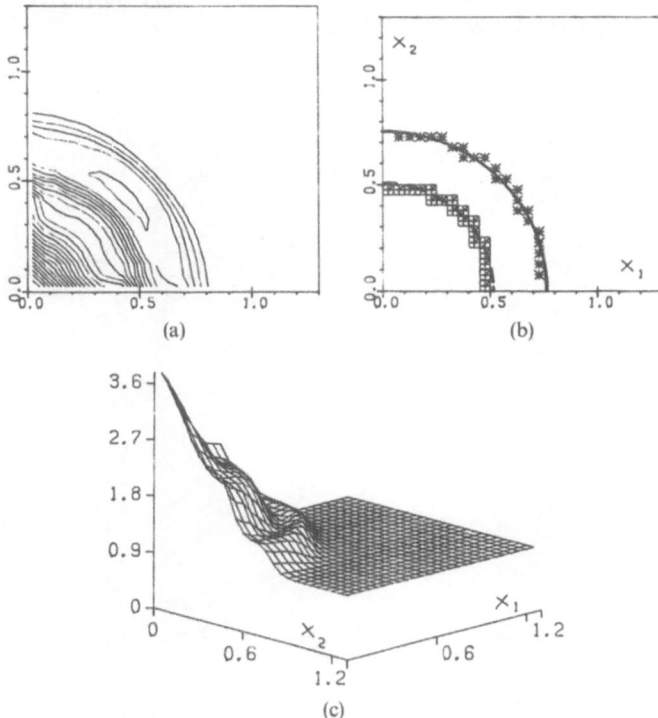

Figure 7.44. The cylindrical shock problem. (a)–(c) $t = 40\tau = 0.2$; (d)–(f) $t = 60\tau = 0.3$. (a), (d) The lines of constant density; (b), (e) the results of singularities classification (∗∗∗)—shock waves; (⊞⊞⊞)—contact discontinuities. (———), (– – – –)—the exact positions of the shock wave and the contact discontinuity, respectively; (c), (f) the isometrics of the surfaces $\rho = \rho(x, y, t)$.

Figure 7.44(e)): this scheme detected spurious rectilinear contact discontinuities near the x- and y-axes which intersect at right angles. We explain this phenomenon by the effect of using the reflection boundary condition at the boundaries $x = 0$ and $y = 0$ in the computations by the Harlow method, which in this case leads to large density gradients on the portions of the above lines which adhere to the coordinate origin $(0, 0)$ (see Figure 7.44(f)). It may be seen in Figures 7.44(b), (e) that the classification scheme I not only correctly classifies the cylindrical discontinuity surfaces into types, but also gives the rather exact position of these discontinuities (with an error within the size of the steps h_1 and h_2). It may be seen in Figures 7.44(c), (f) that the numerical solution gradients have finite nonzero values in the domains behind the shock wave front and in the neighborhood of a contact discontinuity. It may also clearly be seen from Figures 7.44(c), (f) that the cylindrical shock wave front is smeared most intensively in computations by the Harlow method near the x_2-axis and least intensively along a bisector $x_1 = x_2$. This effect was previously predicted by us theoretically and confirmed by numerical computations in [7.183] (see also Section 3.1.2).

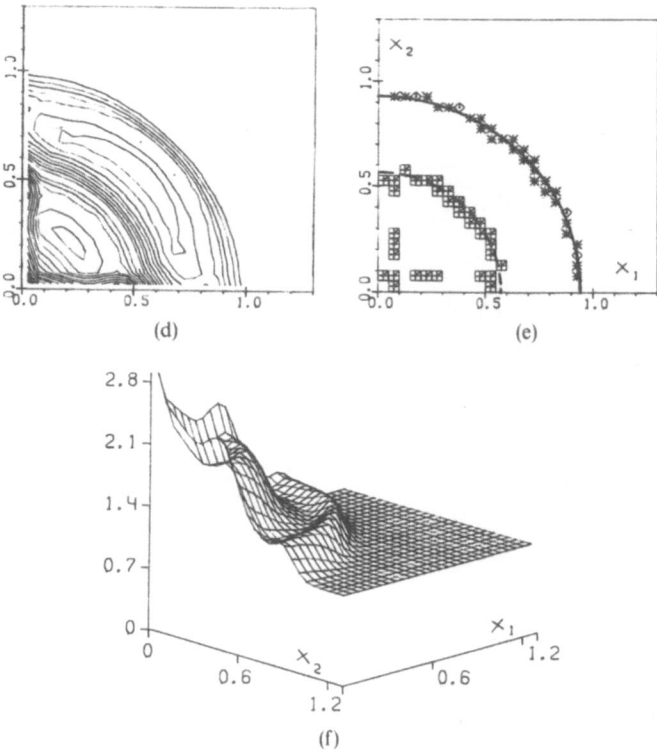

Figure 7.44 (*continued*)

7.6.4. Double Mach Reflection of the Strong Shock Wave

The flow under study can be set up experimentally, for example, in an impinge-ment of a shock wave with a planar front on the wedge. As the result of an interaction of this wave with the wedge surface there occurs a diffraction of a shock wave, and a system of secondary shock waves and contact discontinui-ties is formed. Depending on the values of some basic parameters (the initial angle of the incident shock wave, the shock wave Mach number, the adiabat constant γ, and the atomic composition of a gas in which the shock wave propagates), various types of shock-wave configurations are realized. A survey of some results of the experimental studies of this problem may be found in [7.184]–[7.186]. The numerical studies were carried out by a number of authors (see, for example, [7.187]–[7.195]).

We have used in the computations, the results of which are presented below, the same input parameters as the authors of [7.193]: $\gamma = 1.4$ in the equation of state (1.8), the shock Mach number in air $M_0 = 10$, and the initial angle of the shock wave incidence on the wedge surface which is 60°; the density and pressure of the undisturbed air ahead of the shock are 1.4 and 1, respectively.

The reflecting wall lies along the bottom boundary of the computational domain $\{0 \leq x \leq 4, 0 \leq y \leq 1\}$, beginning at $x = 1/5$. In a short region $0 \leq x \leq 1/5$ along the bottom boundary $y = 0$ the values ρ, u, v, p are set which correspond to the state behind the front of an initially planar shock wave. At the left boundary $x = 0$ the same values of the quantities ρ, u, v, p are set. At the right boundary $x = 4$ the gradients of the functions ρ, u, v, p are set to zero. The values of ρ, u, v, p along the top boundary $y = 1$ are set to describe the exact motion of the Mach 10 shock.

Thus, the computation of the problem under study was carried out in a rectangular domain. This domain was covered by a uniform 100×30 mesh. In the interior nodes the computation was carried out by the MacCormack scheme (4.2.11), (5.2.12) augmented by the smoothing process (7.2.4). In order to avoid a spurious diffusion of the gas across the computation domain boundaries, caused by the application of the smoothing operator (7.2.4), the right-hand side of the relationship (7.2.4) was rewritten in a form similar to (7.6.14). The constant φ in (7.2.4) was chosen empirically to avoid the increase with time of the amplitude of the numerical solution oscillations and to avoid, at the same time, an excessive smearing of shock waves and contact discon-tinuities due to the smoothing (7.2.4). The time step τ_{n+1} was computed at the $(n + 1)$st time step by the solution values at the nth step according to the formula

$$\tau = K \cdot \min(h_1, h_2) \cdot \min_{i,j} \left[1/(|\mathbf{u}_{ij}^n| + c_{ij}^n) \right], \qquad (7.6.19)$$

where

$$|\mathbf{u}_{ij}^n| = \left[(u_{ij}^n)^2 + (v_{ij}^n)^2 \right]^{0.5}, \qquad c_{ij}^n = (\gamma p_{ij}^n/\rho_{ij}^n)^{0.5},$$

the Courant number K was chosen experimentally as a maximum number at which a computation by the MacCormack scheme remained stable.

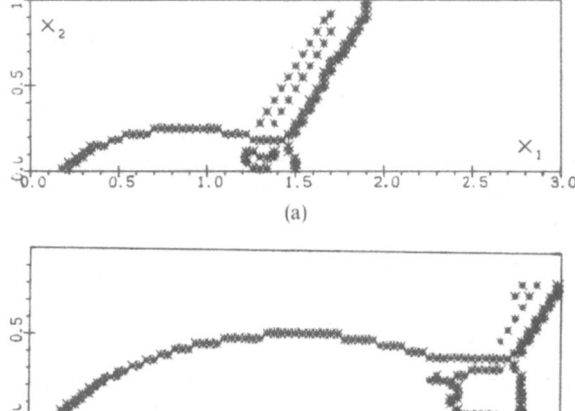

Figure 7.45. The double Mach reflection of a shock. The results of the work of the procedure for the segmentation of the $\rho(x, y, t)$ image: (a) $t = 0.1$; (b) $t = 0.2$.

The input parameters in the problem under study correspond to a double Mach reflection of the shock wave from a wall. In the process of such a reflection two Mach shocks are formed together with two contact discontinuities. The second contact discontinuity is extremely weak, therefore, it is impossible to detect it in the computations on a relatively crude mesh of 100×30 cells. As was pointed out in [7.193], it is generally possible to observe in the 480×120 mesh computations, by the PPMDE scheme developed by the authors of [7.193], the second contact discontinuity by examining the picture of the lines of constant values of the velocity components u and v. The second Mach shock is rather weak, and its amplitude is reduced to zero by the time it reaches the contact discontinuity emanating from the first triple point. The results of a computation of an irregular reflection presented in Figures 7.45–7.47, and 7.49 have been obtained at $K = 0.65$ in (7.6.19), $\varphi = 50$ in (7.2.4). The classification results presented in Figures 7.46(e) and 7.47 are obtained with the aid of the hierarchical classification technique, the scheme of which has been presented in Figure 7.25. The practice of using an image segmentation procedure described in Section 7.3.1 has shown that in cases when there are (in a gas-dynamical flow) closely located discontinuities (more precisely, when the distance between the lines of the discontinuities proves to be less than the size of the zones of shock smearing), it is not reasonable to use the inequalities (7.3.19) in the edge detection. Indeed, there occurs in this case an overlap of the domains of smearing, of the neighboring discontinuities, and the numerical solution errors in the zone of smearing of each of the two neighboring discontinuities become bigger. As a consequence of this, additional errors are introduced into the calculation of the orientation angle of the gradient vector by formula (7.3.9). These errors prove in practice to be so large that a number of pixels belonging to the contours of the shock under

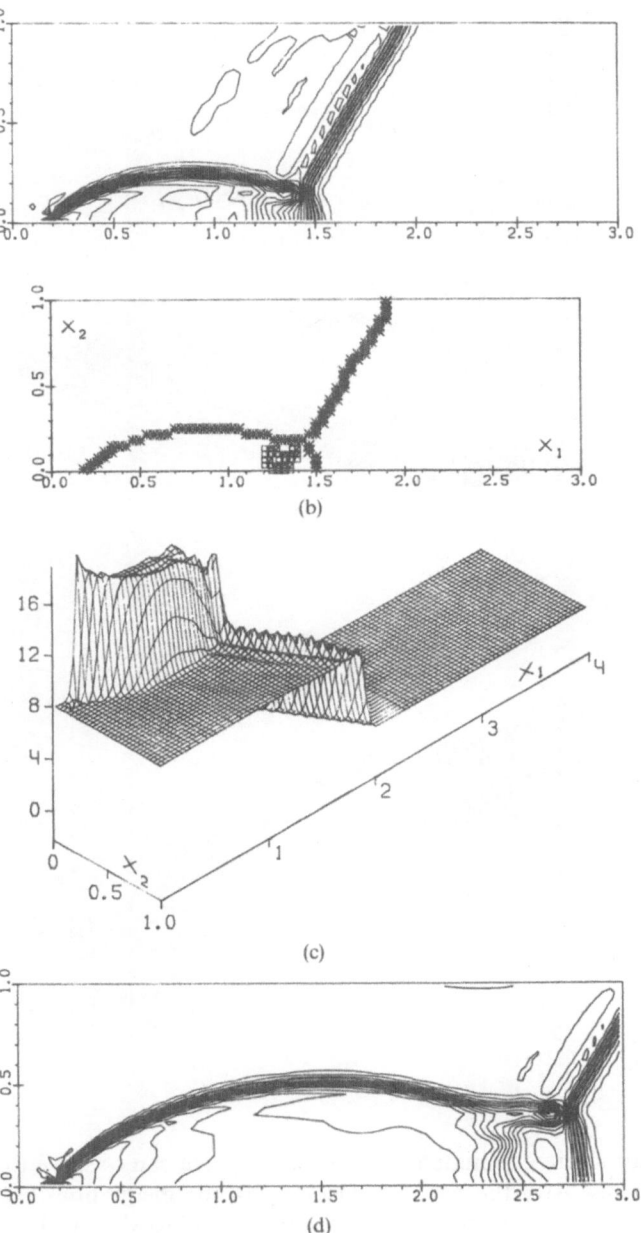

Figure 7.46. Double Mach reflection of a shock. (a)–(c) $t = 0.1$, $n = 75$; (d)–(g) $t = 0.2$, $n = 160$. (a), (d) The lines of constant density; (b), (e) the results of singularities classification: ($***$)—shock waves; (⊞⊞⊞⊞)—contact discontinuities; ($\diamond\diamond\diamond$)—compression waves. (c), (f) The isometrics of the surfaces $\rho = \rho(x, y, t)$; (g) the dimetrics of the surface $\rho = \rho(x, y, t)$, $t = 0.2$. (---)—positions of discontinuities from Woodward and Colella, *J. Comput. Phys.*, **54**, 115–173 (1984).

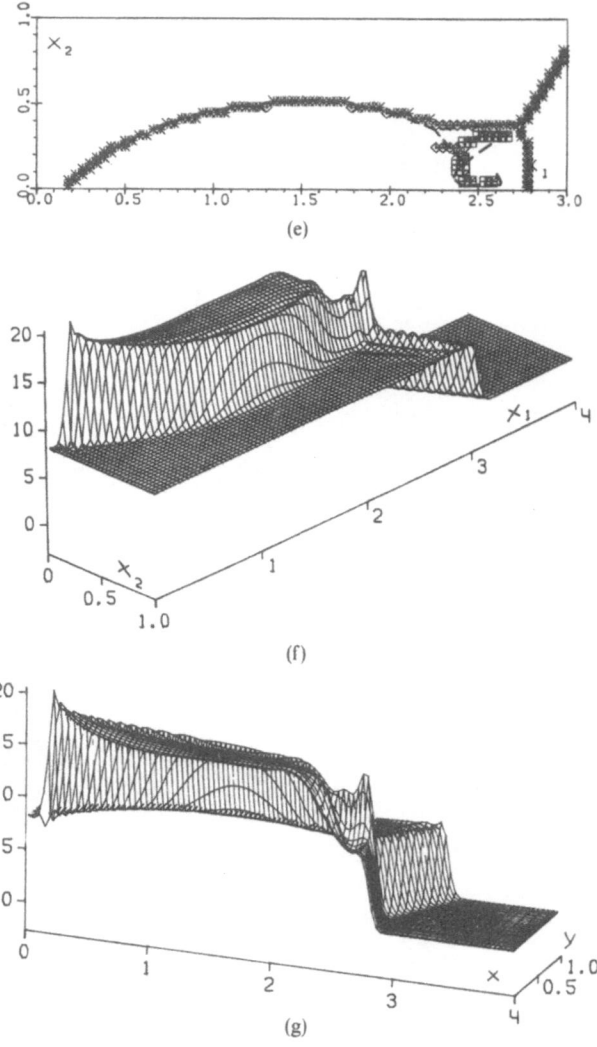

(e)

(f)

(g)

Figure 7.46 (*continued*)

consideration are eliminated from the set $\mathscr{S}_2$. We can observe this in Figure
7.47, obtained with the use of a check-up of the inequalities (7.3.19), at the
segmentation stage. In Figure 7.47 are excluded just the pixels of a reflected
shock wave and of a contact discontinuity which are the neighbors closest to
each other.

It may be seen in Figure 7.46(e) that the first contact discontinuity rolls up
near the reflecting wall, and then follows the direction parallel to the wall and

Figure 7.47. The double
Mach reflection of a shock.
$t = 0.2$.

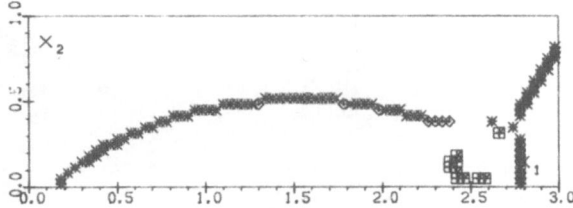

towards the motion of the main Mach shock. In [7.193] this effect has been explained in the following way: near the wall the flow of the denser fluid is deflected by a pressure gradient build up in the region. The result is that a jet of the denser fluid is formed which shoots to the right along the wall. We can also see in Figure 7.46(e) a fragment of the compression wave which, in accordance with known information on the structure of the flow under study, should indeed be a weak Mach wave emanating from the second triple point. The dashed lines in Figures 7.46(e) and 7.49(b) show the positions of the discontinuities which we have found from the pictures of density isolines presented in [7.193], and obtained by the PPMDE method [7.193], [7.196] on a mesh with the steps $h_1 = h_2 = 1/120$. The PPMDE method, as well as the PPMLR method which will be mentioned in Section 7.6.5, represents a second-order extension of the Godunov method of a type first introduced by van LEER in his MUSCL algorithm [7.197].

In Figure 7.46(a), (d) we observe an unphysical structure (an "eye") near the intersection of the three main shock waves. The form of this structure coincides with the one presented in [7.193]. As was shown in [7.193], this structure disappears when a finer spatial mesh is used.

If φ is increased in formula (7.2.4), the amplitude of parasitic oscillations in computations by the MacCormack scheme also increases. For the purpose of comparison with Figure 7.46(a)–(c) we present in Figure 7.48 the results of a classification by the scheme of Figure 7.25 at $\varphi = 200$. Comparing Figures 7.48(a) and 7.46(c), we note a substantial increase in the amplitude of numerical solution oscillations at $\varphi = 200$ in comparison with the case $\varphi = 50$. The elimination of a stage of the check-up of the inequalities (7.3.19) in the process of the segmentation of a digital image enables us to detect the Mach stem (see Fifgure 7.47(c)). At the same time, we observe here a rather large number of artefacts. The artefacts disappear when the check-up of the inequalities (7.3.19) is included, but the Mach stem proves to be almost undetected (Figure 7.48(d)).

The sequential classification scheme presented in Figure 7.26 enables us to obtain more complete information on the properties of the flow singularities in comparison with the scheme of Figure 7.25, since the scheme of Figure 7.26 includes a subdivision of the shock waves class into the subclasses of the normal and oblique shock waves, and a subdivision of the contact discon-

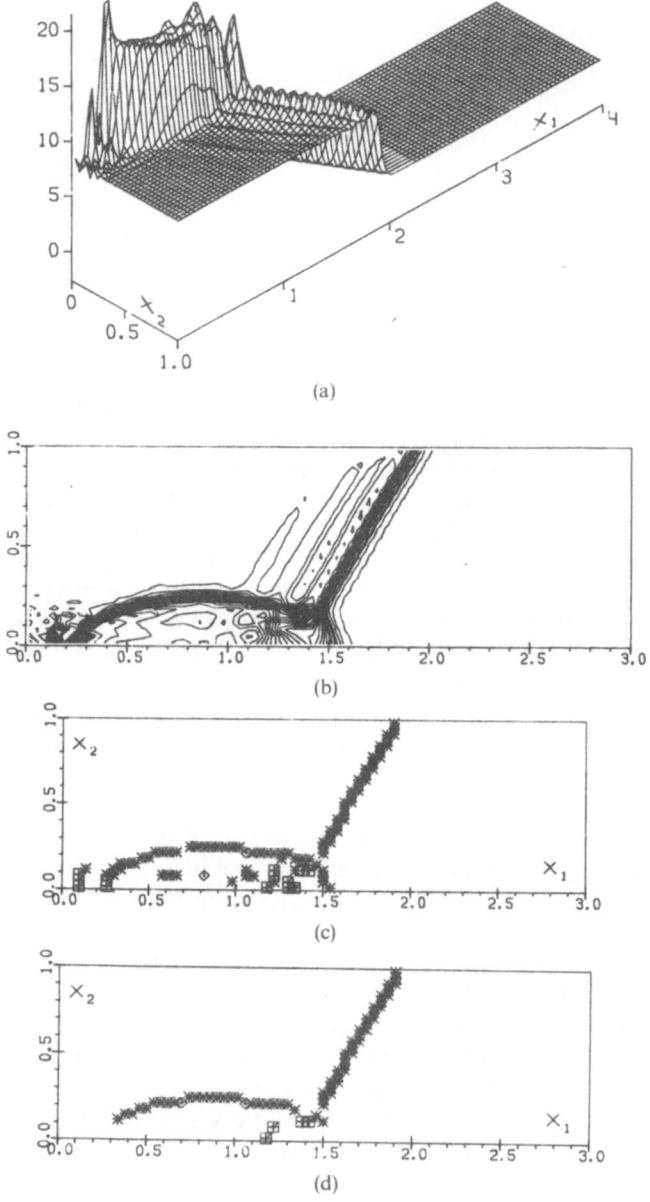

Figure 7.48. The double Mach reflection of a shock. $t = 0.1$, $n = 98$, $K = 0.7$, $\varphi = 200$. (a) The isometrics of the surface $\rho = \rho(x, y, t)$; (b) the lines of constant density; (c) the classification result without checking the inequalities (7.3.19); (d) the classification result in the case of a check-up of the inequalities (7.3.19).

Figure 7.49. Application of
the sequential classification.
(a) $t = 0.1$; (b) $t = 0.2$.
($\times \times \times$)—normal shock
waves; ($***$)—oblique
shock waves; (⊞⊞⊞)—
tangential discontinuities;
($+++$)—purely contact
discontinuities; ($\Diamond \Diamond \Diamond$)—
compression waves. ($---$)—
positions of discontinuities
from Woodward and Colella,
J. Comput. Phys., **54**, 115–
173 (1984).

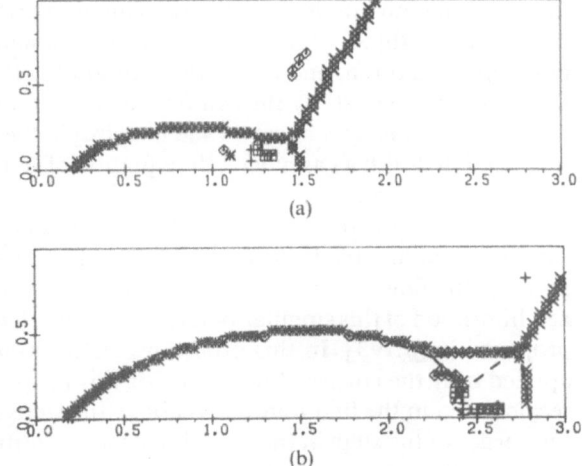

(a)

(b)

tinuities class into the subclasses of tangential and purely contact discontinuities. The results of sequential classification presented in Figure 7.49 are ob-
tained for the following values of the constants δ_1, δ_3, δ_4, δ_6, δ_7 for the features
(7.4.2), (7.4.4), (7.4.20), (7.4.21), (7.5.6), (7.5.7), respectively: $\delta_1 = 0.01$, $\delta_3 = 0.01$,
$\delta_4 = 0.1$, $\delta_6 = 0.05$, $\delta_7 = 0.05$; $N_g = 1$ in (7.3.17). The set of objects (x_m, y_m)
depicted in Figure 7.45, which are subject to classification and are obtained
as a result of the work of the image segmentation procedure, is here the same
as in the case of Figure 7.46. It is interesting to note that the tangential
discontinuity emanating from the main triple point goes over into a purely
contact discontinuity when it rotates towards the flow, but near the reflecting
wall it again becomes a tangential discontinuity. According to [7.193], in a
region limited by the second Mach wave, the curved reflected wave, and the
reflecting wall, there is a very little vertical motion. As may be seen in Figure
7.49, the main Mach stem proved to be a normal shock wave. From this the
conclusion may be drawn that the vertical gas motion is also very small behind
the front of this shock wave. Under the same conditions the incident shock
wave, which initially was a normal shock, is at $t = 0.2$ also classified as a
normal shock wave.

7.6.5. Supersonic Flow in a Wind Tunnel with a Lower-Wall Step

This two-dimensional problem is often used as a test for finite-difference
schemes [7.197], [7.198]. In this problem the initial conditions corresponding
to a uniform Mach 3 flow are given at $t = 0$ in a wind tunnel with a lower-wall
step. The wind tunnel is 1 unit wide and 3 units long. The step is 0.2 units high
and located 0.6 units from the left-hand end of the tunnel. It is assumed that

the tunnel has an infinite width in the direction orthogonal to the plane of the computation (that is, "slab symmetry" is assumed). At the left is a flow-in boundary condition, and at the right all gradients are assumed to vanish (in the problem under study the exit flow is supersonic, therefore, the exit boundary condition has no effect on the flow inside the tunnel). At $t = 0$ the wind tunnel is filled with a gas having the equation of state (1.8) with $\gamma = 1.4$, density 1.4, pressure 1, and velocity 3.

Along the walls of the tunnel reflecting boundary conditions are applied. The corner of the step is the center of a rarefaction fan and hence is a singular point of the flow. To reduce the size of the numerical solution errors in the neighborhood of this singular point, we have implemented a special procedure proposed in [7.193]. In this procedure an additional boundary condition is applied near the corner of the step. For this purpose the flow parameters are recalculated in the first four zones above the step, starting just to the right of the corner of the step; in the row above the quantities are reset in the first two zones. In these six zones the density is reset in such a way that the entropy has the same value as in the zone just to the left and below the corner of the step. The magnitudes of the velocities, not their directions, are also reset, so

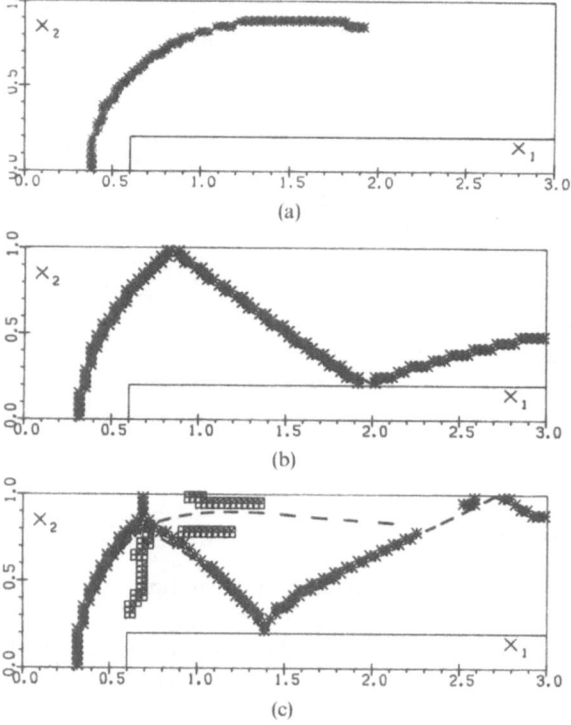

(a)

(b)

(c)

Figure 7.50. (a) $t = 0.5$, $n = 112$; (b) $t = 1.5$, $n = 325$; (c) $t = 3.0$, $n = 632$. (∗∗∗)—shock waves; (⊞⊞⊞)—contact discontinuities. (– – –)—positions of discontinuities from Woodward and Colella, *J. Comput. Phys.*, **54**, 115–173 (1984).

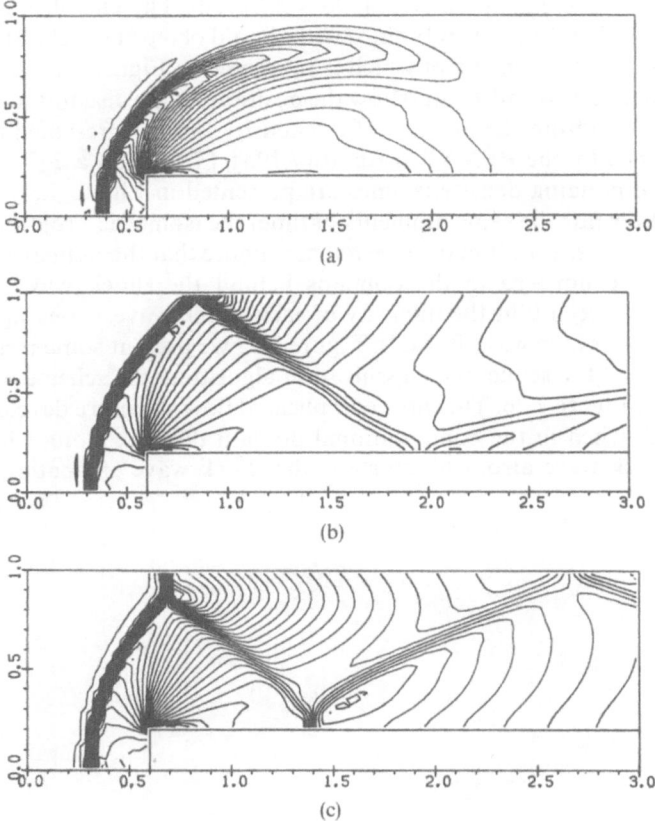

Figure 7.51. Density isolines. (a) $t = 0.5$; (b) $t = 1.5$; (c) $t = 3.0$.

that the sum of enthalpy and kinetic energy per unit mass has the same value as in the same zone used to set the entropy. The above procedure is based on the assumption of a nearly steady flow in the region near the corner. It is clearly inappropriate at the very outset of the calculation. As was noted in [7.193], the above additional boundary conditions remove the grossest errors generated near the corner, but large errors in the flow direction there are bound to remain. These errors may be the cause of an overexpansion of the flow near the corner which was observed in all the runs, although similar effects are observed in experiments using real, viscous gas.

As was shown in [7.197], a steady flow in the problem under study develops by time 12. We shall be interested in the flow for earlier times when the flow structure rapidly changes with time. The computation results presented in Figures 7.50–7.53 are obtained with the aid of the MacCormack scheme (4.2.11), (4.2.12) augmented by the smoothing process (7.2.4). The Courant number $K = 0.6$, $\varphi = 30$ in (7.2.4), 90×30 mesh. The classification results presented in Figure 7.50 are obtained with the aid of the hierarchical classification technique, the scheme of which is presented in Figure 7.25. The dashed lines in Figures 7.50 and 7.53(c) show the positions of the discontinuities which we have found from the pictures of the density isolines presented in [7.193] and obtained by the PPMLR method [7.193], [7.196] on a 240×80 mesh. The corrresponding density isolines are presented in Figure 7.51. In Figure 7.52 we show (for the same moments of time) the isometric projections of the surfaces $\rho = \rho(x, y, t)$. It may be seen in this figure that the numerical solution gradients are nonzero in the domains behind the shock wave fronts. In addition, it is seen that the intensity of each shock wave varies significantly along the line of the wave front. In Figure 7.53 we present some results of the application of the sequential classification algorithm, the scheme of which is depicted in Figure 7.26. The most complicated flow structure develops by the time $t = 3$, when in the computational domain there is, along with the detached shock wave also a Mach stem, the shock wave emanating from the

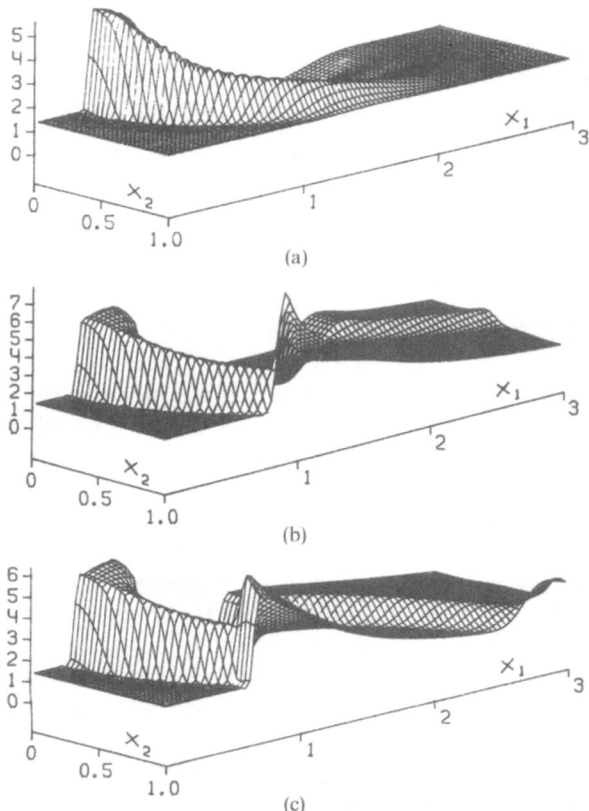

Figure 7.52. (a) $t = 0.5$;
(b) $t = 1.5$; (c) $t = 3.0$.

Figure 7.53. (a) The dimetrics of the surface $\rho = \rho(x, y, t)$ at $t = 3.0$; (b) the result of the work of the procedure of the image $\rho(x, y, t)$ segmentation at $t = 3.0$; (c) the result of sequential classification: ($\times \times \times$)—normal shock waves; ($***$)—oblique shock waves; (⊞⊞⊞)—tangential discontinuities; ($+++$)—purely contact discontinuities. ($---$)—positions of discontinuities from Woodward and Colella, *J. Comput. Phys.*, **54**, 115–173 (1984).

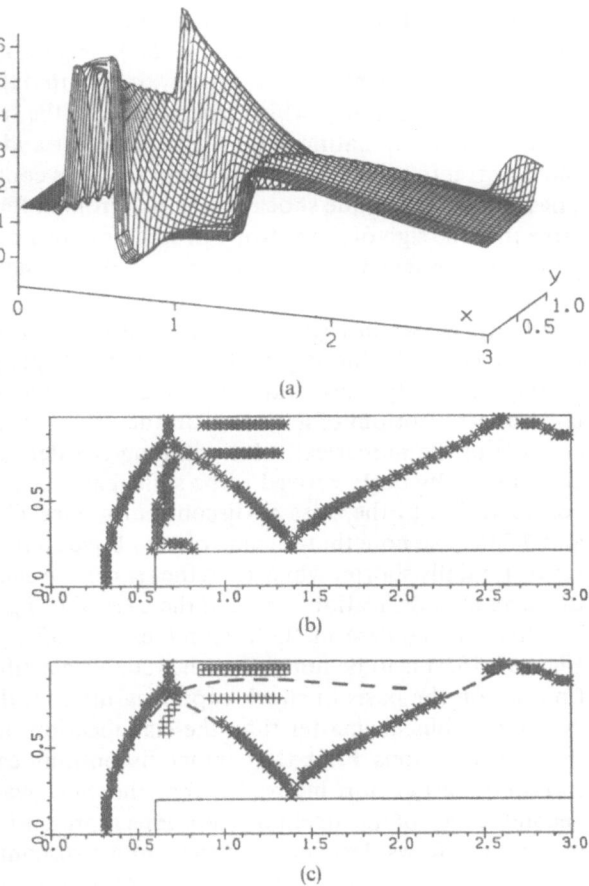

(a)

(b)

(c)

triple point; a secondary shock wave resulting from the reflection of the foregoing shock from the lower wall; and, finally, a shock wave emanating from the upper wall and being the result of a reflection from this wall of the secondary shock just mentioned. In addition, we can see in Figures 7.50(c) and 7.53(c) a contact discontinuity beginning near the triple point. A numerical boundary layer along the upper surface of the step causes a spurious Mach reflection from the step, but the corresponding Mach stem is only two zones high, as may be seen in Figure 7.50(c). This numerical effect, arising in the case of using the MacCormack scheme, was noted previously in [7.193].

We can also observe in Figures 7.50(c) and 7.53(c) the presence of a spurious contact discontinuity near the upper wall of the wind tunnel. This spurious contact discontinuity may be caused for two reasons: first, by the use of simple reflection conditions on the rigid wall in the MacCormack scheme computations; second, the same conditions are used in the computations in the Sobel edge detector at the stage of segmentation of the digital image $\rho(x, y, t)$. We can also see in Figures 7.50(c) and 7.53(c) the gaps in some lines of secondary shock waves. We explain these gaps by the fact that the contact strip containing a contact discontinuity, which emanates from the triple point, intersects with a shock wave emanating from the triple point and with the next secondary shock wave. In the region of intersection of this contact strip with the zone of smearing of an oblique shock wave, the errors in the entropy may become so large that the sign of the entropy production may prove to be incorrect. Such points are assigned to the continuous flow in the classification algorithms used.

Finally, we see in Figure 7.50(c) a spurious contact discontinuity starting near the corner of the step and reaching the triple point. Judging by Figure 7.53(a), this contact discontinuity is located in the domain of the rarefaction fan, near the bottom of a "valley" in the $\rho(x, y, t)$ surface. The change of the entropy in the numerical solution along the detected spurious contact discontinuity obviously proved to be sufficiently large to assign the line under consideration to the class of discontinuity lines. Comparing Figures 7.50(c) and 7.53(c), we note that the line of the above spurious contact discontinuity is substantially shorter while using the sequential classification scheme, so that the level of classification errors in the case of using this scheme proves to be less than in the case of the hierarchical classification scheme presented in Figure 7.25. It is to be noted that the reliable classification of contact discontinuities on the basis of shock-capturing numerical computations proves to be a more difficult matter than the classification of shock waves. We relate this to two reasons. First, the contact discontinuities are smeared in a numerical computation more intensively than the shock waves (see Chapter 4 above). Second, a part of the algebraic equations representing the jump conditions in inviscid gas flows degenerate at the contact discontinuities in the sense that these equations are satisfied identically. As a result of this the number of basic relationships which could be used as the features of contact discontinuities is diminished.

In concluding this section, let us present some data on the computer codes that implement the algorithms presented in this chapter for automatic singularities classification in the two-dimensional inviscid gas flows. If we take the length (in the number of machine words) of the program for solving the two-dimensional problem by the MacCormack scheme as one unit, then the length of the programs realizing the classification schemes of Figures 7.25 and 7.26 is 2.57 units. About 40% of this volume is constituted by the subroutines of the graphical information output (plotting isolines, surfaces, points of discontinuity lines). The length of the program for the classification of singularities in the two-dimensional transonic potential flows (together with graphical subroutines) is approximately equal to the length of the program for computing the transonic potential flow around an airfoil by the AF2-type scheme on a given curvilinear mesh. To economize the core memory of a computer, the magnitudes (7.3.8) of the gradients g_{ij} and of the orientation angles of the gradient vector α_{ij} (7.3.9), needed at the stages of the image segmentation and of the singularities classification, were not stored in the form of two-dimensional numerical arrays, as was done, for example, in [7.89], but were computed each time by the grid values ρ_{ij} of the density with the aid of corresponding subroutines. A comparison with the variant, when g_{ij} and α_{ij} are computed once and then stored in arrays, shows an increase by a factor of about 2 (in the computer time needed for segmentation and classification). The operations of the computation of the quantities g_{ij} and α_{ij} by formulas (7.3.8), (7.3.9) are carried out independently for each pixel (i, j); therefore, the computer time needed for segmentation can be substantially reduced by using the cellular logic devices as well as (to a lesser extent) by using the parallel general-purpose computers.

Let us now present some data on the computer time which was needed for the classification of singularities in various gas-dynamical flows, and for the graphical display of the data (in percent) with respect to the time needed for the solution of a gas-dynamical problem (the computations were carried out on a sequential general-purpose computer):

transonic potential flow—12.5%;
double Mach reflection of a strong shock wave—10.3%;
supersonic flow in a wind tunnel with a lower-wall step—2.7%.

Although there are significant percent differences here, the absolute values (in seconds of computer time) of the time needed for the classification of singularities in the above three problems were close to each other. This is related to the fact that the computer time needed for the digital processing of two-dimensional arrays is determined mainly by the size (that is, by the number of pixels) of the images. In all three problems listed above we have used the number of spatial computing mesh nodes having the same order of magnitude.

It should be noted that the above-presented methods for the localization and classification of singularities in two-dimensional flows can be extended to the case of three-dimensional flows. Indeed, the algorithms of segmentation of three-dimensional digital images have already been considered in the literature (see, for example, [7.199]–[7.204]). As regards the computation of the features needed for the algorithms of the singularities classification, it appears that there are no substantial difficulties to the extension of the corresponding formulas for the case of three space variables, because all the formulas for the features are written in Section 7.4 for the coordinates along a normal and along a tangential direction to the discontinuity surface.

Concluding Remarks

Let us make a few remarks on the applicability of the techniques for the localization of discontinuities in the numerical solutions of gas dynamics problems, with the purpose of helping the reader in his choice of a specific localization technique.

If, at the beginning of a numerical study of some gas-dynamical process, the research worker already has at his disposal sufficiently detailed information on the presence of singularities—on their types (shock waves, contact discontinuities, etc.), on the qualitative behavior of their propagation, and on their orientation with respect to the spatial coordinate axes—then he can use the localization methods presented in Chapters 2–5. The above *a priori* information on singularities may be obtained from the following sources:

(1) the data of physical experiments carried out, for example, in wind tunnels;
(2) the general properties of the mathematical solutions of the class of problems under study; and
(3) the experience and intuition of the research worker.

If the above information is scarce or is absent altogether, then we can employ (for an automatic analysis of the numerical solutions of the two-dimensional problems of gas dynamics) the methods presented in Chapter 7. In the process of development of systems for the classification of singularities (similar to those described in Chapter 7) it is also reasonable to make use of available *a priori* information on the process under study, because the system of *a priori* knowledge of the object creates a system of constraints on the possible interpretation of the object images, and thus helps to avoid ambiguities in the interpretation of objects or processes. In addition, direct use of the *a priori* information (on the process under study in the system of pattern classification) enables us, in many cases, to introduce some simplifications into the recognition algorithms which improve their computational efficiency.

In discussing the merits and shortcomings of methods for the localization of singularities considered in Chapters 2–5, we at first consider a widely accepted technique of a visual determination of the discontinuities location by the concentration of different isolines. As was pointed out in Section 4.1, a maximum concentration of the level lines of some function, for example, the density, takes place in the domains of the largest gradients of this function. It follows from Chapter 2 that, as a rule, the position of the domains of maximum concentration of isolines does not coincide with the exact position of the shock wave fronts. The magnitude of the error arising in such a localization technique is of the order $O(h)$. In addition, the isolines can also be concentrated significantly in the domains of the compression waves, and an additional verification is needed as to whether the places of the most significant concentration of the isolines are indeed a locus of the points of a shock wave front or of a contact discontinuity. It was pointed out in Section 4.1 that a function, such as pressure, remains continuous across the contact discontinuity; therefore, the lines of constant pressure—the isobars—generally do not coalesce in the neighborhood of the contact boundaries, and we can use, for the localization of discontinuities of this kind, the level lines of only those functions which undergo a discontinuity at a contact boundary: these are the functions of density and temperature.

If the first-order difference scheme employed is such that its f.d.a. does not satisfy the conditions of Theorem 4.1 of Section 4.1 in the contact strip, then the position of the regions of maximum concentration of the lines of constant density—the isochors—generally does not coincide with the true position of the contact boundaries. In this case, the corresponding localization error may reach several dozens of the intervals of a spatial computing mesh. Then it is reasonable to use not only the lines of constant density but also the level lines of a certain function $R(p, \rho)$, which is obtained as has been illustrated in Section 4.1, in the example of the Rusanov scheme.

Further, it follows from the results of Section 4.1 that (in the case of using the schemes of the second and higher orders of approximation) the subdomains of maximum concentration of the isochors will also be shifted with respect to the true position of the contact discontinuities. The results of the discontinuities localization in this case may nevertheless prove to be satisfactory in practice, if the computation of a time-dependent problem has been carried out in a bounded interval $0 \leq t \leq T$, where T is a relatively small value. This is related to the fact that the smearing of contact discontinuities occurs substantially less intensely in the case of using the schemes of the second and higher orders of accuracy (see Section 4.1 and Table 4.1).

It is easy to see that the isochors coalesce not only in the flow subdomains containing contact discontinuities, but also in the subdomains containing the shock wave fronts and compression waves. In this connection, there arises an ambiguity in the interpretation of the computation results. In practice, this ambiguity is usually overcome by a wide use of all the available *a priori* information.

Consider now a technique of shock wave front localization on the basis of smeared profiles of the numerical solution, in which the shock location is determined by the points of maximum of the artificial viscosity. We have listed (at the beginning of Chapter 2) a number of works in which this technique was applied for the shock wave localization in a number of complicated two-dimensional nonstationary problems. It follows from the results of Section 2.2 that the localization technique by max q, where q is the artificial viscosity, is applicable in practical computations. If it is necessary that the error in shock localization does not exceed the value h (h is a characteristic dimension of a cell of a spatial grid), then one must, generally speaking, impose some constraints on the nondimensional coefficients entering the expressions for the artificial viscosities (see Table 1.1). If these constraints are not taken into account, then the magnitude of the error in the shock front localization can reach several intervals of h, as has been shown in the computational examples in Section 5.5. The technique of the shock front localization by max q—the differential analyzer—has an advantage over the technique of visual determination of the shock location by the concentration of isolines. This advantage consists of the fact that a computer produces ready information on the shock fronts in the form of the coordinates of their points, or graphical images on the screen display, or on a graph plotter (see Section 3.4). The realization of the differential analyzers described in Section 3.4, which makes use of max q or of a maximum of a scheme viscosity norm, is related to a search for the points of maximum of the above quantities along some beams which are drawn in the (x, y)-plane, in such a way that they intersect the possible shock fronts at angles close to the right angles. This means that a specific strategy of the selection-of-the-beams direction in the differential analyzer subroutines should take into account *a priori* information on the flow under study. The available experimental data (as well as the skill, experience, and intuition of the research worker) are of importance in obtaining the above information—as in the previous localization technique by isolines.

The optimization methods of the strong discontinuities localization, by the shock-capturing computation results presented in Chapter 5, enable us to obtain information directly on the discontinuity line—as in the case of the differential analyzers. The modifications of the original optimization approach of Miranker and Pironneau, proposed in Sections 5.2 and 5.3, have a number of advantages over the differential analyzers of Chapters 2–4. These advantages are listed in Section 5.6. The optimization methods of the discontinuities localization presented in Chapter 5 have the following shortcomings: increased requirements for the computer memory compared with the differential analyzers, and the availability of *a priori* information on the existence of a strong discontinuity in the problem to be solved.

If the information on the discontinuity localization is used for purposes of the difference solution refinement in the neighborhood of a discontinuity, then it is reasonable to use here (as was pointed out in Sections 6.1 and 6.2), the localization algorithms ensuring the best accuracy. This enables us to reduce substantially the computer time needed for the numerical solution of the constrained optimization problems—to which we have reduced, in Chapter 6, the problem of the difference solution refinement.

The solution refinement procedures presented in Chapter 6 are also important, from the point of view that they enable us to restore the true values of the flow parameters on both sides of a strong discontinuity. As was shown in Section 6.4, in a number of computational examples, the refinement of difference solutions in the vicinity of strong and weak discontinuities enables us to calculate correctly (within the framework of the shock-capturing difference methods) the problems with multiple discontinuities in the flow domain, thus broadening the scope of applicability of the shock-capturing difference methods.

The shock-capturing methods, using an adaptive grid whose lines concentrate on the domains of large solution gradients, give the computational aerodynamicist another means of restoring the flow parameters on both sides of the strong discontinuities lines. Indeed, let Δx be a characteristic dimension of a cell of a spatial grid outside the big solution gradient domains. Since the cell sizes in the neighborhood of the discontinuities are typically much smaller than Δx (see Sections 2.5.2, 3.3, and 3.4), the difference solution profiles—in a section which is perpendicular to the orientation of a discontinuity line that is obtained by an interpolation into the points of a rough uniform mesh with the spacing Δx—will have a sharp jump across the discontinuity. Finally, the methods for the restoration of digital images (whose implementation within the framework of computational fluid dynamics was outlined in Section 7.2.2) give the computational aerodynamicist another means of restoring the flow parameters on both sides of the strong discontinuities lines. The results of the computational experiments on the restoration of piecewise constant solutions presented in Section 7.2.2 are, in our opinion, quite encouraging.

If one needs, in the process of the numerical study of some problem, a large number of flow pictures showing the dynamics of the discontinuity surfaces, the problem of computer time needed for obtaining each such picture of the flow (by using some localization technique) becomes important. The known techniques of the discontinuities localization differ substantially with respect to the needed computer time. For some algorithms (such as, for example, [2.14]; see also a survey of the localization techniques at the beginning of Chapter 2), it is comparable to the time for solving the basic two-dimensional problem. To reduce the needed computer time, it is reasonable to employ (in the procedures for the isolines construction as well as in the differential analyzer subroutines described in Section 3.4) the simplest interpolation techniques (for example, linear interpolation). Note that on a computer with parallel processors, the programs for the isolines construction and for the differential analyzer can work simultaneously on the solution of the basic problem.

Methods for the localization and automatic classification of singularities (presented in Chapter 7) are more laborious and complicated than any of the methods for shock localization on the fixed grids presented in Chapters 2–5.

A doubtless advantage of the methods of Chapter 7 is their universality and applicability for an automatic analysis of the numerical solutions of the problems, for which there is no *a priori* information on the singularities.

The variety of existing methods of digital image processing and pattern recognition creates a huge field for research on the applicability of these methods in the systems of localization and automatic classification of the singularities in the numerical solutions of multidimensional gas dynamics problems. We do not want to enumerate these methods here (a review of the most widespread of these methods has already been made in Chapter 7), but we indicate a number of problems which, in our opinion, can be solved successfully with their help. First, we would like to point out the application of the methods for the restoration of digital images to the development of an efficient "nonsmearing" conservative method of a shock-capturing numerical computation on a fixed spatial grid of the convective transfer of arbitrary shocked profiles. The use of the methods of automatic analysis of Chapter 7 in the expert systems for automation of the design of aerodynamic bodies satisfying some optimality requirements also appears to be very promising. Here one can expect a further increase in the computational efficiency of such expert systems by the integration of general-purpose computers and computers based on the cellular logic principle.

References

Chapter 1

[1.1]   J. von Neumann, R. Richtmyer: *J. Appl. Phys.* **21**, 232–237 (1950).
[1.2]   R.D. Richtmyer, K.W. Morton: *Difference Methods for Initial-Value Problems*, 2nd edn (Wiley Interscience, New York, 1967).
[1.3]   P.J. Roache: *Computational Fluid Dynamics* (Hermosa, Albuquerque, New Mexico, 1976).
[1.4]   R. Peyret, T.D. Taylor: *Computational Methods for Fluid Flow* (Springer-Verlag, New York, Heidelberg, Berlin, 1983).
[1.5]   B.L. Rozhdestvensky: "Mathematical Theory of Shock Waves", in *Mathematical Encyclopedia*, Vol. 5 (Soviet Encyclopedia, Moscow, 1985), pp. 468–474.
[1.6]   G.P. Tsybulsky: *Izv. Akad. Nauk SSSR, Ser. Mekhanika Zhidkosti i Gaza*, No. 1, 170–173 (1975).
[1.7]   P.R. Woodward: "Trade-offs in designing explicit hydrodynamical schemes for vector computers", in *Parallel Computations*, ed. by G. Rodrigue (Academic Press, New York, 1982), pp. 153–171.
[1.8]   A.K.M.F. Hussain: *Phys. Fluids* **26**, 2816–2850 (1983).
[1.9]   E. Turkel: *Tellus* **26**, 630–637 (1974).
[1.10] E. Anderson: *Magnetohydrodynamic Shock Waves* (M.I.T. Press, Cambridge, Massachusetts, 1963).
[1.11] A. Jeffrey, T. Taniuti: *Non-Linear Wave Propagation with Applications to Physics and Magnetohydrodynamics* (Academic Press, New York, 1964).
[1.12] J.P. Boris, N.K. Winsor: "Vectorized Computation of Reactive Flow", in *Parallel Computations*, ed. by G. Rodrigue (Academic Press, New York, 1982), pp. 173–215.
[1.13] N.N. Yanenko, R.I. Soloukhin, A.N. Papyrin, V.M. Fomin: *Supersonic Two-Phase Flows Under Conditions of Velocity Non-Equilibrium of Particles* (Nauka, Siberian Division, Novosibirsk, 1980).
[1.14] A.N. Tikhonov: "Mathematical Model", in *Mathematical Encyclopedia*, Vol. 3 (Soviet Encyclopedia, Moscow, 1982), pp. 574–575.
[1.15] S.S. Tong: AIAA Paper 85-0112 (1985).
[1.16] U.B. Mehta, P. Kutler: AIAA Paper 84-1531 (1984).
[1.17] N.E. Kochin, I.A. Kibel, N.V. Rose: *Theoretical Hydromechanics*, Part 1, 6th edn (Fizmatgiz, Moscow, 1963).
[1.18] B.L. Roždestvenskiǐ, N.N. Janenko: *Systems of Quasilinear Equations and Their Application to Gas Dynamics*, 2nd edn (Nauka, Moscow, 1978) [English transl.: *Systems of Quasilinear Equations and Their Applications to Gas Dynamics*, Translations of Mathematical Monographs, Vol. 55 (American Mathematical Society, Providence, Rhode Island, 1983)].
[1.19] Yu.I. Shokin, N.N. Yanenko: *Differential Approximation Method. Application to Gas Dynamics* (Nauka, Siberian Division, Novosibirsk, 1985).
[1.20] L.V. Ovsiannikov: *Lectures on Foundations of Gas Dynamics* (Nauka, Moscow, 1981).
[1.21] E.D. Terentyev, Yu.D. Shmyglevsky: *Zhurn. Vychisl. Matem. i Matem. Fiz.* **15**, 1535–1544 (1975).
[1.22] K.P. Staniukovich: *Transient Motions of a Continuum* (Nauka, Moscow, 1971).

[1.23] N.E. Kochin, I.A. Kibel, N.V. Rose: *Theoretical Hydromechanics*, Part 2, 4th edn (Fizmatgiz, Moscow, 1963).

[1.24] A.J. Chorin, T.J.R. Hughes, M.F. McCracken, J.E. Marsden: *Commun. Pure Appl. Math.* **31**, 205–256 (1978).

[1.25] F.H. Harlow: "The Particle-In-Cell Computing Method for Fluid Dynamics", in *Methods in Computational Physics*, ed. by B. Alder, S. Fernbach, M. Rotenberg, Vol. 3: Fundamental Methods in Hydrodynamics (Academic Press, New York, 1964), pp. 319–343.

[1.26] N.N. Yanenko, N.N. Anuchina, V.E. Petrenko, Yu.I. Shokin: *Numerical Methods in Continuum Mechanics* (in Russian), **1**, No. 1 (A.N. SSSR Siberian Computer Center, Novosibirsk, 1970), pp. 40–62.

[1.27] R.A. Gentry, R.E. Martin, B.J. Daly: *J. Comput. Phys.* **1**, 87–118 (1966).

[1.28] O.M. Belotserkovsky, Yu.M. Davydov: *Coarse Particles Method in Gas Dynamics. Computational Experiment* (Nauka, Moscow, 1982).

[1.29] B.J. Daly: *Math. Comput.* **84**, 346–360 (1963).

[1.30] N.A. Larkin, V.A. Novikov, N.N. Yanenko: *Nonlinear Equations of Variable Type* (Nauka, Siberian Division, Novosibirsk, 1983).

[1.31] P.D. Lax: *Commun. Pure Appl. Math.* **7**, 159–193 (1954).

[1.32] P.D. Lax, B. Wendroff: *Commun. Pure Appl. Math.* **13**, 217–237 (1960).

[1.33] S.K. Godunov, A.V. Zabrodin, M.Ya. Ivanov, A.N. Kraiko, G.P. Prokopov: *Numerical Solution of Multidimensional Problems of Gas Dynamics* (Nauka, Moscow, 1976).

[1.34] A. Harten: "The Method of Artificial Compression: 1. Shocks and Contact Discontinuities"; AEC Research and Development Report C00-3077-50, New York University (1974).

[1.35] A. Harten, J.M. Hyman, P.D. Lax: *Commun. Pure Appl. Math.* **29**, 297–322 (1976).

[1.36] A. Lerat: Lecture Notes in Physics, Vol. 90 (Springer-Verlag, New York, 1979), pp. 345–351.

[1.37] Yu.V. Kasakov, A.V. Feodorov, V.M. Fomin: "Differential Analyser of Shock Waves with Relaxation"; Preprint No. 5 (Institute of Theoretical and Applied Mechanics of the U.S.S.R. Academy of Sciences, Novosibirsk, 1983).

[1.38] M.Ya. Ivanov, V.V. Koretsky, N.Ya. Kurochkina: *Numerical Methods in Continuum Mechanics* (in Russian), **11**, No. 1 (A.N. SSSR Siberian Computer Center, Novosibirsk, 1980), pp. 81–110.

[1.39] M.Ya. Ivanov, V.V. Koretsky, N.Ya. Kurochkina: *Numerical Methods in Continuum Mechanics* (in Russian), **11**, No. 2 (A.N. SSSR Siberian Computer Center, Novosibirsk, 1980), pp. 41–63.

[1.40] M.Ya. Ivanov, V.V. Koretsky, N.Ya. Kurochkina: *Numerical Methods in Continuum Mechanics* (in Russian), **11**, No. 4 (A.N. SSSR Siberian Computer Center, Novosibirsk, 1980), pp. 88–103.

[1.41] M.Ya. Ivanov, V.V. Koretsky: "An Analysis of the Properties of Difference Schemes of Gas Dynamics", in *5th All-Union Congress on Theoretical and Applied Mechanics, Alma-Ata, 1981. Abstracts of the Reports* (Nauka, Alma-Ata, 1981), pp. 169–170.

[1.42] M.Ya. Ivanov: *Zhurn. Vychisl. Matem. i Matem. Fiz.* (in Russian) **22**, 411–417 (1982).

[1.43] S.I. Mukhin, S.B. Popov, Yu.P. Popov: "Dispersion and Dissipative Properties of Difference Schemes For Nonlinear Transport Equations"; Preprint No. 150 (M.V. Keldysh Institute of Applied Mathematics of the U.S.S.R. Academy of Sciences, Moscow, 1981).

[1.44] Yu.N. Sadkov: "Differential Approximations of Difference Schemes and Discontinuous Solutions of Quasilinear Equations", in *Proceedings of the Computer Centre of the Moscow State University*, No. 34 (Moscow State University, Moscow, 1981), pp. 136–144.

[1.45] E.V. Vorozhtsov, N.N. Yanenko: *Comput. Fluids* **8**, 313–326 (1980).

[1.46] S.K. Godunov, A.V. Zabrodin, G.P. Prokopov: *Zhurn. Vychisl. Matem. i Matem. Fiz.* (in Russian) **1**, 1020–1050 (1961).

[1.47] A. Tondl: *Domains of Attraction for Non-Linear Systems. No. 8. Self-Excited Vibrations. No. 9* (National Research Institute for Machine Design Běchovice, Běchovice, 1970).

[1.48] F.R. Gantmacher: *Theory of Matrices*, 3rd edn (Nauka, Moscow, 1967).

[1.49] V.E. Neuvažayev: "Nonadiabatic Motions in an Ideal Gas (Self-Similar Solutions)", in *Difference Methods of Solving Mathematical Physics Problems*, ed. by N.N. Yanenko, Trudy MIAN SSSR, Vol. 122 (Nauka, Moscow, 1973), pp. 24–51.

[1.50] W. Marshall: *Proc. Roy. Soc. London, Ser. A*, **233**, 367–376 (1955).
[1.51] J.D. Jukes: *J. Fluid Mech.* **3**, 275–285 (1957).
[1.52] P. Germain: "Contribution à la théorie des ondes de choc en magnétodynamique des fluides". ONERA publication, No. 97 (Paris, 1959).
[1.53] V.V. Rusanov: Lecture Notes in Physics, Vol. 8 (Springer-Verlag, Berlin, Heidelberg, New York, 1970), pp. 270–278.
[1.54] G. Jennings: *Commun. Pure Appl. Math.* **37**, 25–37 (1974).
[1.55] V.V. Rusanov, I.V. Bezmenov: "On the Limit Profile of the Nonlinear Discontinuity in Difference Schemes for a One-Dimensional Quasilinear Equation"; Preprint, No. 69 (M.V. Keldysh Institute of Applied Mathematics of the U.S.S.R. Academy of Sciences, Moscow, 1980).
[1.56] I.V. Bezmenov: "Formation of a Limit Profile of a Discrete Discontinuity in the Shock Wave Computation by Difference Schemes"; Preprint No. 60 (M.V. Keldysh Institute of Applied Mathematics of the U.S.S.R. Academy of Sciences, Moscow, 1981).
[1.57] V.V. Rusanov, I.V. Bezmenov: *Dokl. Akad. Nauk SSSR* **261**, 817–820 (1981).
[1.58] V.V. Rusanov, I.V. Bezmenov: *Trudy Matem. Inst. Akad. Nauk SSSR* **157**, 178–190 (1981).
[1.59] I.V. Bezmenov: *Dokl. Akad. Nauk SSSR* **277**, 14–16 (1984).
[1.60] V.V. Rusanov, I.V. Bezmenov: "Existence of a Limit Profile of Shock Wave Type for TVD Schemes"; Preprint No. 177 (M.V. Keldysh Institute of Applied Mathematics of the U.S.S.R. Academy of Sciences, Moscow, 1986).
[1.61] M. Morduchov, P. Libby: *J. Aeronaut. Sci.* **16**, 674–684, 704 (1949).
[1.62] E.V. Vorozhtsov: *Numerical Methods in Continuum Mechanics* (in Russian) **8**, No. 2 (A.N. SSSR Siberian Computer Center, Novosibirsk, 1977), pp. 12–27.
[1.63] K. Srinivas, J. Gururaja: *Indian J. Technology* **19**, 405–408 (1981).
[1.64] B. van Leer: "A Choice of Difference Schemes for Ideal Compressible Flow"; Ph.D. Thesis, University of Leiden (1970).
[1.65] V.V. Rusanov: "Difference Schemes of Third-Order Accuracy for "Across"-Computation of Discontinuous Solutions", in *Fluid Dynamics Transactions*, Vol. 4 (Państwowe Wydawnictwo Naukowe, Warszawa, 1969), pp. 285–294.
[1.66] A. Harten: *Commun. Pure Appl. Math.* **30**, 611–638 (1977).
[1.67] A. N. Minailos: *Zhurn. Vychisl. Matem. i Matem. Fiz.* **17**, 1058–1063 (1977).
[1.68] M.Ya. Ivanov, A.N. Kraiko, N.V. Mikhailov: *Zhurn. Vychisl. Matem. i Matem Fiz.* **12**, 441–463 (1972).
[1.69] B. Sturtevant: "Studies of Shock Focusing and Nonlinear Resonance in Shock Tubes", in *Recent Developments in Shock Tube Research* (Stanford, California, 1973), pp. 23–34.
[1.70] A.V. Potapkin: *Numerical Methods in Continuum Mechanics* (in Russian) **14**, No. 3 (A.N. SSSR Siberian Computer Center, Novosibirsk, 1983), pp. 126–139.

Chapter 2

[2.1] J. von Neumann: *Theory of Self-Reproducing Automata* (University of Illinois Press, Urbana, 1966).
[2.2] V.V. Rusanov: Lecture Notes in Physics, Vol. 18 (Springer-Verlag, Berlin, Heidelberg, New York, 1973), pp. 154–162.
[2.3] V.V. Rusanov: Lecture Notes in Physics, Vol. 141 (Springer-Verlag, Berlin, Heidelberg, New York, 1981), pp. 31–43.
[2.4] W.K. Giloi: *Interactive Computer Graphics* (Prentice-Hall, Englewood Cliffs, New Jersey, 1978).
[2.5] O.M. Belotserkovsky, Yu.M. Davydov: *Coarse Particles Method in Gas Dynamics. Computational Experiment* (Nauka, Moscow, 1982).
[2.6] P.G. Buning, J. Steger: "Graphics and Flow Visualization in Computational Fluid Dynamics", in *AIAA 7th Computational Fluid Dynamics Conference, Cincinnati, Ohio, July 15–17, 1985. Collect. Technical Paper* (AIAA, New York 1985), pp. 162–170.
[2.7] R.D. Richtmyer, K.W. Morton: *Difference Methods for Initial-Value Problems*, 2nd edn (Wiley Interscience, New York, 1967).

[2.8] C.W. Hirt, B.D. Nichols: *J. Comput. Phys.* **39**, 201–225 (1981).

[2.9] I.-L. Chern, J. Glimm, O. McBryan, B. Plohr, S. Yaniv: *J. Comput. Phys.* **62**, 83–110 (1986).

[2.10] A.N. Kraĭko, V.E. Makarov, N.I. Tilliaieva: *Zhurn. Vychisl. Matem. i Matem Fiz.* **20**, 716–723 (1980).

[2.11] V.E. Makarov: *Zhurn. Vychisl. Matem. i Matem. Fiz.* **22**, 1218–1226 (1982).

[2.12] V.I. Mileshin, N.I. Tilliaieva: "Supersonic Flow with a Detached Shock Wave Around the Bodies with Annular Channel", in *5th All-Union Congress on Theoretical and Applied Mechanics, Alma-Ata, 1981. Abstracts of the Reports* (Nauka, Alma-Ata, 1981), p. 256.

[2.13] A.V. Potapkin: *Numerical Methods in Continuum Mechanics* (in Russian), **14**, No. 3 (A.N. SSSR Siberian Computer Center, Novosibirsk, 1983), pp. 126–139.

[2.14] P. Lötstedt: *J. Comput. Phys.* **47**, 211–228 (1982).

[2.15] G.E. Weeks, T.L. Cost: *Internat J. Numer. Methods Engrg.* **14**, 441–449 (1979).

[2.16] A.J. Chorin: Lecture Notes in Physics, Vol. 59 (Springer-Verlag, Berlin, Heidelberg, New York, 1976), pp. 129–134.

[2.17] A.J. Chorin: *J. Comput. Phys.* **25**, 253–272 (1977).

[2.18] G.A. Sod: *J. Comput. Phys.* **27**, 1–31 (1978).

[2.19] A.J. Chorin: *J. Comput. Phys.* **35**, 1–11 (1980).

[2.20] R.J. DiPerna: *Commun. Pure Appl. Math.* **35**, 379–449 (1982).

[2.21] G. Marshall, B. Plohr: *J. Comput. Phys.* **56**, 410–427 (1984).

[2.22] J. Glimm, D. Marchesin, O. McBryan: *J. Comput. Phys.* **37**, 336–354 (1980).

[2.23] P. Concus, W. Proskurowski: *J. Comput. Phys.* **30**, 153–166 (1979).

[2.24] G.R. Shubin: *Comput. Fluids* **9**, 299–312 (1981).

[2.25] R.W. MacCormack: AIAA Paper 69-354 (1969).

[2.26] J.B. Bell, G.R. Shubin, J.M. Solomon: *J. Comput. Phys.* **48**, 223–245 (1982).

[2.27] W.L. Miranker, O. Pironneau: Rapport de Recherche No. 123 (IRIA, Domaine de Voluceau Rocquencourt, 78150 Le Chesnay, 1975).

[2.28] W.L. Miranker, O. Pironneau: *Internat J. Comput. Math. Appl.* **2**, 63–71 (1976).

[2.29] A.N. Lyubimov, V.V. Rusanov: *Gas Flows Past Blunt Bodies. Part I. Calculation Method and the Analysis of the Flows* (Nauka, Moscow, 1970) [English transl.: NASA-TT-F715, February, 1973].

[2.30] S.K. Godunov, A.V. Zabrodin, M.Ya. Ivanov, A.N. Kraĭko, G.P. Prokopov: *Numerical Solution of Multidimensional Problems of Gas Dynamics* (Nauka, Moscow, 1976).

[2.31] N.N. Yanenko, V.M. Fomin, E.V. Vorozhtsov: "Differential Analysers of Shock Waves in Shock-Capturing Schemes for Gas Dynamics Problems"; Preprint No. 7 (Institute of Theoretical and Applied Mechanics of the U.S.S.R. Academy of Sciences, Novosibirsk, 1978).

[2.32] F.H. Harlow: "The Particle-In-Cell Computing Method for Fluid Dynamics", in *Methods in Computational Physics*, ed. by B. Alder, S. Fernbach, M. Rotenberg, Vol. 3: Fundamental Methods in Hydrodynamics (Academic Press, New York, 1964), pp. 319–343.

[2.33] V.I. Kosarev: *Zhurn. Vychisl. Matem, i Matem. Fiz.* **11**, 1262–1271 (1971).

[2.34] M.Ya. Ivanov: *Zhurn. Vychisl. Matem. i Matem Fiz.* **15**, 1222–1240 (1975).

[2.35] E.V. Vorozhtsov, N.E. Ermolin, V.M. Fomin: *Numerical Methods in Continuum Mechanics* (in Russian), **10**, No. 2 (A.N. SSSR Siberian Computer Center, Novosibirsk, 1979), pp. 30–39.

[2.36] I.M. Vasenin, A.D. Rychkov: *Numerical Methods in Continuum Mechanics* (in Russian), **1**, No. 2 (A.N. SSSR Siberian Computer Center, Novosibirsk, 1970), pp. 3–9.

[2.37] G.S. Rosliakov, V.P. Suhorukov: "Difference Method for Calculation of Shocked Flows", in *Computational Methods and Programming*, Vol. 19 (Publication of the Moscow State University, Moscow, 1972), pp. 83–96.

[2.38] C.P. Kentzer: *J. Comput. Phys.* **6**, 168–182 (1972).

[2.39] N.N. Anuchina: *Numerical Methods in Continuum Mechanics* (in Russian), **1**, No. 4 (A.N. SSSR Siberian Computer Center, Novosibirsk, 1970), pp. 3–84.

[2.40] *Theoretical Fundamentals and Construction of Numerical Algorithms for Mathematical Physics Problems*, ed. by K.I. Babenko (Nauka, Moscow, 1979).

[2.41] B.L. Roždestvenskiĭ, N.N. Janenko: *Systems of Quasilinear Equations and Their Applications to Gas Dynamics*, 2nd edn (Nauka, Moscow, 1978) [English transl.: *Systems of Quasilinear Equations and Their Applications to Gas Dynamics*, Translations of Mathematical Monographs, Vol. 55 (American Mathematical Society, Providence, Rhode Island, 1983)].

[2.42] M.L. Wilkins: "Calculation of Elastic–Plastic Flow", in *Proceedings of the Section in Numerical Methods in Gas Dynamics of the Second International Colloquium on Gas Dynamics of Explosion and Reacting Systems* (*Novosibirsk, August 19–23, 1969*), Vol. 1 (Publication of the U.S.S.R. Academy of Sciences Computer Center, Moscow, 1971), pp. 408–517.

[2.43] P. Laval: *Rech. Aérospat.* No. 131, 3–16 (1969).

[2.44] P.J. Roache: *Computational Fluid Dynamics* (Hermosa, Albuquerque, New Mexico, 1976).

[2.45] J. Glimm, D. Marchesin, O. McBryan: *J. Comput. Phys.* **39**, 179–200 (1981).

[2.46] E.F. Zhigalko: "Diagnosis of Discontinuities in Self-Similar Nonstationary Gas Flow", in *Flow of a Viscous and Inviscid Gas*, ed. by N.N. Polyahov (Publication of the Leningrad State University, Leningrad, 1981), pp. 77–83.

[2.47] N.N. Yanenko, E.V. Vorozhtsov, V.M. Fomin: *Dokl. Akad. Nauk SSSR* **227**, 50–53 (1976) [English transl.: *Soviet Math. Dokl.* **17**, 358–362 (1976)].

[2.48] G.B. Whitham: *Linear and Nonlinear Waves* (Wiley Interscience, New York, 1974).

[2.49] N.N. Yanenko, E.V. Vorozhtsov: "Differential Analysers of Strong Discontinuities in One-Dimensional Gas Flow", in *Computational Techniques in Transient and Turbulent Flow*, ed. by C. Taylor and K. Morgan (Pineridge Press, Swansea, 1981), pp. 59–96.

[2.50] A.M. Il'in, O.A. Oleinik: *Mat. Sbornik* **51**, 191–216 (1960).

[2.51] L.A. Peletier: "Asymptotic Stability of Travelling Waves", in *IUTAM Symposium on Instability of Continuous Systems* (Springer-Verlag, Berlin, 1971), pp. 418–422.

[2.52] C. Boldrighini, L. Triolo: *Meccanica* **12**, 15–18 (1977).

[2.53] E. Hopf: *Commun. Pure Appl. Math.* **3**, 201–230 (1950).

[2.54] V.M. Fomin, E.V. Vorozhtsov, N.N. Yanenko: *Comput. Fluids* **4**, 171–183 (1976).

[2.55] E.V. Vorozhtsov, V.M. Fomin, N.N. Yanenko: *Numerical Methods in Continuum Mechanics* (in Russian), 7, No. 6 (A.N. SSSR Siberian Computer Center, Novosibirsk, 1976), pp. 8–23.

[2.56] E.V. Vorozhtsov: "Differential Analysers of Shock Waves in Shock-Capturing Schemes for Gas Dynamics Problems"; Thesis (Computer Center of the U.S.S.R. Academy of Sciences, Siberian Division, Novosibirsk, 1976).

[2.57] N.N. Yanenko, V.M. Kovenya, V.D. Lisejkin, V.M. Fomin, E.V. Vorozhtsov: *Comput. Methods Appl. Mech. Engrg.* **17/18**, Part III, 659–671 (1979).

[2.58] E.V. Vorozhtsov, N.N. Yanenko: *Comput. Fluids* **8**, 313–326 (1980).

[2.59] J. von Neumann, R. Richtmyer: *J. Appl. Phys.* **21**, 232–237 (1950).

[2.60] S. Lefschetz: *Differential Equations: Geometric Theory* (Interscience, New York, 1957).

[2.61] N.N. Bautin, E.A. Leontovich: *Methods and Techniques of Qualitative Investigation of Dynamic Systems in the Plane* (Nauka, Moscow, 1976).

[2.62] J. White: *J. Comput. Phys.* **11**, 573–590 (1973).

[2.63] V.I. Arnold: *Additional Chapters of the Theory of Ordinary Differential Equations* (Nauka, Moscow, 1978).

[2.64] I.G. Cameron: *J. Comput. Phys.* **1**, 1–20 (1966).

[2.65] R. Latter: *J. Appl. Phys.* **26**, 955–960 (1955).

[2.66] A.A. Samarskiĭ, Yu.P. Popov: *Difference Schemes of Gas Dynamics* (Nauka, Moscow, 1975).

[2.67] V.P. Nartov, G.G. Chernykh: "On the Numerical Simulation of a Flow Arising During the Collapse of a Mixing Zone in a Stratified Medium"; Preprint No. 15 (Institute of Theoretical and Applied Mechanics of the U.S.S.R. Academy of Sciences, Novosibirsk, 1982).

[2.68] Z. Usakov: *Numerical Methods in Continuum Mechanics* (in Russian), 10, No. 6 (A.N. SSSR Siberian Computer Center, Novosibirsk, 1979), pp. 141–149.

[2.69] E.V. Vorozhtsov: *Numerical Methods in Continuum Mechanics* (in Russian), 8, No. 2 (A.N. SSSR Siberian Computer Center, Novosibirsk, 1977), pp. 12–27.

[2.70] N.N. Yanenko, N.N. Anuchina, V.E. Petrenko, Yu.I. Shokin: *Numerical Methods in Continuum Mechanics* (in Russian), **1**, No. 1 (A.N. SSSR Siberian Computer Center, Novosibirsk, 1970), pp. 40–62.

[2.71] A.A. Samarskiĭ, V.Ya. Arsenin: *Zhurn. Vychisl. Matem. i Matem. Fiz.* (in Russian) **1**, 357–360 (1961).

[2.72] V.I. Paasonen: *Numerical Methods in Continuum Mechanics* (in Russian), **4**, No. 4 (A.N. SSSR Siberian Computer Center, Novosibirsk, 1973), pp. 44–57.

[2.73] R.F. Warming, B.J. Hyett: *J. Comput. Phys.* **14**, 159–179 (1974).

[2.74] R. Liska: *Comput. Phys. Commun.* **34**, 175–186 (1984).

[2.75] L.D. Cloutman, L.W. Fullerton: *J. Comput. Phys.* **29**, 141–144 (1978).

[2.76] A.N. Valiullin, V.G. Ganzha, S.I. Mazurik, F.A. Murzin, V.P. Shapeev, N.N. Yanenko: "Applications of Computers for the Investigation and Construction of Difference Schemes", in *Analytical Calculations on a Computer and Their Applications to Theoretical Physics* (Publication of the Joint Institute of Nuclear Research, Dubna, 1983), pp. 85–96.

[2.77] V.G. Ganzha: *Numerical Methods in Continuum Mechanics* (in Russian), **14**, No. 2 (A.N. SSSR Siberian Computer Center, Novosibirsk, 1983), pp. 39–44.

[2.78] A.N. Valiullin, V.G. Ganzha, V.P. Il'in, V.P. Shapeev, N.N. Yanenko: *Dokl. Akad. Nauk SSSR* **275**, 528–532 (1984).

[2.79] E.V. Vorozhtsov, V.G. Ganzha, V.P. Shapeev: "On Automatic Computer Derivation of Differential Approximations of the Schemes of the Fractional Step Method"; Preprint No. 23 (Institute of Theoretical and Applied Mechanics of the U.S.S.R. Academy of Sciences, Novosibirsk, 1984).

[2.80] L.R. Foy: *Commun. Pure Appl. Math.* **17**, 177–188 (1964).

[2.81] L.S. Pontriagin: *Ordinary Differential Equations*, 3rd edn (Nauka, Moscow, 1970).

[2.82] E. Anderson: *Magnetohydrodynamic Shock Waves* (M.I.T. Press, Cambridge, Massachusetts, 1963).

[2.83] P. Germain: "Contribution à la théorie des ondes de choc en magnétodynamique des fluides"; ONERA publication, No. 97 (Paris, 1959).

[2.84] A.G. Kulikovsky, F.A. Slobodkina: *Prikl. Matem. i Mekhanika* (in Russian) **31**, 593–602 (1967).

[2.85] L.A. Vulis, P.L. Gusika, G.V. Zhizhin: *Zhurn. Prikl. Mekhaniki i Tehnič. Fiziki* (in Russian), No. 5, 143–156 (1972).

[2.86] G.V. Zhizhin: *Inzhenerno-Fizičeskiĭ Zhurn.* (in Russian) **32**, 96–101 (1977).

[2.87] N.N. Yanenko, R.I. Soloukhin, A.N. Papyrin, V.M. Fomin: *Supersonic Two-Phase Flows Under Conditions of Velocity Non-Equilibrium of Particles* (Nauka, Siberian Division, Novosibirsk, 1980).

[2.88] S.K. Godunov, A.V. Zabrodin, G.P. Prokopov: *Zhurn. Vychisl. Matem. i Matem. Fiz.* (in Russian) **1**, 1020–1050 (1961).

[2.89] N.N. Yanenko, A.V. Feodorov, V.M. Fomin: *Dokl. Akad. Nauk SSSR* **254**, 554–559 (1980).

[2.90] A.J. Chorin, T.J.R. Hughes, M.F. McCracken, J.E. Marsden: *Commun. Pure Appl. Math.* **31**, 205–256 (1978).

[2.91] N.A. Larkin, V.A. Novikov, N.N. Yanenko: *Nonlinear Equations of Variable Type* (Nauka, Siberian Division, Novosibirsk, 1983).

[2.92] A. Lerat: Lecture Notes in Physics, Vol. 90 (Springer-Verlag, Berlin, Heidelberg, New York, 1979), pp. 345–351.

[2.93] M.Ya. Ivanov, V.V. Koretsky, N.Ya. Kurochkina: *Numerical Methods in Continuum Mechanics* (in Russian), **11**, No. 2 (A.N. SSSR Siberian Computer Center, Novosibirsk, 1980), pp. 41–63.

[2.94] S.I. Mukhin, S.B. Popov, Yu.P. Popov: "Dispersion and Dissipative Properties of Difference Schemes for Nonlinear Transport Equation"; Preprint No. 150 (M.V. Keldysh Institute of Applied Mathematics of the U.S.S.R. Academy of Sciences, Moscow, 1981).

[2.95] P.D. Lax: *Commun. Pure Appl. Math.* **10**, 537–566 (1957).

[2.96] Yu.I. Shokin: *Differential Approximation Method* (Nauka, Siberian Division, Novosibirsk, 1979) [English transl.: Y.I. Shokin: *The Method of Differential Approximation* (Translated from the Russian by K.G. Roesner) (Springer-Verlag, Berlin, Heidelberg, New York, 1983)].

References   381

[2.97] Yu.I. Shokin, N.N. Yanenko: *Differential Approximation Method. Application to Gas Dynamics* (Nauka, Siberian Division, Novosibirsk, 1985).
[2.98] P.D. Lax: *Commun. Pure Appl. Math.* **7**, 159–193 (1954).
[2.99] V.V. Rusanov: *Zhurn. Vychisl. Matem. i Matem. Fiz.* **1**, 267–279 (1961).
[2.100] R.A. Gentry, R.E. Martin, B.J. Daly: *J. Comput. Phys.* **1**, 87–118 (1966).
[2.101] E.V. Vorozhtsov, N.N. Yanenko: *Comput. Fluids* **9**, 17–32 (1981).
[2.102] M.Ya. Ivanov, O.A. Ryl'ko: *Zhurn. Vychisl. Matem. i Matem. Fiz.* **12**, 1280–1291 (1972).
[2.103] A. Lerat, R. Peyret: *Rech. Aérospat.* No. 2, 61–79 (1975).
[2.104] W. Stark, R. Wojcieczynski: *Z. Angew. Math. Mech.* **60**, S. 383–391 (1980).
[2.105] R. Peyret, T.D. Taylor: *Computational Methods for Fluid Flow* (Springer-Verlag, New York, Heidelberg, Berlin, 1983).
[2.106] E.V. Vorozhtsov: "On Differential Analysers of Shock Fronts", in *Program Complexes of Mathematical Physics*, ed. by V.I. Karnachuk (Publication of the Computer Center of the U.S.S.R. Academy of Sciences, Siberian Division, Novosibirsk, 1972), pp. 98–115.
[2.107] *Shock Tubes*, ed. by Kh. A. Rakhmatulin and S.S. Semeonov (Izdatel'stvo Inostran. Literatury, Moscow, 1962).
[2.108] H. Tijdeman, R. Seebass: *Ann. Rev. Fluid Mechanics* **12**, 181–202 (1980).
[2.109] *Shock Waves. A Bibliography Index of the Home and Foreign Literature* 1960–1969, ed. by N.A. Generalov (Nauka, Moscow, 1979).
[2.110] G. Moretti: Lecture Notes in Physics, Vol. 35 (Springer-Verlag, New York, Heidelberg, Berlin, 1975), pp. 287–292.
[2.111] G. Moretti: "The Importance of Boundary Conditions in the Numerical Treatment of Hyperbolic Equations", in *High-Speed Computing in Fluid Dynamics*, Physics of Fluids, Supplement II (American Institute of Physics, New York, 1969), pp. II-13–II-20.
[2.112] M.D. Salas: *AIAA J.* **14**, 583–588 (1976).
[2.113] Y. Yamamoto, K. Karashima: *AIAA J.* **20**, 9–17 (1982).
[2.114] M.Ya. Ivanov, V.V. Koretsky, N.Ya. Kurochkina: *Numerical Methods in Continuum Mechanics* (in Russian), **11**, No. 1 (A.N. SSSR Siberian Computer Center, Novosibirsk, 1980), pp. 81–110.
[2.115] T.H. Chong: *SIAM J. Numer. Anal.* **15**, 835–857 (1978).
[2.116] B.L. Lohar, P.C. Jain: *J. Comput. Phys.* **39**, 433–442 (1981).
[2.117] J.P. Boris, N.K. Winsor: "Vectorized Computation of Reactive Flow", in *Parallel Computations*, ed. by G. Rodrigue (Academic Press, New York, 1982), pp. 173–215.
[2.118] J.H. Ahlberg, E.M. Nilson, J.L. Walsh: *The Theory of Splines and Their Application* (Academic Press, New York, 1967).
[2.119] V.D. Liseikin: *Numerical Methods in Continuum Mechanics* (in Russian), **12**, No. 1 (A.N. SSSR Siberian Computer Center, Novosibirsk, 1981), pp. 78–81.

Chapter 3

[3.1] Yu.I. Shokin: *Differential Approximation Method* (Nauka, Siberian Division, Novosibirsk, 1979).
[3.2] P.J. Roache: *Computational Fluid Dynamics* (Hermosa, Albuquerque, New Mexico, 1976).
[3.3] F.H. Harlow: "The Particle-In-Cell Computing Method for Fluid Dynamics", in *Methods in Computational Physics*, ed. by B. Alder, S. Fernbach, M. Rotenberg, Vol. 3: Fundamental Methods in Hydrodynamics (Academic Press, New York, 1964), pp. 319–343.
[3.4] R.A. Gentry, R.E. Martin, B.J. Daly: *J. Comput. Phys.* **1**, 87–118 (1966).
[3.5] N.N. Yanenko, V.M. Kovenya, V.D. Lisejkin, V.M. Fomin, E.V. Vorozhtsov: *Comput. Methods Appl. Mech. Engrg.* **17/18**, Part III, 659–671 (1979).
[3.6] P. Lötstedt: *J. Comput. Phys.* **47**, 211–228 (1982).
[3.7] J. Glimm, D. Marchesin, O. McBryan: *J. Comput. Phys.* **39**, 179–200 (1981).
[3.8] P.R. Woodward: "Trade-Offs in Designing Explicit Hydrodynamical Schemes for Vector Computers", in *Parallel Computations*, ed. by G. Rodrigue (Academic Press, New York, 1982), pp. 153–171.

[3.9]   E.V. Vorozhtsov: "On Differential Analysers of Shock Fronts", in *Program Complexes of Mathematical Physics*, ed. by V.I. Karnachuk (Publication of the Computer Center of the U.S.S.R. Academy of Sciences Siberian Division, Novosibirsk, 1972), pp. 98–115.

[3.10] V.M. Fomin, E.V. Vorozhtsov, N.N. Yanenko: *Comput. Fluids* **7**, 109–121 (1979).

[3.11] F.H. Harlow: "The Particle-In-Cell Method for Two-Dimensional Hydrodynamic Problems"; Report No. LAMS-2082 (Los Alamos Scientific Laboratory, Los Alamos, New Mexico, 1956).

[3.12] N.N. Anuchina: *Numerical Methods in Continuum Mechanics* (in Russian) **1**, No. 4 (A.N. SSSR Siberian Computer Center, Novosibirsk, 1970), pp. 3–84.

[3.13] N.N. Yanenko, N.N. Anuchina, V.E. Petrenko, Yu.I. Shokin: *Numerical Methods in Continuum Mechanics* (in Russian) **1**, No. 1 (A.N. SSSR Siberian Computer Center, Novosibirsk, 1970), pp. 40–62.

[3.14] M.D. Taran: "Method of Finite Size Particles for Modeling of Many-Domain Two-Dimensional Gas Dynamics Problems"; Preprint No. 177 (M.V. Keldysh Institute of Applied Mathematics of the U.S.S.R. Academy of Sciences, Moscow, 1979).

[3.15] M.D. Taran, T.V. Taran, A.P. Favorsky: "Algorithm of Numerical Modeling of Hydrodynamic Flows with the Aid of Finite Size Particles"; Preprint No. 114 (M.V. Keldysh Institute of Applied Mathematics of the U.S.S.R. Academy of Sciences, Moscow, 1979).

[3.16] *Theoretical Fundamentals and Construction of Numerical Algorithms for Mathematical Physics Problems*, ed. by K.I. Babenko (Nauka, Moscow, 1979).

[3.17] V.M. Fomin, N.E. Ermolin, E.A. Kroshko, A.V. Fedorov, E.V. Chubarova, R.T. Chernysheva: *Numerical Methods in Continuum Mechanics* (in Russian) **10**, No. 3 (A.N. SSSR Siberian Computer Center, Novosibirsk, 1979), pp. 138–141.

[3.18] T.L. Cook, R.B. Demuth, F.H. Harlow: *J. Comput. Phys.* **41**, 51–67 (1981).

[3.19] G.I. Robul: "Numerical Study of a Cratering Process in a High-Velocity Impact", in *Numerical Methods in Aerodynamics* (Publication of the Moscow State University, Moscow, 1981), pp. 16–21.

[3.20] V.M. Fomin, V.P. Shapeev, N.N. Yanenko: *Comput. Methods Appl. Mech. Engrg.* **32**, 157–197 (1982).

[3.21] Xu Guoyong, Yu Zhilu, Liao Zhen-min, Yuan Xianchun, Zhon Shurong: *Acta Mech. Sinica* **3**, 207–216 (1982).

[3.22] J.U. Brackbill, H.M. Ruppel: *J. Comput. Phys.* **65**, 314–343 (1986).

[3.23] M.W. Evans, F.H. Harlow: "The Particle-In-Cell Method for Hydrodynamic Calculations"; Los Alamos Scientific Laboratory Report No. LA-2139 (1957).

[3.24] Yu.I. Shokin, A.I. Urusov: "On Invariant Difference Splitting-Up Schemes", in *Proceedings of the IVth All-Union Seminar on Numerical Methods and Viscous Fluid Mechanics*, ed. by B.G. Kuznetsov (Computer Center of the U.S.S.R. Academy of Sciences Siberian Division, Novosibirsk, 1973), pp. 192–209.

[3.25] V.E. Petrenko, G.A. Sapozhnikov: "On the Stability and Accuracy of the Particle-In-Cell Method for Viscous Fluid Flows", in *Some Problems of Numerical and Applied Mathematics* (Nauka, Siberian Division, Novosibirsk, 1975), pp. 95–111.

[3.26] R. Collins, H.T. Chen: 'Propagation of a Shock Wave of Arbitrary Strength in Two Half-Planes Containing a Free Surface", in *Proceedings of the Section in Numerical Methods in Gas Dynamics of the Second International Colloquium on Gas Dynamics of Explosion and Reacting Systems (Novosibirsk, August 19–23, 1969)*, Vol. 1 (Publication of the U.S.S.R. Academy of Sciences Computer Center, Moscow, 1971), pp. 1.78–226.

[3.27] R.D. Richtmyer, K.W. Morton: *Difference Methods for Initial-Value Problems*, 2nd edn (Wiley Interscience, New York, 1967).

[3.28] A.I. Tolstykh: *Dokl. Akad. Nauk SSSR* **210**, 48–51 (1973).

[3.29] S.K. Godunov, A.V. Zabrodin, M.Ya. Ivanov, A.N. Kraĭko, G.P. Prokopov: *Numerical Solution of Multidimensional Problems of Gas Dynamics* (Nauka, Moscow, 1976).

[3.30] V.D. Lisejkin, N.N. Yanenko: *Numerical Methods in Continuum Mechanics* (in Russian) **7**, No. 2 (A.N. SSSR Siberian Computer Center, Novosibirsk, 1976), pp. 75–82.

[3.31] N.N. Yanenko, N.T. Danaev, V.D. Lisejkin: *Numerical Methods in Continuum Mechanics* (in Russian) **8**, No. 4 (A.N. SSSR Siberian Computer Center, Novosibirsk, 1977), pp. 157–163.

[3.32] V.M. Kovenya, N.N. Yanenko: *Zhurn. Vychisl. Matem. i Matem. Fiz.* **19**, 174–188 (1979).

[3.33] V.M. Kovenya, N.N. Yanenko: *Comput. Fluids* **8**, 59–70 (1980).

[3.34] N.N. Yanenko, V.M. Kovenya, V.D. Lisejkin, V.M. Fomin, E.V. Vorozhtsov: *Comput. Methods Appl. Mech. Engrg.* **17/18**, Part III, 659–671 (1979).
[3.35] V.M. Kovenya, N.N. Yanenko: *Method of Splitting-Up in Gas Dynamics Problems* (Nauka, Siberian Division, Novosibirsk, 1981).
[3.36] A.G. Zarubin: *Uchenye Zapiski TsAGI* (in Russian) **11**, No. 6, 120–124 (1980).
[3.37] C.M. Ablow: *Appl. Math. Comput.* No. 10–11, 859–863 (1982).
[3.38] J.K. Dukowicz: *J. Comput. Phys.* **54**, 411–424 (1984).
[3.39] A.I. Gulidov, V.M. Fomin: "Modification of the Wilkins Method for Solving the Problems on the Collisions of Bodies"; Preprint No. 49 (Institute of Theoretical and Applied Mechanics of the U.S.S.R. Academy of Sciences, Novosibirsk, 1980).
[3.40] C.W. Hirt, A.A. Amsden, J.L. Cook: *J. Comput. Phys.* **14**, 227–253 (1974).
[3.41] H.G. Horak, E.M. Jones, J.W. Kodis, M.T. Sandford II: *J. Comput. Phys.* **26**, 277–284 (1978).
[3.42] A.I. Sadyrin: "A Development of the Packed Particles Method for Solving the Problems of Impact and of the Penetration of Deformable Bodies", in *Theoretical Fundamentals and the Construction of Numerical Algorithms for the Solution of Mathematical Physics Problems. VI All-Union School, Gorki, September 8–13, 1986. Abstracts of the Reports* (Publication of the Gorki State University, Gor'ky, 1986), p. 134.
[3.43] A.A. Aganin, V.B. Kuznetsov: "Method of Conservative Interpolation of Integral Parameters of the Cells of Arbitrary Meshes", in *Dynamics of Shells in a Flow.* Proceedings of the Seminar, Issue No. 18 (Kazan, 1985), pp. 144–160.
[3.44] N.N. Yanenko, V.M. Kovenya, V.D. Lisejkin, V.M. Fomin, E.V. Vorozhtsov: "On Some Methods for the Numerical Simulation of Flows with Complex Structure", in *6th International Conference on Numerical Methods in Fluid Dynamics*, Tbilisi, 1978, Vol. 2 (Publication of the Institute of Applical Mathematics of the U.S.S.R. Academy of Sciences, Moscow, 1978), pp. 211–224.
[3.45] J.F. Thompson: *AIAA J.* **22**, 1505–1523 (1984).
[3.46] J.F. Thompson, Z.U.A. Warsi, C.W. Mastin: *Numerical Grid Generation. Foundations and Applications* (North-Holland, New York, 1985).
[3.47] J.F. Thompson: *Appl. Numer. Math.* **1**, 3–27 (1985).
[3.48] J.U. Brackbill, J.S. Saltzman: *J. Comput. Phys.* **46**, 342–368 (1982).
[3.49] J. Saltzman, J. Brackbill: *Appl. Math. Comput.* No. 10–11, 865–884 (1982).
[3.50] V.D. Lisejkin: *Numerical Methods in Continuum Mechanics* (in Russian) **9**, No. 6 (A.N. SSSR Siberian Computer Center, Novosibirsk, 1978), pp. 115–118.
[3.51] V.D. Lisejkin, N.N. Yanenko: *Numerical Methods in Continuum Mechanics* (in Russian) **8**, No. 7 (A.N. SSSR Siberian Computer Center, Novosibirsk, 1977), pp. 100–104.
[3.52] G.H. Klopfer, D.S. McRae: "The Nonlinear Modified Equation Approach to Analyzing Finite Difference Schemes", in *AIAA 5th Computational Fluid Dynamics Conference Proceedings* (New York, 1981), pp. 317–332.
[3.53] T.L. Holst, D. Brown: "Transonic Airfoil Calculations Using Solution-Adaptive Grids", in *AIAA 5th Computational Fluid Dynamics Conference Proceedings* (New York, 1981), pp. 136–148.
[3.54] S.M. Bahrakh, Yy. P. Glagoleva, M.S. Samigulin, V.D. Frolov, N.N. Yanenko, Yu.V. Yanilkin: *Dokl. Akad. Nauk SSSR* **257**, 566–569 (1981).
[3.55] S.M. Bahrakh, I.G. Zhidov, V.G. Rogachov, Yu.V. Yanilkin: *Izv. Akad. Nauk SSSR, Ser. Mekhanika Zhidkosti i Gaza*, No. 2, 146–149 (1983).
[3.56] G. Ben-Dor, I.I. Glass: *AIAA J.* **16**, 1146–1153 (1978).
[3.57] G.V. Bazhenova, L.G. Gvozdeva, Yu.P. Lagutov, V.N. Liahov, Yu.M. Faresov, V.P. Fokeev: *Nonstationary Interactions of Shock Waves and Detonation Waves in Gases* (Nauka, Moscow, 1986).
[3.58] M.D. Salas: *AIAA J.* **14**, 583–587 (1976).
[3.59] N.N. Yanenko, Yu.I. Shokin, L.A. Kompaniets, Z.I. Fedotova: "Classification of Difference Schemes of Two-Dimensional Gas Dynamics by the Differential Approximation Method"; Preprint No. 19 (Institute of Theoretical and Applied Mechanics of the U.S.S.R. Academy of Sciences, Novosibirsk, 1982).
[3.60] N.N. Yanenko, Yu.I. Shokin: *Numerical Methods in Continuum Mechanics* (in Russian) **2**, No. 2 (A.N. SSSR Siberian Computer Center, Novosibirsk, 1971), pp. 85–92.
[3.61] N.N. Yanenko: *The Method of Fractional Steps for Solving Multidimensional Problems*

of Mathematical Physics (Nauka, Siberian Division, Novosibirsk, 1967) [English transl.: *The Method of Fractional Steps* (Springer-Verlag, New York, 1971)].

[3.62]   *High Velocity Impact Phenomena*, ed. by R. Kinslow (Academic Press, New York, 1970).

[3.63]   N.A. Zlatin, A.P. Krasilshchikov, G.I. Mishin, N.N. Popov: *Ballistic Devices and Their Application in Experimental Research* (Nauka, Moscow, 1974).

[3.64]   N.N. Yanenko, E.A. Kroshko, V.D. Lisejkin, V.M. Fomin, V.P. Shapeev, Yu.A. Shitov: Lecture Notes in Physics, Vol. 59 (Springer-Verlag, Berlin, Heidelberg, New York, 1976), pp. 454—459.

[3.65]   M.E. Backman, W. Goldsmith: *Internat. J. Engrg. Sci.* **16**, 1–99 (1978).

[3.66]   V.M. Fomin, V.P. Shapeev, N.N. Yanenko: *Comput. Methods Appl. Mech. Engrg.* **32**, 157–197 (1982).

[3.67]   N.N. Yanenko, V.M. Fomin, E.V. Vorozhtsov: "Differential Analysers of Shock Waves in Shock-Capturing Schemes for Gas Dynamics Problems"; Preprint No. 7 (Institute of Theoretical and Applied Mechanics of the U.S.S.R. Academy of Sciences, Novosibirsk, 1978).

[3.68]   V.E. Petrenko, E.V. Vorozhtsov: *Numerical Methods in Continuum Mechanics* (in Russian) **4**, No. 2 (A.N. SSSR Siberian Computer Center, Novosibirsk, 1973), pp. 132–141.

[3.69]   E.V. Vorozhtsov: *Comput. Fluids* **9**, 313–326 (1981).

[3.70]   G. Moretti: *Comput. Fluids* **7**, 191–205 (1979).

[3.71]   R. Peyret, T.D. Taylor: *Computational Methods for Fluid Flow* (Springer-Verlag, New York, Heidelberg, Berlin, 1983).

Chapter 4

[4.1]   B.L. Roždestvenskiĭ, N.N. Janenko: *Systems of Quasi-Linear Equations and Their Applications to Gas Dynamics*, 2nd edn (Nauka, Moscow, 1978) [English transl.: *Systems of Quasi-Linear Equations and Their Applications to Gas Dynamics*, Translations of Mathematical Monographs, Vol. 55 (American Mathematical Society, Providence, Rhode Island, 1983)].

[4.2]   R. Courant, K.O. Friedrichs: *Supersonic Flow and Shock Waves* (Interscience, New York, 1948).

[4.3]   G.V. Bazhenova, L.G. Gvozdeva, Yu.P. Lagutov, V.N. Liahov, Yu.M. Faresov, V.P. Fokeev: *Nonstationary Interactions of Shock Waves and Detonation Waves in Gases* (Nauka, Moscow, 1986).

[4.4]   I. Glass, G.N. Patterson: *J. Aeronaut. Sci.* **22**, 73–100 (1953).

[4.5]   P.D. Lax: *Commun. Pure Appl. Math.* **7**, 159–193 (1954).

[4.6]   K. Förster: *Ingenieur-Archiv* **37**, 45–55 (1968).

[4.7]   V.V. Rusanov: *Zhurn. Vychisl. Matem. i Matem. Fiz.* **1**, 267–279 (1961).

[4.8]   R.P. Fedorenko: *Zhurn. Vychisl. Matem. i Matem. Fiz.* **2**, 1122–1128 (1962).

[4.9]   V.V. Rusanov: "Calculation and Study of Multidimensional Gas Flows by the Method of Finite Differences"; Thesis (Institute of Applied Mathematics of the U.S.S.R. Academy of Sciences, Moscow, 1968).

[4.10]   V.V. Rusanov: "Difference Schemes of Third-Order Accuracy for "Across"-Computation of Discontinuous Solutions", in *Fluid Dynamics Transactions*, Vol. 4 (Państwowe Wydawnictwo Naukowe, Warszawa, 1969), pp. 285–294.

[4.11]   N.N. Kuznetsov: *Zhurn. Vychisl. Matem. i Matem. Fiz.* **12**, 334–351 (1972).

[4.12]   V.V. Rusanov: Lecture Notes in Physics, Vol. 59 (Springer-Verlag, Berlin, Heidelberg, New York, 1976), pp. 378–383.

[4.13]   A. Harten: "The Method of Artificial Compression: 1. Shocks and Contact Discontinuities"; A.E.C. Research and Development Report C00-3077-50 (New York University, New York, 1974).

[4.14]   A. Harten: *Commun. Pure Appl. Math.* **30**, 611–638 (1977).

[4.15]   P. Kutler, L. Sakell, G. Aiello: *AIAA J.* **13**, 361–367 (1975).

[4.16]   V.A. Gridneva, A.I. Korneev, V.G. Trushkov: "On the Numerical Method for Computation of a Flow of Elastic-Plastic Media With Big Deformations", in *Proceedings of the 5th Scientific Conference in Mathematics and Mechanics* (in Russian), Vol. 2 (Publication of the Tomsk State University, Tomsk, 1975), pp. 115–116.

[4.17] J.D. O'Keefe, Th.J. Ahrens: "Shock Effects from a Large Impact on the Moon", in *Proceedings of the 6th Lunar Science Conference*, Houston, Texas, Vol. 3 (Pergamon Press, New York, 1975), pp. 2831—2844.

[4.18] A.G. Awn, D.B. Spalding: "Flow Calculation in Injection Processes", in *Flow, Mixing and Heat Transfer in Furnaces. First Conference on Mechanical Power Engineering, Cairo, February, 1977*, ed. by K.H. Khalil (Pergamon Press, Oxford, 1978), pp. 195–214.

[4.19] G.E. Weeks, T.L. Cost: *Internat. J. Numer. Methods Engrg.* **14**, 441–449 (1979).

[4.20] A.N. Kraïko, V.E. Makarov, N.I. Tilliayeva: *Zhurn. Vychisl. Matem. i Matem. Fiz.* **20**, 716–723 (1980).

[4.21] V.V. Yeriomin: *Trudy NII Mehaniki MGU* (in Russian) No. 30, 176–181 (1973).

[4.22] S.M. Bahrach, Yu.P. Glagoleva, M.S. Samigulin, V.D. Frolov, N.N. Yanenko, Yu.V. Yanilkin: *Dokl. Akad. Nauk SSSR* **257**, 566–569 (1981).

[4.23] F.H. Harlow: "The Particle-In-Cell Computing Method for Fluid Dynamics", in *Methods in Computational Physics*, ed. by B. Alder, S. Fernbach, M. Rotenberg, Vol. 3: Fundamental Methods in Hydrodynamics (Academic Press, New York, 1964), pp. 319–343.

[4.24] S.M. Bahrach. I.G. Zhidov, V.G. Rogachov, Yu.V. Yanilkin: *Izv. Akad Nauk SSSR, Ser. Mekhanika Zhidkosti i Gaza* No. 2, 146–149 (1983).

[4.25] S.M. Bahrach, V.F. Spiridonov, A.A. Shanin: *Dokl. Akad. Nauk SSSR* **276**, 829–833 (1984).

[4.26] N.N. Yanenko, N.N. Anuchina, V.E. Petrenko, Yu.I. Shokin: *Numerical Methods in Continuum Mechanics* (in Russian) **1**, No. 1 (A.N. SSSR Siberian Computer Center, Novosibirsk, 1970), pp. 40–62.

[4.27] N.N. Anuchina: *Numerical Methods in Continuum Mechanics* (in Russian) **1**, No. 4 (A.N. SSSR Siberian Computer Center, Novosibirsk, 1970), pp. 3–84.

[4.28] I.-L. Chern, J. Glimm, P. McBryan, B. Plohr, S. Yaniv: *J. Comput. Phys.* **62**, 83–110 (1986).

[4.29] J. Glimm, O. McBryan, R. Menikoff, D.H. Sharp: *SIAM J. Sci. Statist. Comput.* **7**, 230–251 (1986).

[4.30] I.A. Belov, G.M. Rudakova, V.V. Tsymbalov: "A Numerical Study of Jet Flows Interaction with Obstacles in the Case When Recirculation Flow Domains Arise"; Preprint No. 33 (Computer Center of the U.S.S.R. Academy of Sciences Siberian Division, Krasnoyarsk, 1981).

[4.31] Yu.M. Davydov, M.S. Panteleev: *Zhurn. Prikl. Mekhaniki i Tehnič. Fiziki* (in Russian) No. 1, 117–122 (1981).

[4.32] V.P. Nartov, G.G. Chernykh: "On the Numerical Simulation of a Flow Arising During the Collapse of a Mixing Zone in a Stratified Medium"; Preprint No. 15 (Institute of Theoretical and Applied Mechanics of the U.S.S.R. Academy of Sciences, Novosibirsk, 1982).

[4.33] N.N. Yanenko, E.V. Vorozhtsov: *Dokl. Akad. Nauk SSSR* **247**, 48–52 (1979) [English transl.: *Soviet Math. Dokl.* **20**, 670–675 (1979)].

[4.34] E.V. Vorozhtsov, N.N. Yanenko: *Comput. Fluids* **9**, 1–15 (1981).

[4.35] E.V. Vorozhtsov, N.N. Yanenko: "One-Dimensional Theory of the Contact Strip"; Preprint No. 45 (Institute of Theoretical and Applied Mechanics of the U.S.S.R. Academy of Sciences, Novosibirsk, 1981).

[4.36] N.N. Yanenko, E.V. Vorozhtsov: "Differential Analysers of Strong Discontinuities in One-Dimensional Gas Flow", in *Computational Techniques in Transient and Turbulent Flow*, ed. by C. Taylor and K. Morgan (Pineridge Press, Swansea, 1981), pp. 59–96.

[4.37] E.V. Vorozhtsov: "On the Localisation of Discontinuities in the Difference Solutions by the Sobel Edge Detector"; Preprint No. 12 (Institute of Theoretical and Applied Mechanics of the U.S.S.R. Academy of Sciences, Novosibirsk, 1985).

[4.38] E.V. Vorozhtsov, N.N. Yanenko: *Dokl. Akad. Nauk SSSR* **259**, 18–24 (1981) [English transl.: E.V. Vorožcov, N.N. Janenko: *Soviet Math. Dokl.* **24**, 10–16 (1981)].

[4.39] E.V. Vorozhtsov, N.N. Yanenko: "On the Methods for *K*-Inconsistence Suppression in Difference Schemes of Gas Dynamics"; Preprint No. 44 (Institute of Theoretical and Applied Mechanics of the U.S.S.R. Academy of Sciences, Novosibirsk, 1981).

[4.40] E.V. Vorozhtsov, N.N. Yanenko: *Comput. Fluids* **9**, 17–32 (1981).

[4.41] E.V. Vorozhtsov, N.N. Yanenko: *Comput. Fluids* **11**, 231–249 (1983).

[4.42] E.V. Vorozhtsov, N.N. Yanenko: *Comput. Fluids* **10**, 181–204 (1982).

[4.43] E.V. Vorozhtsov, N.N. Yanenko: *Comput. Fluids* **10**, 205–222 (1982).
[4.44] R.A. Gentry, R.E. Martin, B.J. Daly: *J. Comput. Phys.* **1**, 87–118 (1966).
[4.45] R.D. Richtmyer, K.W. Morton: *Difference Methods for Initial-Value Problems*, 2nd edn (Wiley Interscience, New York, 1967).
[4.46] P.D. Lax: *Commun. Pure Appl. Math.* **10**, 537–566 (1957).
[4.47] C.M. Dafermos: "Quasilinear Hyperbolic Systems Following from the Conservation Laws", in *Nonlinear Waves*, ed. by S. Leibovich and A.R. Seebass (Cornell University Press, Ithaca and London, 1974).
[4.48] M. Morduchov, P. Libby: *J. Aeronaut. Sci.* **16**, 674–684, 704 (1949).
[4.49] A.N. Tikhonov, A.A. Samarskiĭ: *The Equations of Mathematical Physics*, 5th edn (Nauka, Moscow, 1977) [English transl. of 2nd edn: Pergamon Press, Oxford, and Macmillan, New York, 1963: Vols. I, II, Holden-Day, San Francisco, California, 1964, 1967].
[4.50] V.I. Smirnov: *A Course in Higher Mathematics*, Vol. III, Part 2, 8th edn (Nauka, Moscow, 1969) [English transl. of 6th edn: Pergamon Press, Oxford: Addison-Wesley, Reading, Massachusetts, 1964].
[4.51] *Handbook of Mathematical Functions with Formulas, Graphs and Mathematical Tables*, ed. by M. Abramovitz and I. Stegun (National Bureau of Standards Applied Mathematics Series 55, Washington, D.C., 1964).
[4.52] G.D. Yakovleva: *Tables of the Airy Functions and of Their Derivatives* (Nauka, Moscow, 1969).
[4.53] A.V. Lykov: *The Theory of Heat Conduction* (Vysshaya Shkola, Moscow, 1967).
[4.54] V.M. Fomin, E.V. Vorozhtsov, N.N. Yanenko: *Comput. Fluids* **3/4**, 171–183 (1976).
[4.55] S.K. Godunov, A.V. Zabrodin, G.P. Prokopov: *Zhurn. Vychisl. Matem. i Matem. Fiz.* **1**, 1020–1050 (1961).
[4.56] O.M. Belotserkovskiĭ, Yu.M. Davydov: *Coarse Particles Method in Gas Dynamics. Computational Experiment* (Nauka, Moscow, 1982).
[4.57] P.B. Bailey, L.F. Shampine, P.E. Waltman: *Nonlinear Two Point Boundary Value Problems* (Academic Press, New York, 1968).
[4.58] C.F. Lee: *J. Inst. Math. Its Appl.* **10**, 129–133 (1972).
[4.59] B. van Leer: *J. Comput. Phys.* **3**, 473–485 (1969).
[4.60] K. Srinivas, J. Gururaja, P.K. Krishna: *Comput. Fluids* **5**, 87–97 (1977).
[4.61] K. Srinivas, J. Gururaja: *Comput. Fluids* **5**, 139–150 (1977).
[4.62] S. Taki, T. Fujiwara: *AIAA J.* **16**, 73–77 (1978).
[4.63] V.B. Balakin, V.V. Bulanov: *Inzhenerno-Fizičeskiĭ Zhurn.* (in Russian) **21**, 1033–1039 (1971).
[4.64] V.V. Bulanov: *Inzhenerno-Fizičeskiĭ Zhurn.* (in Russian) **32**, 1080–1086 (1977).
[4.65] G.M. Rudakova, A.P. Shashkin: *Izv. Sibirsk. Otdel. Akad. Nauk SSSR, Ser. Tekhn. Nauk* (in Russian) No. 3, 70–77 (1975).
[4.66] B.P. Leonard: *Comput. Methods Appl. Mech. Engrg.* **19**, 59–98 (1979).
[4.67] M.A. Leschziner: *Comput. Methods Appl. Mech. Engrg.* **23**, 293–312 (1980).
[4.68] S.K. Godunov, A.V. Zabrodin, M.Ya. Ivanov, A.N. Kraiko, G.P. Prokopov: *Numerical Solution of Multidimensional Problems of Gas Dynamics* (Nauka, Moscow, 1976).
[4.69] N.N. Yanenko, Yu.I. Shokin: *Dokl. Akad. Nauk SSSR* **182**, 280–281 (1968).
[4.70] N.N. Yanenko, Yu.I. Shokin: "On Approximation Viscosity of Difference Schemes for Hyperbolic Equation Systems", in *Proceedings of the All-Union Seminar on Numerical Methods of Viscous Fluid Mechanics* (in Russian) (Nauka, Siberian Division, Novosibirsk, 1969), pp. 269–282.
[4.71] Yu.I. Shokin: *Differential Approximation Method* (Nauka, Siberian Division, Novosibirsk, 1979).
[4.72] Yu.I. Shokin, N.N. Yanenko: *Differential Approximation Method. Application to Gas Dynamics* (Nauka, Siberian Division, Novosibirsk, 1985).
[4.73] N.N. Yanenko, Yu.I. Shokin, L.A. Tusheva, Z.I. Fedotova: *Numerical Mathematics in Continuum Mechanics* (in Russian) **11**, No. 2 (A.N. SSSR Siberian Computer Center, Novosibirsk, 1980), pp. 123–159.
[4.74] N.N. Yanenko, Z.I. Fedotova, L.A. Tusheva, Yu.I. Shokin: *Comput. Fluids* **11**, 187–206 (1983).
[4.75] R.C.Y. Chin: *J. Comput. Phys.* **18**, 233–247 (1975).
[4.76] A.I. Zhukov: *Uspekhi Mat. Nauk* (in Russian) **14**, 129–136 (1959).

[4.77] S.I. Serdyukova: *Zhurn. Vychisl. Matem. i Matem. Fiz.* **11**, 411–424 (1971).
[4.78] L.D. Landau, E.M. Lifshitz: *Continuum Mechanics* (Gostehizdat, Moscow, 1954).
[4.79] E.V. Vorozhtsov: *Comput. Fluids* **15**, 13–45 (1987).
[4.80] P.J. Roache: *Computational Fluid Dynamics* (Hermosa, Albuquerque, New Mexico, 1976).
[4.81] W.F. Noh, M.H. Protter: *J. Math. Mech.* **12**, 149–191 (1963).
[4.82] R.W. MacCormack: AIAA Paper 69-354 (1969).
[4.83] N.N. Yanenko, Z.I. Fedotova, L.A. Kompaniets, Yu.I. Shokin: *Comput. Fluids* **12**, 93–121 (1984).
[4.84] A.A. Duvanov, F.A. Murzin, V.P. Shapeev: "Analytical Differentiation of Functions on a Computer"; Preprint No. 4 (Institute of Theoretical and Applied Mechanics of the U.S.S.R. Academy of Sciences, Novosibirsk, 1981).
[4.85] K.A. Putilov: *Thermodynamics* (Nauka, Moscow, 1971).
[4.86] F.H. Harlow, A.A. Amsden: *J. Comput. Phys.* **3**, 80–93 (1968).
[4.87] V.F. Lobanov: *Zhurn. Prikl. Mekhaniki i Tehnič. Fiziki* (in Russian) No. 5, 145–149 (1975).
[4.88] G.A. Sapozhnikov: "Influence of a Faced Cavity in the Charge on the Motion of Shells", in *Gas Dynamics of Rapid Processes* (Publication of the Tomsk State University, Tomsk, 1979), pp. 71—78.
[4.89] P.F. Korotkov, V.S. Lobanov: *Zhurn. Prikl. Mekhaniki i Tehnič. Fiziki* (in Russian) No. 4, 156–162 (1973).
[4.90] R.T. Sedgwick, L.J. Hageman, R.G. Herrmann, J.L. Waddell: *Internat. J. Engrg. Sci.* **16**, 859–869 (1978).
[4.91] V.V. Rusanov: "Calculation and Investigation of Multidimensional Gas Flows by the Method of Finite Differences"; Thesis (Institute of Applied Mathematics of the U.S.S.R. Academy of Sciences, Moscow, 1968).
[4.92] P.D. Lax, B. Wendroff: *Commun. Pure Appl. Math.* **13**, 217–237 (1960).
[4.93] A. Lerat, R. Peyret: *Rech. Aérospat.* No. 2, 61–79 (1975).
[4.94] W. Stark, R. Wojcieszynski: *Z. Angew. Math. Mech.* **60**, 383–391 (1980).
[4.95] A. Harten: *Math. Comp.* **32**, 363–389 (1978).
[4.96] G.A. Sod: *J. Comput. Phys.* **27**, 1–31 (1978).
[4.97] V.P. Kolgan: *Zhurn. Vychisl. Matem. i Matem. Fiz.* **18**, 1340–1345 (1978).
[4.98] M.W. Evans, F.H. Harlow: "The Particle-In-Cell Method for Hydrodynamic Calculations"; Los Alamos Scientific Laboratory Report LA-2139 (1957).
[4.99] *Theoretical Fundamentals and Construction of Numerical Algorithms for Mathematical Physics Problems*, ed. by K.I. Babenko (Nauka, Moscow, 1979).
[4.100] V.E. Petrenko, E.V. Vorozhtsov: *Numerical Methods in Continuum Mechanics* (in Russian) **4**, No. 2 (A.N. SSSR Siberian Computer Center, Novosibirsk, 1973) pp. 132–141.
[4.101] E.V. Vorozhtsov: *Comput. Fluids* **9**, 313–326 (1981).
[4.102] J.U. Brackbill, H.M. Ruppel: *J. Comput. Phys.* **65**, 314–343 (1986).
[4.103] T. Mietzner: *Notes Numer. Fluid Mech.* **14**, 175–186 (1986).
[4.104] A. Harten: *J. Comput. Phys.* **49**, 357–393 (1983).
[4.105] T.D. Riney, E.J. Halda: *AIAA J.* **6**, 338–344 (1968).

Chapter 5

[5.1] W.L. Miranker, O. Pironneau: "A Global Shock Fitting Method"; Rapport de Recherche, No. 123 (IRIA, Domaine de Voluceau Rocquencourt, 78150 Le Chesnay, 1975).
[5.2] W.L. Miranker, O. Pironneau: *Internat. J. Comput. Math. Appl.* **2**, 63–71 (1976).
[5.3] O. Pironneau: "Sur les problèmes d'optimisation de structure en mécanique des fluides"; Thèse, Doct. Sci. Math. (University of Paris, Paris, 1976).
[5.4] E.V. Vorozhtsov, V.M. Krepkiĭ, Z. Usakov: *Numerical Methods in Continuum Mechanics* (in Russian) **12**, No. 4 (A.N. SSSR Siberian Computer Center, Novosibirsk, 1981), pp. 30–47.
[5.5] E.V. Vorozhtsov, Z. Usakov: *Numerical Methods in Continuum Mechanics* (in Russian) **14**, No. 3 (A.N. SSSR Siberian Computer Center, Novosibirsk, 1983), pp. 18–32.
[5.6] E.V. Vorozhtsov, N.N. Yanenko: *Internat. J. Numer. Methods Fluids* **4**, 477–496 (1984).

[5.7]   E.V. Vorozhtsov: *Numerical Methods in Continuum Mechanics* (in Russian) **15**, No. 2 (A.N. SSSR Siberian Computer Center, Novosibirsk, 1984), pp. 42–48.

[5.8]   L.Ya. Zlaf: *Variational Calculus and Integral Equations. A Handbook*, 2nd edn (Nauka, Moscow, 1970).

[5.9]   B.L. Roždestvenskiĭ, N.N. Janenko: *Systems of Quasilinear Equations and Their Applications to Gas Dynamics*, 2nd edn (Nauka, Moscow, 1978) [English transl.: *Systems of Quasilinear Equations and Their Applications to Gas Dynamics*, Translations of Mathematical Monographs, Vol. 55 (American Mathematical Society, Providence, Rhode Island, 1983)].

[5.10] R.D. Richtmyer, K.W. Morton: *Difference Methods for Initial-Value Problems*, 2nd edn (Wiley Interscience, New York, 1967).

[5.11] P.J. Roache: *Computational Fluid Dynamics* (Hermosa, Albuquerque, New Mexico, 1976).

[5.12] *Handbook of Mathematical Functions with Formulas, Graphs and Mathematical Tables*, ed. by M. Abramowitz and I. Stegun (National Bureau of Standards Applied Mathematics, Series 55, Washington, D.C., 1964).

[5.13] S.K. Godunov, A.V. Zabrodin, G.P. Prokopov: *Zhurn. Vychisl. Matem. i Matem. Fiz.* **1**, 1020–1050 (1961).

[5.14] V.I. Krylov, V.V. Bobkov, P.I. Monastyrnyĭ: *Computational Methods*, Vol. II (Nauka, Moscow, 1977).

[5.15] G.A. Korn, Th.M. Korn: *Mathematical Handbook for Scientists and Engineers* (McGraw-Hill, New York, 1961).

[5.16] *Modern Numerical Methods for Ordinary Differential Equations*, ed. by G. Hall and J.M. Watt (Clarendon Press, Oxford, 1976).

[5.17] E. Polak: *Computational Methods in Optimization. A Unified Approach* (Academic Press, New York, 1971).

[5.18] N.N. Moiseev, Yu.P. Ivanilov, E.M. Stoliarova: *Optimization Methods* (Nauka, Moscow, 1978).

[5.19] Yu.G. Yevtushenko: *Methods of Solving Extremal Problems and Their Application to Optimization Systems* (Nauka, Moscow, 1982).

[5.20] Ph.E. Gill, W. Murray, M.H. Wright: *Practical Optimization* (Academic Press, New York, 1981).

[5.21] B.T. Poliak: *Introduction in Optimization* (Nauka, Moscow, 1983).

[5.22] I.S. Berezin, N.P. Zhidkov: *Methods of Computations*, Vol. 1, 3rd edn (Nauka, Moscow, 1966).

[5.23] V.M. Fomin, E.V. Vorozhtsov, N.N. Yanenko: *Comput. Fluids* **7**, 109–121 (1979).

[5.24] A.N. Minailos: *Zhurn. Vychisl. Matem. i Matem. Fiz.* **17**, 1058–1063 (1977).

[5.25] E.V. Vorozhtsov, N.N. Yanenko: *Comput. Fluids* **8**, 313–326 (1980).

[5.26] M.L. Wilkins: "Calculation of Elastic-Plastic Flow", in *Proceedings of the Section in Numerical Methods in Gas Dynamics of the Second International Colloquium on Gas Dynamics of Explosion and Reacting Systems* (Novosibirsk, August 19–23, 1969), Vol. 1 (Publication of the U.S.S.R. Academy of Sciences Computer Center, Moscow, 1971), pp. 408–517.

[5.27] D.I. Batishchev: *Search Methods of Optimum Design* (Sovetskoye Radio, Moscow, 1975).

[5.28] N.N. Kalitkin: *Numerical Methods* (Nauka, Moscow, 1978).

[5.29] Ya.B. Zel'dovich, Yu.P. Raizer: *Physics of Shock Waves and High-Temperature Hydrodynamic Phenomena* (Nauka, Moscow, 1966) [English transl.: Vols. 1 and 2, Academic Press, New York, 1967].

[5.30] P.D. Lax, B. Wendroff: *Commun. Pure Appl. Math.* **13**, 217–237 (1960).

[5.31] R. Courant, K.O. Friedrichs: *Supersonic Flow and Shock Waves* (Interscience, New York, 1948).

[5.32] T.Y. Thomas: *J. Math. Mech.* **6**, 455–469 (1957).

[5.33] R. Shankar, M. Prasad: *Internat. J. Engrg Sci.* **17**, 17–21 (1979).

[5.34] V.I. Smirnov: *A Course of Higher Mathematics*, Vol. 4. 3rd Edition (Gos. Izdatel'stvo Tehniko-Teoretičeskoĭ Literatury, Moscow, 1953).

[5.35] A. Harten: *Mathematics of Computation* **32**, 363–389 (1978).

[5.36] R.E. Collins: *Flow of Fluids Through Porous Materials* (Reinhold, New York, 1961).

[5.37] G.P. Tsybulsky: *Izv. Akad. Nauk SSSR, Ser. Mekhanika Zhidkosti i Gaza* No. 1, 170–173 (1975).

[5.38] Z. Usakov: *Numerical Methods in Continuum Mechanics* (in Russian) **10**, No. 6 (A.N. SSSR Siberian Computer Center, Novosibirsk, 1979), pp. 141–149.

[5.39] Z. Usakov: "On One Numerical Experiment in Shock Localisation in the Solution of a Problem on the Filtration of Three Phase Incompressible Fluid", in *Numerical Methods for the Solution of the Filtration Problems for Multiphase Incompressible Fluid. Proceedings of the 4th All-Union Seminar*, ed. by A.N. Konovalov (Publication of the Institute of Theoretical and Applied Mechanics of the U.S.S.R. Academy of Sciences, Novosibirsk, 1980), pp. 241–245.

[5.40] N.E. Kochin: "On the Theory of Discontinuities in the Fluid", in *Collected Works*, Vol. II (Publication of the U.S.S.R. Academy of Sciences, Moscow–Leningrad, 1949), pp. 5–42.

[5.41] V.L. Danilov, A.N. Konovalov, S.I. Yakuba: *Dokl. Akad. Nauk SSSR* **183**, 307–310 (1968).

[5.42] A.N. Konovalov: *Problems of the Filtration of the Multiphase Incompressible Fluid. Lectures for the Students of the Novosibirsk State University* (Publication of the Novosibirsk State University, Novosibirsk, 1972).

[5.43] A.N. Konovalov: *Problems of the Filtration of the Multiphase Incompressible Fluid* (Nauka, Siberian Division, Novosibirsk, 1988).

[5.44] V.P. Il'in: *Difference Methods of Solving Elliptic Equations. Lectures for the Students of the Novosibirsk State University* (Publication of the Novosibirsk State University, Novosibirsk, 1970).

[5.45] A. Harten, J.M. Hyman, P.D. Lax: *Commun. Pure Appl. Math.* **29**, 297–322 (1976).

[5.46] J.B. Bell, G.R. Shubin: *J. Comput. Phys.* **52**, 569–591 (1983).

[5.47] A. Harten: *J. Comput. Phys.* **49**, 357–393 (1983).

[5.48] B.I. Levy, Ya.M. Seidel, V.M. Sankin: *Numerical Methods in Continuum Mechanics* (in Russian) **9**, No. 6 (A.N. SSSR Siberian Computer Center, Novosibirsk, 1978), pp. 105–114.

[5.49] V.L. Danilov, R.M. Kaz: *Hydrodynamic Computations of Fluids Displacement in a Porous Medium* (Nedra, Moscow, 1980).

[5.50] B.N. Pshenichny, Yu.M. Danilin: *Numerical Methods in Extremal Problems* (Nauka, Moscow, 1975).

Chapter 6

[6.1] G. Moretti: Lecture Notes in Physics, Vol. 35 (Springer-Verlag, New York, Heidelberg, Berlin, 1975), pp. 287–292.

[6.2] G. Moretti: "The Importance of Boundary Conditions in the Numerical Treatment of Hyperbolic Equations", in *High-Speed Computing in Fluid Dynamics*, Physics of Fluids, Supplement II (American Institute of Physics, New York, 1969), pp. II-13–II-20.

[6.3] I. Yamamoto, K. Karashima: *AIAA J.* **20**, 9–17 (1982).

[6.4] M.D. Salas: *AIAA J.* **14**, 583–587 (1976).

[6.5] A.N. Liubimov, V.V. Rusanov: *Gas Flows Around Blunt Bodies. Part I. Calculation Method and Analysis of the Flows* (Nauka, Moscow, 1970).

[6.6] G.B. Alalykin, S.K. Godunov, I.L. Kireeva, L.A. Pliner: *Solution of One-Dimensional Gas Dynamics Problems on Moving Meshes* (Nauka, Moscow, 1970).

[6.7] S.K. Godunov, A.V. Zabrodin, M.Ya. Ivanov, A.N. Kraiko, G.P. Prokopov: *Numerical Solution of Multidimensional Problems of Gas Dynamics* (Nauka, Moscow, 1976).

[6.8] V.M. Kovenya, N.N. Yanenko: *The Method of Splitting-Up in Gas Dynamics Problems* (Nauka, Siberian Division, Novosibirsk, 1981).

[6.9] R.J. Gelinas, S.K. Doss, K. Miller: *J. Comput. Phys.* **40**, 202–249 (1981).

[6.10] J.F. Thompson, Z.U.A. Warsi, C.W. Mastin: *Numerical Grid Generation. Foundations and Applications* (North-Holland, New York, 1985).

[6.11] J.F. Thompson: *Appl. Numer. Math.* **1**, 3–27 (1985).

[6.12] J.P. Boris, D.L. Book: *J. Comput. Phys.* **11**, 38–69 (1973).

[6.13] J.P. Boris, D.L. Book: *J. Comput. Phys.* **20**, 397–431 (1976).

[6.14] G.A. Sod: *J. Comput. Phys.* **27**, 1–31 (1978).

[6.15] S.T. Zalesak: *J. Comput. Phys.* **31**, 335–362 (1979).

[6.16] S.P. Popov, Yu.I. Romashkevich: *Zhurn. Vychisl. Matem. i Matem. Fiz.* **19**, 546–550 (1979).
[6.17] G. Adler: "An Iterative Method for Reducing Numerical Diffusion in Problems of Flow Through a Porous Medium"; *Pubbl. Istituto Applic. Calcolo Mauro Picone*, Ser. 3, No. 191 (1979).
[6.18] S.T. Zalesak: *J. Comput. Phys.* **40**, 497–508 (1981).
[6.19] K. Fisher: *Internat. J. Numer. Methods Engrg.* **12**, 931–940 (1978).
[6.20] B.E. McDonald, J. Ambrosiano: *J. Comput. Phys.* **56**, 448–460 (1984).
[6.21] R. Morrow, L.E. Cram: *J. Comput. Phys.* **57**, 129–136 (1985).
[6.22] A. Harten: "The Method of Artificial Compression: 1. Shocks and Contact Discontinuities"; A.E.C. Research and Development Report C00-3077-50 (New York University, New York, 1974).
[6.23] A. Harten: *Commun. Pure Appl. Math.* **30**, 611–638 (1977).
[6.24] A. Harten: *Math. Comp.* **32**, 363–389 (1978).
[6.25] B. van Leer: *J. Comput. Phys.* **32**, 101–136 (1979).
[6.26] P. Woodward, P. Colella: *J. Comput. Phys.* **54**, 115–173 (1984).
[6.27] P. Colella, P.R. Woodward: *J. Comput. Phys.* **54**, 174–201 (1984).
[6.28] B.A. Fryxell, P.R. Woodward, P. Colella. K.-H. Winkler: *J. Comput. Phys.* **63**, 283–310 (1986).
[6.29] S. Eidelman, P. Colella, R. Shreeve: *AIAA J.* **22**, 1609–1615 (1984).
[6.30] A. Harten: *J. Comput. Phys.* **49**, 357–393 (1983).
[6.31] A. Harten: *SIAM J. Numer. Anal.* **21**, 1–23 (1984).
[6.32] P.K. Sweby: *SIAM J. Numer. Anal.* **21**, 995–1011 (1984).
[6.33] S. Osher, S. Chakravarthy: *SIAM J. Numer. Anal.* **21**, 955–984 (1984).
[6.34] S.R. Chakravarthy, K.-Y. Szema, U.C. Goldberg, J.J. Gorski: AIAA Paper 85-0165 (1985).
[6.35] S.R. Chakravarthy, S.Osher: AIAA Paper 85-0363 (1985).
[6.36] A. Harten: *Math. Comput.* **46**, 379–399 (1986).
[6.37] E.V. Vorozhtsov: "On Difference Solution Refinement in the Neighbourhood of a Discontinuity"; Preprint No. 13 (Institute of Theoretical and Applied Mechanics of the U.S.S.R. Academy of Sciences, Novosibirsk, 1983).
[6.38] P.J. Roache: *Computational Fluid Dynamics* (Hermosa, Albuquerque, New Mexico, 1976).
[6.39] K.W. Morton: Lecture Notes in Physics, Vol. 170 (Springer-Verlag, Berlin, Heidelberg, New York, 1982), pp. 77–93.
[6.40] P. Charrier, B. Tessieras: *SIAM J. Numer. Anal.* **23**, 461–472 (1986).
[6.41] S.K. Godunov, A.V. Zabrodin, G.P. Prokopov: *Zhurn. Vychisl. Matem. i Matem. Fiz.* **1**, 1020–1050 (1961).
[6.42] R.A. Gentry, R.E. Martin, B.J. Daly: *J. Comput. Phys.* **1**, 87–118 (1966).
[6.43] O.M. Belotserkovskiĭ, Yu.M. Davydov: *Coarse Particles Method in Gas Dynamics. Computational Experiment* (Nauka, Moscow, 1982).
[6.44] B.L. Roždestvenskii, N.N. Janenko: *Systems of Quasilinear Equations and Their Applications to Gas Dynamics*, 2nd edn (Nauka, Moscow, 1978) [English transl.: *Translations of Mathematical Monographs*, Vol. 55 (American Mathematical Society, Providence, Rhode Island, 1983)].
[6.45] Yu.I. Shokin: *Differential Approximation Method* (Nauka, Siberian Division, Novosibirsk, 1979).
[6.46] A.A. Samarskiĭ, Yu.P. Popov: *Difference Schemes of Gas Dynamics*, 2nd edn (Nauka, Moscow, 1980).
[6.47] A.A. Samarskiĭ, E.S. Nikolaev: *Methods of Solving Grid Equations* (Nauka, Moscow, 1978).
[6.48] J.H. Wilkinson, C. Reinsch: *Handbook for Automatic Computation. Linear Algebra* (Springer-Verlag, Heidelberg, New York, Berlin, 1971).
[6.49] D.K. Faddeev, V.N. Faddeeva: *Computational Methods of Linear Algebra*, 2nd edn (GIFML, Moscow, Leningrad, 1963).
[6.50] J.J. Chattot, J. Guiu-Roux, J. Laminie: "Résolution numérique d'une équation de conservation par une approche variationnelle", in *Proceedings of the 6th International Conference on Numerical Methods in Fluid Dynamics*, Vol. 1 (Publication of the Computer Centre of the U.S.S.R. Academy of Sciences, Moscow, 1978), pp. 32–38.

[6.51] J.J. Chattot, J. Guiu-Roux, J. Lamine: *Internat. J. Numer. Methods Fluids* **2**, 209–219 (1982).
[6.52] L.V. Kantorovich, V.I. Krylov: *Methods of Approximate Solution of Partial Differential Equations* (ONTI NKTP SSSR, Leningrad, Moscow, 1936).
[6.53] G.I. Marchuk: *Methods of Numerical Mathematics* (Nauka, Moscow, 1977).
[6.54] S.G. Mikhlin: *Numerical Realization of Variational Methods* (Nauka, Moscow, 1966).
[6.55] G.J. McRae, W.R. Goodin, J.H. Seinfeld: *J. Comput. Phys.* **45**, 1–42 (1982).
[6.56] S.K. Godunov, V.S. Riaben'kiĭ: *Difference Schemes. Introduction in the Theory* (Nauka, Moscow, 1977).
[6.57] V.I. Paasonen: *Numerical Methods in Continuum Mechanics* (in Russian) **4**, No. 4 (A.N. SSSR Siberian Computer Center, Novosibirsk, 1973), pp. 44–57.
[6.58] A.N. Minailos: *Zhurn. Vychisl. Matem. i Matem. Fiz.* **17**, 1058–1063 (1977).
[6.59] J.H. Ahlberg, E.M. Nilson, J.L. Walsh: *The Theory of Splines and Their Application* (Academic Press, New York, 1967).
[6.60] S.B. Stečkin, Yu.N. Subbotin: *Splines in Numerical Mathematics* (Nauka, Moscow, 1976).
[6.61] Yu.S. Zavialov, B.I. Kvasov, V.L. Miroshnichenko: *Methods of Spline Functions* (Nauka, Moscow, 1980).
[6.62] P.E. Gill, W. Murray: *Numerical Methods for Constrained Optimization* (Academic Press, London, 1974).
[6.63] N.N. Moiseev, Yu.P. Ivanilov, E.M. Stoliarova: *Methods of Optimization* (Nauka, Moscow, 1978).
[6.64] Yu.G. Yevtushenko: *Methods of Solving Extremal Problems and Their Application to Optimization Systems* (Nauka, Moscow, 1982).
[6.65] N.N. Kalitkin: *Numerical Methods* (Nauka, Moscow, 1978).
[6.66] E.D. Eason, C.D. Mote: *Internat. J. Numer. Methods Engrg.* **11**, 641–652 (1977).
[6.67] M. Pappas: *Computers and Structures* **11**, 539–557 (1980).
[6.68] A. Ecer, H.U. Akay: *AIAA J.* **19**, 1174–1182 (1981).
[6.69] E.V. Vorozhtsov, N.N. Yanenko: "One-Dimensional Theory of Contact Strip"; Preprint No. 45 (Institute of Theoretical and Applied Mechanics of the U.S.S.R. Academy of Sciences, Novosibirsk, 1981).
[6.70] E.V. Vorozhtsov, N.N. Yanenko: *Comput. Fluids* **9**, 1–15 (1981).
[6.71] E.V. Vorozhtsov: "On Differential Analysers of Shock Fronts", in *Program Complexes of Mathematical Physics*, ed. by V.I. Karnachuk (Publication of the Computer Center of the U.S.S.R. Academy of Sciences Siberian Division, Novosibirsk, 1972), pp. 98–115.
[6.72] E.V. Vorozhtsov: *Numerical Methods in Continuum Mechanics* (in Russian) **8**, No. 2 (A.N. SSSR Siberian Computer Center, Novosibirsk, 1977), pp. 12–27.
[6.73] B.L. Golovkin: *Parallel Computer Systems* (Nauka, Moscow, 1980).

Chapter 7

[7.1] A.N. Tihonov: "Mathematical Model", in *Mathematical Encyclopedia*, Vol. 3 (Soviet Encyclopedia, Moscow, 1982), pp. 574–575.
[7.2] O.M. Belotserkovskii: "Mathematical Modeling Is the Branch of Informatics", in *Cybernetics. Formation of Informatics* (Nauka, Moscow, 1986), pp. 45–62.
[7.3] V.V. Rusanov: Lecture Notes in Physics, Vol. 18 (Springer-Verlag, Berlin, Heidelberg, New York, 1973), pp. 154–162.
[7.4] A.A. Dorodnitsyn: *Vestnik Akad. Nauk SSSR* No. 2, 85–89 (1985).
[7.5] N.J. Nilson: *Principles of Artificial Intelligence* (Tioga Publishing Company, Palo Alto, California, 1980).
[7.6] A.A. Petrov: "Algorithmical Software for Information and Control Systems of Adaptive Robots (Algorithms of Technical Robot Vision)", in *Itogi Nauki i Tehniki VINITI. Ser. Tehnicheskaya Kibernetika*, No. 17 (VINITI A.N. SSSR, Moscow, 1984), pp. 251–294.
[7.7] L.G. Shapiro: "Computer Vision Systems; Past, Present and Future", in *Pictorial Data Analysis (Proceedings of the NATO Advanced Study Institute on Pictorial Data Analysis, Bonas, France, August 1–12, 1982)*, ed. by R.M. Haralick (Springer-Verlag, Berlin, 1983), pp. 199–237.

[7.8]   *Biomedical Pattern Recognition and Image Processing*, ed. by K.S. Fu, T. Pavlidis (Verlag Chemie GmbH, Weinheim, 1979).

[7.9]   *Digital Image Processing in Medicine*, ed. by K.H. Höhne, Lecture Notes in Medical Informatics, Vol. 15 (Springer-Verlag, New York, 1981).

[7.10] A. Todd-Pokropek: "Medical Image Processing", in *Pictorial Data Analysis (Proceedings of the NATO Advanced Study Institute on Pictorial Data Analysis, Bonas, France, August 1–12, 1982)*, ed. by R.M. Haralick (Springer-Verlag, Berlin, 1983), pp. 295–320.

[7.11] J. Kittler: *Phil. Trans. Roy. Soc. London* **A309**, 325–337 (1983).

[7.12] A.S. Alexeev, V.N. Dement'ev, V.A. Zabelin, V.P. Pyatkin: "Functional Mathematical Software of the Image Processing"; Preprint No. 416 (Siberian Computer Center of the U.S.S.R. Academy of Sciences, Novosibirsk, 1983).

[7.13] *Digital Image Processing for Remote Sensing*, ed. by R. Bernstein (IEEE Press, New York, 1978).

[7.14] A. Cozannet, H. Maitre, J. Fleuret, M. Rousseau: *Optique et Télécommunications* (Eyrolles, Paris, 1981).

[7.15] S.B. Gurevich, V.B. Constantinov, V.K. Sokolov, D.F. Chernykh: *Transfer and Processing of Information by Holographic Methods* (Sovetskoye Radio, Moscow, 1978).

[7.16] *Two-Dimensional Digital Signal Processing I*, ed. by T.S. Huang, Topics in Applied Physics, Vol. 42 (Springer-Verlag, Berlin, Heidelberg, New York, 1981).

[7.17] K. Preston, M.J.B. Duff, S. Levialdi, P.E. Norgren, J.-I. Toriwaki: *Proc. IEEE* **67**, 826–856 (1979).

[7.18] A. Rosenfeld: "Algorithms for Cellular Image Processing", in *ICCD '83: Proc. IEEE Int. Conf. Comput. Des.: VLSI Comput., Port Chester, N.Y., 31 Oct.–3 Nov., 1983* (Silver Spring, Maryland, 1983), pp. 719–722.

[7.19] P.E. Dimotakis, F.D. Debussy, M.M. Koochesfahani: *Phys. Fluids* **24**, 995–999 (1981).

[7.20] M.A. Herman, J. Jimenez: *J. Fluid Mech.* **119**, 323–345 (1982).

[7.21] R. Meynart: *Phys. Fluids* **26**, 2074–2079 (1983).

[7.22] A.K.M.F. Hussain: *Phys. Fluids* **26**, 2816–2850 (1983).

[7.23] T.C. Corke: *AIAA J.* **22**, 1124–1131 (1984).

[7.24] M. Gharib, M.A. Herman, A.H. Yavrouian, V. Sarohia: AIAA Paper 85-0172 (1985).

[7.25] P.G. Buning, J. Steger: "Graphics and Flow Visualization in Computational Fluid Dynamics", in *AIAA 7th Computational Fluid Dynamics Conference, Cincinnati, Ohio, July 15–17, 1985. Collect. Techn. Paper* (New York, 1985), pp. 162–170.

[7.26] P. Kutler, J.L. Steger, F.R. Bailey: AIAA Paper 87-1135 (1987).

[7.27] R.G. Belie: AIAA Paper 87-1179 (1987).

[7.28] P. Kutler: "A Perspective of Computational Fluid Dynamics", in *Proceedings of the Fourth International Conference on Boundary and Interior Layers—Computational and Asymptotic Methods. 7–11 July 1986, Novosibirsk*, ed. by S.K. Godunov, J.J.H. Miller, V.A. Novikov (Boole Press, Dublin, 1986), pp. 332–348.

[7.29] U.B. Mehta, P. Kutler: AIAA Paper 84-1531 (1984).

[7.30] P. Kutler, U.B. Mehta, A. Andrews: Lecture Notes in Physics, Vol. 218 (Springer-Verlag, New York, Heidelberg, Berlin, 1985), pp. 340–345.

[7.31] J.F. Dannenhoffer, III, and J.R. Baron: AIAA Paper 86-0495 (1986).

[7.32] J.F. Dannenhoffer, III, and J.R. Baron: AIAA Paper 87-1111 (1987).

[7.33] A.E. Andrews: *AIAA J.* **26**, 40–46 (1988).

[7.34] L.I. Levkovich-Masliuk: "Processing and Compression of Gas-Dynamical Information with the Aid of Piecewise Polynomial Approximation"; Preprint No. 186 (Institute of Applied Mathematics of the U.S.S.R. Academy of Sciences, Moscow, 1987).

[7.35] I.L. Dobroserdov: "Base Models and Methods for the Construction of Rapid Algorithms and Programs for Computing Non-Isobaric Viscous Jet Flows with Reacting Components", Manuscript dep. VINITI 4.05.1987, No. 4143-V87 (VINITI A.N. SSSR, Moscow, 1987).

[7.36] S.V. Bobyshev, I.L. Dobroserdov: *Modeling in Mechanics* (in Russian) **1** (18), No. 6 (Institute of Theoretical and Applied Mechanics of the U.S.S.R. Academy of Sciences, Novosibirsk, 1987), pp. 3–13.

[7.37] E.V. Vorozhtsov, S.I. Mazurik: "A Method for an Automatic Search for Difference Schemes Having the Largest Volume of the Stability Domain Among the Schemes of a Given Family. 1. Description of the Method and Application to One-Dimensional

Problems". Preprint No. 24 (Institute of Theoretical and Applied Mechanics of the U.S.S.R. Academy of Sciences, Novosibirsk, 1987).

[7.38] E.V. Vorozhtsov, S.I. Mazurik: "A Method for an Automatic Search for Difference Schemes Having the Largest Volume of the Stability Domain Among the Schemes of a Given Family. 2. Application to Two-Dimensional Problems"; Preprint No. 34 (Institute of Theoretical and Applied Mechanics of the U.S.S.R. Academy of Sciences, Novosibirsk, 1987).

[7.39] S.I. Mazurik, E.V. Vorozhtsov: "A Search Method for the Construction of Maximally Stable Difference Schemes for Partial Differential Equations", in *NUMETA '87 Proceedings, 6–10 July 1987, Swansea* (Nijhoff, Dordrecht, The Netherlands, 1987).

[7.40] E.V. Vorozhtsov, S.I. Mazurik: "Construction of Multiply Connected Stability Domains of Difference Schemes by Means of Computer Algebra and Pattern Recognition"; Preprint No. 18 (Institute of Theoretical and Applied Mechanics of the U.S.S.R. Academy of Sciences, Novosibirsk, 1988).

[7.41] E.V. Vorozhtsov, S.I. Mazurik: "Application of Artificial Intelligence Techniques for the Classification of Discontinuities in Gases and for the Stability Analysis of Difference Schemes", in *Soviet Union–Japan Symposium on Computational Fluid Dynamics, U.S.S.R., Khabarovsk, September, 9–16, 1988. Book of Abstracts*, ed. by P.I. Chushkin (Publication of the Far Eastern Division of the U.S.S.R. Academy of Sciences, Vladivostok, 1988), pp. 124–125.

[7.42] E.V. Vorozhtsov, S.I. Mazurik: "Application of Artificial Intelligence Techniques for the Classification of Discontinuities in Gases and for the Stability Analysis of Difference Schemes", in *Soviet Union–Japan Symposium on Computational Fluid Dynamics, U.S.S.R., Khabarovsk, September, 9–16, 1988. Proceedings*, ed. by P.I. Chushkin and V.P. Korobeinikov (Publication of the Computing Center of the U.S.S.R. Academy of Sciences, Moscow, 1989) (in press).

[7.43] S.S. Tong: AIAA Paper 85-0112 (1985).

[7.44] S.S. Tong: AIAA Paper 86-0242 (1986).

[7.45] G.S. Pospelov: "Artificial Intelligence. New Information Technology", in *Cybernetics. Formation of Informatics* (Nauka, Moscow, 1986), pp. 106–121.

[7.46] V.M. Kovenya: "Some Problems of Computational Fluid Dynamics", in *Construction of Numerical Algorithms and the Solution of Mathematical Physics Problems*, ed. by K.I. Babenko (Publication of the Institute of Applied Mathematics of the U.S.S.R. Academy of Sciences, Moscow, 1987), pp. 5–17.

[7.47] E.V. Vorozhtsov: "On the Localisation of Discontinuities in Difference Solutions by the Sobel Edge Detector"; Preprint No. 12 (Institute of Theoretical and Applied Mechanics of the U.S.S.R. Academy of Sciences, Novosibirsk, 1985).

[7.48] E.V. Vorozhtsov: "Extraction of Discontinuity Lines in Difference Solutions by Pattern Recognition Methods", in *Proceedings of the Seminar of Socialist Countries on the Computational Aerohydromechanics* (Publication of the Scientific Council of the U.S.S.R. Academy of Sciences on the Complex Problem "Cybernetics", Moscow, 1985), pp. 30–34.

[7.49] E.V. Vorozhtsov: "Classification of Discontinuities in Gas Flows as the Pattern Recognition Problem"; Preprint No. 23 (Institute of Theoretical and Applied Mechanics of the U.S.S.R. Academy of Sciences, Novosibirsk, 1986).

[7.50] E.V. Vorozhtsov: *Comput. Fluids* 15, 13–45 (1987).

[7.51] E.V. Vorozhtsov: "On the Automatic Classification of Singularities in the Numerical Simulation of Two-Dimensional Gas Flows", in *Image Processing and Remote Sensing. Abstracts of the Reports of a Regional Conference, Novosibirsk, 10–12 November, 1987*, ed. by V.P. Pyatkin (Publication of the Computing Center of the Siberian Division of the U.S.S.R. Academy of Sciences, Novosibirsk, 1987), p. 63.

[7.52] E.V. Vorozhtsov: "Classification of Discontinuities in Two-Dimensional Gas Flows by the Discriminant Analysis Techniques", in *2nd School-Seminar of Socialist Countries "Computational Mechanics and Automation of the Design", 16–23 October 1988, Tashkent* (Publication of the NPO "Kibernetika" Acad. Nauk Uzbekskaya S.S.R., Tashkent, 1988), p. 12.

[7.53] Z.B. Golembo, V.P. Zinkevich: "Mathematical Problems of the Image Analysis and

Recognition", in *Itogi Nauki i Tehniki. Tehnich. Kibernetika*, Vol. 18 (VINITI A.N. SSSR, Moscow, 1985), pp. 123–173.

[7.54] V.A. Kovalevskiĭ: *Optimal Solution Methods in Image Recognition* (Nauka, Moscow, 1976).

[7.55] J.T. Tou, R.C. Gonzalez: *Pattern Recognition Principles* (Addison-Wesley, Reading, Massachusetts, 1974).

[7.56] V.I. Timohin: *Application of the Computers in the Solution of Pattern Recognition Problems* (Publication of the Leningrad State University, Leningrad, 1983).

[7.57] L. Goldfarb: *Pattern Recognition* 17, 575–582 (1984).

[7.58] K.S. Fu: *Syntactic Pattern Recognition and Application* (Prentice-Hall, Englewood Cliffs, New Jersey, 1982).

[7.59] L. Hesselink, J. Helman: AIAA Paper 87-1181 (1987).

[7.60] Yu.N. Vatolin: *Numerical Methods in Continuum Mechanics* (in Russian) 2, No. 3 (A.N. SSSR Siberian Computer Center, Novosibirsk, 1971), pp. 22–37.

[7.61] Yu.N. Vatolin: *Numerical Methods in Continuum Mechanics* (in Russian) 5, No. 2 (A.N. SSSR Siberian Computer Center, Novosibirsk, 1974), pp. 5–6.

[7.62] F. Wahl: *Digitale Bildsignalverarbeitung: Grundlagen, Verfahren, Beispiele* (Nachrichten-technik, Bd. 13) (Springer-Verlag, Berlin, 1984).

[7.63] W.K. Pratt: *Digital Image Processing* (Wiley Interscience, New York, 1978).

[7.64] R.D. Richtmyer, K.W. Morton: *Difference Methods for Initial-Value Problems*, 2nd edn (Wiley Interscience, New York, 1967).

[7.65] N.S. Kokoshinskaya: "Viscous Gas Flow in the Wake of a Blunt Body", in *Proceedings of the 5th All-Union Seminar on Numerical Methods in Viscous Fluid Mechanics*, Part 2, ed. by B.G. Kuznetsov (Publication of the U.S.S.R. Academy of Sciences, Siberian Computer Center, Novosibirsk, 1975), pp. 106–118.

[7.66] W.G. Habashi: "Numerical Methods of Turbomachinery", in *Recent Advances in Numerical Methods in Fluids*, Vol. 1, ed. by C. Taylor, K. Morgan (Pineridge Press, Swansea, 1980), pp. 245–286.

[7.67] K.G. Shkadinsky: "Averaging Method for the Numerical Solution of One-Dimensional Gas Dynamics Problems", in *Numerical Methods of Solving Mathematical Physics Problems* (Supplement to *Zhurn. Vychisl. Matem. i Matem. Fiz.*, 6, No. 4) (Nauka, Moscow, 1966), pp. 200–206.

[7.68] V.N. Liahov: *Numerical Methods in Continuum Mechanics* (in Russian) 5, No. 3 (A.N. SSSR Siberian Computer Center, Novosibirsk, 1974), pp. 69–74.

[7.69] V.B. Balakin, V.V. Bulanov: *Inzhenerno-Fizicheskiĭ Zhurn.* 21, 1033–1039 (1971).

[7.70] S. Levialdi: "Edge Extraction Techniques", in *Fundamentals in Computer Vision*, ed. by O.D. Faugeras (Cambridge University Press, Cambridge, 1983), pp. 117–144.

[7.71] P.J. Roache: *Computational Fluid Dynamics* (Hermosa, Albuquerque, New Mexico, 1976).

[7.72] B.R. Frieden: "Image Enhancement and Restoration", in *Picture Processing and Digital Filtering*, ed. by T.S. Huang (Springer-Verlag, Berlin, New York, 1975), pp. 179–248.

[7.73] R.W. Schafer, R.M. Mersereau, M.A. Richards: *Proc. IEEE* 61, 432–450 (1981).

[7.74] G.I. Vasilenko, A.M. Taratorin: *Image Restoration* (Radio i Sviaz, Moscow, 1986).

[7.75] N.G. Preobrazhenskiĭ, V.V. Pikalov: *Unstable Problems of Plasma Diagnostics* (Nauka, Siberian Division, Novosibirsk, 1982).

[7.76] R.T. Chin, C.L. Yeh: *Computer Vision, Graphics, and Image Processing* 23, 67–91 (1983).

[7.77] L.S. Davis, A. Rosenfeld: *IEEE Trans. Systems, Man Cybernet.* SMC-8, 705–710 (1978).

[7.78] L. Kitchen, M. Pietikäinen, A. Rosenfeld, C.-Y. Wang: *IEEE Trans. Systems, Man Cybernet.* 13, 626–631 (1983).

[7.79] A. Rosenfeld, A.C. Kak: *Digital Picture Processing* (Academic Press, New York, 1976).

[7.80] R. Haralick, L. Shapiro: *Computer Vision, Graphics, and Image Processing* 29, 100–132 (1985).

[7.81] A.K. Griffith: *IEEE Trans. Comput.* C-22, 371–381 (1973).

[7.82] R.C. Gonzalez, R. Safabakhsh: *Computer* 15, No. 12, 17–32 (1982).

[7.83] G.J. Yang, T.S. Huang: *Computer Graphics and Image Processing* 15, 224–245 (1981).

[7.84] R.O. Duda, P.E. Hart: *Pattern Classification and Scene Analysis* (Wiley, New York, 1973).

[7.85]   L.S. Davis: *Computer Graphics and Image Processing* **4**, 248–270 (1975).
[7.86]   *Optico-Structural Machine Analysis of the Images*, ed. by K.A. Yanovskiĭ (Machinos-troyeniye, Moscow, 1984).
[7.87]   L.R. Roberts: "Machine Perception of Three-Dimensional Solids", in *Optical and Electro-Optical Information Processing*, ed. by D.T. Tippett (M.I.T. Press, Cambridge, Massachusetts, 1965), pp. 169–197.
[7.88]   L. Mérö, Z. Vassy: "A Simplified and Fast Version of the Hueckel Operator for Finding Optimal Edges in Pictures", in *Proceedings of the 4th International Conference on Artificial Intelligence, Tbilisi, USSR, September, 1975* (Publication of the Scientific Council in the Complex Problem "Cybernetics" of the U.S.S.R. Academy of Sciences, Moscow, 1975), pp. 650–655.
[7.89]   R. Nevatia, K.R. Babu: *Computer Graphics and Image Processing* **13**, 257–269 (1980).
[7.90]   G.B. Shaw: *Computer Graphics and Image Processing* **9**, 135–149 (1979).
[7.91]   E.S. Deutsch, J.R. Fram: *IEEE Trans. Comput.* **C-27**, 205–213 (1978).
[7.92]   I.E. Abdou, W.K. Pratt: *Proc. IEEE* **67**, 753–763 (1979).
[7.93]   J. Kittler: *Image and Vision Computing* **1**, 37–42 (1983).
[7.94]   E.R. Davies: *Image and Vision Computing* **2**, 134–142 (1984).
[7.95]   G.S. Robinson: *Computer Graphics and Image Processing* **6**, 492–501 (1977).
[7.96]   C.D. Mcllroy, R. Linggard, W. Monteith: *IEE Proc.-E. Computers and Digital Techniques* **13**, Part E, 223–229 (1984).
[7.97]   D.L. Milgram, A. Rosenfeld: "Object Detection in Infrared Images", in *Digital Image Processing Systems*, ed. by L. Bolc, Z. Kulpa (Springer-Verlag, Berlin, 1981), pp. 228–353.
[7.98]   T. Pavlidis: *Algorithms for Graphics and Image Processing* (Springer-Verlag, Berlin, Heidelberg, 1982).
[7.99]   M. Suk, O. Song: *Computer Vision, Graphics, and Image Processing* **26**, 400–411 (1984).
[7.100] F. Murtagh, A.E. Raftery: *Pattern Recognition* **17**, 479–483 (1984).
[7.101] K.C. Gowda: *Pattern Recognition* **17**, 221–237 (1984).
[7.102] D.L. Milgram, M. Herman: *Computer Graphics and Image Processing* **10**, 272–280 (1979).
[7.103] J.C. Bezdek: "Some Recent Applications of Fuzzy C-Means in Pattern Recognition and Image Processing", in *IEEE Workshop Lang. Autom., Chicago, 7–9 November, 1983* (Silver Spring, Maryland 1983), pp. 247–252.
[7.104] *Fuzzy Sets in the Models of Control and Artificial Intelligence*, ed. by D.A. Pospelov (Nauka, Moscow, 1986).
[7.105] M. Dhome, G. Rives, M. Richetin: *Pattern Recognition Lett.* **2**, 101–107 (1983).
[7.106] R.A. Gentry, R.E. Martin, B.J. Daly: *J. Comput. Phys.* **1**, 87–118 (1966).
[7.107] O.M. Belotserkovskii, Yu.M. Davydov: *Coarse Particles Method in Gas Dynamics. Computational Experiment* (Nauka, Moscow, 1982).
[7.108] F. Angrand, A. Dervieux: *Internat. J. Numer. Methods Fluids* **4**, 749–764 (1984).
[7.109] S.M. Bosniakov, V.V. Kovalenko, A.N. Minailos: *Uchenyie Zapiski TsAGI* (in Russian) **15**, No. 2, 20–29 (1984).
[7.110] D.M. Causon, P.J. Ford: *Aeronaut. J.* **89**, No. 886, 226–241 (1985).
[7.111] N.I. Zelinskiĭ, V.A. Sapozhnikov: *Numerical Methods in Continuum Mechanics* (in Russian) **15**, No. 5 (A.N. SSSR Siberian Computer Center, Novosibirsk, 1984), pp. 91–101.
[7.112] B.L. Roždestvenskii, N.N. Yanenko: *Systems of Quasilinear Equations and Their Applications to Gas Dynamics* (Nauka, Moscow, 1968).
[7.113] N. Liron, J. Rubinstein: *SIAM J. Appl. Math.* **44**, 493–511 (1984).
[7.114] *Handbook of Mathematical Functions with Formulas, Graphs and Mathematical Tables*, ed. by M. Abramowitz and I. Stegun (National Bureau of Standards and Applied Mathematics, Series 55, Washington, D.C., 1964).
[7.115] S.K. Godunov, A.V. Zabrodin, G.P. Prokopov: *Zhurn. Vychisl. Matem. i Matem. Fiz.* **1**, 1020–1050 (1961).
[7.116] M. Holt: *Numerical Methods in Fluid Dynamics* (Springer-Verlag, New York, Heidelberg, Berlin, 1977).
[7.117] Yu.I. Shokin, N.N. Yanenko: *Differential Approximation Method. Application to Gas Dynamics* (Nauka, Siberian Division, Novosibirsk, 1985).
[7.118] M.H. Hueckel: *J. Assoc. Comput. Mach.* **18**, 113–125 (1971).

[7.119]   A.J. Tabatabai, O.R. Mitchell: *IEEE Trans. Pattern Anal. Mach. Intell.* **6**, 188–201 (1984).
[7.120]   R. Machuca, A.L. Gilbert: *IEEE Trans. Pattern Anal. Mach. Intell.* **3**, 103–111 (1981).
[7.121]   B. Schucter, A. Rosenfeld: *Commun. Assoc. Comput. Mach.* **21**, 172–176 (1978).
[7.122]   A.I. Zhukov: *Vychislit. Matematika* (in Russian) No. 6, 34–62 (1960).
[7.123]   C.J. Jacobus, R.T. Chien: *IEEE Trans. Pattern Anal. Mach. Intell.* **3**, 581–592 (1981).
[7.124]   A.O. Aboutalib, D. Berkland: "Automatic Edge Detection for Scene Analysis", in *Digital Signal Processing-84*, ed. by V. Cappellini, A.G. Constantinides (Elsevier Science Publishers B.V., North-Holland, Amsterdam, 1984), pp. 513–518.
[7.125]   A. Goshtasby, C.V. Page: "A Multiple Image Segmentation Technique with Subpixel Accuracy", in *Proc. CVPR '83: IEEE Computer Society Conference on Computers, Vision, and Pattern Recognition, Washington, D.C., 19–23 June, 1983* (Silver Spring, Maryland, 1983) pp. 157–158.
[7.126]   P.D. Hyde, L.S. Davis: *Pattern Recognition* **16**, 413–420 (1983).
[7.127]   I.M. Vasenin, A.D. Rychkov: *Numerical Methods in Continuum Mechanics* (in Russian) **1**, No. 2 (A.N. SSSR Siberian Computer Center, Novosibirsk, 1970), pp. 3–9.
[7.128]   G.S. Rosliakov, V.P. Suhorukov: "Difference Method for Calculation of Shocked Flows", in *Computational Methods and Programming*, Vol. 19 (Publication of the Moscow State University, Moscow, 1972), pp. 83–96.
[7.129]   D. Marr, E. Hildreth: *Proc. Roy. Soc. London, Sec. B*, **207**, No. 1167, 187–217 (1980).
[7.130]   D. Marr: *Phil. Trans. Roy. Soc. London* **B275**, 483–524 (1976).
[7.131]   T. Poggio: "Visual Algorithms", in *Physical and Biomedical Processing of Images*, ed. by O.J. Braddick and A.C. Sleigh (Springer-Verlag, Berlin, Heidelberg, New York, 1983), pp. 128–153.
[7.132]   A. Rosenfeld: *Picture Processing by Computer* (Academic Press, New York, 1969).
[7.133]   J.F. Thompson, Z.U.A. Warsi, C.W. Mastin: *Numerical Grid Generation. Foundations and Applications* (North-Holland, New York, 1985).
[7.134]   P.R. Eiseman: *Comput. Methods Appl. Mech. Engrg.* **64**, 321–376 (1987).
[7.135]   A. Jameson: *Trans. ASME: J. Appl. Mech.* **50**, 1052–1070 (1983).
[7.136]   T.J. Baker: "The Computation of Transonic Potential Flow", in *Computational Methods for Turbulent, Transonic, and Viscous Flows*, ed. by J.A. Essers (Hemisphere, Washington, D.C., 1983), pp. 213–289.
[7.137]   L.R. Miranda: *Aircraft J.* **21**, 355–370 (1984).
[7.138]   M.M. Hafez: "Numerical Algorithms for Transonic Inviscid Flow Calculations", in *Advances in Computational Transonics*, ed. by W.G. Habashi (Pineridge Press, Swansea, 1985), pp. 23–58.
[7.139]   J. Flores, J. Barton, T. Holst, T. Pulliam: Lecture Notes in Physics, Vol. 218 (Springer-Verlag, New York, Heidelberg, Berlin, 1985), pp. 213–218.
[7.140]   J.M. Longo, W. Schmidt, A. Jameson: *Z. Flugwiss. Weltraumforsch.* **7**, 47–56 (1983).
[7.141]   D.L. Whitfield, J.L. Thomas: "Transonic Viscous-Inviscid Interaction Using Euler and Inverse Boundary-Layer Equations", in *Computer Methods in Viscous Flows* (Pineridge Press, Swansea, 1984), pp. 451–474.
[7.142]   A. Rizzi, H. Viviand: "Collective Comparison of the Solutions to the Workshop Problems", in *Numerical Methods for the Computation of Inviscid Transonic Flows with Shock Waves. A GAMM Workshop*, ed. by A. Rizzi, H. Viviand (Vieweg, Braunschweig, 1981), pp. 167–221.
[7.143]   G.S. Deiwert: *AIAA J.* **14**, 735–740 (1976).
[7.144]   W.F. Ballhaus, T.L. Holst, J.L. Steger: "Implicit Finite-Difference Simulations of Steady and Unsteady Transonic Flows", in *Proceedings of the Sixth International Conference on Numerical Methods in Fluid Dynamics, Tbilisi, 1978*, Vol. 2 (Publication of the Computer Center of the U.S.S.R. Academy of Sciences, Moscow, 1978), pp. 7–12.
[7.145]   T.L. Holst: "Solution of the Transonic Full Potential Equation in Conservative Form Using an Implicit Algorithm," in *Numerical Methods for the Computation of Inviscid Transonic Flows with Shock Waves. A GAMM Workshop*, ed. by A. Rizzi, H. Viviand (Vieweg, Braunschweig, 1981), pp. 28–36.
[7.146]   J. von Neumann: *Theory of Self-Reproducing Automata* (University of Illinois Press, Urbana, 1966).

[7.147] A. Rizzi: "Computational Mesh for Transonic Airfoils", in *Numerical Methods for the Computation of Inviscid Transonic Flows with Shock Waves. A GAMM Workshop*, ed. by A. Rizzi, H. Viviand (Vieweg, Braunschweig, 1981), pp. 222–263.

[7.148] P. Nepomiastchy: "Propagation de l'erreur dans les schemas à pas fractionnaires pour les équations aux dérivées partielles linéaires et paraboliques"; Thèse (Université de Toulouse, Toulouse, 1969).

[7.149] T.L. Holst: *AIAA J.* **17**, 1038–1045 (1979).

[7.150] J.C. South, M.M. Hafez: AIAA Paper 83-1898-CP, 527–534 (1983).

[7.151] J. Flores, T.L. Holst, R.L. Sorenson: *J. Aircraft* **22**, 50–56 (1985).

[7.152] T.L. Holst, S.D. Thomas: *AIAA J.* **21**, 863–870 (1983).

[7.153] J. Flores, T.L. Holst, D. Kwak, D.M. Batiste: *AIAA J.* **22**, 1027–1034 (1984).

[7.154] T.L. Holst: "Approximate-Factorization Schemes for Solving the Transonic Full-Potential Equation", in *Advances in Computational Transonics*, ed. by W.G. Habashi, Vol. 4 in the series Recent Advances in Numerical Methods in Fluids (Pineridge Press, Swansea, 1985), pp. 59–82.

[7.155] Y.S. Wang, M. Hafez: *AIAA J.* **23**, 808–810 (1985).

[7.156] R.C. Lock: "Test Cases for Numerical Methods in Two-Dimensional Transonic Flow"; AGARD Report R-575-70 (1970).

[7.157] A. Jameson, D.A. Caughey, W.H. Jou, J. Steinhoff: "Accelerated Finite-Volume Calculation of Transonic Potential Flows", in *Numerical Methods for the Computation of Inviscid Transonic Flows with Shock Waves. A GAMM Workshop*, ed. by A. Rizzi, H. Viviand (Vieweg, Braunschweig, 1981), pp. 11–27.

[7.158] M. Hafez, W. Whitlow Jr., S. Osher: *AIAA J.* **25**, 1456–1462 (1987).

[7.159] M.Ya. Ivanov, V.V. Koretsky: *Zhurn. Vychisl. Matem. i Matem. Fiz.* **25**, 1365–1381 (1985).

[7.160] M.Ya. Ivanov, V.V. Koretskiĭ, A.S. Lieberson, D.B. Solovyov: "A High-Accuracy Conservative Approximate Factorization Scheme for the Integration of the Full Equation for the Velocity Potential in a Transonic Range", in *Theoretical Fundamentals and the Construction of Numerical Algorithms for Solving Mathematical Physics Problems. VIth All-Union School, Gor'ky, 8–13 September 1986. Abstracts of the Reports* (Publication of the Gor'ky State University, Gor'ky, 1986), p. 66.

[7.161] P.M. Byvaltsev, M.Ya. Ivanov: "A Rapid Method for the Computation of Potential Transonic Flows in Two- and Three-Dimensional Cascades of Blades", in *Theoretical Fundamentals and the Construction of Numerical Algorithms for Solving Mathematical Physics Problems. VIIth All-Union Seminar, Kemerovo, 18–26 September 1988. Abstracts of the Reports* (Publication of the Kemerovo State University, Kemerovo, 1988), p. 20.

[7.162] A.P. Aralov, Yu.B. Lifschitz, A.A. Shagaev: "Solution of Multidimensional Problems of Potential Transonic Flows", in *Theoretical Fundamentals and the Construction of Numerical Algorithms for Solving Mathematical Physics Problems. VIIth All-Union Seminar, Kemerovo, 18–26 September 1988. Abstracts of the Reports* (Publication of the Kemerovo State University, Kemerovo, 1988), p. 7.

[7.163] V.E. Makarov: "The Use of the Approximate Factorization Method for the Computation of Plane and Spatial Subsonic and Transonic Potential Flows, in *Theoretical Fundamentals and the Construction of Numerical Algorithms for Solving Mathematical Physics Problems. VIth All-Union School, Gor'ky, 8–13 September 1986. Abstracts of the Reports* (Publication of the Gor'ky State University, Gor'ky, 1986), p. 102.

[7.164] A.N. Kraĭko: "Some Questions of the Construction of the Numerical Algorithms for the Computation of Ideal Gas Flows", in *Construction of Numerical Algorithms and the Solution of Mathematical Physics Problems*, ed. by K.I. Babenko (Publication of the Institute of Applied Mathematics of the U.S.S.R. Academy of Sciences, Moscow, 1987), pp. 33–55.

[7.165] O.N. Gus'kov, V.E. Makarov: "Transonic Potential Cascade Flow and Cascade Design, and Simulation of a Separated Incompressible Flow in a Cascade of Blades", in *Theoretical Fundamentals and the Construction of Numerical Algorithms for Solving Mathematical Physics Problems. VIIth All-Union Seminar, Kemerovo, 18–26 September 1988. Abstracts of the Reports* (Publication of the Kemerovo State University, Kemerovo, 1988), p. 38.

[7.166]   M.A. Aizerman, E.M. Braverman, A.I. Rosonoer: *Method of Potential Functions in the Theory of Learning the Machines* (Nauka, Moscow, 1970).

[7.167]   E.B. Hunt: *Artificial Intelligence* (Academic Press, New York, 1975).

[7.168]   S. Watanabe, P.F. Lambert, C.A. Kulikowski, J.L. Buxton, R. Walker: "Evaluation and Selection of Variables in Pattern Recognition", in *Automatic Analysis of Complex Images*, ed. by E.M. Braverman (Mir, Moscow, 1969), pp. 276–295.

[7.169]   B.N. Petrov, G.M. Ulanov, S.V. Ulyanov: "Information Content of the Features and the Compression of Information Processes of the Control", in *Itogi Nauki i Tehniki. Tehničeskaya Kibernetika*, tom 13 (VINITI A.N. SSSR, Moscow, 1980), pp. 3–120.

[7.170]   S.N. Voyakin: "On the Question of Estimating the Usefulness of the Features for Recognition of Natural Objects", in *Proceedings of the 10th Conference for Young Scientists of the Moscow Physico–Technical Institute*, Dolgoprudny, 23 March–7 April 1985, Part 3 (Moscow, 1985), pp. 63–68.

[7.171]   A.L. Gorelik, I.B. Gurevich, V.A. Skripkin: *State-of-the-Art of the Recognition Problem. Some Aspects* (Radio i Sviaz, Moscow, 1985).

[7.172]   G.I. Razorenov, G.A. Poddubskiĭ: *Zavod. Laboratoriya* (in Russian) **51**, No. 7, 48–50 (1985).

[7.173]   A.I. Ivandaev: *Zhurn. Vychisl. Matem. i Matem. Fiz.* **15**, 523–527 (1975).

[7.174]   G.R. Dattatreya, L.N. Kanal: "Decision Trees in Pattern Recognition", in *Progress in Pattern Recognition 2*, ed. by L.N. Kanal, A. Rosenfeld (Elsevier Science Publishers B.V., North-Holland, Amsterdam, 1985), pp. 189–239.

[7.175]   G.R. Dattatreya, V.V.S. Sarma: *IEEE Trans. Pattern Anal. Mach. Intell.* **3**, 293–298 (1981).

[7.176]   J.-J. Chattot: *C.R. Acad. Sci. Paris* **286**, A-111–A-113 (1978).

[7.177]   P. Morice, H. Viviand: *C.R. Acad. Sci. Paris* **292**, B-235–B-238 (1980).

[7.178]   D.E. Knuth: *The Art of Computer Programming*, Vol. 2 (Addison-Wesley, Reading, Massachusetts, 1969).

[7.179]   J. Mace, T.C. Adamson: AIAA Paper No. 0369 (1985).

[7.180]   S.A. Jepps: "A Finite Element Method for Computing Transonic Potential Flow", in *Numerical Methods for the Computation of Inviscid Transonic Flows with Shock Waves. A GAMM Workshop*, ed. by A. Rizzi, H. Viviand (Vieweg, Braunschweig, 1981), pp. 91–97.

[7.181]   S. Abarbanel, M. Goldberg: *J. Comput. Phys.* **10**, 1–21 (1972).

[7.182]   Yu.I. Shokin, L.A. Kompaniets: *Comput. Fluids* **15**, 119–136 (1987).

[7.183]   V.M. Fomin, E.V. Vorozhtsov, N.N. Yanenko: *Comput. Fluids* **7**, 109–121 (1979).

[7.184]   G.V. Bazhenova, L.G. Gvozdeva, Yu.P. Lagutov, V.N. Liahov, Yu.M. Faresov, V.P. Fokeev: *Nonstationary Interaction of Shock Waves and Detonation Waves in Gases* (Nauka, Moscow, 1986).

[7.185]   G. Ben-Dor, I.I. Glass: *J. Fluid Mech.* **92**, 459–496 (1979).

[7.186]   M. Shirouzu, I.I. Glass: *Proc. Roy. Soc. London* **A406**, No. 1830, 75–92 (1986).

[7.187]   Yu. M. Lipnitskii, V.N. Liahov: *Izv. Akad. Nauk SSSR, Ser. Mekhanika Zhidkosti i Gaza*, No. 6, 88–94 (1974).

[7.188]   V.N. Liahov: *Izv. Akad. Nauk SSSR, Ser. Mekhanika Zhidkosti i Gaza*, No. 3, 90–94 (1976).

[7.189]   V.N. Liahov: *Izv. Akad. Nauk SSSR, Ser. Mekhanika Zhidkosti i Gaza*, No. 2, 100–106 (1977).

[7.190]   R.L. Deschambault, I.I. Glass: *J. Fluid Mech.* **131**, 27–57 (1983).

[7.191]   D. Book, J. Boris, A. Kuhl, E. Oran, M. Picone, S. Zalesak: Lecture Notes in Physics, Vol. 141 (Springer-Verlag, New York, Heidelberg, Berlin, 1981), pp. 84–90.

[7.192]   V.P. Goloviznin, A.I. Zhmakin, A.A. Fursenko: *Zhurn. Vychisl. Matem. i Matem. Fiz.* **22**, 484–488 (1982).

[7.193]   P. Woodward, P. Colella: *J. Comput. Phys.* **54**, 115–173 (1984).

[7.194]   J.-H. Lee, I.I. Glass: *Progress Aerospace Sci.* **21**, 33–80 (1984).

[7.195]   P. Colella, H.M. Glaz: Lecture Notes in Physics, Vol. 218 (Springer-Verlag, Berlin, Heidelberg, New York, 1985), pp. 154–158.

[7.196]   P. Colella, P.R. Woodward: *J. Comput. Phys.* **54**, 174–201 (1984).

[7.197]   B. van Leer: *J. Comput. Phys.* **32**, 101–136 (1979).

[7.198]   J.P. Vila: *SIAM J. Numer. Anal.* **23**, 1173–1192 (1986).

[7.199] H.K. Liu: *Computer Graphics and Image Processing* **6**, 123–134 (1977).
[7.200] D.C. Morgenthaler, A. Rosenfeld: *IEEE Trans. Pattern Anal. Mach. Intell.* **3**, 482–486 (1981).
[7.201] S.W. Zucker and R.A. Hummel: *IEEE Trans. Pattern Anal. Mach. Intell.* **3**, 324–331 (1981).
[7.202] J.K. Udupa, S.N. Srihari, and G.T. Herman: *IEEE Trans. Pattern Anal. Mach. Intell.* **4**, 41–50 (1982).
[7.203] J.K. Udupa: *Computer Graphics and Image Processing* **18**, 213–235 (1982).
[7.204] S. Di Zenzo: *Computer Vision, Graphics, and Image Processing* **33**, 116–125 (1986).

Subject Index

Accuracy
  of shock localization 3, 35, 107, 290, 298
  of the image segmentation techniques 289ff
  of the Sobel edge detector 280ff, 290ff
Adams scheme 190, 191
Aerodynamic design 3, 258ff
AF2 scheme 304ff
Airfoil 3, 104, 245, 302ff, 334ff, 350ff
  NACA 0012 302, 351
Airy function 124
Algorithm of processing 1, 254ff
Analog processing 255
Anisotropy of the difference solution 80
Approximation viscosity 15, 17, 50ff, 58, 65, 88, 185ff
A priori information 3, 108, 205, 212, 254, 257, 370ff
Artefacts 285ff, 310
Artificial
  compressibility method 307
  compression method 222, 248ff
  density 307
  intelligence 254ff, 303
  viscosity 15, 17, 39ff, 46ff, 81, 182, 188, 264
    of Neumann and Richtmyer 15, 30
Asymptotic expansion 351
Atmospheric front 1
Automatic design 3, 258
Axisymmetric flow 353ff

Bernoulli integral 95, 114, 303
Boundary conditions 2, 40, 51, 120, 219, 225, 306, 356, 364ff
Breakdown of discontinuity 8

scheme 19ff, 28, 65ff, 135, 151, 166ff, 188, 297
Buckley–Leverett equation 33
Burgers equation 35ff, 181, 240ff

Cellular logic 255, 369, 374
Central line of a contact strip 121ff
Characteristic equation 23
Classes
  of contact discontinuities 313
  of shock waves 312
Classification
  hierarchical 328ff, 343, 348, 355ff
  of discontinuities 255ff
  of singularities 255ff
  parallel 328
  sequential 330ff, 350ff, 362, 366
Cluster 277, 346
  analysis 277, 285ff
Coarse particles method 13, 64ff, 116, 129, 151, 297
Computation control 2, 31, 256
Conservative scheme 14ff
Contact
  discontinuity 7, 8, 11, 113ff, 139, 165ff, 202ff, 224ff, 272, 299, 312, 319ff, 326ff, 338ff, 348ff, 356, 359ff, 367ff
  residual 160ff
  strip 113, 121, 136, 154, 165ff
  surface 7, 113, 139
Contour thinning 284ff
Convergence 12ff, 194, 306
Convolution
  discrete 265
  of the matrix with itself 97
Coordinates
  Cartesian 3, 84, 87, 101, 290, 303, 310
  curvilinear 301ff

Vanishing viscosity   14
Variational difference scheme   240
Vector-matrix
  operator   12
  pseudoviscosity   15

Wave
  compression (zone)   11, 73, 179, 317,
    321, 330, 361
  detonation   82
  equation   135

progressive   16ff, 36, 39, 51, 67, 73,
    93ff
rarefaction (fan)   8, 11, 117, 211,
    364
Width after Prandtl   28ff, 90, 121, 129ff,
    293ff, 307
Wiggles (*see* Oscillations)
Wind tunnel flow   363ff

Zemplen's theorem   8, 61, 82, 189
Zero crossings   299ff